D1171298

INTRODUCTION TO DYNAMICS

INTRODUCTION TO DYNAMICS

BY

L. A. PARS

CAMBRIDGE
AT THE UNIVERSITY PRESS
1953

PUBLISHED BY
THE SYNDICS OF THE CAMBRIDGE UNIVERSITY PRESS

London Office: Bentley House, N.W.1
American Branch: New York
Agents for Canada, India, and Pakistan: Macmillan

Printed in Great Britain by John Wright & Sons Ltd.
at the Stonebridge Press, Bristol

PREFACE

On several occasions during recent years I have given in Cambridge a short course of lectures on elementary dynamics. In these lectures I have suffered some embarrassment from the lack of a suitable text-book to which the class could refer for further information when, from lack of time, I was unable to treat some particular topic as fully as I should have wished. There exist, indeed, many books of something like the right scope, but none that I know has what seems to me the right modern approach to the subject. This book is designed to fill the gap.

The lectures just mentioned were for first-year undergraduates reading for Part I of the Mathematical Tripos, and the arrangement of the lecture-list is such that the class had previously attended a course in elementary statics. This arrangement of the lecture-list is no doubt open to criticism, but it has great practical advantages, and in this book I shall follow the same plan, i.e. I shall assume that the reader already has some knowledge of elementary statics. In theory, no previous knowledge of dynamics is required, though I daresay that in practice some previous acquaintance with the bare elements of the subject might be desirable. The only other equipment required of the reader is a working knowledge of the elements of the calculus.

The scope of the book is fairly clearly defined as the study of motion in two dimensions—particle, rigid body, system—without Lagrange's equations. It is intended as a general introduction to the subject, and I have made no attempt to adhere to the syllabus for any particular examination. I hope that the treatment is such that the reader can proceed naturally and uninterruptedly to the more advanced parts of the subject, that there is nothing he will wish to forget and nothing he will have to unlearn.

There are still some elementary books that seem to suffer from a hangover from the old days of 'mechanics without the calculus'. They use the methods of the calculus as little as may be, and when they do use them it is half-apologetically. I suggest that this attitude is dangerously out-of-date. Nowadays all students starting on the serious study of mathematics at the University have already a pretty good knowledge of the calculus and of mathematical analysis, and an opportunity to use these techniques as a tool for solving the problems of natural philosophy is invaluable. It assures the student (if he needs assurance) of the importance and usefulness of the calculus, and it gives him valuable practice in its applications. I would push the argument even further, and suggest that it is far better to over-estimate the reader's knowledge of analysis than to under-estimate it. If he occasionally meets a theorem he has not seen before, so much the better; he will be all the more readily convinced of its usefulness and importance.

The book is essentially elementary, and this explains the choice of the way in which the subject is developed. One question that arises continually is about 'the general and the particular'—about particular applications of general theories. Is it better to explain the general theory first and pick out the particular case afterwards for special study? Or is it better to treat the special case first as an introduction to the general theory? There is clearly no plain and unequivocal answer to this question; it depends on the topic with which we are dealing and on the audience for whom the exposition is designed. In this book, which is intended to be elementary, I have usually treated the special case before the general theory. This is because I have aimed at finding the most easily comprehended exposition rather than the most concise. Thus (setting aside for a moment the three introductory chapters, which are somewhat apart from the rest of the book) I deal with the motion of a particle on a straight line before motion in a plane. I put the problem of constrained motion in a plane before the problem of free motion. I consider the motion of a rigid body before the motion of a general dynamical system. Even in the motion of a single rigid body I particularize still further, and deal with the special problem of rotation about a fixed axis before the general problem of motion in a plane.

Experience confirms, I think, that this approach is congenial to the students in the early stages of the subject. It does, however, involve a certain amount of repetition. For example, the equation of energy appears several times—for a particle moving on a line, for two particles on a line, for a particle in a plane, for a rigid body, for a system. But this repetitiveness is not altogether a disadvantage. The theorem first appears in a very simple form, and the more recondite cases present themselves as natural generalizations of results already familiar; its reappearance in different contexts serves to impress on the reader its universality and importance.

The vexed question of vectors, or rather of vector notation, has lately been much discussed, and the disputants have been seen maintaining with no little heat their various opinions. I am aware that the *via media* I have chosen in this book will be uncongenial to the extremists of both parties. One party stands for vector notation all the time—even for problems to which it is not particularly well adapted. In some expositions the whole emphasis seems to be on the manipulation of vector formulae rather than on dynamical principles, and we occasionally meet a proof that looks like a stunt or a conjuring trick. This party seems to regard the invention of Cartesian axes as a regrettable accident, a slightly indelicate matter seldom alluded to in polite society. The other party takes the diametrically opposite view. Its members seem always ill at ease with the vectors (just as writers a century ago were ill at ease with the complex variable) and they breathe a sigh of relief when they can get back—which they do as quickly as possible—to the friendly shelter of x, y, z. In this book I have tried to avoid both these extremes. What is of primary importance is a thorough understanding of the dynamical principles, and I have taken what seems to me the common-sense line—that is to say, I have used vector notation freely when it substantially shortens or clarifies the argument. But I have not insisted on it everywhere, because there will be some readers to whom it is unfamiliar, and these readers will make quicker progress if the unfamiliar notation is not used too freely. As I have said, I expect of the reader a fair knowledge of the calculus, which is essential to a proper understanding of the subject. A wide knowledge of the geometrical vector theory is not of comparable importance at this stage; it comes into its own in the field theories of electricity and hydrodynamics. In particular, I do not often use the notation of the vector product. I use the scalar product freely, but in two-dimensional work we can, if we wish, dispense with the notion of the vector product. The reason is, in essence, that the vector product of two vectors lying in a given plane is perpendicular to that plane, and a set of such vectors, all perpendicular to the plane, can be treated effectively as scalars. Thus, for example, I usually write the moment about O of a force (X, Y) at (x, y) as a scalar in the form $xY - yX$. The reader who is expert in the geometrical vector theory is at liberty to write it in the usual form of a vector product if he prefers to do so.

I have included a careful study of the differential equation

$$\tfrac{1}{2}\dot{x}^2 = f(x)$$

which occurs so frequently in dynamics. It is vitally important to be able to infer the general nature of the solution from a glance at the graph of $f(x)$. Then, when the general nature of the solution has been understood (but not sooner) we shall usually wish to determine the explicit relation between x and t. We can attempt to do this by separation of the variables; but I have emphasized that for this equation, and for some others as well, it is often expedient to replace the dependent variable x by a new dependent variable suitably chosen.

A word about the general structure of the book, and about some of its special features. To begin with there are three introductory chapters. They are concerned with the basic notions required in dynamics—vector quantities, vector calculus, and the fundamental ideas of Newtonian dynamics. I expect that some readers will know the content of these chapters sufficiently well already, and for them the book effectively starts with Chapter IV. The first three chapters are to be regarded as an introduction to the subject as a whole, not merely to the present book, and for this reason the space with which we deal in these chapters is three-dimensional. In the following chapters we are mainly concerned with two dimensions—with the motion of a particle in a plane, or of a rigid body or system moving in such a way that each particle moves parallel to a given plane. We begin with particle dynamics.

understanding of the theory. Here we have an example of the pleasing situation (so different from the affairs of everyday life) where the highest principles are also the most profitable.

I have taken particular care to give a proper distribution of emphasis to the various parts of the subject, and I hope I have given a just impression of the relative importance of the various topics. I have put plainness before elegance, and have occasionally admitted infelicities of style in the interests of sheer clarity. I hope that the reader who has mastered this book will be in a good position to pursue the subject into its higher developments—a pilgrimage that will bring him an abounding reward.

A large number of friends and pupils have helped me with various details at different stages, and it is impossible to mention them all here. I am deeply conscious that the book would have been much inferior without their help. In particular, I would express my thanks to Dr J. Bronowski, Mr P. Hall, Mr A. E. Ingham, Dr D. R. Taunt, and Mr A. J. Weir. For help with the diagrams I am indebted to many friends, especially Mr J. H. Halton and Mr G. Matthews. Finally, my best thanks to Dr Taunt and to Dr W. B. Pennington for their assistance in the onerous task of proof-reading.

CAMBRIDGE L. A. PARS
21 *May* 1951

understanding of the theory. Here we have a rare example of the pleasing situation [illegible] where the highest [illegible] are [illegible] of [illegible].

I have taken particular care to give a proper distribution of emphasis to the various parts of the subject, and I hope I have given a just impression of the relative importance of the various topics. I have put certain [illegible] and less [illegible] in the interests of [illegible] clarity.

I hope that the reader who has mastered this book will be in a good position to pursue the subject into its higher developments—a pilgrimage that will bring him an astonishing reward.

A large number of friends and pupils have helped me with various details at different times, and it is impossible to mention them all here. I am deeply conscious that the book would have been much inferior without their help. In particular I would express my thanks to Dr [illegible], Mr [illegible], [illegible] [illegible] [illegible]. I am indebted to many friends, especially [illegible] and [illegible]. Finally, my best thanks to the [illegible] and to [illegible] for their [illegible] in the [illegible] and [illegible] proofs.

CAMBRIDGE, [illegible]
21 August [illegible]

At first sight it may appear that the balance is wrong—that too much attention is given to the rectilinear problem. But this appearance is, I think, illusory. The point is that many of the ideas that we meet in the special case reappear later. They are considered rather fully on their first appearance, so that later the results can be quoted *tout court* without further discussion. I discuss the theory of variable mass for the rectilinear case only; I confess that it seems to me that the importance of this topic has sometimes been over-estimated. I have given rather more space to the theory of orbits than has been usual in books of this scope, but I think this is justified, both by the intrinsic interest of the subject, and by the importance of it in the history of dynamics. In particular, the study of the Newtonian orbit has so permeated the whole subject that a fairly full account of it seems to be essential. I have eschewed the tiresome 'p, r equations' that still find a place in some elementary books.

I give a fairly full account of the motion of two particles in a plane—a simple example of a general dynamical system. The rather detailed discussion of the collision problem occupies more space than is usually devoted to this topic, but this is justified, I think, by the importance of the subject in mathematical physics.

I begin the study of rigid dynamics with a discussion of the motion of a rigid lamina in its plane, an important and elegant theory, packed with new and stimulating ideas. It has been on the whole rather neglected in English text-books.

Finally, I discuss the motion of a system, but without the general analytical theory of the Lagrange equations. Here we are faced with a fundamental difficulty—how far is it worth while to go with only the elementary techniques? I doubt if there is a final answer to this question. There are, of course, plenty of interesting systems, of a fairly simple type, that can be studied without further resources. But the methods are *ad hoc* methods, and the theory lacks the reassuring sense of completeness that we feel in the theory for a particle or in the theory for a single rigid body. The natural object to aim at would be to do as much as is interesting and useful without straining the available tools to do jobs for which they are fundamentally unsuited. I hope my judgment of this delicate matter is not too wide of the mark.

In the matter of notation I have been on the whole conservative. I do not venture to disturb such time-honoured forms as $\mathbf{P} = m\mathbf{f}$, or $T + V = C$. I have aimed at a reasonable standard of consistency of notation, but I would point out that insistence on consistency of notation can be overdone. The student should not become so much the slave of a particular convention that he feels completely at sea when he reads an author who uses a different convention. I have taken the potential V as fundamental in describing a conservative field of force rather than the work-function U, so I usually write $X = -\partial V/\partial x$ rather than $X = \partial U/\partial x$. The advantage of V is its physical significance, as in the equation of energy just mentioned. It has sometimes been maintained that it is expedient to write $X = -\partial V/\partial x$ for the electric field, the minus sign on the right being natural and proper, but that for the mechanical field the same minus sign is an intolerable nuisance. I have never been able to treat this view very seriously.

I have included in the text a number of worked examples to illustrate the application of the theory to concrete problems, and I have added short collections of examples for practice at the ends of the chapters. Most of these are taken from, or are based on examples taken from, Part I of the Mathematical Tripos, or from the examinations for Entrance Scholarships at Cambridge. I have to thank the Cambridge University Press for permission to reprint them here. Among the examples a few of a rather higher standard of difficulty are indicated by an asterisk. The reader should test his mastery of the theory by trying his hand on the examples at the ends of the chapters; but he should guard against the attitude of mind, encouraged by text-books of a certain stamp, that regards the study of dynamics (or of any other branch of mathematics) as preparation for an examination, and puts as the first object the learning of examination tricks. This attitude of mind is not only bad in itself, but, strangely enough, it does not even pay. The candidate who relies on examination tricks is often defeated by a quite simple problem if it lies a little off the beaten track. The best preparation for the examination is a proper

CONTENTS

Chapter X. MOTION OF A PARTICLE IN A PLANE

Chapter XI. THE CONSERVATIVE FIELD; CONSTRAINED MOTION

Chapter XV. THE GENERAL THEORY OF CENTRAL ORBITS

Chapter XVI. OTHER PROBLEMS ON THE MOTION OF A PARTICLE IN A PLANE

Chapter XXII. IMPULSIVE MOTION OF A RIGID BODY

Chapter XXIII. MOTION OF A SYSTEM

Chapter XXIV. THE MOMENTUM OF A SYSTEM

Chapter XXV. IMPULSIVE MOTION OF A SYSTEM

Chapter XXVI. DIMENSIONS

ERRATA

p. 170, Example 3. *Read* $\frac{gT}{2+\lambda}\{2-2^{-(1+\lambda)}\}$.

p. 232, line 4. *After* the possible paths are the *read* parabolas through O with λ as directrix.

p. 243, Example 26. *For* $(q\sqrt{3q}$ *read* $(9\sqrt{39}$

p. 366, last two lines of section **19·7**. *Read* though in one of them the forces, or some of them, act not on the particles but at other points of the imponderable frame.

CHAPTER I

SCALAR QUANTITIES AND VECTOR QUANTITIES

1·1 Scalar quantities and the uniform scale. In Natural Philosophy we study classes of physical quantities that are capable of measurement—classes of quantities such as masses, lengths, velocities. The simplest classes are those for which the members are specified, on an appropriate scale, by a single number—for example, masses, lengths, temperatures. We begin with a sketch of the theory of the measurement.

We consider, then, a class K of such quantities. Our first object is to set up a scale of measurement. The possibility of such a scale depends on the *axiom of congruence*, i.e. the assumption that there is a meaning to the statement that two particular members of K are equal, or congruent, even before any scale has been set up. For example, we assume that there is a meaning to the statement that the mass of a body A is equal to the mass of a body B, and this equality is absolute, something existing prior to the construction of a scale.

The axiom of congruence is all that we need in order to set up a scale. Each member of K is labelled with a measure number, or *measure*, and in the first instance the only restriction on the measures is that congruent members of K have the same measure. Two congruent members of K are considered to be identical and interchangeable physical quantities (so far as the quantity under discussion is concerned) and the measure is taken to be a complete and adequate description. For example, any two particles of equal mass are considered to be identical for purposes of dynamics.

But any useful scale must possess two other properties, which we may call *continuity* and *direction*. We assume that we can recognize when two members of K are nearly equal, and we naturally choose a scale in which nearly equal members have nearly equal measures. Such a scale has the property of continuity. We assume that we can recognize which of two members of K has the greater magnitude, and we naturally choose a scale in which the greater magnitude has the greater measure. Such a scale has the property of direction. We thus set up a primitive or empirical scale of measurement possessing these properties. For example, the scale of temperature on some standard mercury thermometer satisfies our conditions; the bore of the tube need not be uniform, and the graduations need not be equidistant on the stem.

But when we are dealing, as we usually are, with quantities to which the notion of addition is applicable, we are soon led to abandon our empirical scale in favour of the immense advantages of a *uniform scale*, if such a scale exists. The uniform scale has the property that the measure of the sum of two members of K is the sum of the measures of the separate members. *Physical addition is represented by algebraic addition.*

It is not obvious *a priori* that a uniform scale must exist for the class K, but we shall assume that such a scale does exist for each class of quantities with which we wish to deal—for masses and lengths and intervals of time. Such quantities are called *scalar quantities*, and we are thus led to the formal definition:

The class K is a class of scalar quantities if:

(i) A member of K is completely specified by a single number (its measure) on an appropriate scale;

(ii) There is a meaning to the physical sum of two members of K, and this sum is itself a member of K;

(iii) There exists a uniform scale of measurement, i.e. a scale in which the measure of the physical sum of any two members of K is the sum of the measures of the two members.

For example, if K is the class of masses, and we use a uniform scale, a body which is formed by the cohesion of bodies of masses m_1 and m_2 (the symbols are the measures) has mass m_1+m_2. It is this property of mass, when measured on a uniform scale, that is implied in the traditional definition of mass as 'quantity of matter'. If we measure lengths on a uniform scale, and A, B, C are three points in order on a straight line, the length of the segment AC is the sum of the lengths of AB and BC. If we measure time on a uniform scale, and

A, B, C are three instants, B being later than A and C later than B, then the interval AC is the sum of the intervals AB and BC. When we come to temperature, the question is deeper; we can easily construct an empirical scale, but it is not at once obvious how to construct a uniform scale. The difficulty is that there is no immediately obvious process of addition related to the idea of temperature. But the appropriate additive process reveals itself in theoretical thermodynamics, and we construct a uniform scale—the 'absolute scale of temperature'.

1·2 Uniqueness of the uniform scale. The uniform scale, if it exists, is unique, save for a constant multiplier. To prove this, suppose that x is the measure of a member X of K on a uniform scale, and $\phi(x)$ the measure of X on a new uniform scale. (In the first instance we will suppose $x \geqslant 0$, though this restriction will later be lifted.) We may assume that $\phi(x)$ is continuous and monotonic increasing, to preserve the continuity and direction of the scale. We have

$$\phi(a+b) = \phi(a)+\phi(b),$$

for arbitrary positive (or zero) values of a and b. We first observe, by taking $a = 0$, that $\phi(0) = 0$. Next, we easily prove, by induction, that if m is a positive integer,

$$\phi(ma) = m\phi(a).$$

Thus we have $$\phi(ma-nb)+\phi(nb) = \phi(ma),$$

whence $$\phi(ma-nb) = m\phi(a)-n\phi(b),$$

where m and n are positive integers, and $ma \geqslant nb$. If we take $ma = nb$ we deduce

$$\frac{\phi(a)}{a} = \frac{\phi(b)}{b}$$

for arbitrary positive values of a and b whose ratio is rational. Thus, taking now $b = 1$, we have for positive rational values of a

$$\phi(a) = ka,$$

where k is positive since ϕ is monotonic increasing. The completion of the proof for irrational values of a follows from the continuity of ϕ.

The proof is much simpler if we make the additional assumption that ϕ is differentiable. Starting from the equation

$$\phi(a+b) = \phi(a)+\phi(b),$$

we consider variations in a and b such that $a+b$ remains constant. We deduce immediately

$$\phi'(a) = \phi'(b)$$

for arbitrary a and b. Thus

$$\phi'(a) = k,$$

whence

$$\phi(a) = ka+C.$$

Finally, $k>0$, since ϕ is increasing, and $C = 0$, since $\phi(0) = 0$.

1·3 Units. We have seen that the uniform scale is unique, save for choice of the multiplier k. We are thus led to the idea of the *unit* member of K, the member of K whose measure is 1. The unit is at our disposal, but when once it has been chosen, the scale is determined. Every member of K can be thought of as an appropriate multiple of the unit. A scalar quantity can thus be thought of as the synthesis of three elements: (i) the kind of quantity, (ii) the unit, and (iii) the measure, which is the number of units in the given quantity. For example, 'a mass of ten pounds' involves the ideas of kind (mass) and unit (pound) and number of units (ten). Strictly we ought always to distinguish between the measure (ten) and the magnitude (ten pounds) of the scalar quantity (a mass of ten pounds), though in practice a little elasticity of expression is permissible. But we must not lose sight of the fact that in the equations of mathematical physics the symbols that appear relating to scalar quantities—for example, 'a mass m', 'a time t'—denote *numbers*, the measures of the physical quantities.

If we change the unit, the measures change; if the new unit is two of the old units, the measures are reduced to one-half their former values. More generally, if the ratio of the units is $u_1 : u_2$ (i.e. the new unit has measure u_2/u_1 on the old scale) and n_1, n_2 are the measures of the same member of K on the two scales, then $n_1 u_1 = n_2 u_2$. (Cf. Chap. XXVI.)

We can construct measuring instruments with which we can determine the measures of our quantities, on a uniform scale, to a high degree of accuracy. The construction and use of such instruments is a matter that we shall not enter upon in this book.

1·4 Directed lengths. Hitherto we have thought of the measures as positive numbers, but they are not always positive. For example, electric charge may be positive or negative. The measures satisfy the condition (iii) in the definition of scalar quantities; e.g. the sum of charges $+e$ and $-e$ is a charge zero.

When we consider lengths and intervals of time we can introduce negative measures so as to include in the measures not only the idea of size but also the ideas of 'right and left' or 'before and after'. Suppose we are dealing with points on a straight line. It will be convenient to mark a fixed point O of the line as origin, and to fix on a definite sense in the line as the positive sense. If the line is drawn horizontally in the figure, we conventionally take the positive sense to the right; for the present we may treat the phrases 'B is to the right of A' and 'B is on the positive side of A' as synonymous. If X is a point on the line we label the point X with the number x, where x is the measure of the length of OX if X is to the right of (i.e. on the positive side of) O, and x is $-x'$, where x' is the measure of the length of XO, if X is to the left of O. The lengths are of course measured on a uniform scale. If A, B are two points of the line whose labels are a, b, and B is to the right of A, then the measure of the length of AB is $b-a$.

A length is strictly a positive quantity. We now introduce the idea of a *directed length*, which is not necessarily positive. If B is to the right of A the measure of the directed length of AB is simply the measure of the length of AB; if B is to the left of A the measure of the directed length of AB is minus the measure of the length of BA. We denote the measure of the directed length of AB by $\overline{AB}$. If A, B are any two points of the line, $\overline{AB} = -\overline{BA}$, and the measure of the directed length of AB is $b-a$ whether B is to the right of A or to the left. The equation

$$\overline{AB}+\overline{BC} = \overline{AC},$$

or, in symmetrical form,

$$\overline{AB}+\overline{BC}+\overline{CA}=0,$$

is valid for all positions of A, B, C on the line. Notice that our method of labelling the points of the line already involves the essential idea of a directed length. If we wish to distinguish between the directed length of AB and the ordinary length, we may denote the ordinary length by $|AB|$.

We label instants of time in the same sort of way in which we labelled points on the line. We take an instant O as origin. An instant T is labelled with the number t, where t is the measure of the interval (or, briefly, the interval) OT if T is after O, and t is $-t'$, where t' is the interval TO, if T is before O. The intervals are of course measured on a uniform scale. If A, B are two instants we define the measure of the *directed interval* AB as the measure of the interval AB if B is after A, and as minus the measure of the interval BA if B is before A. If A, B are any two instants whose labels are a, b, then $b-a$ is the measure of the directed interval AB, whether B is after A or before A. The measure $\overline{AB}$ of the directed interval AB is positive if B is after A, negative if B is before A.

The directed lengths and intervals take a place in the theory intermediate between the positive scalar quantities and the vector quantities which are the next kind of physical quantities that we consider. The essential importance of the idea of a directed length emerges when we are dealing with a set of vector quantities which are all parallel to a given line.

1·5 Directed quantities and their vectors. The next classes of physical quantities which we must consider have direction (i.e. direction in space) as well as size. Familiar examples are velocity and force. We consider a class K of such directed quantities, and in the first instance we suppose that they are not localized, i.e. size and direction together serve as a complete description of a member of K. Later on (for example, when we deal with forces acting at different points of a rigid body) we shall need to consider the extension of the theory to classes whose members are localized; in that case to describe a member of K we need to specify not only size and direction, but also the point of space with which it is associated. But for the present we deal with classes whose members are completely specified, for the purpose of the theory, by size and direction.

We assume to begin with that size and direction are independent elements in the structure of a member of K. There is a meaning to the magnitude dissociated from the direction, and we can speak of two members of K being of equal magnitude although they are not in the same direction. Thus, we can think of the members of K as quantities of the kind already discussed, which are specified on an appropriate scale by a single positive (or zero) measure, to which the idea of direction has been added. Actually, however, a slightly different analysis is more convenient. We isolate the measure of the non-directed quantity, and we call the abstract directed quantity, which is built up of this measure and the direction, the *vector* of the directed quantity. A vector is a positive (or zero) number, called the *modulus* of the vector, associated with a direction. All vectors of zero modulus are assumed to be identical; any such vector is called a null vector.

The word 'direction' as used above includes sense. In everyday speech the word 'direction' is used ambiguously, sometimes implying 'parallel to a given line', and sometimes implying 'parallel to a given line, and in a definite sense'. It is in the second of these ways that we have used the word 'direction'. This usage involves not merely the idea of a line, but of a line marked with an arrow. With this usage, if we take a sphere, centre O, every point D on the sphere defines a unique direction OD; and conversely, every direction defines a unique point on the sphere.

1·6 The vector diagram. We represent the vectors of the members of K by segments such as AB on an appropriate diagram. The diagram is to be thought of in the first instance as three-dimensional in a Euclidean space; though in most of the applications in this book we shall be concerned with classes of vectors which are all parallel to a plane. In the representation the direction AB is the direction of the vector, and the measure of the length of AB, using any convenient unit of length, is its modulus. The order in which the end-points are named is important, the direction being from A to B. Equal parallel segments in the same direction (and sense) represent the same vector, so every vector can be represented in the diagram in infinitely many ways. The representation in the diagram involves the idea of length, but this is only incidental, and only happens because length is the most convenient representation of the modulus. The notion of a vector does not essentially involve the idea of length, but only the ideas of number and direction.

We denote vectors by symbols in Clarendon type, such as $\mathbf{P}$, $\mathbf{Q}$, $\mathbf{r}$, $\mathbf{v}$, $\mathbf{f}$; or sometimes by symbols such as $\overline{AB}$, where AB is the segment representing the vector in the diagram. The Clarendon type may be regarded as the standard notation, but the other notation is also useful, especially in the early stages, when we wish to refer frequently to the diagram. We denote the modulus of $\mathbf{P}$ by $|\mathbf{P}|$. The nul vector is denoted by $\mathbf{0}$.

1·7 Vector quantities. The most important kind of directed quantities are those called *vector quantities*. Vector quantities satisfy a condition which is related to the process of addition, and is analogous to the requirement of a uniform scale for scalar quantities. We consider now the formal definition.

The class K is a class of vector quantities if:

(i) A member of K is completely specified by its vector;

(ii) There is a meaning to the physical sum of two members of K, and this sum is itself a member of K;

(iii) There exists a scale of measurement such that, if the vectors of two members of K are $\overline{AB}$ and $\overline{BC}$, the vector of their sum is $\overline{AC}$.

The process described in (iii) is called *vector summation*, and the vector formed from two given vectors $\mathbf{P}$ and $\mathbf{Q}$ by this process (i.e. the vector $\overline{AC}$ if $\overline{AB}$ represents $\mathbf{P}$ and $\overline{BC}$ represents $\mathbf{Q}$) is called their vector sum. It is denoted by $\mathbf{P}+\mathbf{Q}$. We can express the definition informally by saying that directed quantities are vector quantities if physical summation is represented by vector summation.

Notice that we speak of the 'vector' of a directed quantity even when this quantity is not a vector quantity. Actually the directed quantities we deal with in this book will all be vector quantities, so the slightly awkward nomenclature will not give rise to any inconvenience.

1·8 Parallel vectors. Some implications of the definition of a vector quantity should be noticed. Suppose first that we have a set of vectors all parallel and all in the same sense. Then the modulus of the sum of two vectors is the sum of their moduli. The vectors reduce effectively to numbers, the vector quantities to scalar quantities. Physical summation is represented by algebraic summation, and we see again that the uniform scale of measurement is fundamental.

Next, consider a set of vectors all parallel to a line λ, but not all in the same sense. Now the given line is associated with *two* directions, if we use the word 'direction' to include sense, as we have done hitherto. But it is usually convenient to fix on one definite sense in λ as the positive sense, and to describe a vector $\overline{AB}$ (where A and B are two points of λ) by the directed length of AB. Indeed, we have already anticipated this representation by using the same notation $\overline{AB}$ both for the vector and for the directed length. If A, B, C are any three points of λ the equation

$$\overline{AB}+\overline{BC} = \overline{AC}$$

is valid whether we interpret a symbol such as $\overline{AB}$ as a vector or as a directed length. With this representation, physical addition corresponds to algebraic addition.

The modulus with the appropriate sign is called the *directed modulus*. Thus the directed modulus is equal to the modulus if the vector is in the positive sense of λ, and to the modulus with the minus sign prefixed if the vector is in the negative sense of λ. A set of vector quantities all parallel to a line λ, and represented by their directed moduli, can be treated in practice as a set of scalar quantities.

When we use directed moduli we sacrifice the advantage of uniqueness in the description of the vectors. Every vector has a definite (positive) modulus, and a definite direction (including sense). But when we use directed moduli we can describe the same vector in two ways, since a vector $\mathbf{P}$ in the negative sense of λ is also described as a vector $-\mathbf{P}$ in the positive sense.

However, no confusion will arise from this, and the notation has great advantages when we deal with a set of vectors all parallel to the same line. Perhaps the advantages are most conspicuous when we use rectangular axes, and we find it convenient to fix on a definite sense in each axis as the positive sense.

1·9 The commutative property of vector summation. It is clear from elementary geometry, since equal parallel segments in the diagram represent the same vector, that vector summation is commutative,

$$\mathbf{P}+\mathbf{Q} = \mathbf{Q}+\mathbf{P}.$$

This equation is illustrated in Fig. 1·9.

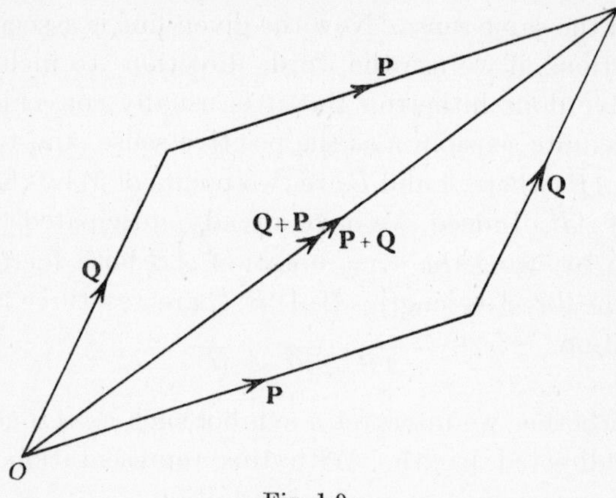

Fig. 1·9

Since $$\overline{AB}+\overline{BA} = \mathbf{0},$$

it is natural to define the negative of a given vector by the equation

$$\overline{AB} = -\overline{BA},$$

and to define subtraction by the equation

$$\mathbf{P}-\mathbf{Q} = \mathbf{P}+(-\mathbf{Q}).$$

We can preserve the analogy with directed lengths, even when the vectors are not all parallel, by writing the rule for vector summation in the symmetrical form

$$\overline{AB}+\overline{BC}+\overline{CA} = \mathbf{0}.$$

1·10 Illustrations. We now illustrate the theory by consideration of some concrete examples.

(1) *Displacements of a particle in space.* We suppose that the space in which the particle moves is Euclidean. A displacement from A to B determines uniquely a directed quantity $\overline{AB}$; the direction is the direction AB, the associated scalar is the length of AB, the modulus is the measure of this length. There is no need to construct a separate vector diagram. We can use the space itself in which the particle moves as the vector diagram, and use the segment AB to represent the vector of the displacement from A to B.

Two displacements which have the same vector (i.e. displacements) from A to B and from A' to B', where the segments AB and $A'B'$ are equal in length, and parallel, and in the same sense, are regarded for our present purpose as equal directed quantities. The physical sum of two displacements is itself a displacement, and the physical sum of the displacement from A to B and the displacement from B to C is the displacement from A to C. Thus the conditions of § 1·7 are fulfilled, and displacements are vector quantities. We refer to the displacement from A to B for brevity as the displacement $\overline{AB}$. Because in this case we do not need a separate vector diagram, displacement is the most fundamental type of vector quantity.

In the dynamics of a particle we choose a base of reference, and we describe the position P of the particle by the displacement, relative to the base, from O to P, where O is an origin fixed in the base. The vector of this displacement $\overline{OP}$ is called the *position vector*, and is denoted by $\mathbf{r}$. In the dynamics of a particle $\mathbf{r}$ is a vector function of t whose properties we wish to study.

The notion of a base of reference is of great importance. The base is any rigid configuration of points, e.g. a rigid body; frequently we use the earth as a base. Usually we employ a rectangular trihedral of reference Ox, Oy, Oz fixed in the base; this usage is so widespread that the trihedral itself is often referred to as the base.

(2) *Statics of a particle.* The notion of force that we use in statics is essentially a refinement of the commonsense notion of force implied in words like *push* and *pull.* In the scientific

applications a force is assumed to act on a particle, or at a definite point of a body or fluid, and to be a directed quantity, completely characterized by its direction and magnitude. Two forces having the same direction and the same magnitude are to be thought of as equivalent and interchangeable physical quantities. For the present we are concerned only with forces acting on a particle, and the position of the particle itself is the point of application of the force. We use a vector diagram in which the vectors of the forces are represented by segments $\overline{AB}$; AB is in the direction of the force, and the measure of the length of AB is the modulus of the force. We recall that the same vector can be represented in the diagram in infinitely many ways, and although we are concerned with forces all acting at the same point the segments representing their vectors in the diagram do not usually all start from the same point.

If two forces act on the particle, they are together equivalent to a single force, their physical sum, or resultant, and this single force is obtained from the two given forces by the vector law of addition. This is the fundamental rule of statics, and in this book we shall assume this rule as a result already known. Thus forces acting on a particle satisfy the conditions in the definition, and constitute a class of vector quantities.

Since we are dealing with statics, we ought strictly to speak not of the resultant of two forces, unless indeed this resultant is nul, but of the third force required to balance the other two. This third force is equal and opposite to the resultant of the first two. The vectors of three forces acting on a particle, and in equilibrium, can be represented by the sides AB, BC, CA of a triangle. This is the famous theorem of the *triangle of forces*. Since forces are vector quantities it is equivalent to the rule for vector summation written in the form already mentioned,

$$\overline{AB}+\overline{BC}+\overline{CA} = \mathbf{0}.$$

(3) *Rotations of a rigid body*. It must be emphasized that the condition (iii) in the definition is an essential part of the idea of a vector quantity. Directed quantities satisfying (i) and (ii) are not vector quantities unless they also satisfy (iii). Perhaps the simplest example of directed quantities which are not vector quantities is afforded by the rotations of a rigid body. Consider

a rigid body, one point of which is attached to a fixed pivot O about which the body can turn. A right-handed rotation of the body about a line OA (from a given initial orientation) through an angle α is a directed quantity; its direction is the direction of OA, its measure (not necessarily positive) is α. The vector of the directed quantity has direction OA and directed modulus α. A rotation α about OA, followed by a rotation β about OB, will bring the body into a new orientation, and we can prove that this new orientation can be derived from the initial orientation by a rotation γ about a line OC. (We omit the proof of this statement, though the proof of the corresponding theorem for the motion of a lamina in a plane will be given in § 18·2.) Thus the conditions (i) and (ii) are satisfied. But (iii) is not. The vector of the physical sum is not obtained from the vectors of the individual rotations by the vector law of addition. Indeed, OC is not even coplanar with OA and OB, and the two operations are not even commutative. Rotations of a rigid body are not vector quantities.

1·11 Retrospect. It may be expedient to pause at this point and glance over the ground we have covered so far.

We have discussed two kinds of physical quantities, those which do not involve direction, and those which do. In both cases we have limited our discussion to special classes of quantities, classes which are additive, and for which the physical sum of two members can be represented in a particularly simple way. The non-directed quantities which satisfy our conditions are called scalar quantities, and the directed quantities which satisfy our conditions are called vector quantities. For scalar quantities there exists a uniform scale of measurement, and this implies that the scalar quantity can be conceived as a multiple of a certain unit, the choice of unit being arbitrary. The multiple, the number of units in the given quantity, is called the measure. For scalar quantities physical addition of quantities corresponds to algebraic addition of their measures.

To describe a scalar quantity we need a unit and a measure. To describe a vector quantity we need a unit and a vector. The unit does not involve the idea of direction; the vector is a synthesis of measure (modulus) and direction. Physical addition of vector quantities corresponds to vector addition of their vectors. This is closely analogous to the algebraic law of addition of measures for scalar quantities. Indeed, from one point of view, it may be said to be the same law; for we shall find that we can regard each vector of a given class as the sum of vectors (called components) parallel to three given lines. The components parallel to one of these lines are, as we have seen, effectively a set of scalar quantities, and the vector law of addition is equivalent to algebraic addition of components.

The symbols denoting scalar quantities in the equations of mathematical physics represent numbers, the measures of the quantities. 'A mass m' is, strictly speaking, 'a mass whose measure is m', though it would be pedantic to insist always on this phraseology. The symbols denoting vector quantities in the equations of mathematical physics represent vectors, the vectors of the vector quantities. 'A force $\mathbf{P}$' is strictly 'a force whose vector is $\mathbf{P}$'. An equation connecting scalar quantities, such as the equation of energy

$$T + V = C, \tag{1}$$

asserts the equality of two numbers. An equation connecting vector quantities, such as the classical equation

$$\mathbf{P} = m\mathbf{f}, \tag{2}$$

asserts the equality (both in direction and modulus) of two vectors. The equations (1) and (2) just quoted will be met with many times in the course of this book.

In the foregoing we have been careful to distinguish between a scalar quantity and its measure, and between a vector quantity and its vector; but when these distinctions have been understood a little laxity of expression is permissible in the interests of brevity. In the literature the word 'scalar' is frequently used both for a scalar quantity and its measure, and the word 'vector' is used both for a vector quantity and its vector. A completely accurate phraseology is apt to be cumbrous, and even stilted. We often read such statements as 'work is the scalar product of force and displacement', and perhaps it would be pedantic to object.

CHAPTER II

VECTOR ALGEBRA AND VECTOR CALCULUS

2·1 Vector algebra. We shall assume in this book that the reader is already familiar with the elements of statics, and therefore with the fundamental ideas of vector algebra. We shall therefore not deal with the algebra of vectors in great detail, but shall be content with an outline of the outstanding ideas.

We begin with a few remarks about vector summation. The two vectors that appear in a vector sum represent vector quantities of the same kind—such as both displacements or both forces. There is no meaning to the physical sum of vector quantities of different kinds. (There is no corresponding restriction for products. Later on we shall meet two kinds of products of vectors, called scalar products and vector products, and the two vectors that occur in the product frequently represent vector quantities of different kinds.)

Vector summation is *associative* and *commutative*. If we have three vectors $\mathbf{P}$, $\mathbf{Q}$, $\mathbf{R}$,

$$(\mathbf{P}+\mathbf{Q})+\mathbf{R} = \mathbf{P}+(\mathbf{Q}+\mathbf{R}),$$

and either member of this equation can be spoken of as the sum $\mathbf{S}$ of the three vectors, and written as $\mathbf{P}+\mathbf{Q}+\mathbf{R}$. Moreover, this sum is the same in whatever order the vectors are arranged. Notice again the parallelism with the properties of scalar quantities measured on a uniform scale. We can arrange the vectors $\mathbf{P}, \mathbf{Q}, \mathbf{R}$ in six ways, and if we illustrate the summation in these six orders in the same figure, we obtain the parallelepiped shown in Fig. 2·1*a*. The figure is flat if the three vectors are coplanar.

In the same way, if we have any finite number of vectors, we can speak unambiguously of their sum. If the vectors are represented in the diagram by $AB, BC, \ldots, JK, KL$, their sum is represented by AL, and we arrive at the same sum if we rearrange the vectors in any order. In the particular case of the statics of a particle we are primarily concerned with the problem of equilibrium; the vector sum is zero, and L coincides with A. The theorem of the 'polygon of forces' can be stated in the form: 'A necessary and sufficient condition for the

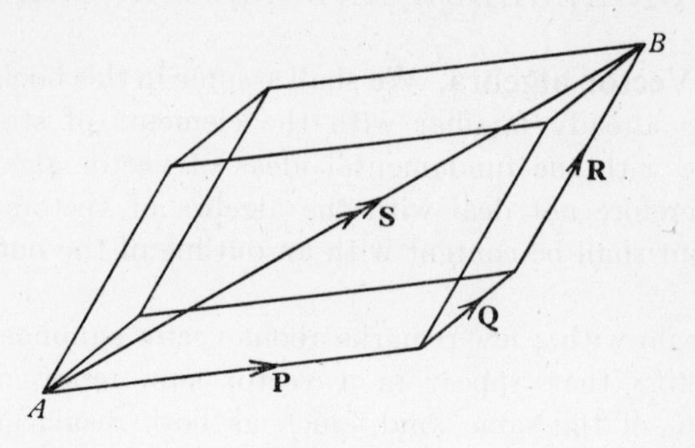

Fig. 2·1*a*

equilibrium of a number of forces acting on a particle is that the vectors of the forces can be represented by the sides AB, $BC, \ldots, JK, KA$ of a polygon $ABC \ldots JKA$.' The polygon is not necessarily plane, and if it is plane it is not necessarily

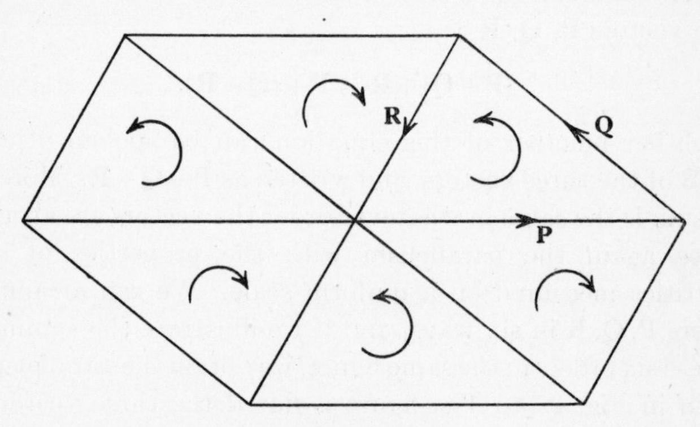

Fig. 2·1*b*

convex. The order is immaterial, and we can construct $n!$ polygons from the same n forces. For $n = 3$ the six possible triangles are shown in Fig. 2·1*b*. This is the form into which Fig. 2·1*a* degenerates when B coincides with A.

2·2 Product of scalar quantity and vector quantity. If $\mathbf{P}$ is a given vector and k a positive number, $k\mathbf{P}$ is defined to be a vector in the same direction (and sense) as $\mathbf{P}$ and of modulus k times that of $\mathbf{P}$. In the special case where k is a positive integer we may alternatively define $k\mathbf{P}$ as the sum of k vectors $\mathbf{P}+\mathbf{P}+\ldots+\mathbf{P}$. This agrees with the definition just given for a general positive value of k. If k is a negative number $k\mathbf{P}$ is a vector in the opposite direction to $\mathbf{P}$, and of modulus $(-k)|\mathbf{P}|$. In all cases $|k\mathbf{P}| = |k||\mathbf{P}|$.

An application of fundamental importance is this: if $\mathbf{P}$ is any vector, and $\mathbf{E}$ is a unit vector (i.e. a vector whose modulus is unity) in the same direction (and sense) as $\mathbf{P}$, then

$$\mathbf{P} = |\mathbf{P}|\mathbf{E}.$$

More generally, if $\mathbf{E}$ is a unit vector, and $\mathbf{P}$ is parallel to $\mathbf{E}$, but not necessarily in the same sense, then

$$\mathbf{P} = p\mathbf{E},$$

where p is the directed modulus of $\mathbf{P}$ for the direction of $\mathbf{E}$.

We naturally define division of a vector by a number k $(\neq 0)$ as multiplication by $1/k$.

If k is the measure of a certain scalar quantity, the vector quantity associated with the product $k\mathbf{P}$ is not usually of the same kind as the vector quantity associated with $\mathbf{P}$. As an illustration, 'the momentum of a particle is the product of its mass and velocity'. If m is the measure of the mass, and $\mathbf{v}$ is the velocity-vector, the momentum-vector is $m\mathbf{v}$. Both $\mathbf{v}$ and $m\mathbf{v}$ are vectors, but the vector quantities to which they belong, velocity and momentum, are of different kinds.

The fundamental properties of the product $k\mathbf{P}$ are represented by the equations:

$$\text{(i)} \qquad a\mathbf{P}+b\mathbf{P} = (a+b)\mathbf{P};$$

$$\text{(ii)} \qquad a(b\mathbf{P}) = b(a\mathbf{P}) = (ab)\mathbf{P};$$

$$\text{(iii)} \qquad a\mathbf{P}+a\mathbf{Q} = a(\mathbf{P}+\mathbf{Q}).$$

In these equations $\mathbf{P}$ and $\mathbf{Q}$ are vectors (belonging of course to vector quantities of the same kind), and a, b are real numbers, positive, negative, or zero. The first two follow easily from the definitions. For the third we consider a diagram (Fig. 2·2) in which AB, BC represent the vectors $\mathbf{P}, \mathbf{Q}$, and $AB', B'C'$ represent the vectors $a\mathbf{P}, a\mathbf{Q}$. (The figure is drawn for the case in which a is positive.) To establish the equation (iii) we have to show that AC' is in the direction AC, and that the length of AC' is a times the length of AC, and these are simple consequences of the properties of similar triangles.

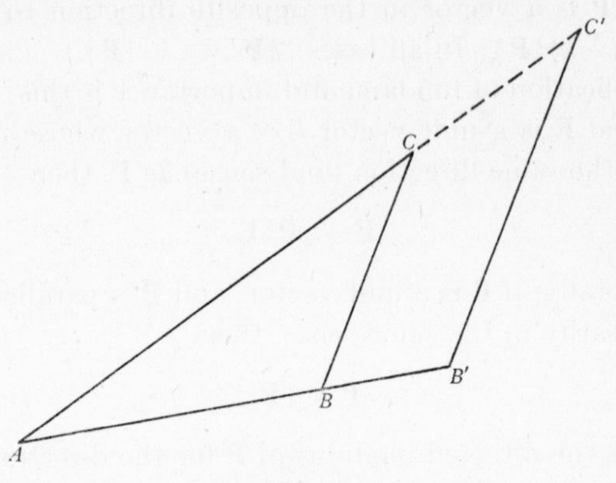

Fig. 2·2

2·3 Resolution. This is the reverse of the process of summation. If AD in the diagram represents a vector, and B is any point in the diagram, then AD can be regarded as the sum of vectors represented by AB and BD. These vectors are spoken of as *components* of the original vector, which is said to be *resolved* into these components. Similarly, if C is any other point, AD can be regarded as the sum of three components $\overline{AB}, \overline{BC}, \overline{CD}$. If the directions of the three components are prescribed (this time the word 'direction' does *not* include sense), and are not all parallel to the same plane, the resolution is always possible, and always unique. If we denote unit vectors in a definite sense in each of the three prescribed directions by $\mathbf{i}, \mathbf{j}, \mathbf{k}$, then any vector can be expressed uniquely in the form $X\mathbf{i} + Y\mathbf{j} + Z\mathbf{k}$. The numbers X, Y, Z are directed moduli, and are not necessarily positive. Usually we employ axes Ox, Oy, Oz in the directions of $\mathbf{i}, \mathbf{j}, \mathbf{k}$, through an origin O in the diagram. The components of the given vector in the prescribed directions are $X\mathbf{i}, Y\mathbf{j}, Z\mathbf{k}$, but we sometimes use a looser form of expression, and say that the components are X, Y, Z; indeed, the **i**-components, for example, of a set of vectors may be thought of as scalar quantities, since they are a set of vectors all parallel to a given line (§ 1·8).

By far the most important case is that in which $\mathbf{i}, \mathbf{j}, \mathbf{k}$ are mutually perpendicular; and we usually suppose them to form a right-handed set. (A rod OA lying along $\mathbf{i}$, and rotated right-handedly through a right angle about $\mathbf{k}$, comes into line with $\mathbf{j}$. In other words, if we look in the direction of $\mathbf{k}$, the rotation of OA through one right angle from $\mathbf{i}$ to $\mathbf{j}$ is clockwise. If the set $\mathbf{i}, \mathbf{j}, \mathbf{k}$ is right-handed, so are $\mathbf{j}, \mathbf{k}, \mathbf{i}$ and $\mathbf{k}, \mathbf{i}, \mathbf{j}$.) If l, m, n are the direction cosines of the direction of $\mathbf{P}$, we have

$$\frac{X}{l} = \frac{Y}{m} = \frac{Z}{n} = |\mathbf{P}|.$$

In the particular case where all the vectors lie in a plane (and this is the case with which we shall be mostly concerned in this book) we take this plane to be the plane Oxy, and

$$X = |\mathbf{P}| \cos\alpha, \quad Y = |\mathbf{P}| \sin\alpha, \quad Z = 0,$$

where α $(0 \leqslant \alpha < 2\pi)$ is the inclination of the direction of $\mathbf{P}$ to Ox. (Alternatively, if we admit directed moduli, we can write

$$X = p\cos\alpha, \quad Y = p\sin\alpha, \quad Z = 0,$$

where $0 \leqslant \alpha < \pi$, and p is not necessarily positive.)

There is a vitally important distinction between the cases where the trihedral is rectangular and where it is not. If the trihedral is rectangular the number X depends only on the direction of Ox, not on the directions of Oy and Oz; if the trihedral is not rectangular, X depends on all three directions, not merely on the direction Ox. Thus if we take a direction Ox there is a definite value of the component of $\mathbf{P}$ in this direction, if we tacitly suppose the trihedral to be rectangular. This definite value is often spoken of as *the* component of $\mathbf{P}$ in the direction Ox, the rectangular axes being understood. When we use non-rectangular axes, and we need to distinguish between the component in the direction Ox that we get when we resolve in the directions Ox, Oy, Oz, and the definite value we should get if the axes were rectangular, we speak of the latter as the *projection* in the direction Ox.

2·4 Components of the vector sum. If $\mathbf{P}_1$ and $\mathbf{P}_2$ are given vectors, and $\mathbf{S} = \mathbf{P}_1+\mathbf{P}_2$, the component of $\mathbf{S}$ in the direction Ox (or, briefly, the x-component of $\mathbf{S}$) is the sum of the x-components of $\mathbf{P}_1$ and $\mathbf{P}_2$. We can prove this by elementary geometry. Alternatively, if

$$\mathbf{P}_1 = X_1\mathbf{i}+Y_1\mathbf{j}+Z_1\mathbf{k},$$

$$\mathbf{P}_2 = X_2\mathbf{i}+Y_2\mathbf{j}+Z_2\mathbf{k},$$

we have $$\mathbf{S} = (X_1+X_2)\mathbf{i}+(Y_1+Y_2)\mathbf{j}+(Z_1+Z_2)\mathbf{k},$$

and it follows that the x-component of $\mathbf{S}$ is X_1+X_2, because the resolution is unique. The axes are not necessarily rectangular, but it is assumed throughout that the prescribed directions are not parallel to a plane.

This theorem we have mentioned already (§ 1·11); it is the connecting link between the theory of measurement of scalar quantities and the theory of vectors. The converse is true, and provides a useful test for the vectorial character of a class K of directed quantities. Suppose that $\mathbf{P}_1$ and $\mathbf{P}_2$ are the vectors of any two members of K, and let us denote the components of $\mathbf{P}_r$ by (X_r, Y_r, Z_r). If the physical sum of the two members of K is itself a member of K, and if the components of its vector are $(X_1+X_2, Y_1+Y_2, Z_1+Z_2)$, then the law of summation is the vector law. The conditions in the definition of a class of vectors are fulfilled.

The theorem is readily extended to the sum of any finite number of vectors, and we deduce the fundamental result in the analytical statics of a particle—the analogue of the polygon of forces in the geometrical theory: 'The necessary and sufficient condition for the equilibrium of a system of forces acting on a particle is that the sums of the components in the three prescribed directions should vanish.'

2·5 Scalar product. The scalar product of the vectors $\mathbf{P}$ and $\mathbf{Q}$, written $\mathbf{P}.\mathbf{Q}$, is defined to be the number $|\mathbf{P}||\mathbf{Q}|\cos\alpha$, where α is the angle between the directions of $\mathbf{P}$ and $\mathbf{Q}$; we may take α in the range $0 \leqslant \alpha \leqslant \pi$. The scalar product is positive if α is acute, negative if α is obtuse; it vanishes if $\mathbf{P}$ is perpendicular to $\mathbf{Q}$. If $\mathbf{P}.\mathbf{Q} = 0$ we can infer $\mathbf{P} = \mathbf{0}$ or $\mathbf{Q} = \mathbf{0}$ or $\alpha = \frac{1}{2}\pi$. If

$\alpha = 0$, $\mathbf{P}.\mathbf{Q} = |\mathbf{P}||\mathbf{Q}|$, and if $\alpha = \pi$, $\mathbf{P}.\mathbf{Q} = -|\mathbf{P}||\mathbf{Q}|$. If we take $\mathbf{P} = \mathbf{Q}$ we get $\mathbf{P}.\mathbf{P} = |\mathbf{P}|^2$.

The vector quantities whose vectors are $\mathbf{P}$ and $\mathbf{Q}$ are not necessarily of the same kind, and the scalar product is the measure (not necessarily positive) of a scalar quantity associated with the two vector quantities with which we started.

We can represent $\mathbf{Q}$ as the vector sum of two vectors $\mathbf{Q}'$ and $\mathbf{Q}''$, of which $\mathbf{Q}'$ is parallel to $\mathbf{P}$ and $\mathbf{Q}''$ is perpendicular to $\mathbf{P}$. ($\mathbf{Q}'$ is the *projection* of $\mathbf{Q}$ in the direction of $\mathbf{P}$, § 2·3.) The directed modulus of $\mathbf{Q}'$ in the direction of $\mathbf{P}$ is $|\mathbf{Q}|\cos\alpha$, and it follows that

$$\mathbf{P}.\mathbf{Q} = \mathbf{P}.\mathbf{Q}'. \tag{1}$$

If $\mathbf{Q}_1$ and $\mathbf{Q}_2$ are two vectors representing vector quantities of the same kind we know from § 2·4 that

$$(\mathbf{Q}_1 + \mathbf{Q}_2)' = \mathbf{Q}_1' + \mathbf{Q}_2', \tag{2}$$

and it follows from equation (1) that

$$\mathbf{P}.(\mathbf{Q}_1 + \mathbf{Q}_2) = \mathbf{P}.\mathbf{Q}_1 + \mathbf{P}.\mathbf{Q}_2. \tag{3}$$

The fundamental properties of the scalar product are represented by the equations:

(i) $$\mathbf{P}.\mathbf{Q} = \mathbf{Q}.\mathbf{P},$$

(ii) $$\mathbf{P}.(\mathbf{Q}_1 + \mathbf{Q}_2 + \ldots + \mathbf{Q}_n) = \mathbf{P}.\mathbf{Q}_1 + \mathbf{P}.\mathbf{Q}_2 + \ldots + \mathbf{P}.\mathbf{Q}_n,$$

(iii) $$\mathbf{P}.k\mathbf{Q} = k(\mathbf{P}.\mathbf{Q}) = k\mathbf{P}.\mathbf{Q}.$$

We notice that the special case of (iii) in which k is a positive integer is already contained in (ii) if we take

$$\mathbf{Q}_1 = \mathbf{Q}_2 = \ldots = \mathbf{Q}_n = \mathbf{Q}$$

and use the alternative definition of $k\mathbf{Q}$ in § 2·2. The proofs of (i) and (iii) are almost immediate, and the proof of (ii) is by induction, starting from equation (3) above.

The concept of the scalar product is useful only when it satisfies a certain fundamental criterion, namely, that the scalar quantity of which it is the measure is thereby measured on a uniform scale. This implies that we understand what is meant by *addition* of the scalar quantity, and that, with the scale of measurement provided by the scalar product, physical addition is represented by algebraic addition. This criterion is satisfied by all the scalar products of interest in physics. The proof of this fact for a particular scalar product involves the interpretation of addition for this quantity, and usually requires an appeal to the property (ii) above, or to the immediate generalization of it contained in the equation

$$(\mathbf{P}_1+\mathbf{P}_2+\ldots+\mathbf{P}_m).(\mathbf{Q}_1+\mathbf{Q}_2+\ldots+\mathbf{Q}_n) = \sum_{r=1}^{m}\sum_{s=1}^{n}\mathbf{P}_r.\mathbf{Q}_s,$$

or to well-known properties of integrals.

The cases we shall first encounter are these: (1) one vector represents a force acting on a particle, and the other a displacement of the particle, and the scalar quantity measured by the scalar product is *work*; (2) one vector represents a force acting on a moving particle, and the other the velocity of the particle, and then the scalar quantity measured by the scalar product is *power*. We shall see later (§§ 3·6 and 3·7) that these scalar products do in fact satisfy the criterion just mentioned.

Finally, let us introduce rectangular axes, and write

$$\mathbf{P} = X\mathbf{i}+Y\mathbf{j}+Z\mathbf{k},$$

where $\mathbf{i}, \mathbf{j}, \mathbf{k}$ are unit vectors, mutually perpendicular, as before. We have, using the property (ii),

$$\mathbf{i}.\mathbf{P} = \mathbf{i}.(X\mathbf{i}+Y\mathbf{j}+Z\mathbf{k}) = X\mathbf{i}.\mathbf{i}+Y\mathbf{i}.\mathbf{j}+Z\mathbf{i}.\mathbf{k} = X,$$

since $\mathbf{i}.\mathbf{i} = 1$, $\mathbf{i}.\mathbf{j} = \mathbf{i}.\mathbf{k} = 0$. It follows that the projection of $\mathbf{P}$ in the direction of a given unit vector $\mathbf{E}$ is $(\mathbf{E}.\mathbf{P})\mathbf{E}$. Its directed modulus is $(\mathbf{E}.\mathbf{P})$, and, as we have already noticed, this can also be written as $|\mathbf{P}|\cos\alpha$, where α is the angle between the directions of $\mathbf{P}$ and of $\mathbf{E}$. We thus recover the fact that the scalar product of two vectors is the product of the modulus of one vector and the directed modulus of the projection on it of the other.

Again, if we have two vectors $\mathbf{P}_1$ and $\mathbf{P}_2$, and we consider them resolved into components along the axes, we have

$$\mathbf{P}_1.\mathbf{P}_2 = (X_1\mathbf{i}+Y_1\mathbf{j}+Z_1\mathbf{k}).(X_2\mathbf{i}+Y_2\mathbf{j}+Z_2\mathbf{k})$$
$$= X_1X_2+Y_1Y_2+Z_1Z_2.$$

Since the scalar product is independent of the choice of axes, we see that the expression $X_1 X_2 + Y_1 Y_2 + Z_1 Z_2$ is an *invariant*, i.e. an expression whose value does not change when we change the axes.

2·6 Vector calculus. In dynamics we deal largely with vector quantities whose vectors are not constants but functions of the time. For example, in the dynamics of a particle, we describe the position of the particle P by the position vector $\mathbf{r}$, which is the vector of the displacement OP relative to the base of reference. Here $\mathbf{r}$ is a vector function of the independent variable t, $\mathbf{r} = \mathbf{r}(t)$.

We also study vector quantities that are functions of position in space. If we describe a point of space by its position vector $\mathbf{r}$, and $\mathbf{P}$ is a function of position, we write $\mathbf{P} = \mathbf{P}(\mathbf{r})$. Observe at once that the symbol $\mathbf{r}$ now has a different status, since $\mathbf{r}$ is now the independent variable. The dependent variable $\mathbf{P}$ and the independent variable $\mathbf{r}$ are both vectors. When the vector $\mathbf{P}$ is the vector of a force we have a 'field of force'; the notion of a field of force is fundamental in the theory of electricity. It is clear that the idea of a vector function of a vector variable is a rather more complex idea than that of a vector function of a scalar variable.

To begin with we consider a vector function of the time, $\mathbf{P}(t)$. We represent the vector $\mathbf{P}$ by a segment OA in the vector diagram. As $\mathbf{P}$ varies the point A moves in the diagram. The line of thought follows very closely the familiar ideas of the theory of a function of one variable. We naturally begin with a definition of *continuity*: $\mathbf{P}(t)$ is continuous at t_0 if $\mathbf{P}(t) - \mathbf{P}(t_0) \to \mathbf{0}$ (i.e. if $|\mathbf{P}(t) - \mathbf{P}(t_0)| \to 0$) as $t \to t_0$. The vector $\mathbf{P}(t) - \mathbf{P}(t_0)$ is represented by the segment A_0A in the diagram, and the definition is merely the precise form of the statement that, if $\mathbf{P}(t)$ is continuous at t_0, then the length $|A_0A|$ is small when $|t - t_0|$ is small.

Next, we are inevitably led to the idea of a *derivative* or differential coefficient of the vector function. The line of thought is almost identical with that in the ordinary differential calculus. We consider an instant t_0, and values of t in the neighbourhood of t_0. Then $\mathbf{P}(t) - \mathbf{P}(t_0)$ is the vector $\overline{A_0A}$

in the diagram (Fig. 2·6), and the vector

$$\frac{1}{t-t_0}\{\mathbf{P}(t)-\mathbf{P}(t_0)\}$$

is the vector $\overline{A_0B}$, where $\overline{A_0A}$ is produced to B, and the directed length $\overline{A_0B}$ is $\dfrac{1}{t-t_0}$ times the length $\overline{A_0A}$; $\overline{A_0B}$ is in the same direction as $\overline{A_0A}$ if $t-t_0>0$. As $t\to t_0$ the point B may tend to a limiting position L. If this happens $\overline{A_0L}$ is defined

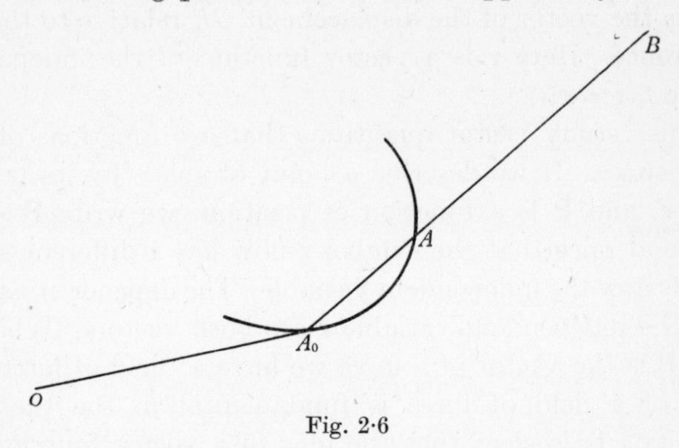

Fig. 2·6

to be the derivative of $\mathbf{P}$ at t_0. It is denoted by $(d\mathbf{P}/dt)_{t=t_0}$ or by $\dot{\mathbf{P}}(t_0)$. We can express our definition by the equation

$$\dot{\mathbf{P}}(t_0)=\lim_{t\to t_0}\frac{\mathbf{P}(t)-\mathbf{P}(t_0)}{t-t_0},$$

if this limit exists. (If the limit does not exist $\mathbf{P}$ has no derivative at t_0.) We can also write the equation in the form

$$\dot{\mathbf{P}}(t_0)=\lim_{\delta t\to 0}\frac{\mathbf{P}(t_0+\delta t)-\mathbf{P}(t_0)}{\delta t},$$

or, even more concisely,

$$\dot{\mathbf{P}}(t_0)=\lim_{\delta t\to 0}\frac{\delta\mathbf{P}}{\delta t}.$$

The derivative is also called the 'differential coefficient' and the 'rate of change', a usage familiar in the differential calculus. If $\dot{\mathbf{P}}(t_0)$ exists, $\mathbf{P}(t)$ is said to be differentiable at t_0. The derivative of a vector quantity whose vector is $\mathbf{P}(t)$ is the directed quantity whose vector is $\dot{\mathbf{P}}(t)$.

There is one point which should be noticed here, though discussion of it may conveniently be postponed (to § 3·4). We have just spoken of the directed quantity whose vector is $\dot{\mathbf{P}}$, and this implies the use of a particular scale of measurement for this new quantity. This scale of measurement is the so-called derived or theoretical scale. It is automatically determined so soon as the scales of measurement for $\mathbf{P}$ and for t are prescribed.

2·7 The vectorial character of the derivative. Is the derivative of a vector quantity itself a vector quantity? This question is fundamental. We have to inquire whether the derivative satisfies the three conditions laid down in the definition of a vector quantity (§ 1·7). The first is clearly satisfied—the derivative is a directed quantity. How about the second and third?

Now the second and third are concerned with the notion of the physical sum, so to deal with them we must envisage the possibility of *two simultaneous rates of change*. This means that in the interval from t_0 to $t_0+\delta t$, $\mathbf{P}$ receives two independent increments, $\delta_1\mathbf{P}$ and $\delta_2\mathbf{P}$. Since vector addition is commutative, there is no ambiguity in the interpretation of the total increment $\delta\mathbf{P}$, which is the vector sum of $\delta_1\mathbf{P}$ and $\delta_2\mathbf{P}$. Thus

$$\delta\mathbf{P} = \delta_1\mathbf{P}+\delta_2\mathbf{P},$$

and therefore, by (iii) of § 2·2,

$$\frac{\delta\mathbf{P}}{\delta t} = \frac{\delta_1\mathbf{P}}{\delta t}+\frac{\delta_2\mathbf{P}}{\delta t},$$

where of course the addition is vectorial. We assume that, as $\delta t\to 0$, $\delta_1\mathbf{P}/\delta t$ and $\delta_2\mathbf{P}/\delta t$ tend to limits $[\dot{\mathbf{P}}(t_0)]_1$ and $[\dot{\mathbf{P}}(t_0)]_2$. Now the theorem which states that the limit of the sum of two functions is the sum of their limits, which is familiar in analysis, is also true for vector functions. Thus, letting $\delta t\to 0$, $\delta\mathbf{P}/\delta t$ tends to a limit which is the vector sum of $[\dot{\mathbf{P}}(t_0)]_1$ and $[\dot{\mathbf{P}}(t_0)]_2$. Thus the third condition in the definition of vector quantities is satisfied. The derivative of a vector quantity is itself a vector quantity.

2·8 Properties of the derivative. Two simple properties of the derivative are easily established. (i) If $\mathbf{P} = \theta\mathbf{E}$, where $\mathbf{E}$ is a fixed unit vector, and θ is a differentiable function of t, then $\dot{\mathbf{P}}$ exists, and is equal to $\dot{\theta}\mathbf{E}$. To prove this, we have

$$\frac{\delta\mathbf{P}}{\delta t} = \frac{\delta\theta}{\delta t}\mathbf{E},$$

so
$$\left|\frac{\delta\mathbf{P}}{\delta t} - \dot{\theta}\mathbf{E}\right| = \left|\left(\frac{\delta\theta}{\delta t} - \dot{\theta}\right)\mathbf{E}\right| = \left|\frac{\delta\theta}{\delta t} - \dot{\theta}\right|,$$

which $\to 0$ with δt, and the result follows.

(ii) If $\mathbf{S} = \mathbf{P} + \mathbf{Q}$, and $\mathbf{P}$ and $\mathbf{Q}$ are differentiable, then $\mathbf{S}$ is differentiable, and

$$\dot{\mathbf{S}} = \dot{\mathbf{P}} + \dot{\mathbf{Q}}.$$

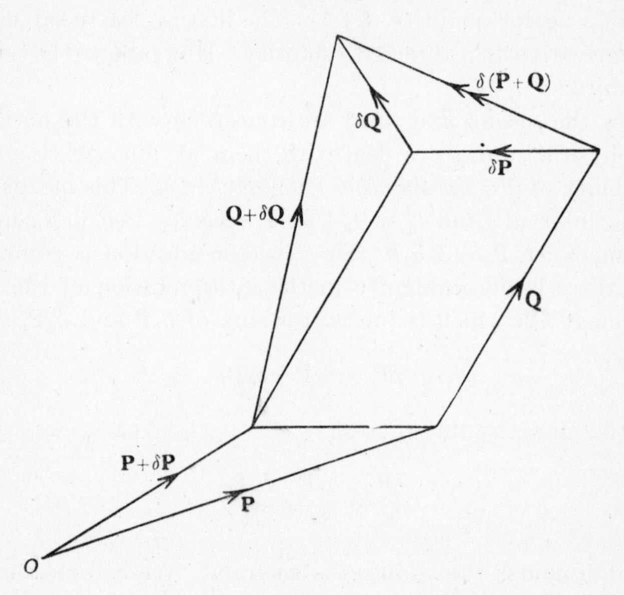

Fig. 2·8

The proof is simple; if δ denotes an increment that occurs in the interval t_0 to $t_0 + \delta t$, we have

$$\begin{aligned}\delta\mathbf{S} &= \delta(\mathbf{P} + \mathbf{Q}) \\ &= (\mathbf{P} + \delta\mathbf{P} + \mathbf{Q} + \delta\mathbf{Q}) - (\mathbf{P} + \mathbf{Q}) \\ &= \delta\mathbf{P} + \delta\mathbf{Q}.\end{aligned}$$

Fig. 2·8 illustrates this equation (in general it is not plane). Thus

$$\frac{\delta \mathbf{S}}{\delta t} = \frac{\delta \mathbf{P}}{\delta t} + \frac{\delta \mathbf{Q}}{\delta t},$$

and the result follows on letting $\delta t \to 0$.

We easily prove, by induction, a similar theorem for the sum of any number of differentiable vector functions of t,

$$\frac{d}{dt}(\mathbf{P}_1 + \mathbf{P}_2 + \ldots + \mathbf{P}_n) = \dot{\mathbf{P}}_1 + \dot{\mathbf{P}}_2 + \ldots + \dot{\mathbf{P}}_n.$$

If we introduce a fixed trihedral of reference, not necessarily rectangular, and resolve $\mathbf{P}$ into its components in the directions of the axes, we have

$$\mathbf{P} = X\mathbf{i} + Y\mathbf{j} + Z\mathbf{k}.$$

If X, Y, Z are differentiable functions of t we have

$$\begin{aligned} \frac{d\mathbf{P}}{dt} &= \frac{d}{dt}(X\mathbf{i} + Y\mathbf{j} + Z\mathbf{k}) \\ &= \frac{d}{dt}(X\mathbf{i}) + \frac{d}{dt}(Y\mathbf{j}) + \frac{d}{dt}(Z\mathbf{k}) \\ &= \dot{X}\mathbf{i} + \dot{Y}\mathbf{j} + \dot{Z}\mathbf{k}, \end{aligned}$$

and $\dot{\mathbf{P}}$ exists, and its components are $\dot{X}, \dot{Y}, \dot{Z}$. Indeed, it is clear that a necessary and sufficient condition for $\mathbf{P}$ to be differentiable is that its components X, Y, Z should be differentiable functions of t.

2·9 Velocity. The primary application of the idea of differentiation is to the position vector $\mathbf{r}(t)$ of a moving particle A (§ 1·10). The derivative, if it exists, is the vector $\dot{\mathbf{r}}$ or $\mathbf{v}$ of the *velocity*. Usually the path of the particle is a smooth, rectifiable curve. Suppose in the first instance that the particle moves in such a way that the arc s, measured from a fixed point on the curve to A, increases with t. Then (referring to Fig. 2·6) we notice first that the direction of $\overline{A_0B}$ tends to that of the tangent at A_0, drawn in the direction of increasing s, so the direction of the velocity is the direction of this tangent.

Also $$|A_0 B| = \frac{|A_0 A|}{|t-t_0|} = \frac{|s-s_0|}{|t-t_0|}\frac{|A_0 A|}{|s-s_0|},$$

whence $$|A_0 L| = \dot{s},$$

since the ratio of arc to chord tends to 1. If now we denote by $\mathbf{T}$ a unit vector in the direction of the tangent to the path and in the sense of increasing s, we have

$$\mathbf{v} = \dot{s}\mathbf{T}.$$

(In the proof given we supposed $\dot{s} > 0$, but it is easy to see that the result is true also if $\dot{s} < 0$.)

2·10 Differentiation of a product.

(i) We have seen already that, if $\mathbf{E}$ is a fixed unit vector,

$$\frac{d}{dt}(\theta\mathbf{E}) = \dot{\theta}\mathbf{E}.$$

We now consider the differentiation of a product $\theta\mathbf{P}$, where θ is a differentiable scalar function of t, and $\mathbf{P}$ is a differentiable vector function of t. If δ denotes increments which occur in the interval from t_0 to $t_0 + \delta t$, we have

$$\begin{aligned}\delta(\theta\mathbf{P}) &= (\theta+\delta\theta)(\mathbf{P}+\delta\mathbf{P}) - \theta\mathbf{P} \\ &= \theta\delta\mathbf{P} + \delta\theta\mathbf{P} + \delta\theta\delta\mathbf{P},\end{aligned}$$

and $$\frac{1}{\delta t}\delta(\theta\mathbf{P}) = \theta\frac{\delta\mathbf{P}}{\delta t} + \frac{\delta\theta}{\delta t}\mathbf{P} + \frac{\delta\theta}{\delta t}\frac{\delta\mathbf{P}}{\delta t}\delta t.$$

Now as $\delta t \to 0$, $\theta\frac{\delta\mathbf{P}}{\delta t} \to \theta\dot{\mathbf{P}}$, since

$$\left|\theta\frac{\delta\mathbf{P}}{\delta t} - \theta\dot{\mathbf{P}}\right| = \left|\theta\left(\frac{\delta\mathbf{P}}{\delta t} - \dot{\mathbf{P}}\right)\right|$$

which $\to 0$ with δt. Also $\frac{\delta\theta}{\delta t}\mathbf{P} \to \dot{\theta}\mathbf{P}$ (as in the proof already given for the case when $\mathbf{P}$ is constant). Finally

$$\frac{\delta\theta}{\delta t}\frac{\delta\mathbf{P}}{\delta t}\delta t \to \mathbf{0},$$

since $$\left|\frac{\delta\theta}{\delta t}\frac{\delta\mathbf{P}}{\delta t}\delta t\right| = \left|\frac{\delta\theta}{\delta t}\right|\left|\frac{\delta\mathbf{P}}{\delta t}\right||\delta t| \to 0.$$

We see, therefore, that $\theta\mathbf{P}$ is itself differentiable, and

$$\frac{d}{dt}(\theta\mathbf{P}) = \theta\dot{\mathbf{P}} + \dot{\theta}\mathbf{P}.$$

(ii) The proof for the differentiation of a scalar product is almost as simple. Let

$$R = \mathbf{P}.\mathbf{Q},$$

where $\mathbf{P}$ and $\mathbf{Q}$ are differentiable vector functions of t. Then

$$\begin{aligned} \delta R &= (\mathbf{P}+\delta\mathbf{P}).(\mathbf{Q}+\delta\mathbf{Q}) - \mathbf{P}.\mathbf{Q} \\ &= (\mathbf{P}+\delta\mathbf{P}).\mathbf{Q} + (\mathbf{P}+\delta\mathbf{P}).\delta\mathbf{Q} - \mathbf{P}.\mathbf{Q} \\ &= \delta\mathbf{P}.\mathbf{Q} + \mathbf{P}.\delta\mathbf{Q} + \delta\mathbf{P}.\delta\mathbf{Q}. \end{aligned}$$

Therefore $$\frac{\delta R}{\delta t} = \frac{\delta\mathbf{P}}{\delta t}.\mathbf{Q} + \mathbf{P}.\frac{\delta\mathbf{Q}}{\delta t} + \left(\frac{\delta\mathbf{P}}{\delta t}.\frac{\delta\mathbf{Q}}{\delta t}\right)\delta t.$$

Now $$\frac{\delta\mathbf{P}}{\delta t}.\mathbf{Q} \to \dot{\mathbf{P}}.\mathbf{Q},$$

since $$\left|\frac{\delta\mathbf{P}}{\delta t}.\mathbf{Q} - \dot{\mathbf{P}}.\mathbf{Q}\right| = \left|\left(\frac{\delta\mathbf{P}}{\delta t} - \dot{\mathbf{P}}\right).\mathbf{Q}\right| \leqslant \left|\frac{\delta\mathbf{P}}{\delta t} - \dot{\mathbf{P}}\right| |\mathbf{Q}|,$$

which tends to zero with δt; and similarly

$$\mathbf{P}.\frac{\delta\mathbf{Q}}{\delta t} \to \mathbf{P}.\dot{\mathbf{Q}}.$$

The last term on the right

$$\left(\frac{\delta\mathbf{P}}{\delta t}.\frac{\delta\mathbf{Q}}{\delta t}\right)\delta t \to 0,$$

since $$\left|\frac{\delta\mathbf{P}}{\delta t}.\frac{\delta\mathbf{Q}}{\delta t}\right| \leqslant \left|\frac{\delta\mathbf{P}}{\delta t}\right|\left|\frac{\delta\mathbf{Q}}{\delta t}\right| \to |\dot{\mathbf{P}}||\dot{\mathbf{Q}}|.$$

Finally, then, we see that R is differentiable, and

$$\dot{R} = \mathbf{P}.\dot{\mathbf{Q}} + \dot{\mathbf{P}}.\mathbf{Q}.$$

One application of this result, of particular importance, is that in which $\mathbf{P} = \mathbf{Q} = \mathbf{v}$. We have

$$\frac{d}{dt}(\tfrac{1}{2}|\mathbf{v}|^2) = \frac{1}{2}\frac{d}{dt}(\mathbf{v}.\mathbf{v}) = \mathbf{v}.\dot{\mathbf{v}}.$$

2·11 Second derivative. If a vector function of t, $\mathbf{P}(t)$, has a derivative $\dot{\mathbf{P}}(t)$ at and in the neighbourhood of t_0, this vector may itself have a derivative at t_0, which is called the second derivative of $\mathbf{P}$, and denoted by $\left[\frac{d^2}{dt^2}\mathbf{P}(t)\right]_{t=t_0}$ or $\ddot{\mathbf{P}}(t_0)$. If we introduce a fixed trihedral of reference, and the components of $\mathbf{P}$ are X, Y, Z, then the components of $\ddot{\mathbf{P}}$ are $\ddot{X}, \ddot{Y}, \ddot{Z}$. Indeed, it is clear that a necessary and sufficient condition for $\mathbf{P}$ to have a second derivative at t_0 is that X, Y, Z should each have a second derivative at this instant. The second derivative is the vector of a new vector quantity which is itself a function of t.

CHAPTER III

THE FUNDAMENTAL IDEAS OF NEWTONIAN MECHANICS

3·1 Newtonian mechanics. We begin with a discussion of the motion of a single particle or point-mass. We describe the position of the particle relative to the chosen base (§ 1·10) by the position vector $\mathbf{r}(t)$. The vector $\mathbf{r}$ is always a continuous function of t. For all but a finite number of exceptional values of t it possesses a first derivative $\dot{\mathbf{r}}$ or $\mathbf{v}$, and a second derivative $\ddot{\mathbf{r}}$ or $\dot{\mathbf{v}}$ or $\mathbf{f}$. The vector quantity whose vector is $\dot{\mathbf{r}}$ is the velocity, and the vector quantity whose vector is $\ddot{\mathbf{r}}$ is the acceleration. The positive scalar quantity $|\dot{\mathbf{r}}|$ or $|\mathbf{v}|$ is sometimes called the *speed*; this is a useful terminology, but in practice the word 'velocity' is often used not only for the vector quantity $\mathbf{v}$ but also for $|\mathbf{v}|$, i.e. for the speed; the context makes it clear which usage is intended.

From the definition of acceleration we have

$$\mathbf{f}(t_0) = \lim_{\delta t \to 0} \frac{1}{\delta t}\{\mathbf{v}(t_0+\delta t) - \mathbf{v}(t_0)\}.$$

We remind the reader that in this definition velocities are thought of as non-localized vector quantities, and the fact that the particle is at a different position at the instant $t_0+\delta t$ from its position at t_0 is here irrelevant. (Later on, in different circumstances, we shall find that velocities must sometimes be thought of as localized. For example, in the theory of central orbits we shall sometimes speak of the moment of momentum about a point O; in that case the momentum $m\mathbf{v}$ must be thought of as localized in the particle. For our present purpose, however, velocities are non-localized.)

If we introduce a trihedral of reference, fixed in the base, we have

$$\mathbf{r} = x\mathbf{i} + y\mathbf{j} + z\mathbf{k},$$

$$\dot{\mathbf{r}} = \dot{x}\mathbf{i} + \dot{y}\mathbf{j} + \dot{z}\mathbf{k},$$

$$\ddot{\mathbf{r}} = \ddot{x}\mathbf{i} + \ddot{y}\mathbf{j} + \ddot{z}\mathbf{k}.$$

The components x, y, z of the position vector $\mathbf{r}$ (§ 2·3) are the Cartesian coordinates of P, referred to the axes Ox, Oy, Oz. The components of velocity are $\dot{x}, \dot{y}, \dot{z}$, and the components of acceleration are $\ddot{x}, \ddot{y}, \ddot{z}$. The axes are not necessarily rectangular. But it is usually simplest to have rectangular axes, and it is frequently taken for granted, without explicit statement, that the axes are rectangular and right-handed.

In statics we consider a particle or a body or a system of bodies at rest (i.e. permanently at rest, not merely instantaneously at rest), and the forces that maintain this state of rest. In dynamics there is no longer equilibrium, and we study the motion resulting from the action of force. We begin with the simplest problem, the motion of a particle.

The corner-stone of Newtonian dynamics is the realization that the property of motion that is immediately linked up with force is the *acceleration*. Given the mass m of a particle and the force $\mathbf{P}$ acting on it the acceleration $\mathbf{f}$ is determined. This was a far-reaching discovery—certainly one of the most far-reaching ever made. Until the time of Galileo no precise idea of the relation between force and the consequent motion had emerged. The Greeks had held that force is necessary to maintain motion, a belief which might be supposed to imply that the essential connexion is between force and velocity. Newton held that *the velocity of a particle remains constant when no force acts on it*; this is the First Law of Motion. The essential connexion is between force and acceleration, or, more generally, if we consider not one single particle but a number of different particles, between force and *kineton*. The kineton is the vector quantity whose vector is $m\mathbf{f}$; using the looser phraseology we may say that kineton is mass times acceleration.

Newton's Second Law of Motion states that the motion of a particle under the action of a force $\mathbf{P}$ is determined by the rule that *the kineton is in the same direction* (*and sense*) *as the force, and the magnitude of the kineton is proportional to the magnitude of the force.* In the classical form

$$\mathbf{P} = km\mathbf{f},$$

where k is a positive constant (i.e. an absolute constant, independent of m).

What is the status of this law? One possible approach to the subject would be to regard Newton's theory as axiomatic, and to work out as an exercise in abstract mathematics the consequences of the theory, without claiming any relationship between the results of the theory and the events of the actual world. But this would be a sadly mistaken policy. It would rob the theory of the greater part of its interest, and it would deprive us of the immensely valuable help provided by our knowledge of what does in fact happen in the real world. It is therefore altogether preferable to regard Newtonian mechanics as a science based upon experiment. But the material that we handle in theoretical mechanics is a simplification of the material of our actual experience; we deal with an idealized and simplified conceptual model of the real world. Thus, we consider first the theory of force acting on a particle; a particle, or point-mass, is the idealization of a small massive body, so small that we may think of all its mass as concentrated in a point. When we come to deal with bodies of finite size, we consider them in the first instance to be rigid, i.e. not subject to deformation under stress. Needless to say, no actual body is rigid in the strict theoretical sense. But many bodies are such that the change of shape which they suffer is very small under the stresses which they ordinarily bear, and the theoretical rigid body is an idealization of such nearly rigid bodies. Moreover, we go further, and derive the theory of the motion of the rigid body from the theory for particles, by regarding the rigid body as an aggregate of particles set in a rigid and imponderable frame. For example, we can conceive of the frame as built up of light rigid rods, freely hinged together at their ends, the particles being attached at the joints. The purpose of these simplifications is of course to obtain problems more tractable than those provided by the highly complex systems of the actual world. In a similar way, the rest of the apparatus of theoretical mechanics is a simplification of the systems met with in nature. We speak of a rigid rod—an idealization of a beam so thin that we can, for the purpose of our calculations, regard the matter composing it as lying on a straight line, and so stiff and inelastic that it does not stretch or bend; an inelastic string—an idealization of a cord, thin enough to be regarded as linear, and inelastic enough to be sensibly of invariable length; elastic strings (that exactly obey Hooke's law), perfectly flexible strings (that offer no resistance whatever to bending), bodies with perfectly smooth surfaces, and so on. With these we build up our simplified model of the real world, simplified enough to be amenable to calculation, but sufficiently accurate to give us results which are nearly true in the real world. It is hardly necessary to point out that different simplifications are required in different circumstances, and that a model valid in one context is invalid in another. A conspicuous example is the idea of a rigid body described above; it is adequate for our immediate purpose, but would be inadequate in another branch of physics, such as the theory of heat.

3·2 The second law of motion. We return to our discussion of the dynamics of a single particle. As we have already remarked, we shall regard the subject as one founded on experiment. Indeed, the most convincing justification of the theory is its almost complete agreement with the observed results in astronomy and in a diversity of terrestrial experiments. But it will be useful, in order to provide a concise exposition of the content of Newton's second law, to imagine a series of primitive experiments, experiments rather simpler than any that are practicable in the laboratory. This series of experiments consists simply of observing the motion of particles when acted on first by a single force, and then simultaneously by two or more forces. Any practicable experiments would be more complex than these that we describe, but if we suppose by way of illustration that such a series of experiments could be performed, they would reveal the results enumerated below. In experiments (i) to (iv) a single force **P** acts on a particle of mass m, in experiment (v) two forces act simultaneously. (We have spoken here of a force **P** and of a mass m. We recall that, strictly speaking, **P** is a *vector*, the vector of the force, and m is a *number*, the measure of the mass.)

The results are these:

(i) The acceleration of the particle is in the direction of the force acting on it, and in the same sense.

(ii) The magnitude of the acceleration depends only on the mass of the particle and the magnitude of the force. In other words, there exists a uniform (i.e. single-valued) function $\phi(m, |\mathbf{P}|)$ such that $|\mathbf{f}| = \phi(m, |\mathbf{P}|)$.

(iii) For a given particle (m fixed) $|\mathbf{f}| \propto |\mathbf{P}|$.

(iv) For a given force ($\mathbf{P}$ fixed) $|\mathbf{f}| \propto 1/m$.

We can sum up the results so far in the statement

$$\mathbf{f} \propto \frac{1}{m}\mathbf{P},$$

or in the classical form already mentioned

$$\mathbf{P} = km\mathbf{f},$$

where k is an absolute positive constant.

(v) If two forces $\mathbf{P}, \mathbf{P}'$ act simultaneously on the particle its acceleration is the vector sum of the accelerations it would acquire under the actions of the two forces separately.

We are thinking of this result as one based on experimental observation like the others, but it should be noticed that it has a rather different status from the others because it would result as a *deduction* from either of two highly plausible assumptions. (a) We might assume that the two forces acting together are equivalent to their vector sum just as in statics. With this assumption the acceleration $\mathbf{f}''$ is given by the equation

$$\mathbf{P} + \mathbf{P}' = km\mathbf{f}''.$$

If $\mathbf{P} = km\mathbf{f}$, $\mathbf{P}' = km\mathbf{f}'$, we see that

$$\mathbf{f}'' = \mathbf{f} + \mathbf{f}',$$

which is the result observed in experiment (v). (b) We might assume that each force produces its own individual effect, just as if the other force were absent (the hypothesis of 'the physical independence of forces'). With this assumption the acceleration produced is the vector sum $\mathbf{f} + \mathbf{f}'$, and again we obtain the result observed in experiment (v). For the development of the theory it is indifferent whether we regard (v) as an experimental result, or whether we prefer to accept one or other of these hypotheses.

In theoretical mechanics we choose our unit of force such that $k = 1$, and we obtain the fundamental equation

$$\mathbf{P} = m\mathbf{f}.$$

If we use this derived or theoretical unit of force, Newton's second law asserts that the kineton created by a force is equal vectorially (i.e. both in magnitude and direction) to the force. But we must be on our guard against such a form of words as 'force *is* kineton' which would reduce the second law of motion to a mere tautology.

3·3 The Newtonian base. The description given above of the idealized experiments from which we deduced the second law of motion is incomplete until we specify the base to which the measurements are referred. If the law is valid for one base it will not in general be valid for a second base which is in motion relative to the first. The essence of Newton's mechanics is the assumption that *there exists a base of reference for which the second law of motion holds*. We shall assume this to be true, and we shall call such a base a *Newtonian base*.

Assuming that a Newtonian base exists, the next step is to identify it in the actual world. A precise determination would be an almost unattainable ideal, but we can be certain that for a base determined by the so-called fixed stars the Newtonian theory is valid to a very high degree of accuracy.

The Newtonian base is clearly not unique. If we have a Newtonian base A, and a second base B moves uniformly (i.e. with constant velocity) and without rotation relative to A, then B is also a Newtonian base.

Again, suppose we have a Newtonian base A, and a second base B moves with uniform acceleration $\mathbf{f}_0$, and without rotation, relative to A. If we study the motion of a particle relative to the second base B, the motion of the base is compensated for if we add a *uniform field* (i.e. a force constant in magnitude and direction) $-\mathbf{f}_0$ per unit mass. In other words, the motion relative to B (as measured by an observer who takes B as his base of reference) is given by Newton's second law, just as if B were a Newtonian base, if a fictitious force $-m\mathbf{f}_0$ is superposed on the forces already acting on the particle. The proof is simple. If $\mathbf{f}$ is the acceleration of the particle relative to B, its acceleration relative to the Newtonian base A is $\mathbf{f}+\mathbf{f}_0$ (cf. § 10·5), and, by the second law of motion,

$$m(\mathbf{f}+\mathbf{f}_0) = \mathbf{P},$$

where $\mathbf{P}$ is the resultant of the forces acting on the particle. Hence

$$m\mathbf{f} = \mathbf{P} - m\mathbf{f}_0,$$

which proves the theorem. It turns out that an analogous result is valid for dynamical systems of a more general type, not merely for a single particle.

If the motion of B relative to A is more complicated, the problem of compensating for the motion of the base is correspondingly more difficult. But in certain cases, if the relative motion is small enough, the requisite compensation may be negligible for some purposes, and for these purposes we may form equations of motion just as if B were a Newtonian base. The most conspicuous case occurs when we take the base B to be the earth. Often we are concerned with phenomena taking place in a neighbourhood of a point O on the earth's surface, the scale of the problem (i.e. the size of the part of the orbit considered) being small in comparison with the earth's radius. We take axes Ox, Oy, Oz through O, fixed relative to the earth. As a first approximation we may neglect the motion of the earth, and treat the earth itself, or the trihedral $Oxyz$, as a Newtonian base. (We shall find, as a second approximation, that we can partially compensate for the fact that the base is not Newtonian, by adding a fictitious uniform field, as in the simpler case just considered; but this time the compensation is only approximate, and only valid for phenomena in the neighbourhood of O. Cf. § 10·14.)

3·4 Units. The measure of the speed of a moving particle is $|\dot{\mathbf{r}}|$, which implies that the unit of speed is determined when the units of length and time have been chosen. If the particle moves in the positive direction along the axis Ox, so that $\mathbf{r} = x\mathbf{i}$, the speed is $\dot{x}$. If $\dot{x}$ has the constant value 1, $x - x_0 = t - t_0$. The unit of speed is the speed in which unit length is covered in unit time. This unit is the *derived* or *theoretical* unit. If the units of length and time are the centimetre and the second, the derived unit of speed is 1 cm./sec.

But, as we have already remarked (§ 1·3), the choice of unit of a scalar quantity is arbitrary. There is no obligation on us to adopt this unit of speed, except as a matter of convenience. If we used a unit n times as large, the velocity would be $\frac{1}{n}\dot{\mathbf{r}}$.

In the same way, we have defined the acceleration vector as $\ddot{\mathbf{r}}$, which implies a unit of acceleration (or, more precisely, of the associated scalar quantity $|\ddot{\mathbf{r}}|$) of 1 cm./sec./sec. If the particle moves along Ox with constant unit acceleration, its speed increases by 1 cm./sec. every second.

The unit of force is chosen in theoretical mechanics, as we have seen, in such a way that Newton's second law of motion takes the form $\mathbf{P} = m\mathbf{f}$. The unit force produces unit acceleration in unit mass. This again is merely a matter of convenience; it is the simplest unit to use in theoretical work, but for practical purposes other units are commonly used.

If we measure time in seconds, length in centimetres, and mass in grams, the theoretical unit of force is called a *dyne*; it is the force which gives a mass of 1 g. an acceleration of 1 cm./sec./sec. It is easy to see what is the relation between this theoretical unit and the more practical unit of the weight of a gram. If a body—small enough to be thought of as a particle—be allowed to fall freely from rest, it falls vertically downwards with an acceleration which is constant, and the same for all bodies. (We suppose here that air resistance is negligible, or that the motion takes place in a vacuum.) A historic test of this important observation was made in the famous experiment of Galileo at Pisa. This constant acceleration is always denoted by g; its actual value is about 981 cm./sec./sec. If we suppose, for simplicity, that the body has a mass of 1 g., so that the force on it is the weight of 1 g., then this force is, by Newton's law, also equal to g dynes. Thus a force of 1 g. weight is equal to a force of g dynes. The theoretical unit of force is therefore about 1/981 times the weight of 1 g.

If the scale of the problem is small in comparison with the size of the earth, the weight and the direction in which it acts (referred to axes moving with the earth) are sensibly constant, and we have an example of a 'uniform field of force'. (But we must remember that the base is not Newtonian and the uniform field that we observe is not strictly the effect of gravity alone. As we have noticed above (§ 3·3) we can partially compensate for the fact that the base is not Newtonian by adding a fictitious correcting field. What we usually call the weight is the force of gravity augmented by this fictitious correcting field. We shall see later on (§ 10·14) that for points near the Earth's surface the correcting field is very small in comparison with gravity. For problems on a sufficiently small scale both the field of gravity and the correcting field are sensibly uniform, and the resultant field—only slightly different from the field of gravity—is also uniform. This is true for practical purposes, for example in the theory of the simple pendulum, and even in the theory of projectiles.)

If we measure time in seconds, length in feet, and mass in pounds, the theoretical unit force is called a poundal. We use the same symbol as before, g, to denote the acceleration of a body falling freely, but its numerical value is of course different in the different units. In foot-pound-second units the value of g is about 32·2 ft./sec./sec. (32 is a conventional approximation). The force on a mass of m pounds falling freely is m pounds weight; it is also mg poundals, so one pound weight is equal to g poundals. The poundal is thus about equal to the weight of half an ounce.

The value of g differs slightly at different places, increasing with latitude and decreasing with increase of height above sea-level; but the greatest variation is only of the order of 0·5 per cent, and is negligible for many purposes. As a first approximation the weight of a body is the same at all places on the earth's surface.

3·5 Work. If a constant force $\mathbf{P}_0$ (i.e. a force constant both in magnitude and direction) acts on a particle, and the particle suffers a displacement $\mathbf{Q}$, the work W done by the force is defined to be *the scalar product of force and displacement*,

$$W = \mathbf{P}_0 . \mathbf{Q}.$$

Here $\mathbf{P}_0$ is the vector of the force, $\mathbf{Q}$ the vector of the displacement, and W is a number, the measure of the scalar quantity *work*.

We can express this also in other ways. We may write

$$W = |\mathbf{P}_0|\,|\mathbf{Q}|\cos\alpha,$$

where α is the angle between the directions of $\mathbf{P}_0$ and $\mathbf{Q}$; and this is equivalent to the statement that the work is equal to the product of the magnitude of the displacement, and the projection of the force on the direction of the displacement.

If we introduce rectangular axes, and denote the components of $\mathbf{P}_0$ by X_0, Y_0, Z_0 and the components of $\mathbf{Q}$ by $\delta x, \delta y, \delta z$, we have

$$W = X_0\,\delta x + Y_0\,\delta y + Z_0\,\delta z.$$

More generally, suppose the particle moves from a point A to a point B along a curve of simple type (Fig. 3·5*a*), and that the force $\mathbf{P}$ acting on the particle is not constant, but varies continuously as the particle moves. The work done by the force $\mathbf{P}$ during the journey from A to B is defined as the line-integral

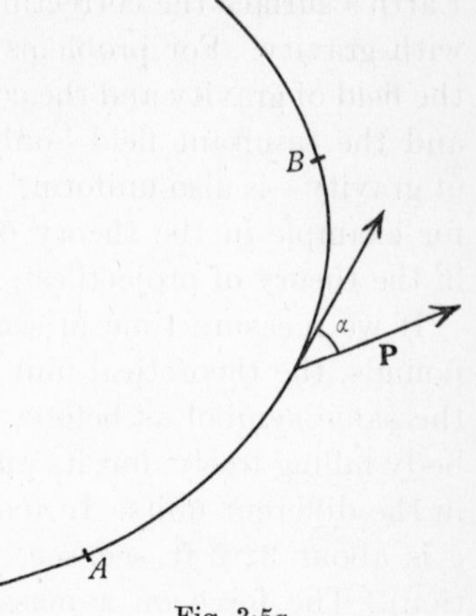

Fig. 3·5*a*

$$W = W_{AB} = \int_A^B (X dx + Y dy + Z dz)$$

taken along the curve. We observe at once that this is consistent with the definition already given for the special case where $\mathbf{P}$ is constant.

The line-integral can be expressed in other ways. If we think of the coordinates of a point on the curve as expressed in terms of the arc s, measured from any convenient fixed point on the curve, and we denote by $\mathbf{T}$ a unit vector in the direction of the tangent to the curve, drawn in the direction of increasing s, the components of $\mathbf{T}$ are $\dfrac{dx}{ds}, \dfrac{dy}{ds}, \dfrac{dz}{ds}$.

Thus $$W_{AB} = \int_A^B \left(X\frac{dx}{ds} + Y\frac{dy}{ds} + Z\frac{dz}{ds} \right) ds = \int_A^B \mathbf{P}.\mathbf{T}\,ds, \tag{1}$$

or, if we write $d\mathbf{r}$ for $\mathbf{T}\,ds$,

$$W_{AB} = \int_A^B \mathbf{P}.d\mathbf{r}.$$

We can also write (1) in the form

$$W_{AB} = \int_A^B |\mathbf{P}| \cos\alpha\, ds,$$

where α is the angle between $\mathbf{P}$ and $\mathbf{T}$.

If the particle moves in a field of force (i.e. the force is a function of position in space, $\mathbf{P} = \mathbf{P}(\mathbf{r})$) the work done by the field depends on the termini A and B of the path, and also in general on the path from A to B. It does not depend on the speed at which the path is described.

Example. *If $X = -ky$, $Y = kx$, $Z = 0$, to find the work done when the particle moves in the plane $z = 0$ from O to B (a, a), (i) by the path OAB, (ii) by the path OCB (Fig. 3·5b).*

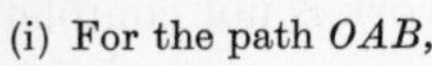

(i) For the path OAB,

$$W = \int_O^A X\,dx + \int_A^B Y\,dy$$
$$= 0 + \int_0^a ka\,dy$$
$$= ka^2.$$

(ii) For the path OCB,

$$W = \int_O^C Y\,dy + \int_C^B X\,dx$$
$$= 0 - \int_0^a ka\,dx$$
$$= -ka^2.$$

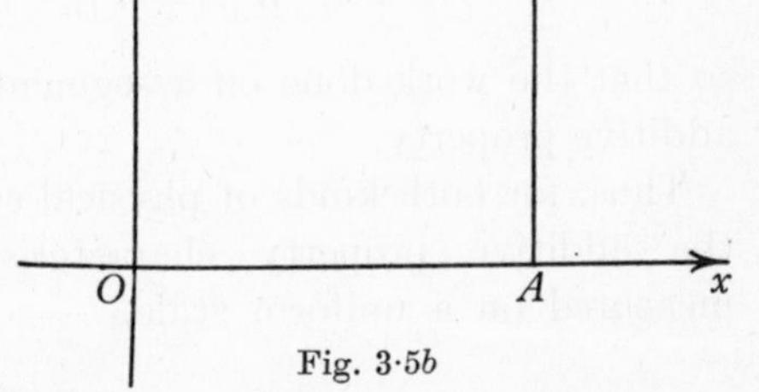

Fig. 3·5b

The work done by the field when the particle moves right round the square $OABCO$ is $2ka^2$. More generally, if the particle moves round any simple closed curve

$$W = \oint X\,dx + Y\,dy$$
$$= k \oint x\,dy - y\,dx$$
$$= 2kS,$$

where S is the area enclosed by the curve.

3·6 The scalar character of work. So far we have tacitly assumed that the physical quantity work which we have defined is in fact a scalar quantity measured on a uniform scale. It is clearly a non-directed quantity, and our definition includes the determination of the measure on a certain scale. There remains the task postponed from § 2·5, namely, to establish the additive property which is the third distinctive property in the definition (§ 1·1).

To begin with we notice that if $\mathbf{P}$ is a displacement of a particle, and $\mathbf{Q}_1, \mathbf{Q}_2, \ldots, \mathbf{Q}_n$ are (constant) forces acting on the particle, then the property (ii) of § 2·5 ensures that the work done by the vector sum of the forces is the algebraic sum of the works done by the separate forces. More generally, if the particle moves along a curve, and a number of forces, not necessarily constant, act simultaneously on the particle at each point of its path, the work done by the resultant of the forces on any segment of the path is the algebraic sum of the works done by the separate forces.

But this account of the scalar quantity work is not complete, for there is a different way in which the notion of addition arises. Suppose that a particle moves on a curve under the action of a variable force. If A, B, C are three points on the path, we have, by a well-known property of the line-integral,

$$W_{AC} = W_{AB} + W_{BC},$$

so that the work done on a segment of the path also has the additive property.

Thus, for both kinds of physical summation, work possesses the additive property characteristic of scalar quantities measured on a uniform scale.

3·7 Power. Power is rate of doing work, i.e. the work done per unit of time. Suppose a particle, acted on by a force $\mathbf{P}$, moves from a point A at $t = \alpha$ to a point B at $t = \beta$. The work done by the force is, by equation (1) of § 3·5,

$$W = \int_A^B \mathbf{P}.\mathbf{T}\,ds,$$

where the integral is taken along the path of the particle.

Hence $$W = \int_\alpha^\beta \mathbf{P}.\mathbf{v}\,dt,$$

and therefore the power R is given by

$$R = \mathbf{P}.\mathbf{v}. \tag{1}$$

Power is a scalar quantity, measured by the scalar product of force and velocity. We can write this also in the equivalent form

$$R = |\mathbf{P}|\,|\mathbf{v}|\cos\alpha, \tag{2}$$

where α is the angle between $\mathbf{P}$ and $\mathbf{v}$. Finally, if we introduce rectangular axes, denoting the components of $\mathbf{P}$ by X, Y, Z, we have

$$R = X\dot{x} + Y\dot{y} + Z\dot{z}. \tag{3}$$

It is clear (from the property (ii) of § 2·5) that if a number of forces act simultaneously on a particle the power of the vector sum of the forces is the algebraic sum of the powers of the separate forces. Power is a genuine scalar quantity measured on a uniform scale.

3·8 Units of work and power. If we use c.g.s. units, with the centimetre as unit of length and the dyne as unit of force, the theoretical unit of *work* is the work done when a force of one dyne acts on a particle, and the particle moves through a distance of 1 cm. in the direction of the force. This unit is called an *erg*. It is inconveniently small for practical purposes, and in electrical engineering the unit of work generally adopted, called a joule, is 10^7 times as large: one joule is 10^7 ergs.

If we use the foot as unit of length, and the poundal as unit of force, the corresponding unit is the *foot-poundal*. Engineers more often use the *foot-pound* as unit, which is the work done by a force of one pound-weight when the particle on which it acts moves through a distance of one foot in the direction of the force. One foot-pound is equal to g (roughly 32) foot-poundals.

The theoretical unit of *power* in c.g.s. units is one erg per second. This is the rate of working by a force of one dyne when the particle on which it acts has a velocity of 1 cm./sec. in the direction of the force. This unit also is inconveniently small, and the practical unit, the watt, is 10^7 times as large: a watt is one joule per second.

In foot-pound-second units the natural unit of power is one foot-poundal per second, but this again is inconveniently small for practical work. Engineers usually take as the unit of power a horse-power, which is 550 foot-pounds/sec., i.e. 550 times the rate of working of a pound-weight, when the particle on which it acts has a velocity of 1 ft./sec. in the direction of the force. Since the pound-weight is g poundals, the unit horse-power is 550 g foot-poundals/sec.

3·9 Gradient. Let ϕ be a uniform function of position in space. We can use the notation $\phi(P)$, where P is an arbitrary point, or $\phi(\mathbf{r})$, where $\mathbf{r}$ is the position vector of P, or $\phi(x, y, z)$, where x, y, z are the rectangular Cartesian coordinates of P, i.e. the components of $\mathbf{r}$. We assume that $\phi(x, y, z)$ has continuous first derivatives in the domain of definition.

Consider in particular the value of ϕ at points on a simple curve Γ. If we denote the arc-length of Γ by s, we can express the rectangular coordinates x, y, z of points on Γ as functions of s. Moreover, we may regard ϕ at points on Γ as a function of s, and we have

$$\frac{d\phi}{ds} = \frac{\partial\phi}{\partial x}\frac{dx}{ds} + \frac{\partial\phi}{\partial y}\frac{dy}{ds} + \frac{\partial\phi}{\partial z}\frac{dz}{ds}. \tag{1}$$

Now $\frac{dx}{ds}, \frac{dy}{ds}, \frac{dz}{ds}$ are, as we have seen, the components of $\mathbf{T}$, the unit vector along the tangent to Γ in the direction of increasing s; $\frac{\partial\phi}{\partial x}, \frac{\partial\phi}{\partial y}, \frac{\partial\phi}{\partial z}$ are the components of a vector called the *gradient* of ϕ, and written grad ϕ. Equation (1) can be written

$$\frac{d\phi}{ds} = \operatorname{grad}\phi\,.\,\mathbf{T}; \tag{2}$$

or

$$\frac{d\phi}{ds} = |\operatorname{grad}\phi|\cos\alpha, \tag{3}$$

where α is the angle between $\operatorname{grad}\phi$ and $\mathbf{T}$.

Let us consider more closely the significance of the vector $\operatorname{grad}\phi$. We have so far tacitly assumed that the directed quantity whose 'components' are $\dfrac{\partial\phi}{\partial x}, \dfrac{\partial\phi}{\partial y}, \dfrac{\partial\phi}{\partial z}$ satisfies the criteria for a vector quantity, and this is easily established. The relevant physical summation here is the addition of two scalar functions ϕ_1 and ϕ_2. Now the gradient of $\phi_1+\phi_2$ is obtained from the gradients of ϕ_1 and ϕ_2 by the vector law of addition, because $\dfrac{\partial}{\partial x}(\phi_1+\phi_2) = \dfrac{\partial\phi_1}{\partial x}+\dfrac{\partial\phi_2}{\partial x}$, with similar results for y and z. The result now follows from § 2·4.

Equation (2) expresses the fundamental property of $\operatorname{grad}\phi$. Let P be the point x, y, z, and denote the value of ϕ at this point by C. If $\mathbf{T}$ lies in the tangent plane at P to the surface $\phi = C$, we have $d\phi/ds = 0$, so that

$$\operatorname{grad}\phi\,.\,\mathbf{T} = 0;$$

it follows that $\operatorname{grad}\phi$ is perpendicular to all lines in the tangent plane. Thus $\operatorname{grad}\phi$ is in the direction of the normal at P to the surface $\phi = C$.

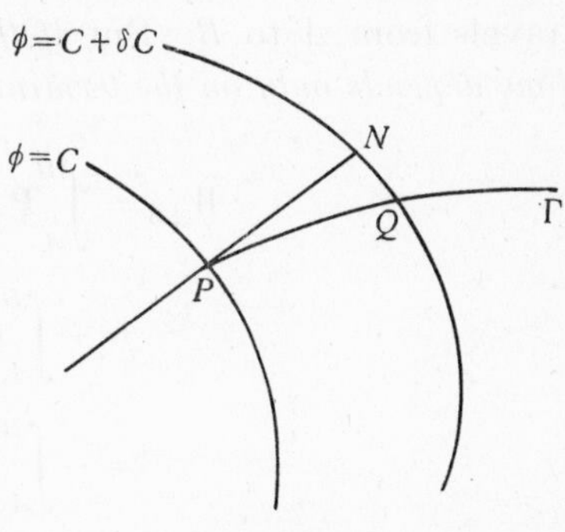

Fig. 3·9

Next, if we take $\mathbf{T}$ normal to the surface $\phi = C$, and in the direction of increasing ϕ, we have

$$|\operatorname{grad}\phi| = \frac{d\phi}{ds};$$

thus if the normal at P to $\phi = C$ meets the surface $\phi = C+\delta C$ in N, we have

$$|\operatorname{grad}\phi| = \lim_{PN\to 0}\frac{\delta C}{PN}.$$

We can now see that $\operatorname{grad}\phi$ is a definite physical quantity, defined at each point of space, and in no way dependent on the choice of the Cartesian axes that we used above. Also we can see more clearly the geometrical significance of the equation (3). For if the curve Γ through P meets $\phi = C+\delta C$ in Q, we have

$$\frac{\delta C}{PQ} = \frac{\delta C}{PN}\frac{PN}{PQ}. \tag{4}$$

As $\delta C \to 0$, $\dfrac{\delta C}{PN} \to |\operatorname{grad} \phi|$ and $\dfrac{PN}{PQ} \to \cos\alpha$, so equation (3) is the limiting form of equation (4).

3·10 Conservative field. Suppose that V is a uniform function of position in space. We assume that V is defined and possesses a continuous gradient in some domain; in some cases this domain is the whole space. A field of force $\mathbf{P} = \mathbf{P}(\mathbf{r})$ defined by the equation

$$\mathbf{P} = -\operatorname{grad} V \tag{1}$$

is said to be *conservative*, and $V = V(\mathbf{r}) = V(x, y, z)$ is called the *potential function*, or simply the *potential*, of the field.

Now we saw (§ 3·5) that the work done when a particle moves from A to B in a field of force depends on the termini A and B, and also in general on the path Γ by which the particle travels from A to B. But *if the field is conservative, the work done depends only on the termini and not on the path.* For

$$\begin{aligned} W_{AB} &= \int_A^B \mathbf{P}.\mathbf{T}\,ds \\ &= -\int_A^B \operatorname{grad} V.\mathbf{T}\,ds \\ &= -\int_A^B \frac{dV}{ds}\,ds \\ &= V(A) - V(B). \end{aligned} \tag{2}$$

This vitally important result is the characteristic property of the conservative field.

Another way of expressing this result is that the work done when the particle traverses a closed curve, returning to the starting point, is zero. This we see by taking a closed path (so that $B \equiv A$) and remembering that V is a uniform function.

The converse is true; *if the work done when the particle travels round an arbitrary closed curve is zero, then the field is conservative.* To prove this, we observe first that the work done when the particle moves from A to B is independent of the path. For if we take two paths joining A to B, say the paths ACB

and ADB in the figure (Fig. 3·10), we may regard the path $ACBDA$ as a closed curve, and then

$$W_{ACB} + W_{BDA} = 0,$$

i.e.

$$W_{ACB} = -W_{BDA} = W_{ADB}.$$

If therefore we take an arbitrary origin O in the plane as a point of zero potential, we can define unambiguously the function V by the equation

$$V(P) = -W_{OP}. \quad (3)$$

Then $W_{OA} + W_{AB} + W_{BO} = 0$,

whence

$$W_{AB} = -W_{OA} - W_{BO} = V(A) - V(B), \quad (4)$$

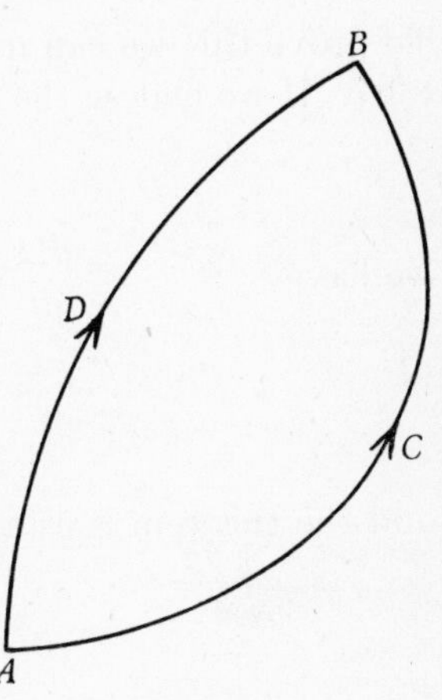

Fig. 3·10

and the field possesses the characteristic property, defined by equation (2), of the conservative field. It follows also that the function V defined by (3) has the property (1); if we take A as the point (x_0, y_0, z_0) and B as the point (x, y_0, z_0), and the path from A to B is a straight line, we have, from (4),

$$W_{AB} = \int_A^B X\, dx = V_A - V_B,$$

whence

$$X = -\frac{\partial V}{\partial x}.$$

Similarly

$$Y = -\frac{\partial V}{\partial y} \quad \text{and} \quad Z = -\frac{\partial V}{\partial z}.$$

The fact that the work done by the field when the particle moves round a closed curve is zero is sometimes taken as the property defining a conservative field. It is now evident that this definition is equivalent to that given at the beginning of this paragraph.

It should be noticed that the point chosen as the zero of V is arbitrary. We get the same field if we replace the function V by $V + K$, where K is constant.

The surfaces $V =$ constant are called *level surfaces*. The field at any point is normal to the level surface through that point.

Finally, we establish a simple formula for the power when a particle moves in a conservative field. We denote by $\Omega(t)$ the function of t formed by substituting for x, y, z in V the coordinates of the particle at time t in its motion. Then

$$R = -\frac{d\Omega}{dt}. \tag{5}$$

To prove this we can use either equation (1) or equation (3) of § 3·7. If we choose the latter, since

$$\Omega(t) = V(x, y, z),$$

we have
$$\begin{aligned}\frac{d\Omega}{dt} &= \frac{\partial V}{\partial x}\dot{x} + \frac{\partial V}{\partial y}\dot{y} + \frac{\partial V}{\partial z}\dot{z} \\ &= -(X\dot{x} + Y\dot{y} + Z\dot{z}) \\ &= -R,\end{aligned}$$

and the theorem is established.

CHAPTER IV

MOTION ON A STRAIGHT LINE, THE SIMPLEST PROBLEMS

4·1 Motion of a particle on a straight line. If the resultant force acting on a particle is, at each instant, in a given fixed direction (but not necessarily always in the same sense), and if further its initial velocity is also in this direction, the particle moves on a straight line. This is the simplest problem of the dynamics of a particle. The underlying base of reference is assumed to be a Newtonian base, and the adjective *fixed* here means fixed relative to this base.

It may be helpful to enumerate some of the more familiar types of problem which fall into this class.

(i) A free particle, i.e. a particle moving with no force acting on it, or under the action of forces in equilibrium.

(ii) A particle moving in a uniform field of force, when the direction of projection is parallel to the field. (We speak of a field of force when the force is a function of position in space, and of a uniform field when the force has the same value, both in direction and magnitude, at all points of space.)

(iii) A particle moving under the action of a force which is fixed in direction, the initial velocity being also in this direction. One important special case occurs when the force is a function of position on the line. This is the one-dimensional analogue of a field of force, or, simply, it is a one-dimensional field. A simple illustration of such a field is a force which is directed towards a fixed point of the line, and is in magnitude proportional to the distance from this point (the harmonic oscillator).

There are problems in which bodies of finite size, and even systems of bodies, can be treated, to a first approximation, as particles—as though the whole mass were concentrated in one point. As examples we may mention:

(iv) A ball thrown vertically upwards, either in a vacuum or in a resisting medium.

(v) A bead which slides on a fixed straight wire. In this problem the forces acting on the bead fall into two distinct classes, (*a*) the forces other than the reaction of the wire on the bead, and (*b*) the reaction of the wire on the bead. If the wire is smooth only the component of the resultant of the forces of type (*a*) along the wire is relevant, the component normal to the wire being balanced by the reaction of the wire on the bead. If the wire is rough, the reaction is no longer normal to the wire; the normal component of reaction balances the normal component of the forces of type (*a*), and the tangential component of the reaction is in the sense opposite to that of the velocity and is assumed to be (like the limiting friction in statics) proportional to the normal component. The factor of proportionality μ is a physical quantity depending on the material surfaces in contact; it is usually assumed to be the same at all points of the wire. So long as motion continues, the reaction of the wire on the bead acts in a direction making an angle $\frac{1}{2}\pi + \lambda$ with the direction of motion, where $\tan\lambda = \mu$.

(vi) Similar remarks hold for a small body sliding on one side of a fixed plane ϖ if all the forces on the body and the initial velocity lie in a given plane perpendicular to Q. But in this case there is the additional complication that the plane can only supply an *outward* reaction. Let us denote the normal component of the reaction of the plane on the body, measured away from the plane, by N. (If the plane is smooth N is the whole reaction.) We need the condition $N \geqslant 0$. If a rectilinear motion on the plane requires $N < 0$ for $t > t_1$, then the particle leaves the plane at $t = t_1$, and the rectilinear motion ceases.

(vii) A car moving on a straight road, level or inclined.

In these and similar problems the earth is taken to be, as a first approximation, a Newtonian base. This is sometimes denoted by a phrase such as 'neglecting the rotation of the earth'.

4·2 Statement of the problem. Consider then a particle, or point-mass, A, which moves on a straight line. We describe the position of the particle on the line by the directed length (§ 1·4) of the displacement OA from a fixed origin O on the line (Fig. 4·2). The positive sense is taken to the right in the figure.

Thus x is a uniform (i.e. one-valued) function of the time t, $x = x(t)$, and our first object is to study the properties of this function.

The function $x(t)$ is always continuous. Usually (i.e. for all but isolated exceptional values of t) it possesses a first derivative, the velocity, which is denoted by dx/dt or $\dot{x}$ or v; and a second derivative, the acceleration, which is denoted by d^2x/dt^2 or $\ddot{x}$ or $\dot{v}$ or f. (We are here using derived units of velocity and acceleration; see § 3·4.) The velocity $\dot{x}$ is positive if x increases with t, i.e. if the particle is moving to the right in the figure.

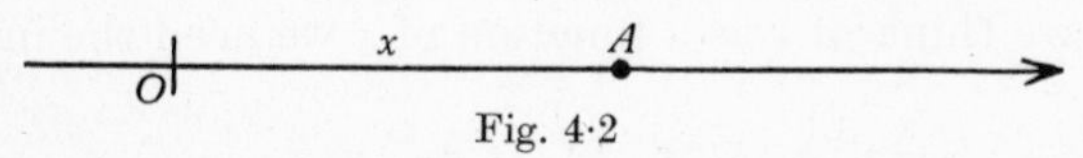

Fig. 4·2

The acceleration $\ddot{x}$ is positive if $\dot{x}$ increases with t; it should be noticed that $\ddot{x}$ may be positive when $\dot{x}$ is negative, namely, when the particle is moving to the left and $|\dot{x}|$ is decreasing.

We have here an example of classes of vector quantities all parallel to a line, and, as we saw (§ 1·8), these reduce effectively to scalar quantities, not necessarily positive. The scalars replace the vectors of the general theory.

There may be exceptional values of t at which $\dot{x}$ and $\ddot{x}$ do not exist. This phenomenon arises most commonly in a problem involving the action of an impulse, say at the instant $t = t_0$. Then the velocity $\dot{x}$ does not exist at t_0; it does exist in a neighbourhood of t_0, except at t_0 itself, but the limits of $\dot{x}$ as $t \to t_0 + 0$ and as $t \to t_0 - 0$ are different. The velocity is discontinuous at t_0. But such values of t are exceptional, and in general the function $x(t)$ possesses both first and second derivatives. Moreover, in general the second derivative is itself a continuous function of t.

Since x and v are both functions of t, v is a function of x (though not in general a uniform function), and in some problems it is advantageous to consider primarily the dependence of v upon x. It is important to observe that, when we adopt this point of view, the role of the variable x is changed. Instead of thinking of x as the dependent variable, x being a function of t, we must now regard x as the independent variable, v being thought of as a function of x. Explicitly, suppose that

$$x = \phi(t), \quad v = \phi'(t),$$

ϕ and ϕ' being uniform functions. The first of these equations defines t as a function of x, but in general as a many-valued function; so when we substitute for t in terms of x in the second equation we find that v is a many-valued function of x. For example, in the simple problem in which

$$x = \tfrac{1}{2}gt^2, \quad v = gt,$$

t and v are *two-valued* functions of x,

$$t = \pm\sqrt{(2x/g)}, \quad v = \pm\sqrt{(2gx)}.$$

When we think of v as a function of x we need the important formula

$$f = \frac{d}{dx}(\tfrac{1}{2}v^2). \tag{1}$$

If $v \neq 0$ this is equivalent to

$$f = v\frac{dv}{dx}, \tag{2}$$

but the formula (1) is more general than (2), since it is valid even when $v = 0$. The proof of these formulae that is usually given may be written compactly as follows:

$$f = \frac{dv}{dt} = \frac{dv}{dx}\frac{dx}{dt} = v\frac{dv}{dx} = \frac{d}{dx}(\tfrac{1}{2}v^2).$$

This fails in the case where $v = 0$ and dv/dx does not exist. A proof covering also this case needs a little more care. We consider an instant t_0, and we assume that there is a neighbourhood of t_0 in which v does not vanish except possibly at t_0 itself. We assume also that f is a continuous function of t at t_0. Then, using Cauchy's form of the mean value theorem, we have for $t \neq t_0$

$$\frac{\tfrac{1}{2}v^2(t) - \tfrac{1}{2}v^2(t_0)}{x(t) - x(t_0)} = \frac{v(\theta)\,f(\theta)}{v(\theta)},$$

where θ lies between t and t_0. Since $v(\theta) \neq 0$ the second member is equal to $f(\theta)$, and tends to $f(t_0)$ as $t \to t_0$. The result follows.

To determine the motion we need Newton's second law (§ 3·2). The vector $\mathbf{P}$ and the position vector $\mathbf{r}$ and its derivatives are now all in the same direction. We denote the force in the direction Ox by P or by X, not necessarily positive. The position vector $\mathbf{r}$ is replaced by the scalar x, and the vectors $\dot{\mathbf{r}}$ and $\ddot{\mathbf{r}}$ by $\dot{x}$ (or v) and $\ddot{x}$ (or $\dot{v}$ or f). The second law of motion now takes the simpler form

$$mf = P, \tag{3}$$

and this determines f when P is given. We must understand clearly what is meant by the statement that 'P is given'. It means that P is given in terms of t and x and v. The problem is reduced to the integration of a differential equation of the second order—a differential equation of the particular type in which $\ddot{x}$ is expressed explicitly as a function of t, x and $\dot{x}$.

In the simplest problems the equation of motion is obtained immediately in the form (3). In other problems (for example, in problems of types (v) and (vi) of § 4·1 above, and in Example 4 of § 4·9 below) the equation in the form (3) is only obtained after some preliminary discussion.

The simplest problem is that in which P is constant, and other special cases occur when P is a function of t only, or of x only, or of v only. The case where $P = P(x)$ is of paramount importance (motion in a field of force). The case where P is a function of v occurs in problems with resisting media. We shall find it convenient to break up the discussion into five parts, considering first the four special cases mentioned, and finally the general case. We thus obtain the scheme: (*a*) f constant; (*b*) f a given function of t; (*c*) f a given function of x; (*d*) f a given function of v; (*e*) the general case, $f = \phi(t, x, v)$.

4·3 Graphical representation of the motion. In the study of rectilinear motion it will often be useful to consider the graphs representing (i) v as a function of t; (ii) x as a function of t; and (iii) v as a function of x. The first two of these are very simply related; if $v = \phi(t)$, we have $x - x_0 = \int_{t_0}^{t} \phi(\theta)\, d\theta$, where x_0 is the value of x at $t = t_0$. The distance travelled in the interval from t_0 to t is the area bounded by the graph and the ordinates for t_0 and t. This is a directed length, the sign showing whether the displacement is to the right or to the left. Again, if we take the graph (ii), the slope of this graph represents velocity, since $v = \dot{x}$.

Consider now (iii), representing the relation between x and v. When the representative point lies in the lower half-plane $v < 0$, and the particle is moving to the left; when it lies in the upper half-plane $v > 0$, and the particle is moving to the right. If PN is the perpendicular from a point P of the graph on to Ox, and the normal at P meets Ox in G, then the directed length NG represents the acceleration f, since $f = v\,dv/dx$ (Fig. 4·3). In

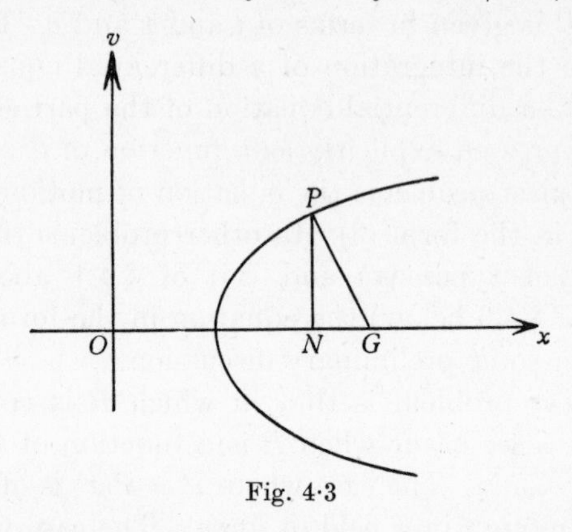

Fig. 4·3

general the graph crosses the x-axis normally, for a finite value of dv/dx at a point on Ox would imply $f = 0$ at this point as well as $v = 0$, and this can happen only in exceptional cases. (There is a concrete example of the exceptional case in § 6·11, case (ii).) We shall find that for motion in a field of force v^2 is a uniform function of x, and then the (x, v) graph is symmetrical about Ox.

4·4 Motion with constant acceleration. We turn now to the simplest of the problems enumerated in § 4·2, case (a), in which the particle moves under the action of forces whose vector sum is constant (both in direction and in magnitude). We have

$$mf = P,$$

$$f = P/m = c, \text{ say.}$$

The determination of the motion in this case is very simple. Since

$$\ddot{x} = c, \tag{1}$$

we have

$$v = \dot{x} = u + ct, \tag{2}$$

and

$$x = a + ut + \tfrac{1}{2}ct^2. \tag{3}$$

The significance of the constants u and a is evident; u is the value of v, and a is the value of x, at the instant $t = 0$. Figs. 4·4a, b show the graph of v as a function of t, which is a straight line, and the graph of x as a function of t, which is a parabola. The figures are drawn for the case in which c and a and u are positive. The equation (3) is the expression of x as a Taylor series in powers of t; the coefficient of t^n in this series is $\frac{1}{n!}\left(\frac{d^n x}{dt^n}\right)_{t=0}$, and this vanishes for $n > 2$.

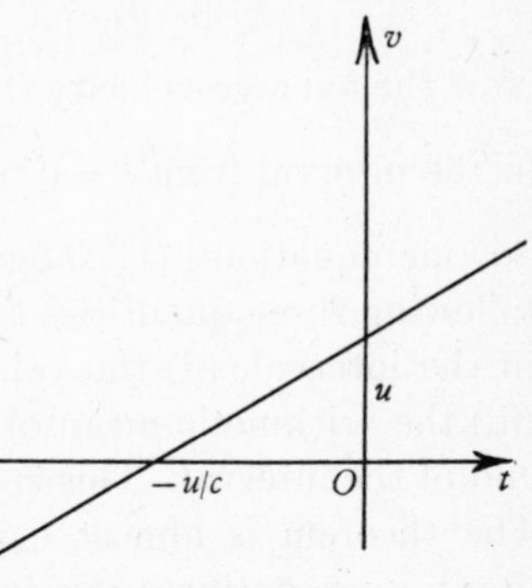

Fig. 4·4a

If we eliminate t between (2) and (3) we obtain immediately the (x, v) relation

$$v^2 = u^2 + 2c(x - a). \tag{4}$$

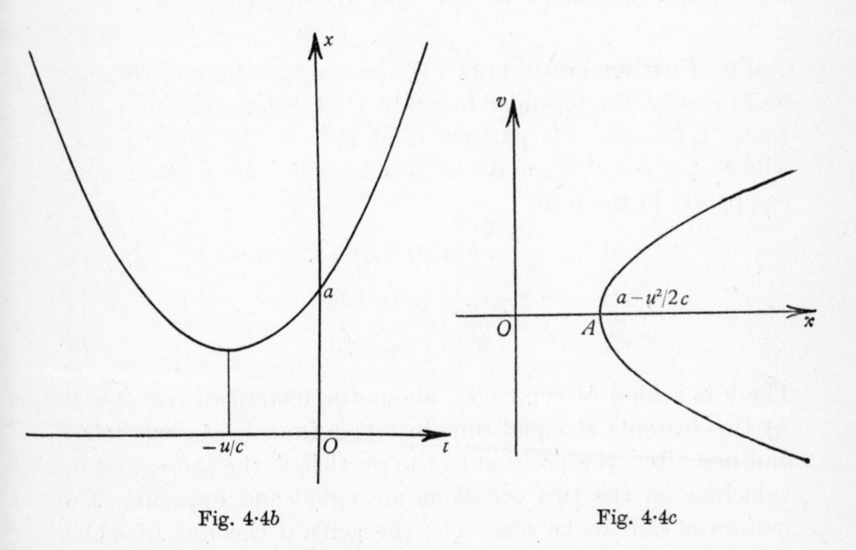

Fig. 4·4b

Fig. 4·4c

The graph representing this relation is a parabola (Fig. 4·4c), as we could have anticipated from the property that NG is constant. We notice that dv/dx does not exist at the point A, and the graph cuts the x-axis normally.

4·5 Theorem. We now prove a useful and important theorem. Consider a particular instant t_1, and let x_1, v_1, denote the corresponding values of x, v. Then from equations (2) and (3) of § 4·4, we see that

$$\frac{x_1 - a}{t_1} = u + c(\tfrac{1}{2}t_1) = \frac{u + v_1}{2}. \tag{1}$$

Now the average velocity (i.e. the average with respect to time) in the interval from $t = 0$ to $t = t_1$ is $\left(\int_0^{t_1} v(t)\,dt\right)\Big/ t_1 = (x_1 - a)/t_1$. So our equations (1) show that in any interval of time the following three quantities are all equal: (i) the average velocity in the interval, (ii) the velocity at the middle of the interval, (iii) the arithmetic mean of the velocities at the beginning and end of the interval. This is the theorem we wished to establish. The theorem is almost evident from the linear form of the graph representing v as a function of t if we remember that the area under the graph represents the distance travelled.

4·6 Further comments. We now return to, and consider more closely, the formulae found in § 4·4. At the instant $t = t_0$, where $t_0 = -u/c$, the particle is at rest at the point $x = x_0$, where $x_0 = a - u^2/2c$, and our formulae can be written more compactly in the form

$$v = c(t - t_0), \tag{1}$$

$$x - x_0 = \tfrac{1}{2}c(t - t_0)^2, \tag{2}$$

$$v^2 = 2c(x - x_0). \tag{3}$$

There is a kind of symmetry about the instant of rest, $t = t_0$. At two instants at equal time intervals from $t = t_0$, one before and one after, the position of the particle is the same, and its velocities on the two occasions are equal and opposite. The motion is said to be *reversible*; the general theorem of which this is a special case will appear later (§ 5·4).

Of course if we think of the motion as *started* at the instant $t = 0$, then our formulae only determine the motion for positive values of t, whereas t_0 will be negative if u and c have the same sign; in this case the particle never comes to rest in the actual motion. But this of course does not invalidate the formulae (1), (2), (3) above. These formulae represent a motion with constant acceleration *for all values of* t, though it is only positive values of t in which we are immediately interested.

A case that frequently occurs is that in which $u > 0$ and $c < 0$, say $c = -g$, where $g > 0$. (A negative acceleration is called a *retardation*.) This is the problem of a ball thrown vertically upwards. It comes to rest at a height $u^2/2g$ above the point of projection; this is the highest point reached, and the time taken to reach it is u/g; the time taken to return to the point of projection is $2u/g$. These results are needed frequently.

4·7 Another method. It may be of interest to reconsider the problem of motion with constant acceleration from another point of view. If we start from the equation

$$\frac{d}{dx}(\tfrac{1}{2}v^2) = c,$$

we obtain, on integration,

$$\tfrac{1}{2}(v^2 - u^2) = c(x-a),$$

which is equivalent to equation (4) of § 4·4. To find the relation between x and t we have now to integrate the differential equation

$$\tfrac{1}{2}\dot{x}^2 = \tfrac{1}{2}u^2 + c(x-a). \tag{1}$$

This is a differential equation of a general type which we shall discuss more fully in Chapter VI, but a few preliminary comments on this special case may be useful. To begin with, there is an ambiguity of sign for $\dot{x}$; this we could have foreseen, because, as we have noticed already in § 4·6, two values of v, equal in magnitude but opposite in sign, are associated with any value of x. Next, the equation (1) would have the same form if $\dot{x}$ had the value $-u$ instead of $+u$ when $t = 0$, and there will be a corresponding ambiguity in the (t, x) relation. For example, if we integrate the equation by separation of variables, we find

$$\pm ct = \int_a^x \frac{c\,d\theta}{\sqrt{[u^2 + 2c(\theta - a)]}}$$

$$= \sqrt{[u^2 + 2c(x-a)]} - u,$$

giving

$$x = a \pm ut + \tfrac{1}{2}ct^2.$$

The ambiguity is easily resolved, because $\dot{x} = +u$ at $t = 0$.

We can also derive the (t, x) relation from equation (1) by a slightly different method, and it may be of interest to do so as a trivial example of a technique which will turn out to be important later on. We introduce in place of x an auxiliary variable θ defined by the equation

$$u^2 + 2c(x-a) = \theta^2.$$

Then

$$c\dot{x} = \theta\dot{\theta},$$

and on substituting in equation (1) we obtain the equation

$$\dot{\theta}^2 = c^2.$$

The usefulness of the method arises from the simplicity of this result. We have

$$\dot{\theta} = \pm c, \quad \theta = \pm c(t-t_0),$$

and the (t, x) relation becomes

$$u^2 + 2c(x-a) = c^2(t-t_0)^2;$$

this is equivalent to equation (2) of § 4·6. The constant t_0 that appears here and in similar cases is called an *epoch constant*. In this case $t = t_0$ is the instant at which $v = 0$, $t_0 = -u/c$.

4·8 Momentum and kinetic energy. Our discussion of motion with constant acceleration started from the equation

$$mf = P,$$

where P is constant. Now we have seen that f is represented by either of the formulae dv/dt or $d(\frac{1}{2}v^2)/dx$, so we have

$$\frac{d}{dt}(mv) = \frac{d}{dx}(\tfrac{1}{2}mv^2) = P.$$

The quantity mv in the first term is the *momentum* of the particle—properly a vector quantity, but here enabled to pose as a scalar quantity, because its direction is the same throughout the motion. The scalar quantity $\frac{1}{2}mv^2$ is called the *kinetic energy* of the particle. The reason for the importance of the kinetic energy will emerge later (§ 5·3), and a discussion of its significance may conveniently be postponed. We notice, however, at this point that, for motion under a *constant* force P, Pt measures the increase of momentum in any interval of time of duration t, and Px measures the increase of kinetic energy in any interval of distance of length x. These observations include in particular the facts, already familiar, that if the particle starts from rest at the origin at $t = 0$, then $v \propto t$ and $v^2 \propto x$.

The fact that the increase of kinetic energy is equal to Px is a special case of a famous principle, the 'equation of energy', which we shall meet frequently in the sequel (see, for example, § 5·3 and §7·7).

It may be of interest to anticipate at this point the generalization of the results just found for the case when the force is not constant. If the force is given as a function of t, $X(t)$, the increase in momentum in the time-interval from t_1 to t_2 is equal to

$$\int_{t_1}^{t_2} X(t)\,dt.$$

If the force is given as a function of $x, X(x)$, the increase in kinetic energy when the particle moves from x_1 to x_2 on the x-axis is equal to

$$\int_{x_1}^{x_2} X(x)\,dx.$$

4·9 Examples. We now consider the explicit solution of some problems on motion with constant acceleration.

Example 1. *A ball is thrown up with velocity u_1, and after an interval of time t_0 a stone is thrown up from the same point with velocity u_2. Will they ever meet?*

Taking the instant of projection of the stone as the zero of the time-scale the heights of the ball and stone at time t are

$$x_1 = a + u_3 t - \tfrac{1}{2}gt^2,$$
$$x_2 = \quad u_2 t - \tfrac{1}{2}gt^2,$$

where $a = (u_1 - \tfrac{1}{2}gt_0)t_0$, $u_3 = u_1 - gt_0$. They meet when $(u_2 - u_3)t = a$, and the question is, does this equation give a positive value of t?

(i) $0 < t_0 < u_1/g$. The ball is still rising when the stone is projected; $a > 0, u_3 > 0$. They meet if $u_2 > u_3$. The result is of course evident without the calculation.

(ii) $u_1/g < t_0 < 2u_1/g$. The ball is falling when the stone is projected, but has not yet returned to the origin; $a > 0, u_3 < 0$. They do meet.

(iii) $2u_1/g < t_0$. In this case $a < 0, u_3 < 0$, and they never meet.

Example 2. *A stone is dropped down a shaft, and the impact on the bottom is heard t seconds later. Given that the velocity of sound is a ft./sec., find the depth of the shaft in feet. Examine the numerical case $t = 5{\cdot}94$, $a = 1100$.*

Let h be the depth of the shaft, g the acceleration of gravity, using feet and seconds as units of length and time. The time for the stone to reach the bottom is $\sqrt{(2h/g)}$, and the time for the sound to reach the top is h/a, so we have

$$\sqrt{\frac{2h}{g}} + \frac{h}{a} = t.$$

Hence
$$\left(\sqrt{\frac{h}{a}} + \sqrt{\frac{a}{2g}}\right)^2 = t + \frac{a}{2g},$$

giving
$$\sqrt{\frac{h}{a}} = \sqrt{\left(t + \frac{a}{2g}\right)} - \sqrt{\frac{a}{2g}}.$$

In the numerical case, taking $g = 32$, we find

$$\sqrt{\left(t + \frac{a}{2g}\right)} = 4{\cdot}809, \quad \sqrt{\frac{a}{2g}} = 4{\cdot}146, \quad \text{giving } h = 484.$$

Example 3. *A particle moves with constant retardation on a straight line. It travels distances AB, BC, of 28 ft. and 32 ft., in successive intervals of 2 sec. and 4 sec. Find the distance from C to the point D where the particle comes to rest, and find the time taken to travel from C to D.*

We use the theorem of § 4·5. Taking seconds and feet as units, and taking $t = 0$ at A, we see that the velocity at $t = 1$ is 14 ft./sec., since this is the average velocity in the interval from $t = 0$ to $t = 2$. Similarly, the velocity at $t = 4$ is 8 ft./sec. The velocity falls from 14 ft./sec. to 8 ft./sec. in 3 sec., so the retardation is 2 ft./sec./sec. The velocity at $C(t = 6)$ is therefore 4 ft./sec. The time from C to rest is therefore 2 sec. The particle comes to rest at $t = 8$.

Also the average velocity in the interval from $t = 6$ to $t = 8$ is the mean of the velocities at the ends of the interval, i.e. $\frac{1}{2}(4+0) = 2$ ft./sec., so the distance CD is 4 ft.

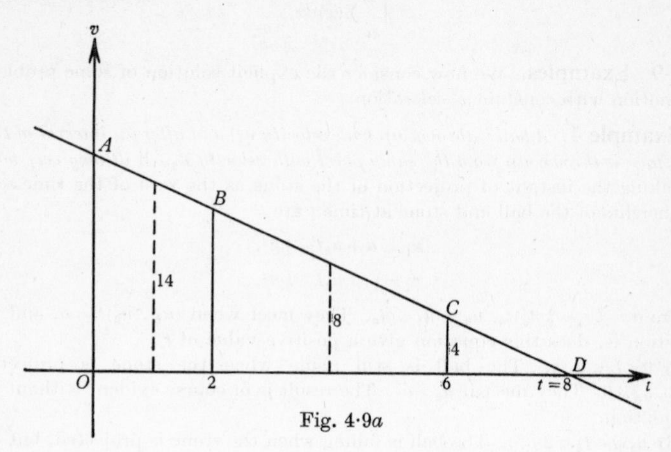

Fig. 4·9*a*

There are many other methods of tackling problems of this kind. For example, we can use the (t, v) graph, which, since f is constant, is a straight line (Fig. 4·9*a*). This line goes through the points (1, 14) and (4, 8), and its equation is therefore

$$2t+v = 16.$$

Thus the particle comes to rest when $t = 8$. The value of v at C, when $t = 6$, is 4, and the distance CD is the area under the graph from $t = 6$ to $t = 8$, namely, 4.

Of course the second of these methods is not fundamentally distinct from the first; cf. the last sentence of § 4·5.

Example 4. *A particle is projected, with velocity u, down the line of greatest slope of a rough plane inclined at an angle α to the horizontal. The surface is uniform, the angle of friction being λ, which is greater than α. To discuss the motion.*

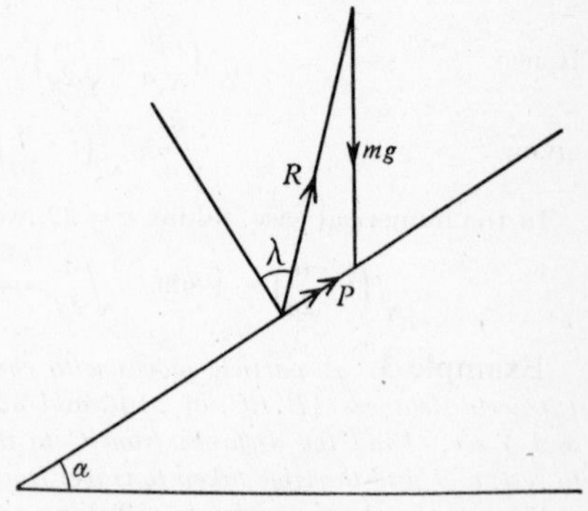

Fig. 4·9*b*

The friction which is called into play opposes the motion, and, so long as motion continues, the reaction makes an angle λ with the normal to the plane (Fig. 4·9*b*). Since $\lambda > \alpha$ the acceleration is directed up the plane. The resultant of the reaction R (of the plane on the particle) and the weight mg is up the plane, and its magnitude P is given by

$$\frac{P}{\sin(\lambda-\alpha)} = \frac{mg}{\cos\lambda}.$$

If we take the positive direction down the incline we have

$$f = -g \sin(\lambda - \alpha) \sec \lambda,$$

and the particle moves with constant retardation. It comes to rest after an interval $u/(-f)$. After this the particle remains at rest on the plane. The limiting friction is only called into play so long as motion continues, and after the particle comes to rest the reaction of the plane on it is just equal and opposite to its weight.

Another method is to resolve the reaction into tangential and normal components F, N. Since the resultant of these and the weight is along the plane,

$$N = mg \cos \alpha$$

and the force opposing the motion is

$$F - mg \sin \alpha.$$

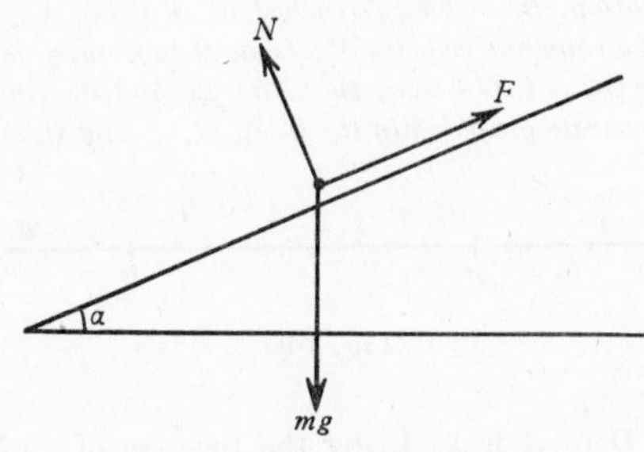

Fig. 4·9c

Now $F = \mu N$, where $\mu = \tan \lambda$, and therefore

$$f = -g(\mu \cos \alpha - \sin \alpha) = -g \sin(\lambda - \alpha) \sec \lambda,$$

as before.

(Observe carefully that the equation for motion *up* the plane is different from that for motion down the plane, since the friction acts always so as to oppose the motion. The reader will easily verify that if the particle moves up the plane the retardation is $g \sin(\lambda + \alpha) \sec \lambda$.)

Example 5. *A train can be accelerated by a force of* 55 *lb. per ton weight and when steam is shut off can be braked by a force of* 440 *lb. per ton weight. Find the least time between stopping stations* 3850 *ft. apart and the greatest velocity of the train.*

In this context a force of 55 lb. means a force of 55 lb. weight; this usage is not uncommon, though it is not strictly accurate; a pound is a unit of mass, a pound weight is a unit of force.

If the mass of the train is k tons, which is $2240k$ pounds, the accelerating force is $55gk$ poundals. The train travels with its maximum acceleration for say t_1 sec., covering x_1 ft., and then with its maximum retardation for say t_2 sec., covering x_2 ft. If the maximum velocity is V ft./sec., we have (§ 4·8)

$$55gx_1 = \tfrac{1}{2} \times 2240 V^2 = 440gx_2,$$

giving

$$x_1 + x_2 = \frac{1120}{32}\left(\frac{1}{55} + \frac{1}{440}\right) V^2 = \frac{1120 \times 9}{32 \times 440} V^2.$$

Since $x_1+x_2 = 3850$, we have

$$V^2 = \frac{3850\times 32\times 440}{1120\times 9} = \left(\frac{220}{3}\right)^2,$$

$$V = \tfrac{220}{3} = 73\tfrac{1}{3}.$$

This is the maximum velocity in feet per second; it is equivalent to 50 m.p.h.

Further (§ 4·5)

$$\frac{V}{2} = \frac{x_1}{t_1} = \frac{x_2}{t_2} = \frac{x_1+x_2}{t_1+t_2},$$

giving

$$t_1+t_2 = 105.$$

This is the time of the journey in seconds, equivalent to $1\frac{3}{4}$ min.

Example 6. *A tram moves from rest at O to rest at C. From O to A there is constant acceleration, the velocity reached at A being V; from A to B the tram moves with the constant velocity V; from B to C there is constant retardation. The distance* $OC = l$, $OA = a$, $BC = b$. *To find the time for the journey, and to find and illustrate graphically the* (t,v), (x,v), *and* (t,x) *relations.*

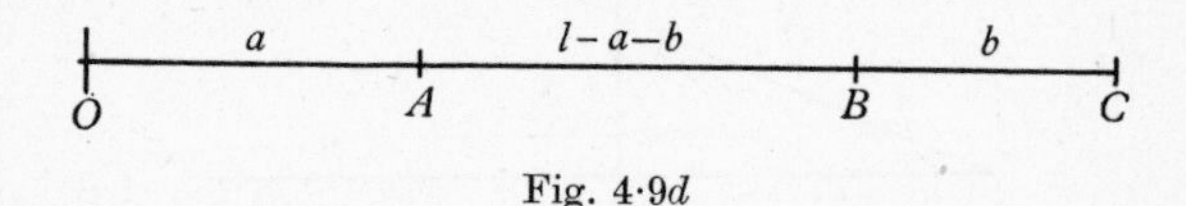

Fig. 4·9*d*

The time from O to A is $2a/V$, by the theorem of § 4·5; the time from A to B is $(l-a-b)/V$; the time from B to C is $2b/V$. Thus the total time for the journey is $t_0 = (l+a+b)/V$.

The time-velocity graph is shown in Fig. 4·9*e*. On OA the constant acceleration is $V^2/2a$; this we can see from the graph, since $f = \tan\theta$, or from equation (4) of § 4·4. Similarly, on BC the retardation is $V^2/2b$. The equations of the three lines are

$$v = \frac{V^2}{2a}t, \quad v = V, \quad v = \frac{V^2}{2b}(t_0-t).$$

The distance-velocity graph is shown in Fig. 4·9*f*. For the first run $v^2 \propto x$ (equation (4) of § 4·4), and the equations for the three pieces are

$$\frac{v^2}{V^2} = \frac{x}{a}, \quad v = V, \quad \frac{v^2}{V^2} = \frac{l-x}{b}.$$

(In drawing the figure we notice that the tangent to the first parabola at $x = a$ cuts Ox at $x = -a$.)

The time-distance graph (Fig. 4·9*g*) is obtained by integration from the time-velocity graph (as explained in § 4·3), or (except for the middle run) by elimination of v from the two preceding. The results are easily found to be

$$x = \frac{V^2}{4a}t^2, \quad x = Vt-a, \quad x = \frac{V^2}{4b}(t_0-t)^2.$$

This graph has a continuous slope, since v is continuous; but f is not continuous, and the other two graphs have slopes which are not continuous at the change from one run to the next.

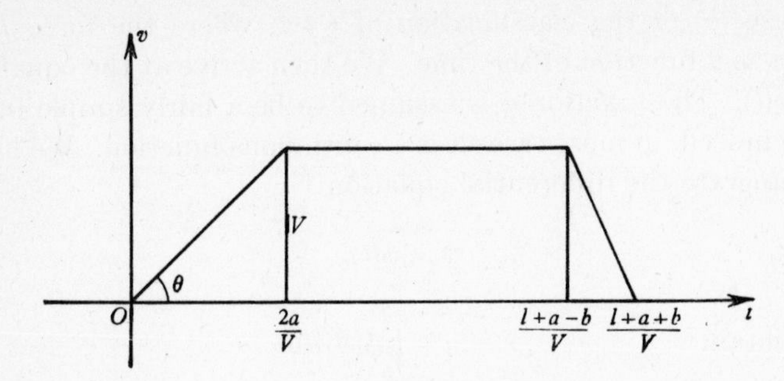

Fig. 4·9e

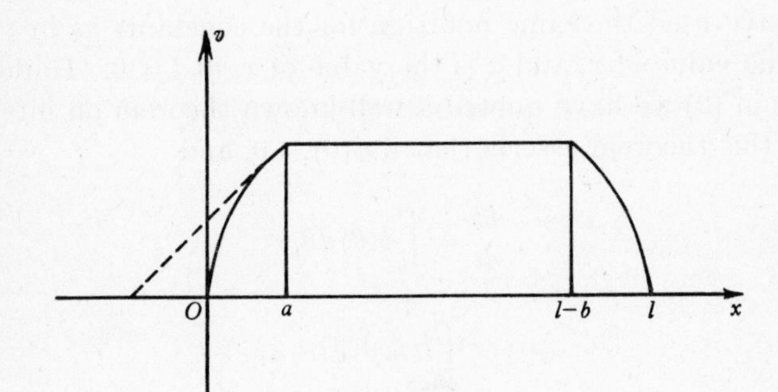

Fig. 4·9f

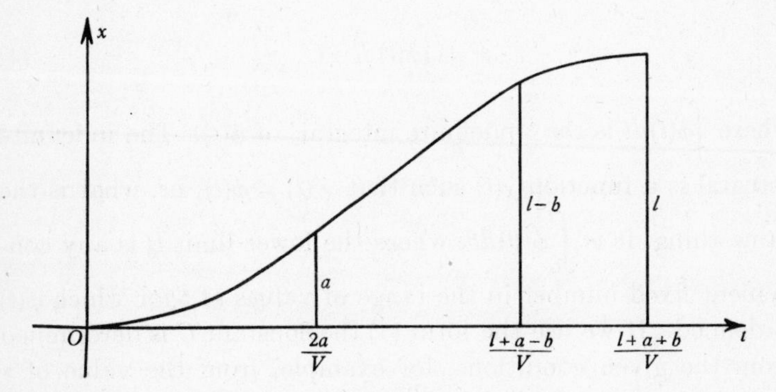

Fig. 4·9g

4·10 Acceleration a function of the time. We now turn to case (*b*) of the classification of § 4·2, where the force P is given as a function of the time. We then arrive at the equation $f = \phi(t)$. The function ϕ is assumed to be a fairly simple function; indeed, in most cases it is a continuous function. We have to integrate the differential equation

$$\ddot{x} = \phi(t). \tag{1}$$

We obtain
$$\dot{x} = u + \int_0^t \phi(\theta)\, d\theta, \tag{2}$$

and
$$x = a + ut + \int_0^t (t-\theta)\,\phi(\theta)\, d\theta. \tag{3}$$

We have used the same notation for the constants as in § 4·4; u is the value of v, and a is the value of x, at $t = 0$. To derive (3) from (2) we have quoted a well-known theorem on integration; this theorem asserts that if $\psi(0) = 0$, and

$$\frac{d\psi}{dt} = \int_0^t \phi(\theta)\, d\theta,$$

then
$$\psi(t) = \int_0^t (t-\theta)\,\phi(\theta)\, d\theta.$$

In the concrete applications we can avoid the appeal to this theorem by direct integration of (2).

The equation (2) is sometimes written

$$\dot{x} = \int \phi(t)\, dt + C, \tag{4}$$

where $\int \phi(t)\,dt$ is the 'indefinite integral' of $\phi(t)$. The indefinite integral is a function $\chi(t)$ such that $\chi'(t) = \phi(t)$; or, what is the same thing, it is $\int_{t_0}^t \phi(\theta)\,d\theta$, where the lower limit t_0 is any convenient fixed number in the range of values of t for which $\phi(t)$ is defined. If we use the form (4) the constant C is determined from the given conditions, for example, from the value of v at $t = 0$.

Example. *The acceleration f is given by the formulae*

$$f = c(\alpha - t) \quad (0 \leqslant t \leqslant \alpha),$$

$$f = 0 \qquad (t > \alpha),$$

where c and α are positive constants; find and illustrate the (t,x) and (t,v) relations if the particle starts from rest at the origin at $t = 0$. Find also and illustrate the (x,v) relation.

We have, for $0 \leqslant t \leqslant \alpha$,

$$\ddot{x} = c(\alpha - t),$$

whence
$$v = \dot{x} = -\tfrac{1}{2}c(\alpha - t)^2 + \tfrac{1}{2}c\alpha^2,$$

and
$$x = \tfrac{1}{6}c(\alpha - t)^3 - \tfrac{1}{2}c\alpha^2(\alpha - t) + \tfrac{1}{3}c\alpha^3,$$

remembering that $v = x = 0$ at $t = 0$. (It is somewhat more convenient to write the formulae in terms of powers of $(\alpha - t)$ rather than of t.) We can write the relations more simply in the form

$$u - v = \tfrac{1}{2}c\eta^2, \tag{5}$$

$$a - x = u\eta - \tfrac{1}{6}c\eta^3, \tag{6}$$

where $a(= \frac{1}{3}c\alpha^3)$ and $u(= \frac{1}{2}c\alpha^2)$ are the values of x and of v at $t = \alpha$, and $\eta = \alpha - t$.

If we use instead the equation (3) we have

$$x = c\int_0^t (t - \theta)(\alpha - \theta)\,d\theta$$

$$= c\int_{\alpha - t}^{\alpha} \{\xi - (\alpha - t)\}\xi\,d\xi,$$

giving again the same result.

If we wish to construct the (x, v) curve for $0 \leqslant t \leqslant \alpha$ we can think of the equations (6) and (5), which are of the forms

$$x = x(\eta), \quad v = v(\eta),$$

as giving the parametric representation of the curve. Or we can eliminate η and find the explicit relation between x and v; we can rewrite (6), in virtue of (5),

$$a - x = u\eta - \tfrac{1}{3}\eta(u - v)$$

$$= (2u + v)\,\eta/3,$$

and now the elimination is immediate, giving

$$2(v - u)(v + 2u)^2 + 9c(x - a)^2 = 0.$$

Of course it is only the piece of this curve in the rectangle $0 \leqslant x \leqslant a$, $0 \leqslant v \leqslant u$, with which we are concerned in the dynamics problem.

For $t > \alpha$ the particle moves with constant velocity u. Thus $v = u$, and $x = a + u(t - \alpha)$.

In this problem f is a continuous function of t, and all the graphs (Figs. 4·10*a*, *b*, *c*) have continuous gradients.

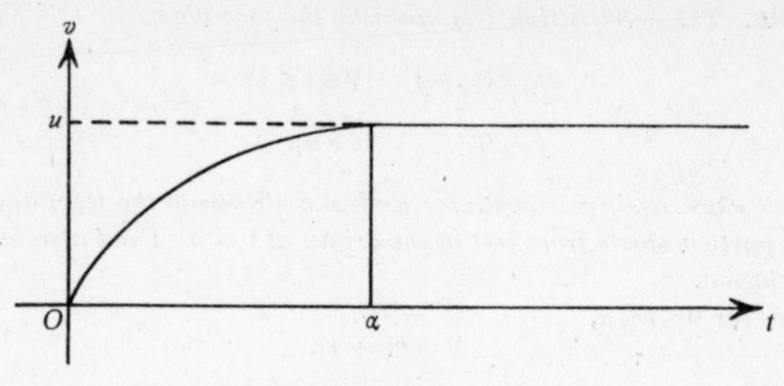

Fig. 4·10*a*

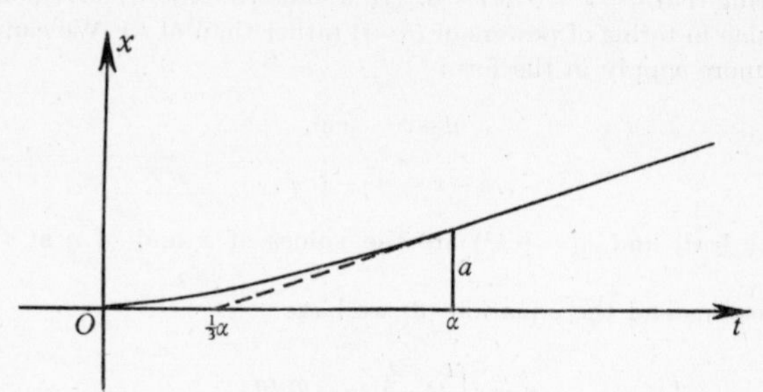

Fig. 4·10*b*

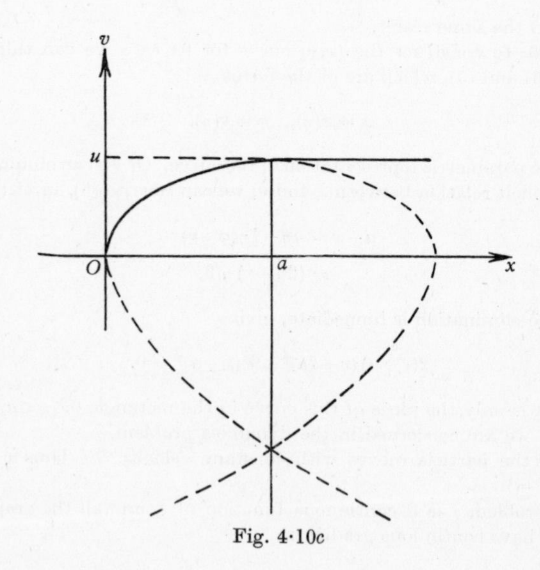

Fig. 4·10*c*

EXAMPLES IV

1. Two trains take 6 sec. to pass one another when going in opposite directions, but only 5 sec. if the speed of one is increased by one-half. How long will one train take to pass the other if they travel in the same direction with their original speeds?

2. One of two motor cars, travelling over the same route in the same direction, passed two places on the route at 2.15 and 3.57 respectively; the other car passed the same two places at 2.28 and 3.40. (i) Supposing the speed of each car to have been uniform, find at what time the former car was passed by the latter. (ii) Supposing the speed of each car to have been uniform, except that owing to a breakdown the former car was stationary for 6 min., show that, if the breakdown occurred between 2.43 and 3.7, the former car was stationary when it was passed by the latter.

Indicate how a graphical method can be applied to solve these problems.

3. A body moving in a straight line travels distances AB, BC, CD, of 153 ft., 320 ft., 135 ft. respectively in three successive intervals of 3 sec., 8 sec., and 5 sec. Show that these facts are consistent with the hypothesis that the body is subjected to uniform retardation. On this hypothesis find the distance from D to the point where the velocity vanishes, and the time occupied in describing this distance.

4. A point moves with uniform acceleration. Explain what is meant by the mean velocity with respect to (a) distance, (b) time, and show that the difference between the means is equal to

$$\frac{1}{6}\frac{(U-u)^2}{U+u},$$

where u and U are the initial and final velocities.

5. A tram moves from one stopping-place to another, the first part of the journey with uniform acceleration, the rest with uniform retardation. Prove that the average speed is one-half the maximum speed.

6. If g is increased effectively by air resistance to $g+f$ in upward motion, and reduced to $g-f$ in downward motion, find the time a ball is in the air when projected upwards with velocity u and caught at the same point on the descent.

For what value of f is the time the same as for motion in a vacuum?

7. A particle is placed on an inclined plane of inclination α; the coefficient of friction between the particle and the plane at all points of the plane is μ. Prove that the particle cannot rest on the plane unless $\mu \geqslant \tan\alpha$.

If $\mu < \tan\alpha$, and the particle is projected with speed u up a line of greatest slope, prove that when it returns to the starting-point its speed is

$$u\sqrt{\left(\frac{1-k}{1+k}\right)},$$

where $k = \mu/\tan\alpha$.

8. Two stopping-points of an electric tramcar are 440 yards apart. The maximum speed of the car is 20 m.p.h. and it covers the distance between stops in 75 seconds. If both acceleration and retardation are uniform and the latter is twice as great as the former, find the value of each of them, and also how far the car runs at its maximum speed.

9. A particle is projected in a straight line with a certain velocity and a constant acceleration. One second later another particle is projected after it with half the velocity and double the acceleration. When it overtakes the first the velocities are 22 and 31 ft./sec. respectively. Prove that the distance traversed is 48 ft.

10. A train passes a station A at 40 m.p.h. and maintains this speed for 7 miles, and is then uniformly retarded, stopping at B, which is 8 miles from A. A second train starts from A just as the first train passes A, and being uniformly accelerated for part of the journey and uniformly retarded for the rest stops at B at the same time as the first train. What is the greatest speed on the journey of the second train?

11. The weight of a train is 400 tons, the part of the weight of the engine supported by the driving wheels is 30 tons, and the coefficient of friction μ between the driving wheels and the rails is 0·16. Prove that at the end of a minute after starting on the flat the velocity will be less than 15·8 m.p.h. (Assume that the tangential component of the reaction between a wheel and the rail is not greater than μ times the normal component.)

12. If a constant force P acting upon a body of mass M for a time t gives it a velocity v, establish the relation $Pt = Mv$.

A tug-boat A of 400 tons is attached to a ship B of 4000 tons by means of a cable fixed to A and passing round a capstan on B, the arrangement being such that a pull of 5 tons causes the cable to slip round the capstan. The propulsive force on the tug-boat exerted by its propeller is 3 tons and it starts off with the cable slack. At the instant when the cable becomes taut the velocity of A is 5 ft./sec., and B is stationary. Prove that the common velocity of A and B when the cable has just ceased slipping is 1 ft./sec. Determine also the time during which slip occurs, and the length of cable which passes over the capstan. (The resistance of the water may be neglected.) [For a comment on the nomenclature—'a force of 3 tons', meaning a force of 3 tons weight—see Ex. 5 of § 4·9.]

13. The total resistance to a car with the brakes on is $\frac{1}{16}$ of the weight of the car; show that the car, when running at 8 m.p.h., can be stopped in about $11\frac{1}{2}$ yards.

Show also that, if the ordinary resistance is $\frac{1}{64}$ of the weight, each stoppage by the brakes and recovery of the previous speed of 8 m.p.h. adds to the work of traction an amount approximately equal to 181 foot-pounds/cwt. of the car mass.

14.* A particle moving in a straight line leaves A with velocity u and comes to rest at B, the length of AB being a. The distance AB is divided into n equal intervals at $A_1, A_2, \ldots, A_{n-1}$. At A_1 the particle receives a certain retardation, which is increased by equal amounts at $A_2, A_3, \ldots, A_{n-1}$, the motion between two points of subdivision being with constant retardation.

Prove that the retardation r of the particle when coming to rest at B is independent of n.

If the $(n-1)$ equal increases of retardation occur at equal time-intervals instead of at equal space-intervals, prove that the retardation at B is not independent of n, and tends to $\frac{4}{3}r$ as n tends to infinity.

15. A train starts from a station A with acceleration 1 ft./sec./sec., and this acceleration decreases uniformly for $2\frac{4}{9}$ minutes, at the end of which time the train has attained full speed which is maintained for another 5 minutes. Then brakes are applied and produce a constant retardation $2\frac{1}{5}$ ft./sec./sec. and bring the train to rest at B. Find the maximum speed and the distance AB.

16. In starting a tram of mass 3200 lb. the pull exerted by a horse is initially 200 lb., and this pull decreases uniformly with the time until at the end of 10 sec. it has fallen to 40 lb., an amount just sufficient to overcome the frictional resistance of the tram. Determine, and illustrate by means of a graph, the variation of velocity with time, and find the distance run during this period of 10 sec.

Show that the horse-power is a maximum at the end of 5 sec., and find its maximum value, taking g to be 32 f.s. units.

17. A train of mass 300 tons is originally at rest on a level track. It is acted on by a horizontal force F which uniformly increases with the time, in such a way that $F = 0$ when $t = 0$, $F = 5$ when $t = 15$; F being measured in tons weight, t in seconds. When in motion, the train may be assumed to be acted on by a frictional force of 3 tons weight, independent of the speed of the train. Find the instant of starting, and show that at $t = 15$, the speed of the train is 0·64 ft./sec., whilst the horse-power required at this instant is about 13.

18. By proper choice of units the curve on a time base representing the acceleration of an electric tram is a quadrant of a circle, whose centre is the origin. The initial acceleration is 2·5 ft./sec./sec., and the acceleration falls to zero in 20 sec. Calculate the velocity acquired and the distance described in that time.

19. A boat of mass m is moving with a velocity V which varies periodically and which can be expressed, in terms of the time t, as $V = u + v \sin nt$, u and v being constants. The resistance to motion is λV^2. Write down expressions giving the values of the propelling force and of the rate of working, at any instant, and show that, if $u = 4v$, the work done during a complete period is about $9\frac{1}{2}$ per cent greater than would be done in the same time if the boat were propelled uniformly at the same average speed.

20. A particle moves in a straight line under the action of a given (variable) force. What physical quantity is represented by the area lying under the curve and bounded by two ordinates in the several cases where the abscissae and ordinates represent graphically (1) time and velocity, (2) time and force, (3) distance and force?

A cable is used for raising loads, the greatest tension that the cable will bear being W tons weight. Show, by consideration of the time-velocity graph, that the least time in which a load of W' tons can be raised through h feet from rest to rest by means of the cable is $\sqrt{\{2hW/g(W-W')\}}$ seconds. The weight of the cable itself is negligible.

21. A particle moves on a straight line, the relation between time and distance being

$$t = ax + bx^2,$$

where a and b are constants. Determine the relations between (i) distance and acceleration, (ii) distance and velocity, (iii) time and velocity.

If the particle travels 2000 ft. in 1·9 sec., and its velocity is then 1000 ft./sec., show that its initial velocity was about 1111 ft./sec.

22. Power is transmitted from one pulley to another by means of a belt round both pulleys. Show that if the tensions in the tight and slack sides of the belt are T_1, T_2 lb. weight, and the speed of the belt is v ft./sec., the horse-power transmitted is $(T_1 - T_2)v/550$.

Power is delivered by an engine through ropes passing round the flywheel of the engine and over pulleys in the mill. If the flywheel is 30 ft. in diameter and turns at 90 revolutions per minute, determine how many ropes of $1\frac{1}{2}$ in. diameter will be required to transmit 2000 horse-power. Assume the tension on the tight side to be equal to twice the tension on the slack side, and allow not more than 300 lb./sq.in. pull in the ropes.

CHAPTER V

MOTION ON A STRAIGHT LINE, THE FIELD OF FORCE

5·1 Motion in a field. We now turn to case (*c*) of the classification of § 4·2. We consider a (one-dimensional) field of force; a particle of mass m moves on a straight line under the action of a force $P = X(x)$, where $X(x)$ is a given function of x. This function is assumed to be of simple type; usually it is a continuous function.

The first thing to notice is the change in the role of the variable x at this point (cf. § 4·2). Frequently we have used x to denote the position of the particle at time t; then x was a dependent variable, a function of t to be determined. Here, however, for the moment, x is the independent variable, and $X(x)$ is a given function of this independent variable.

When we study the field of force, we are led almost inevitably to introduce the *potential function* $V = V(x)$. This function is defined by the equation

$$V = -\int_a^x X(\theta)\,d\theta, \tag{1}$$

where the lower limit a in the integral is any convenient fixed number in the range in which $X(x)$ is defined. An alternative form of the definition is contained in the equation

$$\frac{dV}{dx} = -X. \tag{2}$$

The two definitions are equivalent, the arbitrary constant of integration in (2) corresponding to the arbitrary lower limit of integration in (1).

The work done by the field when the particle moves from $x = x_1$ to $x = x_2$ is

$$W = \int_{x_1}^{x_2} X(x)\,dx = V_1 - V_2, \tag{3}$$

where V_1 is written for $V(x_1)$, and V_2 for $V(x_2)$. This equation (3) represents the fundamental property of the potential function.

It will be observed that, since a is arbitrary, the potential function is not precisely defined; it involves an arbitrary additive constant. The difference between the values of the potential function at two points of the line is, however, precisely defined, and it is only such a difference that appears in the fundamental equation (3).

The work done by the field when the particle moves from A to B is $V_A - V_B$. We can look at this result from another point of view. Suppose an agent or carrier takes the particle at constant speed from A to B. Since the speed is constant, the forces on the particle are in equilibrium at each instant, and the force exerted by the carrier on the particle is $-X$. The work done by the carrier when the particle is taken from A to B is thus precisely opposite to the work done by the field, i.e. the work done by the carrier is $V_B - V_A$. We can thus define the potential function at B as the work done by the carrier to take the particle at constant speed *to* B, starting from an arbitrary fixed point O. The choice of O, the point at which V vanishes, merely affects the additive constant in V.

The equation (3) above is of course a special case of the equation (2) of § 3·10, but there is an important difference between the two cases. When we deal with the general three-dimensional field, such an equation holds only when the field is conservative. When we deal with the one-dimensional problem, such an equation holds always. We can express this in another way by saying that, when we are dealing with motion in a straight line, any field is conservative; a potential function always exists, and it is in fact the function defined by equation (1) above.

5·2 Examples of the calculation of V. We now consider the function V for some of the classical cases; we can use either of the definitions (1) or (2) of § 5·1, or the fundamental property (3).

(i) *The uniform field.* If the field is $-mg$ (so that the positive sense in Ox is opposite to the direction of the field), the fundamental equation (3) gives us

$$V_1 - V_2 = -\int_{x_1}^{x_2} mg\,dx = mg(x_1 - x_2),$$

so that
$$V_1 - mgx_1 = V_2 - mgx_2 = C,$$

whence
$$V = mgx + C.$$

Or, even more simply, from equation (2)

$$\frac{dV}{dx} = -X = mg,$$

whence $$V = mgx + C.$$

It may be noticed that such a formula holds as a first approximation in a more general problem, where X is not constant, if the scale of the motion is so small that the variation of X is negligible. The approximation is then that obtained by retaining only the first power of $(x-a)$ in a Taylor series for V in powers of $(x-a)$,

$$\begin{aligned} V(x) &= V(a) + (x-a)\,V'(a) \\ &= V(a) - (x-a)\,X(a). \end{aligned}$$

(ii) *Harmonic attraction.* The particle is attracted towards O with a force $mn^2\,|\,x\,|$, where n is a positive constant. Here

$$X = -mn^2x,$$

a formula which is valid both for positive and negative values of x. If we take $V = 0$ at O we have (equation (1) of § 5·1)

$$V = \int_0^x mn^2\,\theta\,d\theta = \tfrac{1}{2}mn^2x^2.$$

This case arises when the particle is attached to a fixed point A by a light spring obeying Hooke's law (Fig. 5·2). Measuring from the equilibrium position of the particle the restoring force

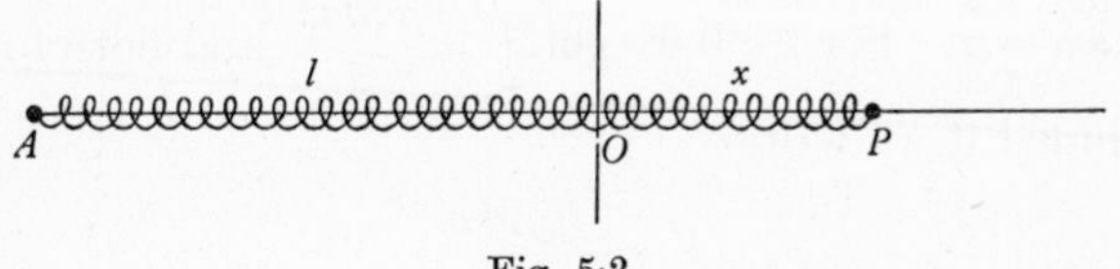

Fig. 5·2.

is $k\,|\,x\,|$, or $(\lambda/l)\,|\,x\,|$, where l is the natural length of the spring, and λ the tension that would be required to double its length. (In practice Hooke's law is obeyed to a reasonably high degree of approximation only when $|\,x\,|$ is small; it is clear that our assumption becomes invalid before x reaches the value $-l$.) Here

$$V = \tfrac{1}{2}kx^2 = \frac{\lambda}{2l}x^2,$$

and $$n^2 = k/m = \lambda/(lm).$$

If we consider an elastic *string* instead of a spring the formulae for the restoring force and the potential function are valid only for stretching.

It is just a matter of convenience which of the constants, k or λ, we choose to employ in a particular context. Perhaps k is more generally useful, but λ, called the *modulus*, has the advantage that its value depends only on the material, supposed uniform, of the string or spring, not on its length; λ is the tension needed to double the length.

(iii) *Gravitational field.* A particle moves on the line Ox under the gravitational attraction of a fixed particle, of mass M, at O. Newton's great discovery of the law of gravitation, which we shall consider more fully later on, tells us that the attraction between the particles is $\gamma Mm/x^2$, where γ is a universal constant. Thus, for $x > 0$,

$$X = -\gamma Mm/x^2,$$

and $$V_1 - V_2 = -\gamma Mm \int_{x_1}^{x_2} \frac{1}{x^2}\,dx = -\gamma Mm\left(\frac{1}{x_1} - \frac{1}{x_2}\right).$$

Thus $$V = -\frac{\gamma Mm}{x} + C.$$

We always give C, which is arbitrary, the value zero, so that $V \to 0$ as $x \to \infty$. For $x < 0$ we get $V = \dfrac{\gamma Mm}{x}$, and both formulae are included if we write

$$V = -\frac{\gamma Mm}{r},$$

where $r = |x|$. But in practice we should not often be dealing with the two regions, $x > 0$ and $x < 0$, in the same problem, because these two regions are separated by the discontinuity at O; $V \to -\infty$ as $r \to 0$.

5·3 The equation of energy. We return now to the general problem of motion in a field whose potential is V. If we substitute for x in $V(x)$ the value of $x(=x(t))$ at time t during the motion we obtain a function of t, say $\Omega(t)$. Since

$$\Omega(t) = V(x),$$

we have
$$\frac{d\Omega}{dt} = \frac{dV}{dx}\frac{dx}{dt} = -vX.$$

Now the equation of motion can be written

$$m\frac{dv}{dt} = X,$$

whence
$$mv\frac{dv}{dt} = vX = -\frac{d\Omega}{dt},$$

and this equation can be written in the form

$$\frac{d}{dt}(\tfrac{1}{2}mv^2 + \Omega) = 0.$$

We deduce
$$\tfrac{1}{2}mv^2 + \Omega = C, \tag{1}$$

where C is constant. This is the famous 'equation of energy' or 'integral of energy'. The term $\frac{1}{2}mv^2$ is called the *kinetic energy* of the particle, and is almost universally denoted by T. Kinetic energy is, like work, a scalar quantity, and the formula $\frac{1}{2}mv^2$ gives its measure on a uniform scale.

The equation of energy

$$T + \Omega = C \tag{2}$$

can also be written in the form

$$T + V = C \tag{3}$$

if we remember that the symbol V has here a meaning slightly different from that which it had when originally introduced. Originally $V(x)$ was a function of the independent variable x, defined for all values of x, or at least for all values of x in some domain. Here (equation (3)) V means the value of this function, not at a general point x, but at the point actually occupied by the particle at time t, an instant at which its velocity is v. The equation (3) is, of course, the (x, v) relation for the problem.

It is a differential equation of the first order

$$\tfrac{1}{2}m\left(\frac{dx}{dt}\right)^2 + V(x) = C,$$

in which x is the dependent variable, and $V(x)$ is a given function of x. The equation of energy is the dominating principle for the whole theory of motion in a field.

The equation of energy expresses the fact that the sum of two terms remains constant throughout the motion, the first term T depending on the velocity of the particle, the second term V depending on its position. In the form (2) we are thinking primarily of t as the independent variable; $T+\Omega$ has the same value for all values of t. In the form (3) we are thinking primarily of x as the independent variable; $T+V$ has the same value for all values of x.

The equation (3) shows us that, for motion in a field, v^2 is a uniform function of x, and this remark suggests a different approach to the matter. Originally we took t as independent variable, and thought of x as a function of t,

$$x = \phi(t),$$

whence

$$v = \phi'(t).$$

The elimination of t between these two equations leads, as we have noticed already (§ 4·2), to a functional relation between v and x, though not in general a simple relation. In the case under discussion, however, we know that the relation is of a fairly simple type, namely, one in which v^2 is a uniform function of x.

If we think from the outset of x as the independent variable we have (§ 4·2)

$$f = \frac{d}{dx}(\tfrac{1}{2}v^2),$$

and our equation of motion can be written

$$\frac{d}{dx}(\tfrac{1}{2}mv^2) = -\frac{dV}{dx},$$

or

$$\frac{d}{dx}(T+V) = 0;$$

we recover immediately the equation of energy,

$$T+V = C. \qquad (3)$$

An equivalent form of (3) is

$$T_2 - T_1 = V_1 - V_2 = \int_{x_1}^{x_2} X\,dx,$$

and this exhibits the result in a slightly different form; the increase in kinetic energy in any stretch of the motion is equal to the work done by the field. We have noticed one special case of this result already (§ 4·8).

The equation of energy, together with its analogues in more complicated cases, is of far-reaching importance. Its implications extend beyond ordinary dynamics into the whole realm of physics. We conceive of energy as a concrete physical reality. For example, we think of the work done in stretching a string as energy stored up in the string. If one end of the string is fixed, and the other attached to a particle, this store is drawn upon, and converted into kinetic energy of the particle, when the string is allowed to relax. The generalization of the *conservation of energy* occupies so central a position in our thought about the physical world, that when we encounter a dynamical problem in which the energy is not conserved, we prefer to say, not that energy has been destroyed but that it has been converted into a new form, i.e. a form (such as heat) other than the kinetic and potential forms that we deal with in dynamics. Actually if we examine more closely the microscopic structure of matter we discover that heat is itself a form of kinetic energy; but this kinetic energy is not usually significant for classical dynamics. For example, a rigid body at rest has zero kinetic energy from the point of view of classical dynamics; but it does contain heat, and this is a form of energy.

Example. *A particle is tied to a fixed point A by a light elastic string. The natural length of the string is a, and the stretched length, when the particle hangs in equilibrium, is $(a+b)$. To find the depth c below the equilibrium position at which the particle comes to rest, when it falls under gravity from rest at A.*

There is no discontinuity of velocity in this problem, and the energy equation is valid. Also $T = 0$, both initially and finally, so we have only to equate (i) the loss of potential energy due to the downward displacement in the field of gravity and (ii) the work done in stretching the string from length a to

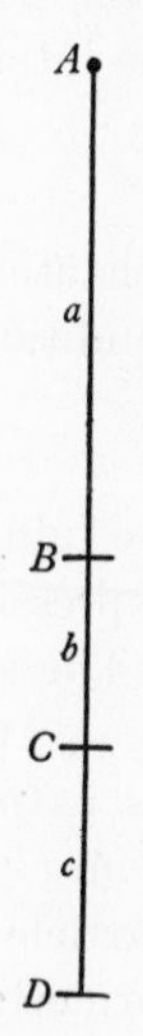

Fig. 5·3

$(a+b+c)$, where $(a+b+c)$ is the depth below A at which the particle comes to rest. (In Fig. 5·3 AB is the unstretched length of the string, and C the equilibrium position when the particle hangs at rest.) We have, therefore, if the tension is k times the stretch (as in § 5·2 (ii)),

$$mg(a+b+c) = \tfrac{1}{2}k(b+c)^2,$$

while

$$mg = kb.$$

Thus

$$(b+c)^2 = 2b(a+b+c),$$

$$c^2 = b^2+2ba = (a+b)^2-a^2.$$

5·4 Reversibility. One other property which is very characteristic of the problem of motion in a field of force is that of reversibility. This property is almost evident from the fact that v^2 is a uniform function of position, so that, with a fixed value for the energy constant C, the speed is the same at a given point whether the particle moves to the right or to the left.

We now put this idea into a more precise form. The equation of motion is

$$m\ddot{x} = X(x). \tag{1}$$

Suppose a solution is $x = \phi(t)$ in the range $t = 0$ to $t = t_0$ $(t_0 > 0)$, and that, at $t = t_0$, the velocity is reversed. We may suppose that at this instant the particle collides with a perfectly elastic wall. Then the energy constant is unchanged. The solution in the interval from $t = t_0$ to $t = 2t_0$ is

$$x = \phi(2t_0 - t), \tag{2}$$

because this satisfies the differential equation (1), and the conditions

$$x(t_0+0) = x(t_0-0), \quad \dot{x}(t_0+0) = -\dot{x}(t_0-0) \tag{3}$$

are fulfilled. The equation (2), valid for $t_0 < t < 2t_0$, is the precise expression of the property of reversibility.

The simplest case occurs when the particle comes to rest at $t = t_0$ at a point B. After coming to rest the particle retraces its path. The time from B to a point C is equal to the time from C to B in the original motion, and the velocity when the particle returns to C is the same as when the particle was formerly at C, but with the sign changed. An example has already appeared in the problem of the uniform field (§ 4·6).

5·5 Harmonic motion. The general theory of motion in a field of force is based on the equation of energy, and this theory we shall consider in detail in the next chapter. But before doing so, it will be convenient to make a preliminary study of the important special case of harmonic motion. (Another special case, that of the uniform field, we have discussed already in § 4·4.)

Consider then a particle moving on a line, and subject to an attraction towards a fixed point of the line (the centre of attraction) and proportional to the distance from this point. A particle whose motion is controlled by this law is called a *harmonic oscillator*, and its motion is called *harmonic motion* or *simple harmonic motion*.

Taking the origin at the centre of attraction we have

$$X = -mn^2x, \quad V = \tfrac{1}{2}mn^2x^2,$$

where n is a positive constant. The equation of motion is

$$\ddot{x}+n^2x = 0, \tag{1}$$

valid for all values of x, not merely for positive values. The solution of this is

$$x = a\cos nt+\frac{u}{n}\sin nt, \tag{2}$$

where a is the value of x, and u is the value of v, at $t = 0$. We shall find that this formula, giving the position of the particle at any time during the motion, is often needed.

It is sometimes convenient to suppress the constants a and u, representing the position and velocity at $t = 0$, in favour of new constants c, α defined by $c\cos\alpha = a$, $c\sin\alpha = u/n$ (Fig. 5·5a); here $c > 0$, and $0 \leqslant \alpha < 2\pi$. It is clear that $c^2 = a^2+(u/n)^2$, and that $\tan\alpha = u/na$, α being between 0 and π if $u > 0$; in particular α is acute if a and u are both positive. In terms of these new constants equation (2) becomes

$$x = c\cos(nt-\alpha). \tag{3}$$

Again, if we write $\alpha = nt_0$, this becomes

$$x = c\cos n(t-t_0), \tag{4}$$

whence also

$$v = \dot{x} = -nc\sin n(t-t_0). \tag{5}$$

It is clear now that the motion is the projection on Ox of the motion of a point Q which moves uniformly in a circle with centre O and radius c, the angular velocity being n. The motion is confined to the stretch from $x = -c$ to $x = c$, and t_0 is a value of t when $x = c$. (Cf. Fig. 5·5b.)

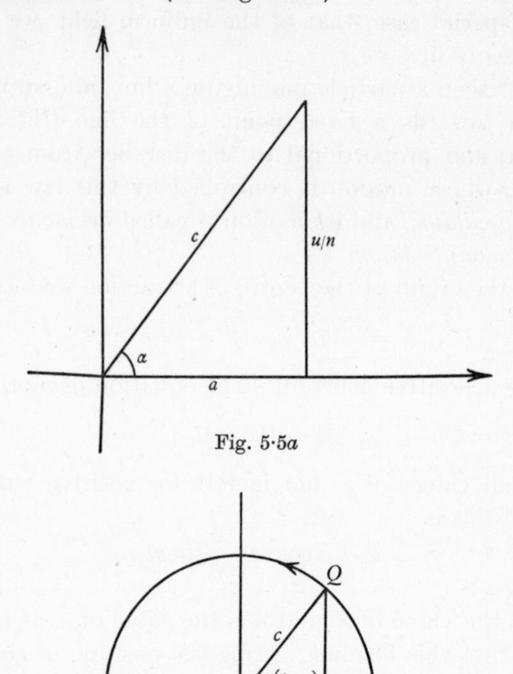

Fig. 5·5a

Fig. 5·5b

The motion is *periodic* with the period $\sigma = 2\pi/n$, that is to say

$$x(t+\sigma) = x(t), \tag{6}$$

for all values of t; a corresponding property holds for the velocity

$$\dot{x}(t+\sigma) = \dot{x}(t). \tag{7}$$

The period σ is the time taken for Q to traverse the complete circle. To prove (6) we have only to observe that the position of Q, and therefore also of P, at time $t+\sigma$ is the same as its position at time t. The results of (6) and (7) are evident also from (4) and (5) in virtue of the fundamental property of the functions $\cos\theta$ and $\sin\theta$, that they are periodic with period 2π.

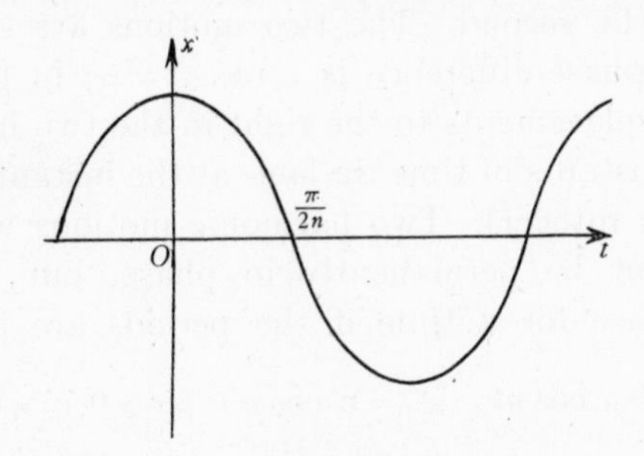

Fig. 5·5c

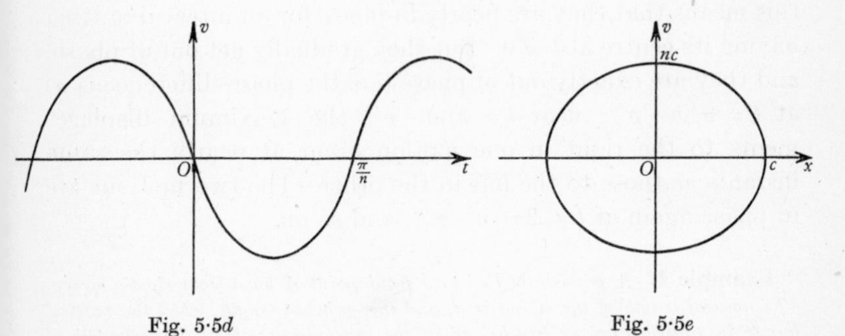

Fig. 5·5d

Fig. 5·5e

The motion that takes place in the interval from $t = 0$ to $t = \sigma$ is repeated in the interval from $t = \sigma$ to $t = 2\sigma$, and generally in any interval from $t = r\sigma$ to $t = (r+1)\sigma$, where r is any integer.

It is clear of course that 2σ, for example, is also a period; but when we speak of the period we usually mean the smallest period, and that is, in this case, $\sigma = 2\pi/n$.

It is easy to construct the (t, x), (t, v), and (x, v) graphs for the motion. The first two are given by (4) and (5), and the (x, v) relation is

$$v^2 + n^2x^2 = n^2c^2, \tag{8}$$

which is the equation of an ellipse. Equation (8) is simply the equation of energy.

In the formula (3) the length c is called the *amplitude* of the oscillation and the angle $(nt-\alpha)$ is called the *phase*. If we have two harmonic motions with the same period

$$x = c\cos(nt-\alpha), \quad x' = c'\cos(nt-\alpha') \quad (c>0, c'>0),$$

there is said to be a phase-difference $\alpha'-\alpha$ between the first motion and the second. The two motions are said to be *in phase* if the phase-difference is zero, $\alpha' = \alpha$; in that case the maximum displacements to the right in the two motions occur at the same instants of time (in fact, at the instants $(\alpha+2r\pi)/n$, where r is an integer). Two harmonic motions with different periods cannot be permanently in phase, but they can be nearly in phase for a time if the periods are nearly equal. Thus if

$$x = c\cos nt, \quad x' = c'\cos n't \quad (c>0, c'>0),$$

where $|n-n'|$ is small, the two motions are 'in phase at $t=0$'; this means that they are nearly in phase for an interval of time having its centre at $t=0$. But they gradually get out of phase, and they are exactly out of phase (i.e. the phase-difference is π) at $t=\pi/|n-n'|$: near $t=\pi/|n-n'|$ the maximum displacements to the right in one motion occur at nearly the same instants as those to the left in the other. The two motions are in phase again at $t=2\pi/|n-n'|$, and so on.

Example 1. *A particle is tied to a fixed point A by a light elastic string. The natural length of the string is a, and the stretched length, when the particle hangs in equilibrium, is $(a+b)$, as in the example of § 5·3. The particle is allowed to fall from rest at A. To discuss the subsequent motion, and, in particular, to find the time that elapses before the particle first comes to rest.*

We denote by B the point at depth a below A, i.e. the point at which the string becomes taut, and by C the point at which the particle would hang in equilibrium. If the tension in the string is $k\times$ (extension) we have $kb = mg$. If the depth of the particle below B is ξ we have for the equation of motion, so long as the string is taut $(\xi>0)$,

$$m\ddot{\xi} = mg-k\xi = -k(\xi-b),$$

i.e.

$$\frac{d^2}{dt^2}(\xi-b)+n^2(\xi-b) = 0,$$

where $n^2 = k/m = g/b$. If we write $\xi = b+x$, so that x is the depth below C, the equation becomes

$$\ddot{x}+n^2x = 0,$$

and the motion is a harmonic motion, with C as centre.

It is perhaps worth while to look at this part of the argument again from a slightly different point of view. When the particle is at C, the tension in the string is equal to the weight of the particle; when the particle is at a depth x below C, the *increase* in tension is $k\times$ (*increase* in length), so that the restoring force (excess of tension over weight) is kx. Thus the equation of motion is

$$m\ddot{x} = -kx,$$

i.e. $\ddot{x} = -n^2 x$, as before.

During the motion from A to B (while the string is slack) the particle has constant acceleration g. The time from A to B is $\sqrt{(2a/g)}$, and the velocity at B is $\sqrt{(2ga)}$. In the subsequent harmonic motion, now measuring time from the instant of reaching B, we have

$$x = -b\cos nt + \frac{\sqrt{(2ga)}}{n}\sin nt$$

$$= -b\cos nt + \sqrt{(2ab)}\sin nt$$

$$= -c\cos(nt+\alpha),$$

where

$$c^2 = b^2 + 2ab \quad \text{and} \quad \tan^2\alpha = 2a/b \quad (0 < \alpha < \tfrac{1}{2}\pi).$$

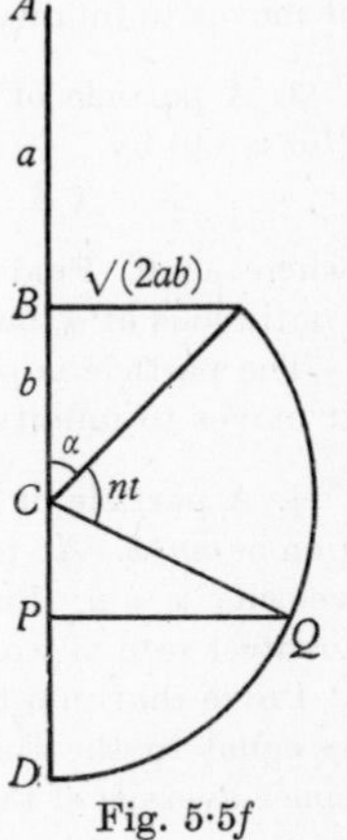

Fig. 5·5f

The motion is represented by the projection of a uniform circular motion, as in the figure (Fig. 5·5f). The particle reaches the lowest point D when $nt+\alpha = \pi$,

$$t = (\pi - \alpha)/n.$$

The total time from A to D is

$$\sqrt{\left(\frac{2a}{g}\right)} + (\pi-\alpha)\sqrt{\left(\frac{b}{g}\right)}.$$

After the particle comes to rest at D the harmonic motion continues until the particle reaches B, and when it does so it has the speed with which it originally reached B, but now upwards instead of downwards. After this point it moves freely under gravity, and comes to rest again at A. Here we have an illustration of the theory of reversibilty (§ 5·4), and the time from D to A is equal to the time taken from A to D. When the particle reaches A it rests there instantaneously, and then the cycle repeats itself.

Example 2. *The eccentric circular cam.* A circular disc is made to rotate in its own plane about a point of itself not its centre, the angular velocity being maintained at the constant value n. A piston rod, free to move along its length, is kept in contact with the disc by a spring. Then the motion of the piston is harmonic. This follows easily, because the projection of the motion of the centre of the disc is harmonic.

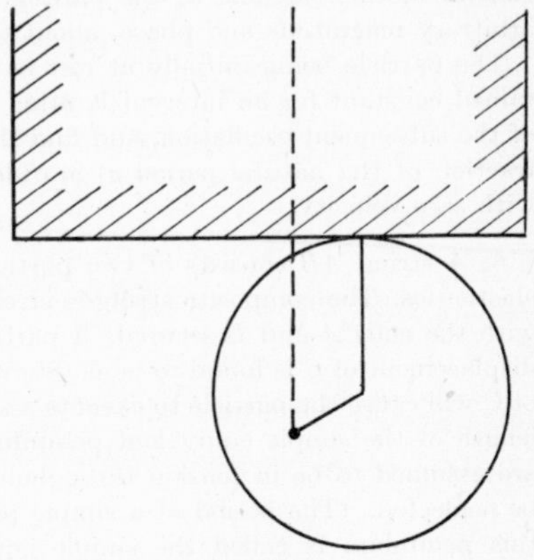
Fig. 5·5g

EXAMPLES V

1. A horse pulls a wagon of 10 tons from rest against a constant resistance of 50 lb. weight. The pull exerted is initially 200 lb. weight, and decreases uniformly with the distance covered until it falls to 50 lb. weight at a distance of 167 ft. from the start. Show that the resulting velocity of the wagon is very nearly 6 ft./sec.

2. A particle of mass m moves on the axis Ox in the field of force $X = -mab \sin bx$. It is projected from the origin with velocity u. Prove that it moves to infinity if $u^2 > 4a$.

3. A particle of mass m moves on the axis Ox in the field of force given (for $x > 0$) by

$$X = -\mu m x/a^3 \quad (x \leqslant a), \qquad X = -\mu m/x^2 \quad (x > a),$$

where $\mu > 0$. Find the potential function V and illustrate by a graph. (V is continuous at a, and tends to zero as $x \to \infty$.)

The particle is projected from the origin with velocity u (>0). Prove that it moves to infinity if $au^2 > 3\mu$.

4. A particle of mass m moves on the line Ox under the harmonic attraction $-mn^2x$. At the instant $t = 0$ it is projected from the point $x = a$ with velocity $\dot{x} = u$. Find the amplitude c of the oscillation, and prove that the greatest rate of working of the force is $\frac{1}{2}mn^3c^2$.

Prove that in a harmonic oscillation the time average of the kinetic energy is equal to the time-average of the potential energy. Prove also that the space-average of the kinetic energy is twice the space-average of the potential energy.

5. A particle is constrained to move along a straight line, and is attracted towards a fixed point O in that line by a force proportional to its distance from O. It is subjected in addition to a constant force X, acting in the same straight line away from O, and of magnitude sufficient to hold the particle in equilibrium at another point A at distance a from O. Show that the most general motion possible to the particle is a simple harmonic oscillation, of arbitrary magnitude and phase, about the point A as centre.

The particle being initially at rest at O, the force X is applied and maintained constant for an interval θ, after which it ceases. Find the amplitude of the subsequent oscillation, and find the smallest value of θ (expressed as a fraction of the natural period σ) in order that the particle shall arrive at A with zero velocity.

6. A string AB consists of two portions AC, CB, of unequal lengths and elasticities. The composite string is stretched and held in a vertical position with the ends A and B secured. A particle is attached to C, and the steady displacement of C is found to be δ. Show that a further vertical displacement of C will cause the particle to execute a simple harmonic motion, and that the length of the simple equivalent pendulum is δ. Both portions of the string are assumed to be in tension throughout, and the weight of the string may be neglected. (The period of a simple pendulum of length l is $2\pi\sqrt{(l/g)}$, and this pendulum is called the simple equivalent pendulum for any particle oscillating with this period.)

7. A particle is attached to the mid-point of a light elastic string of natural length a. The ends of the string are attached to fixed points A and B, A being at a height $2a$ vertically above B, and in equilibrium the particle rests at a depth $5a/4$ below A. The particle is projected vertically downwards from this position with velocity $\sqrt{(ga)}$. Prove that the lower string slackens after a time $\frac{\pi}{12}\sqrt{\left(\frac{a}{g}\right)}$, and that the particle comes to rest after a further time $\beta\sqrt{\left(\frac{a}{2g}\right)}$, where β is the acute angle defined by the equation $\tan\beta = \sqrt{(3/2)}$.

8. A number of strings, AB, BC, ..., JK, of different lengths and elasticities, are tied together at B, C, ..., J, and the combined string is hung from A and a particle is tied to K. In equilibrium the stretch of the combined string is c. Show that for small vertical oscillations the length of the simple equivalent pendulum is c.

9. A mass is suspended from a ceiling by an elastic string of natural length a. When the mass hangs in equilibrium the length of the string is $a+c$. The mass is started off from this position with downward vertical velocity v. If in the subsequent motion the string never becomes slack, show that $v^2 < cg$.

If $v^2 = 4cg$ and $3c < 2a$ and the particle is started as before, find the time from the lowest point of its path to the highest.

10. A heavy particle of mass m is attached to one end of an elastic string of natural length a, whose other end is fixed at O. The particle is let fall from rest at O. Show that part of the motion is simple harmonic, and that, if the greatest depth of the particle below O is $a\cot^2\frac{1}{2}\theta$, the modulus of elasticity of the string is $\frac{1}{2}mg\tan^2\theta$, and that the particle attains this depth in time

$$\sqrt{\left(\frac{2a}{g}\right)}\{1+(\pi-\theta)\cot\theta\},$$

where θ is a positive acute angle.

11. Explain what is meant by simple harmonic motion, and obtain an expression for the acceleration at any point of the motion in terms of the displacement and the number of oscillations per second.

A horizontal shelf is given a horizontal simple harmonic motion. The amplitude of the motion is a feet, and n complete oscillations are performed per second. A particle of mass m lb. is placed on the shelf at the instant when it is at the extremity of its motion. Prove that, if μ is less than $4\pi^2n^2a/g$, slipping between the particle and shelf will occur for a period t, where t is found from the equation

$$\frac{\sin 2\pi nt}{2\pi nt} = \frac{\mu g}{4\pi^2 n^2 a}.$$

Show that, if for a particular case this value of t is $1/6n$, the distance through which the particle moves relatively to the shelf in this time is

$$\frac{a}{2}\left(1-\frac{\pi\sqrt{3}}{6}\right).$$

12. A particle of mass m is attached to one end B of a light elastic string AB, the other end A being fixed. When the particle hangs in equilibrium the length of the string exceeds the unstretched length by a. Prove that

if the particle executes small vertical oscillations about the position of equilibrium, the period is the same as that of a simple pendulum of length a.

A similar string BC, carrying a particle of mass $\frac{2}{3}m$ at C, is now attached to the first particle at B, and the system hangs in equilibrium from A. If now the particles oscillate vertically, the downward displacements of B and C from their equilibrium positions being x and y, establish the equations of motion

$$\ddot{x} = n^2(y-2x), \quad \tfrac{2}{3}\ddot{y} = n^2(x-y),$$

where $n^2 = g/a$. It is assumed that the strings remain taut.

Prove that a motion is possible in which $x/2 = y/3$ throughout.

CHAPTER VI

THE GENERAL THEORY OF MOTION IN A FIELD OF FORCE

6·1 Motion in a field. To begin with we make a small change of notation. We write mX, mV in place of X, V (so that X now represents the force per unit mass and V the potential per unit mass) and mC in place of C. The equation of energy now takes the form

$$\tfrac{1}{2}\dot{x}^2 = C - V.$$

This, as we have remarked already, is the (x, v) relation for the problem, and v^2 is a uniform function of x. Corresponding to a given x there are two values of v, equal in magnitude but of opposite signs. It will be convenient to consider primarily not the (x, v) graph itself but the curve

$$y = C - V.$$

From this curve we can obtain the values of $\tfrac{1}{2}v^2$ and of f for the particle when its position is x; these values are given by the formulae

$$\tfrac{1}{2}v^2 = y \quad \text{and} \quad f = \frac{dy}{dx}.$$

The values of y and dy/dx determine the kinetic energy per unit mass and the acceleration at the point x.

The motion that we are to study is controlled by a differential equation of the form

$$\tfrac{1}{2}\dot{x}^2 = \phi(x), \qquad (1)$$

where ϕ is a given function of x. In our case $\phi(x)$ has the form $C - V$, but it will be more convenient to deal first with the general case rather than with this particular form. The equation (1) is of vital importance in dynamics, and we shall meet it again and again. Our discussion of the motion will be based on a study of the graph

$$y = \phi(x).$$

6·2 Libration motion. We notice first that motion occurs on a stretch of the x-axis for which $y = \phi(x) \geqslant 0$. In the simplest (and commonest) case x lies initially between consecutive simple zeros a, b $(a < b)$ of $\phi(x)$; the relevant part of the graph is of the general form illustrated in Fig. 6·2a, with $y > 0$ for $a < x < b$. Since a and b are simple zeros, dy/dx does not vanish at a or at b; indeed, it is clear that $dy/dx > 0$ at a, and $dy/dx < 0$ at b.

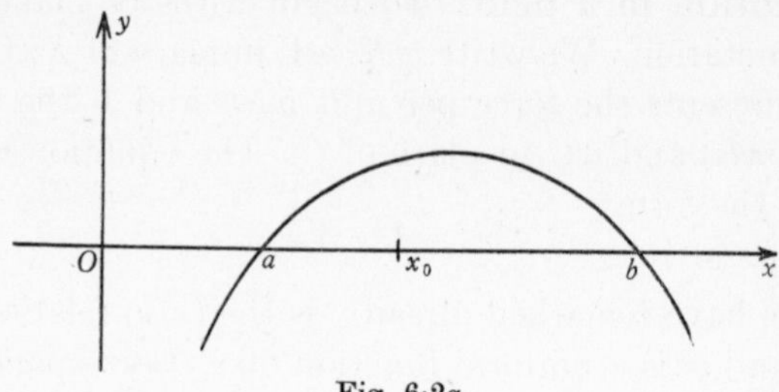

Fig. 6·2a

The particle oscillates between $x = a$ and $x = b$, and the motion is periodic. We notice first that $v = 0$ at a and at b, $f > 0$ at a, and $f < 0$ at b. Suppose, for definiteness, that $x = x_0$ at $t = 0$ $(a < x_0 < b)$ and that $v > 0$ at $t = 0$. (The actual value of v is of course $\sqrt{[2\phi(x_0)]}$.) Then for the first part of the motion $\dot{x} > 0$, and

$$t = \int_{x_0}^{x} \frac{d\theta}{\sqrt{[2\phi(\theta)]}}.$$

Now the integrand tends to infinity as θ tends to b, but the integral is nevertheless *convergent* at b, i.e. the integral tends to a limit as x tends to b. This is true whenever b is a *simple* zero of ϕ. Since the integral is convergent at b the particle reaches b in a finite time. At b it rests instantaneously, v passing through the value zero from positive to negative since $f < 0$ at b. Then the particle moves towards a, and, by a similar argument, reaches a in a finite time, rests there instantaneously, and then moves towards b. It reaches x_0 with the original (positive) velocity after a time

$$\sigma = 2\int_{a}^{b} \frac{d\theta}{\sqrt{[2\phi(\theta)]}}. \tag{1}$$

But now, at $t = \sigma$, both x and v have the same values that they had at $t = 0$. The motion that occurs in the interval from

$t = \sigma$ to $t = 2\sigma$ is a repetition of the motion that occurs in the interval from $t = 0$ to $t = \sigma$. The same is true for any interval $r\sigma$ to $(r+1)\sigma$, where r is any positive integer. The motion is periodic with period σ. Such a motion, in which the particle continues to oscillate between two points of the line, is called a *libration motion.*

We notice also, from the theory of the reversibility of the motion, that the graph of x as a function of t is symmetrical about the ordinates representing the instants of rest at a and b. The general form is as shown in Fig. 6·2b.

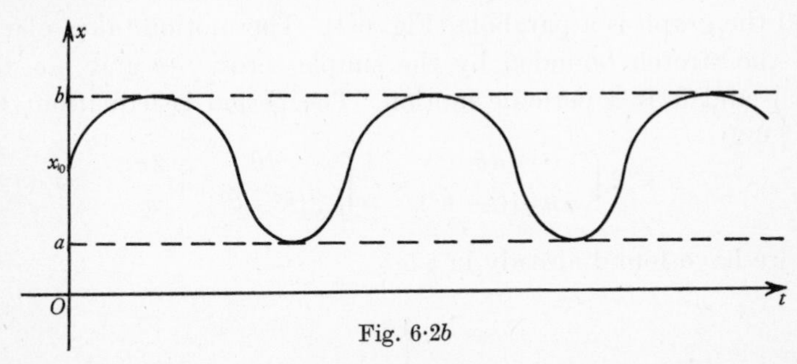

Fig. 6·2b

6·3 Convergent and divergent integrals. We have used in the last paragraph the idea of a convergent integral, and it may be useful at this point to remind the reader, in a general way, of what is involved. Suppose we have a function of x, $f(x)$, which is continuous for $0 \leqslant x < b$, but which tends to ∞ as x tends to b. For example, we might take for $f(x)$ the function $\dfrac{1}{\sqrt{(b^2-x^2)}}$, or the function $\dfrac{1}{b-x}$. Suppose now we consider

$$F(y) = \int_0^y f(x)\,dx,$$

where y is any number in the range $0 \leqslant y < b$. There is no doubt about the existence of $F(y)$, since the integrand is continuous in the range $0 \leqslant x \leqslant y$. The question is, how does $F(y)$ behave as $y \to b$?

In some cases $F(y)$ tends to a limit as y tends to b, and then the integral is said to be *convergent* at b. In other cases $F(y)$ tends to infinity as y tends to b, and then the integral is said to be *divergent* at b.

The simple examples mentioned will serve to illustrate both cases. If

$$f(x) = \frac{1}{\sqrt{(b^2-x^2)}}, \qquad F(y) = \sin^{-1} y/b,$$

and $F(y) \to \tfrac{1}{2}\pi$ as $y \to b$. The integral is convergent at b. If $f(x) = \dfrac{1}{b-x}$,

$$F(y) = \log\frac{b}{b-y},$$

and $F(y) \to \infty$ as $y \to b$. The integral is divergent at b.

For a general account of the theory, the reader may consult Hardy, *A Course of Pure Mathematics*, 7th edition, p. 364.

6·4 An example of libration motion. One example of a libration motion we have discussed already in § 5·5, the harmonic oscillator. Here $V = \frac{1}{2}n^2x^2$, and

$$\phi(x) = C - \tfrac{1}{2}n^2x^2.$$

It is clear at once that no motion is possible with $C < 0$. If $C = 0$ the particle simply rests at O. So we suppose $C > 0$, and write $C = \frac{1}{2}n^2c^2$, where $c > 0$. Then

$$y = \phi(x) = \tfrac{1}{2}n^2(c^2 - x^2), \tag{1}$$

and the graph is a parabola (Fig. 6·4). The motion takes place on the stretch bounded by the simple zeros, $-c$ and $+c$, of $\phi(x)$, and it is a periodic motion. The period is (equation (1) of § 6·2)

$$\sigma = 2\int_{-c}^{c} \frac{d\theta}{n\sqrt{(c^2-\theta^2)}} = \frac{4}{n}\int_0^c \frac{d\theta}{\sqrt{(c^2-\theta^2)}} = \frac{2\pi}{n}, \tag{2}$$

as we have found already in § 5·5.

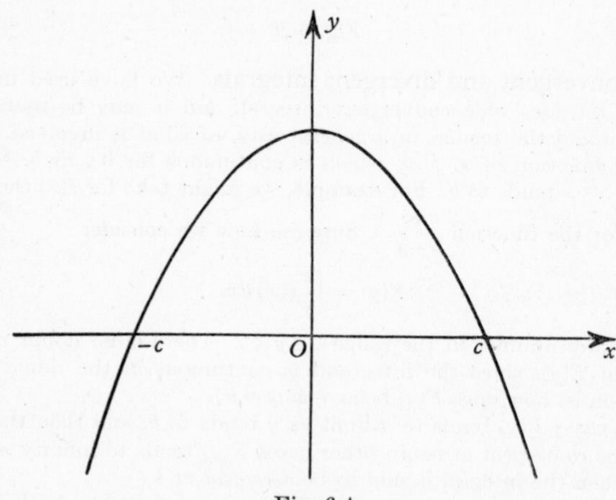

Fig. 6·4

There is another point that should be noticed before leaving this problem. Suppose we start from the energy equation

$$\dot{x}^2 = n^2(c^2 - x^2), \tag{3}$$

and wish to derive from it the relation between t and x, a relation which we have already found by another method.

One way would be to separate the variables and integrate the equation

$$n\,dt = \pm \frac{dx}{\sqrt{(c^2-x^2)}}. \tag{4}$$

There is, however, a better technique for dealing with equation (3), a technique which we have already mentioned in § 4·7. We introduce an auxiliary variable θ in place of x in such a way that we obtain a differential equation, relating t and θ, of very simple form. If in equation (3) we substitute

$$x = c\cos\theta, \tag{5}$$

we get

$$\dot{\theta}^2 = n^2,$$

whence

$$\theta = \pm n(t-t_0). \tag{6}$$

Since the cosine is an even function we can take the plus sign without loss of generality, and the solution of equation (3) is

$$x = c\cos n(t-t_0). \tag{7}$$

The particle is at c at $t = t_0$. This result (7) we have found already (equation (4) of § 5·5).

There is nothing particularly recondite about the substitution (5), for we have only to look ahead a little from equation (3) to equation (4), and consider what substitution would be appropriate for the integration of the second member of equation (4). All we have done is to introduce this substitution at an earlier stage.

6·5 Further study of libration motion. We return to the general case of a libration motion studied in § 6·2. The relation between t and x is

$$dt = \pm \frac{dx}{\sqrt{[2\phi(x)]}}, \tag{1}$$

and we must consider how to fix the ambiguous sign. This is very simple; the particle continually oscillates from a to b and back again, and the sign is plus while it moves to the right, minus when it moves to the left. Expressed even more concisely, the sign on the right in equation (1) is such that, as x oscillates, the second member is always positive.

But this ambiguous sign is occasionally an embarrassment, and it is sometimes convenient to avoid it by the use of an auxiliary variable θ *which continually increases with t.* One way of achieving this is to think of the libration as the projection of motion in a circle having the points $x = a$ and $x = b$ as ends of a diameter. As Q (Fig. 6·5) moves round the circle, always in the same sense, P oscillates between the points a and b.

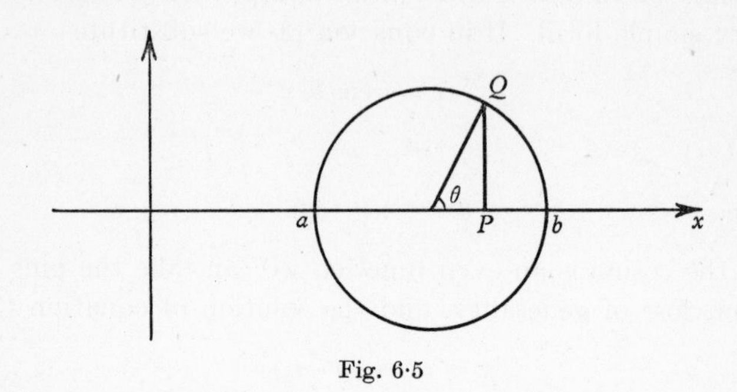

Fig. 6·5

The relation between x and θ is

$$x = \alpha + \beta \cos \theta, \tag{2}$$

where $$\alpha + \beta = b, \quad \alpha - \beta = a;$$

$$\alpha = \frac{b+a}{2}, \quad \beta = \frac{b-a}{2}.$$

The relation (2) can also be written in the form (though this is usually less convenient than (2))

$$x = b \cos^2 \tfrac{1}{2}\theta + a \sin^2 \tfrac{1}{2}\theta.$$

We obtain the relation between t and θ by substituting for x, from equation (2), in the equation of energy,

$$\tfrac{1}{2}\dot{x}^2 = \phi(x). \tag{3}$$

In the simple case of harmonic motion, where we have already used this technique, the velocity of Q on its circle was constant, but this of course is not true in the general case.

Now a and b are simple zeros of ϕ, so we can write

$$\phi(x) = (b-x)(x-a)\psi(x),$$

where $\psi(x) > 0$ for $a \leqslant x \leqslant b$, and equation (3) becomes

$$\tfrac{1}{2}\dot{x}^2 = (b-x)(x-a)\psi(x).$$

We now substitute for x in terms of θ from equation (2). We have

$$\dot{x} = -(\beta \sin\theta)\dot{\theta},$$

and (by direct substitution, or from the geometry of the figure)

$$b-x = \beta(1-\cos\theta), \quad x-a = \beta(1+\cos\theta),$$

$$(b-x)(x-a) = \beta^2\sin^2\theta,$$

giving $$\dot{\theta}^2 = 2\psi(\alpha+\beta\cos\theta),$$

whence $$\dot{\theta} = \sqrt{[2\psi(\alpha+\beta\cos\theta)]}. \tag{4}$$

There is no ambiguity of sign in equation (4), since $\dot{\theta}$ is positive. Equation (4) exhibits the relation between t and θ, and the period of the motion is

$$\sigma = 2\int_0^{\pi} \frac{d\theta}{\sqrt{[2\psi(\alpha+\beta\cos\theta)]}}.$$

(We observe, for future reference, that the substitution (2) has the property

$$d\theta = \mp \frac{dx}{\sqrt{[(b-x)(x-a)]}}.\Bigg)$$

6·6 Limitation motion and motion to infinity.

In the preceding discussion we supposed that x lay originally between consecutive simple real zeros a, b of $\phi(x)$; the relevant part of the graph of $\phi(x)$ was of the form shown in Fig. 6·2a. We now consider what happens when the particle approaches a multiple zero b of $\phi(x)$, i.e. a point at which both y and dy/dx vanish. At such a point the graph touches the axis of x (Fig. 6·6). If, for example, $\phi(x)$ has a double zero at $x = b$, then

$$\phi(x) = (b-x)^2\psi(x),$$

where $\psi(x)$ is positive in the neighbourhood of b.

We suppose, to fix the ideas, that at $t = 0$ we have $x = x_0 < b$ and $\dot{x} > 0$. Then (as in § 6·2)

$$t = \int_{x_0}^{x} \frac{d\theta}{\sqrt{[2\phi(\theta)]}}. \tag{1}$$

This time the integral is divergent as $x \to b$, and $x \to b$ as $t \to \infty$. The particle continues to approach the point b, but never actually reaches it. Such a motion is called a *limitation motion*. Of course a similar result holds if x is decreasing and approaching a multiple zero a of $\phi(x)$; then $x \to a$ as $t \to \infty$.

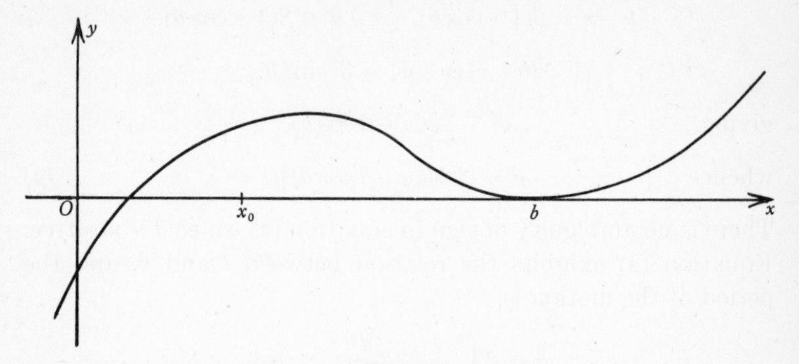

Fig. 6·6

Finally, we consider a problem in which the particle moves to infinity. If $\phi(x) > 0$ for $x \geqslant a$, and $\dot{x}(= \sqrt{[2\phi(a)]})$ is positive at a, then $\dot{x}$ remains positive always, and the particle moves to infinity in the positive direction. The relation between position and time (taking $t = 0$ when $x = a$) is

$$t = \int_{a}^{x} \frac{d\theta}{\sqrt{[2\phi(\theta)]}}. \tag{2}$$

The same relation holds if a is the greatest real zero of $\phi(x)$, and if $\phi'(a) > 0$ and $\phi(x) > 0$ for $x > a$: the particle starts from rest at $x = a$ at $t = 0$. In this case the integral is improper, since the integrand tends to infinity as x tends to a, but the integral converges at a since a is a simple zero of $\phi(x)$.

In most of the familiar problems the integral on the right in equation (2) diverges as $x \to \infty$: in this case $x \to \infty$ as $t \to \infty$.

Examples of this occur in the problem of a particle repelled from O by a force proportional to x (see case (i) of § 6·11) and in the problem of a particle attracted to O by a force inversely proportional to x^2 (see cases (ii) and (iii) of § 6·12). But in other problems, for example in the problem of a particle repelled from O by a force proportional to x^2, the integral on the right in equation (2) converges as $x \to \infty$, and the particle moves to infinity in a finite time. (There is a concrete example of this phenomenon in No. 3 of the examples at the end of this chapter.)

There are, of course, similar cases in which the particle moves to $-\infty$ instead of to $+\infty$.

6·7 Classification.

We have seen in the preceding paragraphs that the motion represented by the differential equation

$$\tfrac{1}{2}\dot{x}^2 = \phi(x)$$

depends primarily on the real zeros of $\phi(x)$, and that we can predict the general nature of the motion from the graph of $\phi(x)$. The case in which we are interested at the moment is that in which

$$\phi(x) = C - V,$$

and it is now clear that all possible types of motion in the given field, for all values of the energy constant C, can be enumerated by considering the graph of $-V$. We get the graph of $C-V$ by raising the graph bodily (if $C>0$) or lowering it (if $C<0$); or, even more simply, by drawing the graph of $-V$ once for all, and lowering or raising the x-axis. And when we have the graph of $C-V$ for a given value of C, the general nature of the motions possible with that value of C can be read off from the graph.

If, for a particular value of C, there is a stretch of the x-axis between consecutive simple real zeros of $C-V$, with $C-V>0$ between them, then a possible motion with this value of C is a libration between the two zeros. If we choose a value of C such that the graph of $C-V$ touches the x-axis, then, if the particle approaches the multiple zero, we have a limitation motion.

If, for a particular value of C, $C-V>0$ for $x \geqslant a$, and if $\dot{x}>0$ when $x=a$, the particle moves to infinity in the positive direction.

6·8 Points of equilibrium. Suppose that $V'(a) = 0$, so that the graph of $-V$ has a horizontal tangent at $x = a$. Then if we choose $C = V(a)$ the graph of $C - V$ touches the x-axis at $x = a$. One possibility with this value of C is that the particle rests in equilibrium at $x = a$.

Supposing, to begin with, that $V''(a) \neq 0$, there are two different cases to consider. If (i) $V''(a) > 0$ the curve lies below the x-axis near a (Fig. 6·8a), and if (ii) $V''(a) < 0$ the curve lies above the x-axis (Fig. 6·8b).

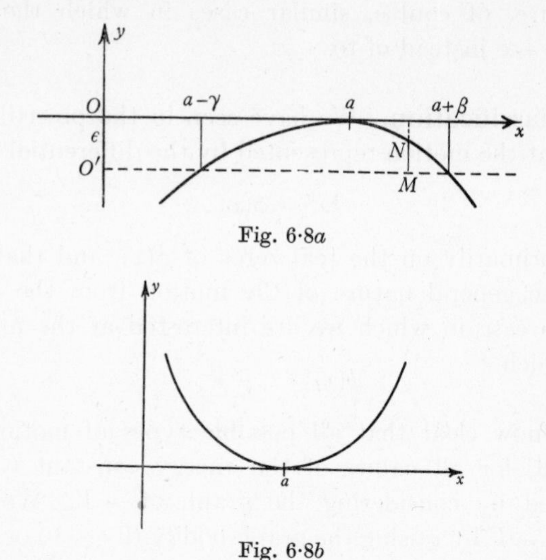

Fig. 6·8a

Fig. 6·8b

(i) In the first case rest at a is the only possibility in the neighbourhood of a.

Consider now what happens when the equilibrium is disturbed; the particle is projected from a point not too remote from a with not too great a speed, so that C is increased, say from $V(a)$ to $V(a)+\epsilon$. The graph is raised (or the x-axis is lowered, to the position shown by the dotted line in Fig. 6·8a) and, provided ϵ is not too large, the disturbed motion is a libration motion between $a+\beta$ and $a-\gamma$, where $a+\beta$ and $a-\gamma$ are the relevant zeros of $V(a)+\epsilon-V(x)$ as shown in the figure. The disturbed motion is a periodic motion in the neighbourhood of a. Equilibrium of this type is said to be *stable*. The equilibrium at a is stable if V has a minimum value at a.

The (x, v) graph for the disturbed motion is important. Its equation is

$$\tfrac{1}{2}v^2 = V(a) + \epsilon - V(x), \tag{1}$$

and the second member of this equation is represented by the intercept MN in Fig. 6·8a. The curve defined by equation (1) is a closed oval curve, symmetrical about Ox, cutting Ox orthogonally at $a+\beta$ and at $a-\gamma$, and having horizontal tangents where $x = a$ (Fig. 6·8c).

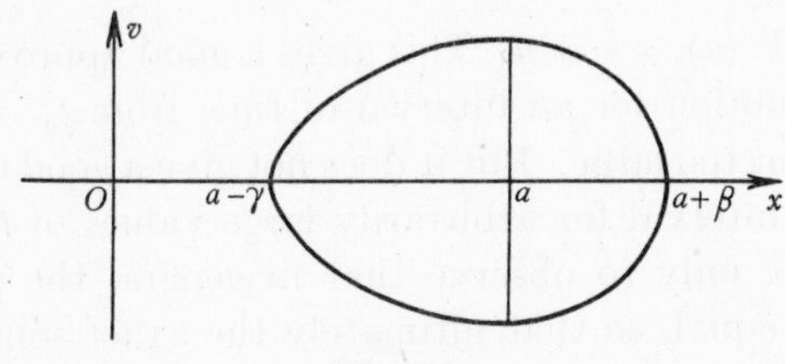

Fig. 6·8c

It is easy to find an approximation to the period σ of the oscillation in the case when ϵ is small. If ϵ is small, β and γ are small (of order $\sqrt{\epsilon}$), and $|x-a|$ remains small in the disturbed motion for all time. If we write $x = a + \xi$ in the equation of energy

$$T = C - V,$$

we find (remembering $V'(a) = 0$, $V''(a) > 0$)

$$\begin{aligned}\tfrac{1}{2}\dot{\xi}^2 &= V(a) + \epsilon - V(a+\xi)\\ &= \epsilon - \tfrac{1}{2}\xi^2 V''(a) + O(\xi^3)\\ &= \tfrac{1}{2}n^2(\alpha^2 - \xi^2) + O(\xi^3),\end{aligned}$$

where $n^2 = V''(a)$, $\alpha^2 = 2\epsilon/V''(a)$. Now, since $|\xi|$ remains small throughout the motion, we obtain a first approximation by neglecting the term $O(\xi^3)$, and the equation so obtained represents a harmonic oscillation of period $2\pi/n$ and amplitude α. If ϵ is small a first approximation to the period σ of the disturbed motion is $2\pi/n$, and the approximation is the closer the smaller the value of ϵ. The approximation we have found is independent of the value of ϵ, provided only that ϵ is small.

We have found that if ϵ is small a first approximation to the disturbed motion is a harmonic motion of period $2\pi/n$; but in general the period σ of the actual oscillation is not exactly equal to $2\pi/n$, and the precise significance of the first approximation deserves a moment's consideration. If we suppose, to fix the ideas, that t_0 is an instant at which the particle passes through a, so that $\xi = 0$, $\dot{\xi} = u = \pm\sqrt{(2\epsilon)}$ at $t = t_0$, then the first approximation we have found is

$$\xi = \alpha \sin n(t-t_0),$$

where $n^2 = V''(a)$, $\alpha = u/n$. This gives a good approximation to the actual motion for an interval of time from $t_0 - t_1$ to $t_0 + t_1$, where t_1 is not too large. But it does not give a good approximation in this interval for arbitrarily large values of t_1; to prove this we have only to observe that in general the periods are not exactly equal, so that ultimately the exact solution is out of phase with the solution given by the first approximation.

Consider as a concrete illustration the potential function

$$V = \tfrac{1}{2}n^2\left(x^2 + \frac{1}{p}x^3\right),$$

where p is a fixed positive number. The origin is a point of stable equilibrium; in the notation used above $a = 0$ and $\xi = x$. If p is large we have a close approximation to the problem in which the disturbed motion is accurately harmonic. Consider a disturbance in which the energy constant C is increased from zero to $\frac{1}{2}n^2\alpha^2$. Provided α^2 is sufficiently small the disturbed motion is a libration; explicitly, we need $\dfrac{\alpha^2}{p^2} < \dfrac{4}{27}$, as is easily verified. (If $\dfrac{\alpha^2}{p^2} = \dfrac{4}{27}$ the disturbed motion is a limitation motion to $x = -\frac{2}{3}p$.) The graph of $C - V$ is shown in Fig. 6·8d, and the equation of energy is

$$\begin{aligned}\tfrac{1}{2}\dot{x}^2 &= C - V\\ &= \tfrac{1}{2}n^2\left(\alpha^2 - x^2 - \frac{1}{p}x^3\right)\\ &= \frac{1}{2}\frac{n^2}{p}(\beta - x)(x+\gamma)(x+\kappa).\end{aligned}$$

The zeros of the cubic $C - V$ are β, $-\gamma$, $-\kappa$; if α is small β and γ are nearly equal to α, and κ is nearly equal to p. If we put

$$\beta = \alpha(1-\theta), \quad \gamma = \alpha(1+\phi),$$

we easily verify the formulae

$$\theta = \tfrac{1}{2}\lambda - \tfrac{5}{8}\lambda^2 + \lambda^3 - \dots,$$

$$\phi = \tfrac{1}{2}\lambda + \tfrac{5}{8}\lambda^2 + \lambda^3 + \dots,$$

where $\quad \lambda = \alpha/p, \quad$ and $\quad \kappa = p\{1 - \lambda(\theta+\phi)\} = p(1 - \lambda^2 - 2\lambda^4 - \dots).$

The disturbed motion is a libration between β and $-\gamma$, and its period, determined by the substitution introduced in § 6·5, namely,

$$x = \frac{\beta-\gamma}{2} + \frac{\beta+\gamma}{2}\cos\psi,$$

is

$$\sigma = \frac{2}{n}\int_0^\pi \frac{d\psi}{\sqrt{(\mu+\nu\cos\psi)}}, \tag{2}$$

where

$$\mu = \frac{1}{p}\left(\kappa + \frac{\beta-\gamma}{2}\right),$$

and

$$\nu = \frac{1}{2p}(\beta+\gamma).$$

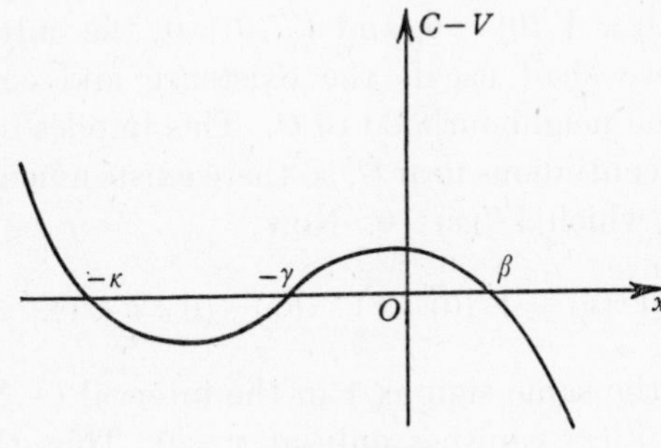

Fig. 6·8*d*

We can express (2) in the form

$$\sigma = \frac{4}{n\sqrt{(\mu+\nu)}}\int_0^{\frac{1}{2}\pi} \frac{d\theta}{\sqrt{(1-k^2\sin^2\theta)}}, \tag{3}$$

where

$$k^2 = \frac{2\nu}{\mu+\nu} = \frac{\beta+\gamma}{\kappa+\beta} < 1.$$

The integral

$$K = \int_0^{\frac{1}{2}\pi} \frac{d\theta}{\sqrt{(1-k^2\sin^2\theta)}}$$

is called *the complete elliptic integral of the first kind*, and its value, for different values of k^2 (<1), is given in the tables.

As we know, a first approximation to σ, when $\lambda(=\alpha/p)$ is small, is $2\pi/n$. To find a better approximation we notice that

$$\mu = 1-\tfrac{3}{2}\lambda^2+O(\lambda^4), \quad \nu = \lambda+O(\lambda^3),$$

which suffice to give the approximation

$$\sigma = \frac{2\pi}{n}\left(1+\frac{15}{16}\lambda^2\right).$$

This is obtained easily from (2) without troubling to transform first to (3).

Let us now return to the consideration of the exact solution for the disturbed motion. It will simplify the further discussion a little if we now take $a = 0$, $V(a) = 0$ (as in the example just discussed). This involves no loss of generality. We have merely chosen the point of equilibrium as the origin from which x is measured, and we have chosen suitably the arbitrary constant in the definition of V.

We have seen that the disturbed motion is a periodic motion through an interval about $x = 0$ 'provided ϵ is not too large'. We now find an upper bound for ϵ, i.e. an explicit value ϵ_0 with the property that the disturbed motion is of this type if $\epsilon < \epsilon_0$. We have $V(0) = V'(0) = 0$, and $V''(0) > 0$; the only additional assumption we shall use is the existence and continuity of $V'''(x)$ in some neighbourhood of O. This implies in particular that $V''(x)$ is continuous near O, so there exists a neighbourhood $-\delta < x < \delta$ in which $V''(x) > 0$. Now

$$V'(x) = V'(0) + xV''(\theta x) \quad (0 < \theta < 1),$$

so $V'(x)$ has the same sign as x in the interval $(-\delta, \delta)$, and in this interval $V'(x)$ vanishes only at $x = 0$. Thus the function $-V(x)$ strictly increases in the range $-\delta < x < 0$ and strictly decreases in the range $0 < x < \delta$.

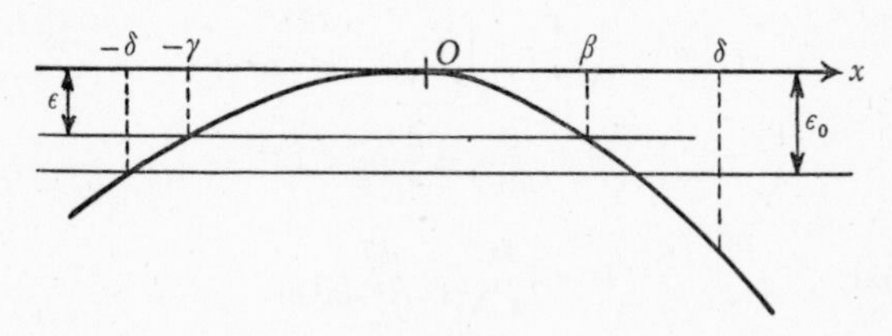

Fig. 6·8e

We now take ϵ_0 as the smaller of $V(-\delta)$ and $V(\delta)$ (Fig. 6·8e). Then if $\epsilon < \epsilon_0$ the equation

$$\epsilon - V(x) = 0$$

has precisely one root in each of the ranges $-\delta < x < 0$ and $0 < x < \delta$. If these roots are $-\gamma, \beta$ the disturbed motion is a libration motion between β and $-\gamma$.

Before leaving this case ($V''(a) > 0$) there is a final observation of great importance. If

$$V'(x) = n^2x + O(x^2),$$

then $V(x)$ has a minimum at $x = 0$, and the theorem (that the disturbed motion is a libration if $\epsilon < \epsilon_0$) is valid. If therefore we wish to exhibit the theory in terms of the equation of motion instead of the equation of energy we only need that $X(x)$ shall be of the form $-n^2x + O(x^2)$. The solution of the differential equation

$$\ddot{x} + n^2x = O(x^2) \tag{4}$$

represents a libration in the neighbourhood of $x = 0$ if the initial values of $|x|$ and $|\dot{x}|$ are not too large. And if these initial values are small we get a good approximation to the motion by omitting the term $O(x^2)$. This approximation is a harmonic motion with period $2\pi/n$.

Notice that we could not afford to neglect the term $O(x^2)$, even for a first approximation, unless we had some prior assurance that x remains small. In the problem before us we have such assurance. It might happen, and in some problems of a somewhat similar type it does happen, that in the actual motion the effect of the second-order terms is cumulative, so that x does not remain small in the actual motion. The equation obtained by neglecting the term $O(x^2)$ would not in such a case represent the motion even approximately.

The equation (4) will reappear in the sequel (for example in § 13·7).

(ii) We now turn to the second case, $V''(a) < 0$, and we find that matters are quite different. To begin with, there are now *two* possibilities with the same value of C. The particle may rest in equilibrium at a, or it may suffer a limitation motion tending to a from above or from below.

Consider first the case of equilibrium. If we increase C slightly, so as to raise the graph, we have a motion which is not confined to the immediate neighbourhood of a. If we decrease C slightly, so as to lower the graph, we have a motion which is confined to a region wholly to the right of a or wholly to the left, and in either case not confined to the immediate neighbourhood of a. Whether C is increased or decreased, the acceleration at points near a is away from a. Such a position of equilibrium is called *unstable*. The equilibrium at a is unstable if V has a maximum at a.

If we start from the limitation motion, not from rest in equilibrium, and make a small increase or a small decrease in C, the resulting motion is clearly the same as if we had started from the state of equilibrium. The motion that takes place after the disturbance is radically different in character from the original limitation motion. The idea of stability can be extended to include motion as well as equilibrium. The limitation motion is said to be unstable, this nomenclature in this context implying no more than the fact that a slight disturbance gives rise to a motion of entirely different type.

(iii) So far we have supposed $V''(a) \neq 0$, but the case (iii) $V''(a) = 0$, $V'''(a) \neq 0$ is very easy to deal with. The graph of $C - V$ has a horizontal inflexion at a. There may be rest in equilibrium at a, or a limitation motion towards a (from above if $V'''(a) < 0$, from below if $V'''(a) > 0$). In the sense already described the equilibrium and the limitation motion are unstable.

6·9 The theory of small oscillations. We return to the important special case (case (i) of § 6·8) in which V has a minimum value at $x = a$. We have seen that if the particle is projected from a point sufficiently near to a with a sufficiently small speed, then the motion is a periodic motion in the neighbourhood of a. Moreover, if the disturbance is small the period is approximately $2\pi/n$, where the positive number n can be found in either of two ways: (i) $n^2 = V''(a)$, (ii) the restoring force when the particle is displaced through a small distance ξ from the position of equilibrium is approximately $mn^2\xi$, the approximation being correct to the first order in ξ.

We now consider some concrete applications of this theory.

Example 1. *A particle of mass m is attached to the middle point of a stretched string, of negligible mass, whose ends are attached to fixed points A and B at a distance $2a$ apart. To find the period of the small oscillation which results when the particle is displaced through a small distance at right angles to the string and released.*

We can see from the symmetry that the particle moves on a straight line at right angles to AB, and it is evident on physical grounds that if we release it from rest at a distance c from AB the motion is a small oscillation of amplitude c. Explicitly, if we denote the displacement at time t by x (Fig. 6·9a) the equation of motion is of the type discussed above

$$\ddot{x}+n^2x = O(x^2),$$

and the approximation we are concerned with is that obtained by neglecting the term $O(x^2)$. Now the restoring force is $2Tx/r$, where T is the tension in either half of the string, and $r^2 = a^2+x^2$. But for our present purpose we only need the restoring force correct to order x, so for $2T/r$ we can write the value which it has when $x = 0$, i.e. $2T_0/a$, where T_0 is the tension of the stretched string before the equilibrium is disturbed. Thus $n^2 = 2T_0/ma$.

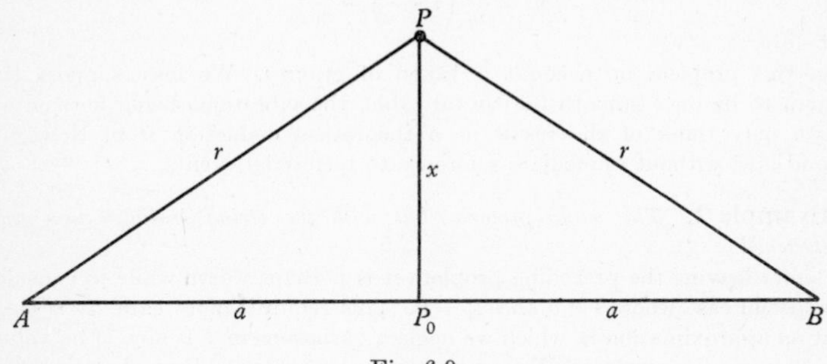

Fig. 6·9a

This result is independent of any assumption about the elastic properties of the string. Indeed, to this order of approximation the tension is sensibly unchanged during the motion.

It may be of interest to recover the result from the formula $n^2 = V''(0)$, though for this problem the method just given is simpler. We now suppose the string to obey Hooke's law, so that the force on the particle, in the direction of increasing x, is

$$X = -2T\frac{x}{r} = -2k(r-l)\frac{x}{r},$$

where $2l$ is the unstretched length of the string, and $k = \lambda/l$. Now

$$r^2 = a^2+x^2, \quad r\frac{dr}{dx} = x,$$

so if V denotes as usual the potential per unit mass

$$m\frac{dV}{dx} = 2k(r-l)\frac{dr}{dx},$$

whence $$mV = k(r-l)^2.$$

(This is the potential energy of the stretched string. We can start with this formula if we are content to anticipate a little about the theory of the motion of a dynamical system, which will be expounded in Chapter XXIII. We need to assume that the equation of energy holds for the system we are considering, and that the work done in stretching a string depends only on the initial and final lengths, not on the path of one end relative to the other during the stretching.) To solve our problem we need the value of $V''(x)$ for $x = 0$. It is not particularly difficult to find the value of $V''(x)$ for a general value of x, but a simpler way of achieving our immediate object is to expand $V(x)$ in powers of x in the form

$$V(x) = V(0) + \tfrac{1}{2}n^2x^2 + O(x^4).$$

Here n^2 is the required value of $V''(0)$. Now

$$mV(x) = k(r^2 - 2rl + l^2)$$
$$= k\left\{a^2 + x^2 - 2la\left(1 + \frac{x^2}{a^2}\right)^{\frac{1}{2}} + l^2\right\},$$

and the coefficient of x^2 is $k(1 - l/a)$. Thus

$$n^2 = \frac{2k}{m}\left(1 - \frac{l}{a}\right) = \frac{2T_0}{ma}$$

as before.

In this problem no account is taken of gravity. We may suppose the system to lie on a smooth horizontal table, the vibrations being horizontal; or we may think of the result as a theoretical deduction from Newton's second law, without immediate reference to terrestrial events.

Example 2. *The same problem, but with the string initially taut and unstretched.*

Before leaving the preceding problem it is perhaps worth while to consider the special case where $l = a$ and $T_0 = 0$. This requires more care. It is clear that an approximation in which we neglect variations in T is now of no value. The first approximation no longer gives us a harmonic motion, for the restoring force is now of order x^3: in fact, $V''(0) = V'''(0) = 0$, and the theory given above is no longer applicable. The period of the libration is not now independent of the amplitude, even in a first approximation.

Suppose then that the particle is pulled aside through a small distance c and released from rest. The motion will be an oscillation of amplitude c. We have the equation of motion

$$m\ddot{x} = -2\frac{\lambda}{a}(r-a)\frac{x}{r},$$

whence

$$\ddot{x} = -\frac{2\lambda}{ma}\left(1 - \frac{a}{r}\right)x.$$

Now

$$\frac{r^2}{a^2} = 1 + \frac{x^2}{a^2},$$

so that

$$1 - \frac{a}{r} = 1 - \left(1 + \frac{x^2}{a^2}\right)^{-\frac{1}{2}} = \frac{x^2}{2a^2} + O\left(\frac{x^4}{a^4}\right).$$

The equation of motion is therefore, to a first approximation,

$$\ddot{x} = -\frac{\lambda}{ma^3}x^3 = -2b^2x^3, \quad \text{say},$$

whence, on integrating,

$$\dot{x}^2 = b^2(c^4 - x^4).$$

In the equation of motion we have neglected a term $O(x^5)$ in comparison with the term $-2b^2x^3$. This is legitimate for a first approximation provided that $|x|$ remains small during the motion, and this is true in this case since $V(= \tfrac{1}{2}b^2x^4 + O(x^6))$ has a minimum at 0. The motion is a libration between $+c$ and $-c$ as we expect. (Of course this statement holds for the exact solution, not merely for the approximate solution.)

To find an approximation to the period we can either (i) separate the variables to obtain

$$\sigma = \frac{4}{b}\int_0^c \frac{dx}{\sqrt{(c^4-x^4)}} = \frac{4}{bc}\int_0^1 \frac{dy}{\sqrt{(1-y^4)}};$$

or (ii) we can use the method of § 6·5, introducing an auxiliary variable θ defined by the equation

$$x = c\cos\theta;$$

this leads to
$$\dot\theta = bc\sqrt{(2-\sin^2\theta)},$$

and the period is
$$\sigma = \frac{2\sqrt{2}}{bc}\int_0^{\frac{1}{2}\pi} \frac{d\theta}{\sqrt{(1-k^2\sin^2\theta)}}, \quad k^2 = \tfrac{1}{2}.$$

We have already met an integral of this form in § 6·8. Replacing the value of b, and taking the value of the integral from the tables, we find

$$\sigma = \left(7{\cdot}416\sqrt{\frac{ma^3}{\lambda}}\right)\frac{1}{c}.$$

The period is inversely proportional to c; it tends to infinity as $c \to 0$ as we should anticipate.

Example 3. *A heavy particle is attached to the middle point of a light elastic string. The natural length of the string is $2a$, and its ends are attached to two points A and B at the same level and at a distance $2a$ apart. When in equilibrium the particle hangs at a depth h below the line AB. Prove that when the particle executes small oscillations vertically the period is $2\pi/n$, where*

$$n^2 = \frac{g}{h}\left(1 + \frac{a}{b} + \frac{a^2}{b^2}\right),$$

where $b = \sqrt{(a^2+h^2)}$.

It will be illuminating to examine this problem in various ways.

(i) We use the equation $n^2 = V''(h)$. We denote the depth of the particle below AB by z, and the length of each string AP, BP by r. Then, as in Example 1,

$$mV = k(r-a)^2 - mgz,$$

and
$$V' = \frac{2k}{m}(r-a)\frac{z}{r} - g.$$

We have used the equations $r^2 = a^2+z^2$, $dr/dz = z/r$. Since $V' = 0$ in the position of equilibrium

$$\frac{2k}{m}(b-a)\frac{h}{b} = g,$$

an equation which we verify immediately by resolving vertically. Differentiating again we find

$$V'' = \frac{2k}{m}\left(1 - \frac{a}{r} + \frac{az^2}{r^3}\right),$$

and since n^2 is the value of V'' in the position of equilibrium

$$\begin{aligned} n^2 &= \frac{2k}{m}\left(1 - \frac{a}{b} + \frac{ah^2}{b^3}\right) \\ &= \frac{bg}{(b-a)\,h}\left(1 - \frac{a}{b} + \frac{ah^2}{b^3}\right) \\ &= \frac{g}{h}\left(1 + \frac{a}{b} + \frac{a^2}{b^2}\right). \end{aligned}$$

(ii) We denote the depth of P below the equilibrium position by x, and we find the restoring force correct to order x. If T_0 is the tension in the position of equilibrium we have, resolving vertically,

$$2T_0h/b = mg.$$

If T is the tension for the displaced position the equation of motion is

$$m\ddot{x} = mg - 2T\frac{h+x}{r},$$

and we can write this equation in the form

$$m\ddot{x} = mg\left\{1 - \frac{T}{T_0}\frac{b}{h}\frac{h+x}{r}\right\}.$$

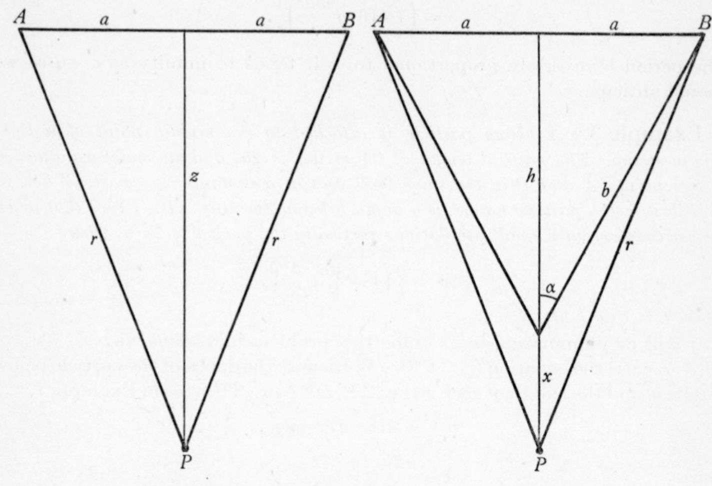

Fig. 6·9*b* Fig. 6·9*c*

Now
$$\frac{T}{T_0} = \frac{r-a}{b-a},$$

giving
$$\ddot{x} = -\frac{g}{(b-a)\,hr}\{ah(r-b)+b(r-a)x\}.$$

So far the result is exact. But for our present purpose we only need the second member correct to order x. To this order

$$r-b = x\cos\alpha = \frac{h}{b}x,$$

and in the rest we can write b for r, giving the equation

$$\ddot{x}+n^2x = 0$$

with
$$n^2 = \frac{g}{b^2h}\frac{b^3-a^3}{b-a} = \frac{g}{h}\left(1+\frac{a}{b}+\frac{a^2}{b^2}\right).$$

There are many other ways of finding the required approximation for $r-b$; perhaps the simplest is to write $r = b+\xi$, and then

$$(b+\xi)^2 = a^2+(h+x)^2,$$

giving
$$2b\xi+\xi^2 = 2hx+x^2.$$

This is exact; to the order of approximation required for our present purpose $b\xi = hx$ as before.

(iii) We can use the method of (ii) and write $b+\xi$ for r from the start. We bear in mind that ξ is small, of the same order as x, and we retain only the first power of x and of ξ. To a sufficient approximation $\xi = (h/b)x$. We have

$$\frac{T}{T_0} = 1+\frac{\xi}{b-a} = 1+\frac{h}{b(b-a)}x,$$

and
$$\frac{h+x}{r} = \frac{h+x}{b+\xi} = \frac{h}{b}\left(1+\frac{x}{h}-\frac{\xi}{b}\right) = \frac{h}{b}\left(1+\frac{a^2}{hb^2}x\right),$$

the last formula in each case being correct to order x. If we substitute in the equation of motion we recover the same result as before. As a general working rule, however, it is better not to approximate too early.

(iv) We can use the energy method, as in (i) above, and find n^2 from the approximation

$$V = \text{constant}+\tfrac{1}{2}n^2x^2.$$

But if we use this attack on the problem we must remember that we shall require an approximation for ξ correct to order x^2, not merely to order x. We have

$$2b\xi = 2hx+x^2-\xi^2,$$

and to get the approximation we need we can replace ξ^2 in the last term by $(hx/b)^2$, giving

$$2b\xi = 2hx+(a^2/b^2)x^2.$$

Now
$$\begin{aligned} mV &= k(b+\xi-a)^2-mg(h+x) \\ &= k(b-a)^2+2k(b-a)\,\xi+k\xi^2-mgh-mgx \\ &= \text{constant}+\left\{2k(b-a)\frac{h}{b}-mg\right\}x \\ &\qquad +\frac{k}{b}\left\{(b-a)\frac{a^2}{b^2}+\frac{h^2}{b}\right\}x^2+O(x^3). \end{aligned}$$

The coefficient of x is zero, as we expect, and replacing k by $mgb/2(b-a)h$ we have

$$V = \text{constant}+\frac{g}{2h}\left(\frac{a^2}{b^2}+\frac{b+a}{b}\right)x^2+O(x^3),$$

giving again
$$n^2 = \frac{g}{h}\left(1+\frac{a}{b}+\frac{a^2}{b^2}\right).$$

6·10 Study of some special fields.

We now consider the application of the preceding theory to some particular fields. The essential point is that we can predict all the motions that are possible in the given field by reference to the graph $y = -V(x)$. From this graph itself we can see what motion is possible with $C = 0$, and for other values of C we have only to consider instead the graph of $C - V$ which is obtained by raising or lowering the graph of $-V$. In practice we usually leave the graph unchanged and lower or raise the x-axis. When we have determined the general nature of the motion in each particular case (but not sooner) we consider the explicit formula giving x as a function of t.

Two special fields have been discussed already, and it only remains to point out how aptly they illustrate the general theory. (i) For the uniform field (§§ 4·4–4·8) mg we have (measuring x in the direction of the field) $V = -gx$, and the graph

$$y = C - V = g(x - x_0) \quad (C = -gx_0),$$

shows that there is only one type of motion whatever the value of C. If $\dot{x} < 0$ initially the particle moves to the left until it reaches the point $x = x_0$, where it rests instantaneously, and then moves always to the right, x tending to infinity with t. (ii) For the harmonic oscillator (§§ 5·5 and 6·4) $V = \frac{1}{2}n^2x^2$, and the graph of $-V$ is a parabola touching the x-axis at $x = 0$, and lying below the axis. For $C < 0$ no motion is possible, and there remain two possibilities, $C = 0$ and $C > 0$. For $C = 0$ the particle rests in stable equilibrium at O. For $C > 0$ we have a libration motion between two points equidistant from O to right and left. The explicit forms of the (t, x), (t, v) and (x, v) relations we have already found and illustrated graphically in § 5·5.

If we superpose the two fields just mentioned, so that the particle is subject to a harmonic attraction and also to a uniform field, we obtain a quadratic form for V,

$$X = -n^2x + g, \quad V = \tfrac{1}{2}n^2x^2 - gx + c.$$

This is equivalent to a harmonic attraction to a new centre of attraction, namely, the point $x = g/n^2$. We can see this by writing V in the form

$$\tfrac{1}{2}n^2\left(x - \frac{g}{n^2}\right)^2 + \text{constant}.$$

Alternatively, the equation of motion

$$\ddot{x} = -n^2x + g$$

can be written in the form

$$\ddot{y} + n^2y = 0,$$

where $y = x - (g/n^2)$, and this represents a harmonic oscillation about $y = 0$. A concrete example of these superposed fields is provided by the problem of a particle hanging from an elastic thread (Ex. 1 of § 5·5).

In the following paragraphs we consider two new problems, the first that of a repulsion proportional to distance, the second that of an attraction proportional to the inverse square of the distance.

6·11 Repulsion proportional to distance. Here $X = n^2x$, $V = -\frac{1}{2}n^2x^2$, and the graph of $-V$ is a parabola touching the x-axis at the origin, and lying above the axis (Fig. 6·11a). We will suppose for definiteness that $x = a$ and $\dot{x} = -w$ at $t = 0$, where $a > 0$ and $w > 0$. Then $C = \frac{1}{2}(w^2 - n^2a^2)$. It is clear from the theory that there are three types of motion, according as $C < 0$, $C = 0$, $C > 0$. The three cases are illustrated in the figure, the origin being placed at $O_1 (C < 0)$ or $O_2 (C = 0)$ or $O_3 (C > 0)$.

(i) $C < 0, w < na$. We introduce the positive constant c defined by $c^2 = a^2 - w^2/n^2$. The particle moves to the left until $x = c$. At $x = c$ it rests instantaneously, then moves to the right with increasing velocity.

The equation of motion is $\ddot{x} - n^2x = 0$, and the solution is

$$x = a\cosh nt - \frac{w}{n}\sinh nt.$$

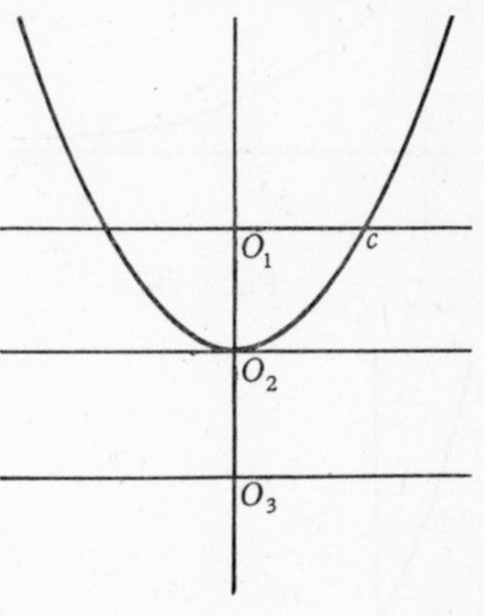

Fig. 6·11a

We introduce the positive constant t_0 defined by the equations

$$\frac{\cosh nt_0}{a} = \frac{\sinh nt_0}{w/n} = \frac{1}{c}.$$

With this notation $$x = c\cosh n(t - t_0),$$

and $t = t_0$ is the instant at which x has its minimum value c. The (t, x) graph is the *catenary* shown in Fig. 6·11b. This curve is the curve assumed by a uniform flexible string hanging freely under gravity. The relation connecting x and v is

$$\frac{x^2}{c^2} - \frac{v^2}{n^2c^2} = 1,$$

and this is the equation of a branch of a hyperbola as shown in Fig. 6·11c.

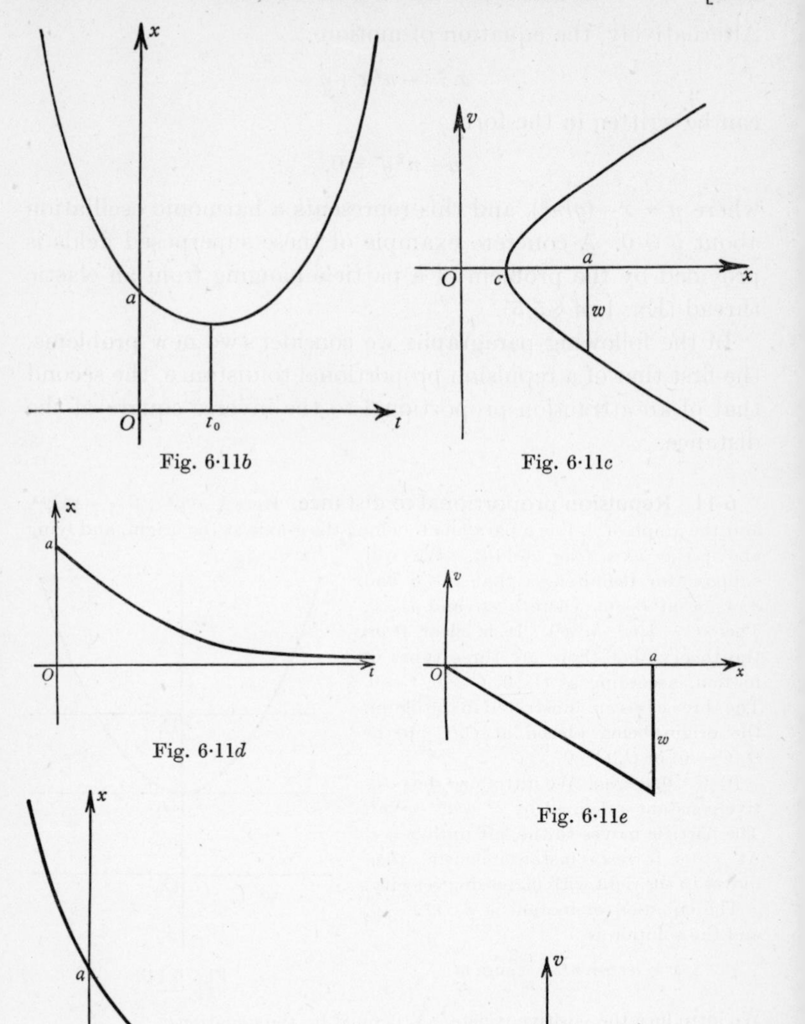

Fig. 6·11*b*

Fig. 6·11*c*

Fig. 6·11*d*

Fig. 6·11*e*

Fig. 6·11*f*

Fig. 6·11*g*

(ii) $C=0$. Here $w=na$, and the solution is $x=ae^{-nt}$. We have, as is evident from the graph of $-V$, a limitation motion, in which $x\to 0$ as $t\to\infty$. The (x,v) graph is a portion of the straight line $v=-nx$. We have here an example of the exceptional case, where the (x,v) graph does not cut the x-axis normally. This is possible because f vanishes as well as v at $x=0$. The (t,x) and (x,v) graphs are shown in Figs. 6·11*d* and 6·11*e*.

(iii) $C>0$. Here $w>na$, and the particle moves always to the left. Write $w^2/n^2-a^2=b^2$. The solution is

$$x = a\cosh nt - \frac{w}{n}\sinh nt,$$

and if we write
$$\frac{\cosh nt_0}{w/n} = \frac{\sinh nt_0}{a} = \frac{1}{b},$$

we have $x=-b\sinh n(t-t_0)$.

The (x,v) relation is $\dfrac{v^2}{n^2b^2}-\dfrac{x^2}{b^2}=1$, a branch of a hyperbola lying in the lower half-plane. The (t,x) and (x,v) graphs are shown in Figs. 6·11*f* and 6·11*g*.

6·12 Newtonian attraction.

For $x>0$ we have $X=-\mu/x^2$, $V=-\mu/x$, where μ is a positive constant. Suppose for definiteness $x=c$ and $\dot{x}=u$, where $c>0$ and $u>0$, at $t=0$. The energy equation is

$$\tfrac{1}{2}\dot{x}^2 = C+\frac{\mu}{x}, \quad (1)$$

where

$$C = \tfrac{1}{2}u^2-\frac{\mu}{c}. \quad (2)$$

The graph of the second member of equation (1) is a rectangular hyperbola, and it is clear that there are three cases to consider, $C<0$, $C=0$, $C>0$. (See Fig. 6·12*a*, where these three cases are illustrated, the origin being placed either at $O_1(C<0)$, or $O_2(C=0)$, or $O_3(C>0)$.)

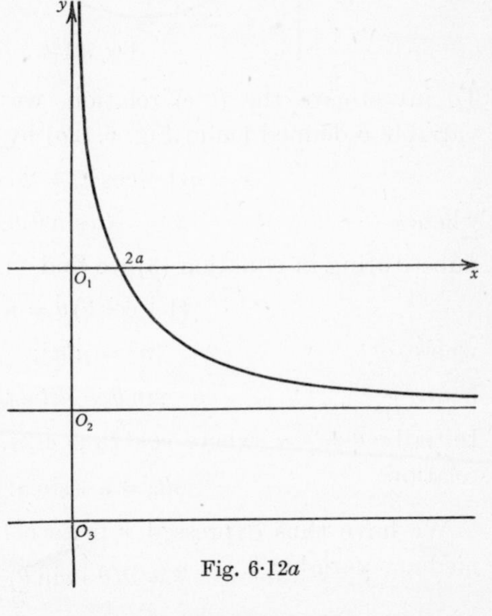

Fig. 6·12*a*

The sign of C is linked up with the important notion of the *velocity from infinity*. If the particle fell freely from rest at infinity to $x=c$ the velocity it would acquire would be $\sqrt{(2\mu/c)}$;

this velocity is called the velocity from infinity. Now $C<0$ means $u<\sqrt{(2\mu/c)}$; C is negative if the velocity of projection is less than the velocity from infinity.

We now consider the motion in the three cases.

(i) $C<0$, say $C=-\mu/2a$. The particle comes to rest at $x=2a$, and then falls into the centre of force. The equation of energy is

$$\dot{x}^2=\mu\left(\frac{2}{x}-\frac{1}{a}\right). \tag{3}$$

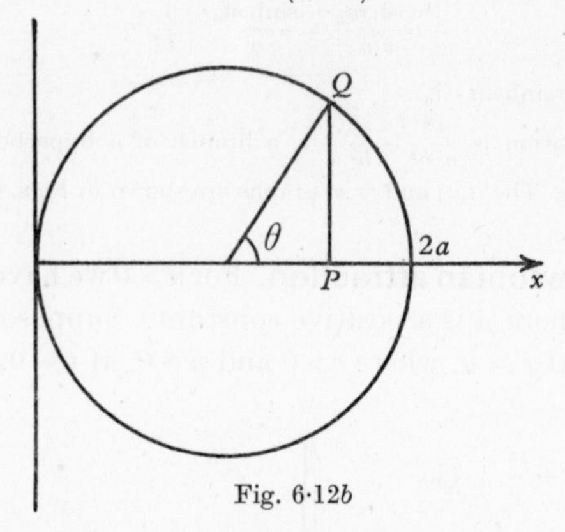

Fig. 6·12*b*

To investigate the (t, x) relation, we introduce an auxiliary variable θ defined (as in Fig. 6·12*b*) by

$$x=a(1+\cos\theta)=2a\cos^2\tfrac{1}{2}\theta, \tag{4}$$

whence $$\dot{x}=-(a\sin\theta)\dot{\theta}.$$

Substituting in equation (3) we find

$$(1+\cos\theta)\,\dot{\theta}=n,$$

where $$n^2=\mu/a^3;$$

hence $$\theta+\sin\theta=n(t-t_0). \tag{5}$$

Initially $\theta=-\alpha$, where $\cos^2\frac{1}{2}\alpha=c/2a$, and t_0 is given by the relation

$$nt_0=\alpha+\sin\alpha.$$

We have thus expressed x in terms of t through the intermediary variable θ,

$$\xi=a(\theta+\sin\theta), \tag{6}$$

$$x=a(1+\cos\theta), \tag{7}$$

where $\xi=an(t-t_0)$. The parametric form is the simplest expression of the relation between ξ and x.

The curve represented by the equations (6) and (7) is called a *cycloid*; it is the path of a point on the rim of a circular disc which rolls without slipping on a straight line. We shall study this curve more fully later, in § 11·12. The representation in terms of the parameter θ is shown in Fig. 6·12*c*.

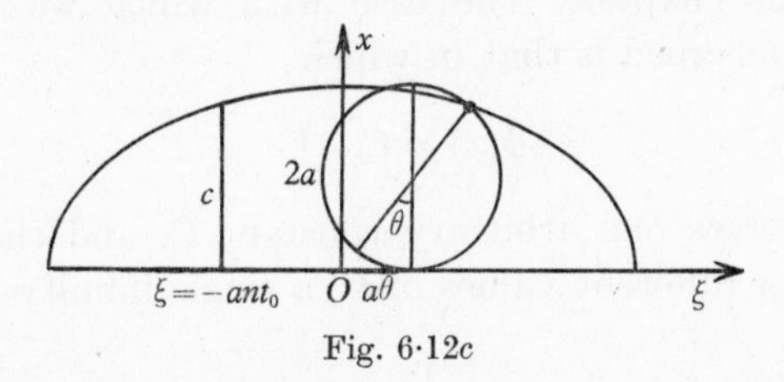

Fig. 6·12*c*

One result of special interest is the time required to fall from rest at $x = 2a$ into the centre of force. This is the interval from $\theta = 0$ to $\theta = \pi$, corresponding to the time-interval π/n.

(ii) $C = 0$. The velocity of projection is equal to the velocity from infinity. In this case x continually increases and tends to infinity with t. Explicitly

$$\dot{x}^2 = 2\mu/x,$$

from which we easily find

$$x^{\frac{3}{2}} = c^{\frac{3}{2}} + nt, \quad n^2 = \tfrac{9}{2}\mu.$$

(iii) $C > 0$. The velocity of projection is greater than the velocity from infinity, and $x \to \infty$ with t. If we write $C = \mu/2a$ we have the energy equation

$$\dot{x}^2 = \mu\left(\frac{2}{x} + \frac{1}{a}\right).$$

We introduce a parameter θ defined by

$$x = a(\cosh\theta - 1)$$

which leads to

$$\sinh\theta - \theta = n(t + \beta), \quad n^2 = \mu/a^3.$$

Initially $\theta = \alpha$, where $\sinh^2 \frac{1}{2}\alpha = c/2a$, and β is given by the relation $n\beta = \sinh\alpha - \alpha$. If we write $\xi = an(t+\beta)$ the (ξ, x) relation is

$$\begin{cases} \xi = a(\sinh\theta - \theta), \\ x = a(\cosh\theta - 1). \end{cases}$$

6·13 Final comments. The theory of motion in a field of force has led us to the differential equation

$$\tfrac{1}{2}\dot{x}^2 = \phi(x), \tag{1}$$

and the study of this differential equation has been the main topic of this chapter. The case with which we have been primarily concerned is that in which

$$\phi(x) = C - V.$$

Here ϕ involves one arbitrary constant C, and the equation describes, for different values of C, a single infinity of possible motions.

The differential equation (1) is of great importance in dynamics, and reappears in many important problems. A conspicuous example is the theory of central orbits, where the relation between r and t is of the form (1). In that case ϕ involves two arbitrary constants, and the equation describes a double infinity of possible motions. The relation between r and θ is of the same type (see § 15·1).

If we think of the equation (1) as defining the motion of a particle on a straight line, the particle can rest in equilibrium at a point a at which $\phi(a) = \phi'(a) = 0$. The equilibrium is stable if $\phi''(a) < 0$, and the period of a small oscillation about this position of stable equilibrium is $2\pi/n$, where $n^2 = -\phi''(a)$.

A more general form is

$$\tfrac{1}{2}\psi(x)\,\dot{x}^2 = \chi(x), \tag{2}$$

where $\psi(x) > 0$ for all values of x. The graph of $\chi(x)$ alone will suffice to give us a general idea of the motion. Equilibrium is possible at a point a at which $\chi(a) = \chi'(a) = 0$, and the equilibrium is stable if $\chi''(a) < 0$. This is true because, if we write $\phi(x)$ for $\chi(x)/\psi(x)$,

$$\phi''(a) = \frac{\chi''(a)}{\psi(a)},$$

and $\chi''(a) < 0$ implies $\phi''(a) < 0$. The period of a small oscillation about the position of stable equilibrium is $2\pi/n$, where

$$n^2 = -\frac{\chi''(a)}{\psi(a)}.$$

EXAMPLES VI

1. A particle moves on a straight line under the action of a force which is a function of position on the line. Establish the equation of energy

$$\tfrac{1}{2}v^2 = C - V,$$

where x is the distance of the particle from a fixed point of the line, v its velocity, $V(=V(x))$ the potential per unit mass, and C is constant.

If $V = k\,|\,x\,|$, where k is constant, prove that, if $C > 0$ and $k > 0$, the motion is periodic with period $4\sqrt{(2C)}/k$.

2. The acceleration of a particle P moving in a straight line is given by $n^2x - 3n^2x^2/2a$, where x is the distance of P from a fixed point O of the line. If the velocity of P is zero when $x = a$, describe the general nature of the subsequent motion and show that the time elapsed when $x = 3a/4$ is $(\log 3)/n$.

3. The acceleration of a particle P moving in a straight line is given by $-n^2x + 3n^2x^2/2a$, where x is the distance of P from a fixed point O of the line.

(i) Find the period of a small oscillation about the position of stable equilibrium.

(ii) If the particle starts from rest at $x = a$ at the instant $t = 0$, prove that in the subsequent motion $x = a\sec^2(\tfrac{1}{2}nt)$; show that $x \to \infty$ as $t \to \pi/n$.

4. A particle is placed at a point P of a rough plane inclined at an angle α to the horizontal. Prove that the particle will remain at rest at P if the coefficient of friction μ at P satisfies the condition

$$\mu \geqslant \tan\alpha.$$

A particle slides on a plane inclined at an angle α to the horizontal. The coefficient of friction μ is not constant, but is proportional to the distance r from a point O of the plane, $\mu = kr$. A particle is projected from O with velocity u up the line of greatest slope. Prove that the particle will remain at rest when it first comes to rest if

$$u^2 \geqslant 3g\sin^2\alpha/(k\cos\alpha).$$

5. A and B are two equal circular cylinders with axes horizontal at the same level and at distance l apart. A is smooth, and B, which is rough, is kept rotating with constant peripheral speed v, the highest point of B moving away from A. A long straight plank is gently lowered into contact with the cylinders and released, the plank being at right angles to the axes of the cylinders, and with its centre of gravity G at a distance kl $(k < 1)$ from the vertical through the axis of A. Show that slipping of the plank on B will cease before G comes over the axis of B, provided that $l > v^2/\mu g(1-k^2)$, where μ is the coefficient of friction between B and the plank. Sketch the form of the speed-time curve for the plank under these conditions, and prove that the total distance through which the plank slips relatively to the surface of B is

$$kl[\alpha\sinh^{-1}\alpha + 1 - \sqrt{(1+\alpha^2)}],$$

where

$$\alpha^2 = v^2/\mu g k^2 l.$$

6. A light elastic string is stretched between two points in the same vertical line, distant l apart. The tension in the string is F. A body, whose weight is small compared with F, is attached to the midpoint of the string, causing it to sink a distance d. Show that the periodic time, T_1, of small vertical oscillations of the body is the same as that of a simple pendulum of length d.

If the periodic time of small horizontal oscillations of the body is T_2, show that the mass of the body is approximately

$$4\frac{Fd}{gl}\left(\frac{T_2}{T_1}\right)^2.$$

(See the note at the end of Ex. 6, p. 84.)

7. A light endless elastic string of unstretched length $4a$ passes over two small smooth pegs at the same level distant $2a$ apart. A particle is attached to a point on the string and when the particle is in equilibrium the string forms the three sides of an equilateral triangle. Prove that the period of vibration of the particle in a vertical line is the same as that of a pendulum of length $4\sqrt{3}a/7$.

8. A uniform bar AB of mass M and length $2a$ is suspended from a point O by two equal elastic strings OA, OB. In the position of equilibrium the inclination of each string to the vertical is α. The increase in the tension of either string when its length is increased by x is Mgx/b. The bar vibrates vertically, remaining horizontal. Prove that the period of a small oscillation is $2\pi/n$, where

$$n^2 = \frac{g}{\cos\alpha}\left(\frac{2}{b}\cos^3\alpha + \frac{1}{a}\sin^3\alpha\right).$$

9. A particle P of mass m moves in a plane under the influence of n fixed centres of attraction $A_1, A_2, \ldots, A_n$ in the plane. The magnitude of the attraction to A_s is $m\lambda_s r_s$, where λ_s is a positive constant and r_s is the distance PA_s. If G is the centre of gravity of masses $\lambda_1, \lambda_2, \ldots, \lambda_n$ at $A_1, A_2, \ldots, A_n$, prove that G is a position of stable equilibrium, and that the period of an oscillation about it is $2\pi/n$, where $n^2 = \lambda_1 + \lambda_2 + \ldots + \lambda_n$.

*10. A particle of mass m moves in a plane under the influence of N fixed centres of attraction equally spaced round a circle of radius a. The magnitude of the attraction to each centre of attraction is a constant force P. Prove that the centre of the circle is a position of stable equilibrium, and that the period of a small oscillation about it is $2\pi/n$, where $n^2 = NP/2ma$.

*11. A light inextensible string has particles of mass $5m$ attached to its ends and a particle of mass $8m$ attached to its middle point. It hangs over two smooth pegs, fixed at the same lèvel at a distance $2a$ apart, with the mass $8m$ between the pegs. The system executes small oscillations in which the masses move vertically. Prove that the length of the equivalent simple pendulum is $20a/3$. (Assume that the energy remains constant—which will be proved in Chapter XXIII—and use the theorem of § 6·13.)

CHAPTER VII

MOTION ON A STRAIGHT LINE, THE REMAINING PROBLEMS

7·1 Acceleration a function of the velocity. We now consider problems (case (*d*) of our classification in § 4·2) in which the acceleration is a given function of the velocity, $f = \phi(v)$. The most familiar case is that of motion in a resisting medium; in that case $\phi(v)$ has the opposite sign to v. If we start from the equation

$$\frac{dv}{dt} = \phi(v)$$

one integration will determine the relation connecting t and v; if $v = u$ at $t = 0$ we have

$$dt = \frac{1}{\phi(v)}\,dv$$

giving
$$t = \int_u^v \frac{1}{\phi(\theta)}\,d\theta.$$

This is the relation connecting v, or $\dot{x}$, and t. A second integration will determine the relation connecting t and x.

Another attack on the problem starts from the equation

$$v\frac{dv}{dx} = \phi(v),$$

giving
$$x - a = \int_u^v \frac{\theta}{\phi(\theta)}\,d\theta,$$

where $x = a$, and $v = u$, at $t = 0$. This is the relation connecting x and v. It could have been found also by elimination of t from the (t, v) and (t, x) relations already mentioned.

7·2 Resistance proportional to the speed. We begin with the problem of a particle moving in a medium that offers a resistance proportional to the speed. The force on the particle is $-kmv$, where k is a positive constant, and we have the equation of motion

$$\frac{dv}{dt} = -kv; \tag{1}$$

this equation is valid whether the particle moves to the right or to the left. We can use the method of separation of variables, as in § 7·1, and obtain

$$t = -\frac{1}{k}\int_u^v \frac{d\theta}{\theta} = -\frac{1}{k}\log\frac{v}{u},$$

$$v = ue^{-kt}. \tag{2}$$

(Of course, if u is positive, we can anticipate that v decreases as t increases.) But it is simpler to multiply the equation (1),

$$\frac{dv}{dt} + kv = 0,$$

by the integrating factor e^{kt}, giving

$$ve^{kt} = \text{constant} = u,$$

which is equivalent to equation (2).

Thus
$$\dot{x} = ue^{-kt}, \tag{3}$$

and therefore
$$x - a = \frac{u}{k}(1 - e^{-kt}). \tag{4}$$

Eliminating t from (2) and (4) gives us the (x, v) relation

$$u - v = k(x - a). \tag{5}$$

The decrease in velocity is proportional to the distance covered. Of course we can also derive this last relation from the equation

$$v\frac{dv}{dx} = -kv.$$

7·3 Vertical motion. We now discuss the motion when a particle is thrown vertically upwards in a medium offering a resistance $km\,|\,v\,|$. We measure x vertically upwards, and suppose $x = 0$, $\dot{x} = u$, at $t = 0$.

We have $$\ddot{x}+k\dot{x} = -g. \tag{1}$$

We could derive from this equation the first integral

$$\dot{x}+kx = u-gt,$$

but it turns out as before to be a little simpler to start from equation (1), multiply through by e^{kt}, and integrate; this leads to

$$\dot{x}e^{kt} = u-\frac{g}{k}(e^{kt}-1),$$

whence $$\dot{x} = \left(u+\frac{g}{k}\right)e^{-kt}-\frac{g}{k}. \tag{2}$$

We derive at once the expression of x in terms of t,

$$x = \left(\frac{u}{k}+\frac{g}{k^2}\right)(1-e^{-kt})-\frac{g}{k}t. \tag{3}$$

The formula (3) takes a rather simpler form if we introduce the idea of the *terminal velocity*. The terminal velocity c is the constant speed at which the particle would fall if the resistance just balanced the weight,

$$kmc = mg, \quad c = g/k.$$

In terms of c equation (3) takes the simpler form mentioned, namely,

$$kx = (c+u)(1-e^{-kt})-c(kt). \tag{4}$$

The particle returns to the point of projection after a time $t_2(=\phi/k)$ given by the equation

$$1-e^{-\phi} = \lambda\phi, \tag{5}$$

where λ denotes the positive constant $c/(c+u)$. We can find an approximate solution of equation (5) by plotting the graphs of the two members and finding the value of ϕ where they intersect; we notice that if $0<u<c$, $\frac{1}{2}<\lambda<1$. Or we can write equation (5) in the form

$$\frac{e^{-\phi}-1+\phi}{\phi} = \frac{u}{c+u},$$

and, still supposing $0<u<c$, we can write this as

$$\frac{1}{2!}\phi-\frac{1}{3!}\phi^2+\frac{1}{4!}\phi^3-\ldots = \alpha-\alpha^2+\alpha^3-\ldots,$$

where $\alpha = u/c = ku/g$. If α is small a first approximation for ϕ is 2α (i.e. $t_2 = 2u/g$) as we expect; and if we express ϕ as a power-series in α, we find the first terms, by successive approximation, to be

$$\phi = 2\alpha(1 - \tfrac{1}{3}\alpha + \tfrac{2}{9}\alpha^2 - \ldots),$$

i.e.

$$t_2 = \frac{2u}{g}\left(1 - \frac{1}{3}\frac{ku}{g} + \frac{2}{9}\frac{k^2u^2}{g^2} - \ldots\right).$$

For a small resistance the particle returns to the origin sooner than it would if there were no resistance.

The graph showing the relation between t and x is very similar in its general appearance to Fig. 7·4 (which actually refers to a different problem, namely, one in which the resistance is proportional to the square of the speed). In our present problem the equation of the asymptote is

$$kx = c + u - ckt.$$

The greatest height is reached when $\dot{x} = 0$, i.e. at time

$$t_1 = \frac{1}{k}\log\left(1 + \frac{u}{c}\right),$$

and its value h is given by

$$kh = u - c\log\left(1 + \frac{u}{c}\right).$$

We can, of course, derive this result as simply from the equation of motion

$$v\frac{dv}{dx} = -kv - g = -k(v + c),$$

giving

$$kh = \int_0^u \frac{v}{v+c}\,dv = u - c\log\left(1 + \frac{u}{c}\right)$$

as before.

7·4 Resistance proportional to the square of the speed. This problem needs more care. If the resistance is kmv^2, the equation $f = -kv^2$ is valid only if $v > 0$; if $v < 0$ the equation of motion takes the form $f = kv^2$. We can write, if we wish,

$$f = -kv\,|\,v\,|,$$

which holds generally, but this is not convenient to work with.

In practice it is usually advisable to measure distances in the direction of motion, so that $v>0$; in a problem in which the direction of motion changes, we consider the two parts separately.

Consider in particular the problem of a particle thrown vertically upwards with velocity u in such a medium; to find the subsequent motion. During the ascent, measuring x upwards,

$$f = -g-kv^2 = -k(c^2+v^2),$$

where $c = \sqrt{(g/k)}$, is the terminal velocity. We can start either with the (t, v) relation or the (x, v) relation. Taking the former

$$\frac{dv}{dt} = -k(c^2+v^2). \tag{1}$$

We can integrate this by separation of variables, or, more simply, by making the substitution

$$v = c\tan\theta;$$

this leads to

$$\theta = \alpha - kct,$$

where α is the acute angle defined by

$$u = c\tan\alpha.$$

Thus

$$v = c\tan(\alpha - kct).$$

We notice that v decreases as t increases, and reaches the value zero at $t = t_1$, where $t_1 = \alpha/(kc)$. The solution we have found is valid for $0 \leqslant t \leqslant t_1$.

To find the (t, x) relation we have

$$v = \frac{dx}{dt} = c\tan kc(t_1-t), \tag{2}$$

whence

$$k(h-x) = \log\sec kc(t_1-t), \tag{3}$$

where $h = (\log\sec\alpha)/k$ is the greatest height attained.

To find the (x, v) relation during the ascent we must eliminate (t_1-t) between the equations (2) and (3), which gives us the relation

$$(c^2+v^2) = (c^2+u^2)e^{-2kx}.$$

We can, of course, easily find this relation directly by integration of the equation

$$v\frac{dv}{dx} = -k(c^2+v^2);$$

if we write z for c^2+v^2 this equation is

$$\frac{dz}{dx}+2kz = 0,$$

whence
$$z = z_0 e^{-2kx}.$$

Now consider the descent. It will be convenient to measure x downwards from the point of rest, and we may as well shift the origin of time also, and take $t = 0$ at the instant of rest. We have

$$\frac{dv}{dt} = g-kv^2 = k(c^2-v^2).$$

As before, we can solve this equation either by separation of variables, or by the introduction of an auxiliary variable. If we write

$$v = c\tanh\theta$$

we find
$$\theta = kct, \quad \text{and} \quad v = c\tanh kct.$$

A further integration gives us

$$kx = \log\cosh kct.$$

The elimination of t gives us the (x, v) relation during the descent

$$v^2 = c^2(1-e^{-2kx}),$$

which again we can easily find directly. The speed with which the particle returns to the starting-point is

$$c\sqrt{(1-e^{-2kh})} = c\sin\alpha.$$

Let us now return to the original notation. For the (t, x) relation in the two parts of the problem we have

$$k(h-x) = \log\sec kc(t_1-t) \quad (0\leqslant t\leqslant t_1),$$

$$k(h-x) = \log\cosh kc(t-t_1) \quad (t\geqslant t_1).$$

The (t, x) graph is shown in Fig. 7·4. The instant t_2 at which the particle returns to the point of projection is given by

$$kct_2 = \alpha + \log(\sec\alpha + \tan\alpha).$$

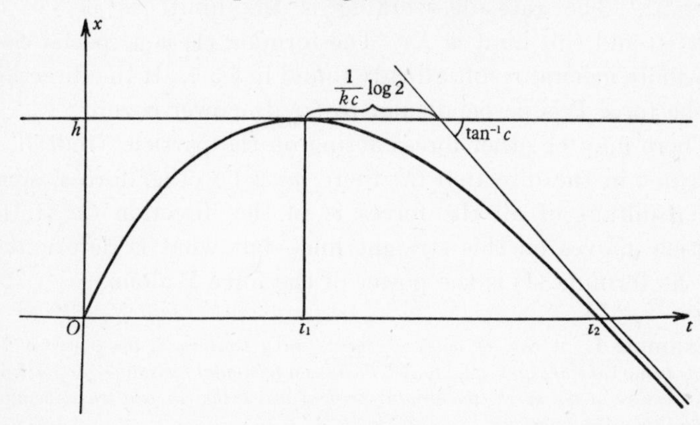

Fig. 7·4

If k is small, t_2 will differ only slightly from the value $2u/g$ which we should obtain if there were no resistance. To exhibit the magnitude of the deviation we expand

$$\beta = \alpha + \log(\sec\alpha + \tan\alpha)$$

in powers of $p = \tan\alpha = u\sqrt{(k/g)}$. Now

$$\frac{d\beta}{dp} = \frac{d\beta}{d\alpha}\bigg/\frac{dp}{d\alpha} = \frac{1}{\sec^2\alpha} + \frac{1}{\sec\alpha}$$

$$= (1+p^2)^{-1} + (1+p^2)^{-\frac{1}{2}}$$

$$= 2 - \tfrac{3}{2}p^2 + \tfrac{11}{8}p^4 - \dots$$

To find β (which vanishes with p) we integrate the power-series term by term, giving

$$\beta = 2p - \tfrac{1}{2}p^3 + \tfrac{11}{40}p^5 - \dots.$$

Thus

$$t_2 = \frac{2p}{\sqrt{(kg)}}\left(1 - \frac{1}{4}p^2 + \frac{11}{80}p^4 - \dots\right)$$

$$= \frac{2u}{g}\left(1 - \frac{1}{4}\frac{ku^2}{g} + \frac{11}{80}\frac{k^2u^4}{g^2} - \dots\right).$$

7·5 Power. The power of a force, or of the agent supplying the force, is the rate at which it works, i.e. the work done per unit of time (§ 3·7). If a particle moves on the line Ox the power of a force $\mathbf{P}$ acting on the particle is given by the formula

$$R = Xv, \tag{1}$$

where X is the component (projection) of $\mathbf{P}$ in the direction Ox.

It is easy to establish this result. If the particle moves from x to $x+\delta x$ in the interval t to $t+\delta t$ the work done by the force in this interval is $\bar{X}\delta x$, where $\bar{X}$ is a mean value of X in the interval. The rate of working is the limit, as $\delta t \to 0$, of $\bar{X}\delta x/\delta t$, and this limit is Xv. The formula (1) is a special case of a more general result already found in § 3·7. If the direction of the force **P** is perpendicular to Ox its power is zero.

There may be other forces acting on the particle—indeed, if **P** is not in the direction Ox there *must* be other forces, since the resultant of all the forces is in the direction Ox if the particle moves on this straight line—but what is determined by the formula (1) is the power of the force **P** alone.

Example 1. *A car, of mass m, moves on a level road, the power of the engine being constant and equal to K. There is a frictional resistance proportional to the square of the speed, the steady speed at which the car can travel being c. To find the (x, v) relation.*

We treat the car as a particle of mass m acted on by a propulsive, or tractive, force X and a resistance F, where

$$Xv = R = \text{constant} = K,$$

and $F = kv^2$. Then the equation of motion is

$$mf = X - F = \frac{K}{v} - kv^2,$$

and since $f = 0$ when $v = c$, $k = K/c^3$. Thus the equation of motion can be written

$$mf = K\left(\frac{1}{v} - \frac{v^2}{c^3}\right).$$

The speed c plays a part somewhat similar to that of the terminal velocity in the problems of resisting media, and in most problems it is not actually attained.

Writing now $f = v\dfrac{dv}{dx}$ we find

$$v\frac{dv}{dx} = \frac{K}{m}\frac{c^3 - v^3}{c^3 v},$$

whence

$$dx = \frac{mc^3}{K}\frac{v^2\,dv}{c^3 - v^3}.$$

If $v = u$ when $x = a$ we have therefore

$$x - a = \frac{mc^3}{3K}\int_u^v \frac{3\theta^2\,d\theta}{c^3 - \theta^3} = \frac{mc^3}{3K}\log\frac{c^3 - u^3}{c^3 - v^3}.$$

This is the required (x, v) relation. It can also be written in the form

$$v^3 = c^3 - (c^3 - u^3)\exp\left[-3K/mc^3(x-a)\right].$$

We find, as we should expect, that $v \to c$ as $x \to \infty$.

Notes on Example 1

(i) This is a problem in which a system of bodies is treated, to a first approximation, as a single particle. This procedure can be justified in two ways at different levels.

(*a*) We shall see later (§ 23·3) that the motion of the centre of gravity G of a system is given by the same law as the motion of a single particle if we suppose all the mass and all the external forces concentrated at G. This principle provides one way of justifying the procedure in the problem just solved. From this point of view $X-F$ is the aggregate of the external forces on the car, X, the propulsive force, can be interpreted as acting at the points of contact of the wheels with the road, the resistance F as acting on the body of the car; we can suppose, to simplify, that it acts at G.

(*b*) There is another, and more penetrating, way of looking at the equation of motion

$$mf = X - F.$$

It is more natural to think of the propulsive force X as related to the driving couple on the back wheels, rather than as the actual reaction on the wheels at the point of contact with the road; and to include in the symbol F the effect of friction generally, in particular the effect of the brakes. We shall find (§ 23·5) that if we give these meanings to the symbols X and F *the equation of motion*

$$mf = X - F$$

is still valid, but the symbol m now represents a fictitious mass, constant, but larger than the actual mass of the car.

(ii) The hypothesis of constant power is not plausible if the car starts from rest. For the formula $X = K/v$ implies that $X \to \infty$ as $v \to 0$, and the hypothesis is therefore not tenable in the neighbourhood of $v = 0$. Nevertheless, if we overlook the fact that the hypothesis is indefensible on physical grounds, the motion determined on the basis of it is quite reasonable, even though the car starts from rest. If, to isolate the difficulty, we suppose for the moment that there is no resistance, the equation of motion is

$$mf = K/v.$$

If $v = 0$ and $x = 0$ at $t = 0$, this equation implies $v^2 \propto t$, $v^3 \propto x$, and $x^2 \propto t^3$, and these results are not too unreasonable.

A more realistic assumption than that of constant power is that the tractive force X varies as $1/(v+a)$, where a is a positive constant (in practice small compared with the terminal velocity c), say

$$X = K/(v+a).$$

With this form for X the power R increases steadily from zero as v increases, and $R \to K$ as $v \to \infty$ (though usually values of v beyond a certain terminal value c are not relevant).

A more accurate representation of the resistance to the motion of a car is given by the formula $A + Bv^2$, where A and B are positive constants; the term A is called the *rolling resistance*, and the term Bv^2 the *wind resistance*. It is clear that the rolling resistance is of dominating importance when v is small, and the wind resistance when v is large. There is a certain value v_1 of v where the change-over occurs, and $A = Bv_1^2$, so the resistance can be written in the form $B(v^2 + v_1^2)$. (For an ordinary car v_1 is in the neighbourhood of $\frac{1}{3}c$.)

If we use these new forms, both for tractive force and resistance, in place of the simpler forms in Example 1, the equation of motion becomes

$$mf = \frac{K}{v+a} - (A+Bv^2).$$

We assume that $K > aA$, and then the terminal velocity c is the (unique) positive real root of the equation

$$(v+a)(A+Bv^2) = K.$$

The equation of motion can now be written in the form

$$mf = (c-v)\left\{B(c+v) + \frac{A+Bc^2}{v+a}\right\}.$$

Example 2. *Numerical example. If in Example* 1 *the mass of the car is* 1 *ton, the horse-power is* 10, *and the steady speed is* 60 *m.p.h., in what distance will the velocity increase from* 15 *to* 45 *m.p.h.?*

Here, using foot-pound-second units, we have $m = 2240$, $K = 10 \times 550 \times 32$ (taking $g = 32$), $c = 88$ (since 60 m.p.h. is 88 ft./sec.). The required distance, in feet, is

$$\frac{2240 \times (88)^3}{3 \times 10 \times 550 \times 32} \log \frac{1-\frac{1}{64}}{1-\frac{27}{64}} = \frac{2240 \times (88)^3}{3 \times 10 \times 550 \times 32} \log \frac{63}{37} = 1538, \text{ approximately.}$$

It is perhaps hardly necessary to warn the reader that the logarithm which occurs in this result is a logarithm to the base e; to find this we must use a table of Naperian logarithms, or, if this is not available, we must multiply the logarithm to the base 10 obtained from the ordinary tables by $\log_e 10$, which is 2·303 approximately.

7·6 The general case of rectilinear motion.

The last case (e) of rectilinear motion in our classification in § 4·2 is the general case

$$f = \phi(t, x, v). \tag{1}$$

It is clear that this problem is of a higher degree of complexity than the special cases we have previously dealt with, and it is unlikely that we can make much progress with the solution of the completely general problem. There is, however, one classical problem of this type for which the solution, with given initial values of x and v, can be found. This is the problem in which the function ϕ has the simple form

$$\psi(t) + Ax + Bv,$$

where A and B are constants. When ϕ has this form the equation (1) becomes a linear differential equation of the second order with constant coefficients, and an equation of this type can be solved by rule.

If A and B are both negative we have the problem of a harmonic oscillator, when the motion is resisted by a frictional force proportional to the speed, and there is (as well as the harmonic attraction) an additional, or applied, force which is a given function of t. The general equation of motion for this problem is

$$\ddot{x} + 2k\dot{x} + n^2 x = \psi(t), \tag{2}$$

where $n > 0$, $k > 0$. The most important cases occur when $\psi(t)$ is constant, or of the form $c \cos(qt + \alpha)$. It will be convenient to break up the problem, considering first some particular cases.

(i) $k = 0$, $\psi(t) = $ constant. This is the problem of harmonic motion with a uniform field superposed, with which we have dealt already (§ 6·10).

(ii) $k = 0$, $\psi(t) = c \cos qt$. The differential equation is

$$\ddot{x} + n^2 x = c \cos qt. \tag{3}$$

(a) If $q \neq n$ the solution is

$$x = \frac{c}{n^2 - q^2} \cos qt + \left(a - \frac{c}{n^2 - q^2}\right) \cos nt + \frac{u}{n} \sin nt,$$

where as usual $x = a$ and $\dot{x} = u$ at $t = 0$. The first term in this solution is called the *forced oscillation*; it has the same period $2\pi/q$ as the applied force. The remaining terms, of period $2\pi/n$, constitute the *free oscillation*. The amplitude of the forced oscillation $\left(\dfrac{c}{n^2 - q^2}\right)$ is large when q is nearly equal to n, i.e. when the forced oscillation has nearly the same period as the undisturbed harmonic motion. This effect, the production of a forced oscillation of amplitude which is large in comparison with the amplitude of the applied force, is called *resonance*; though the term is sometimes used in a stricter sense, namely, when q is exactly, and not merely approximately, equal to n. We shall consider this case (of exact equality) in a moment.

An interesting special case occurs when $a = u = 0$, and we get

$$\begin{aligned} x &= \frac{c}{n^2 - q^2} (\cos qt - \cos nt) \\ &= \frac{2c}{n^2 - q^2} \sin\left(\frac{n-q}{2} t\right) \sin\left(\frac{n+q}{2} t\right). \end{aligned}$$

This may be thought of as a harmonic oscillation of period $4\pi/(n+q)$, with an amplitude which itself varies harmonically with period $4\pi/|n-q|$. In the special case of resonance, when $|n-q|$ is small, the period $4\pi/(n+q)$ is nearly equal to the period of the free oscillation $2\pi/n$; the amplitude varies slowly with a period $4\pi/|n-q|$, which is large. The (t, x) relation is shown in Fig. 7·6*a*.

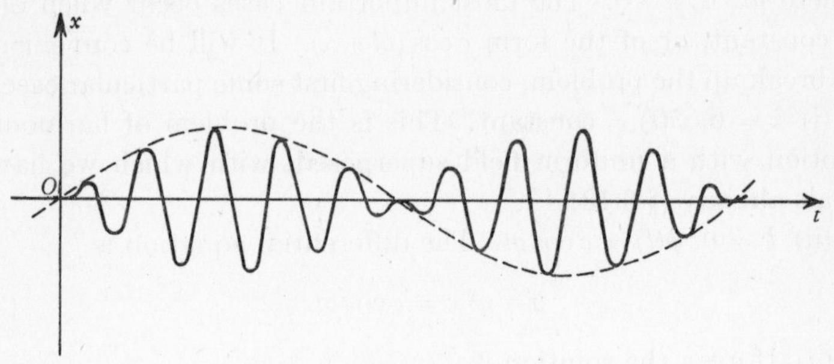

Fig. 7·6*a*

(*b*) If $q = n$ the solution is

$$x = \frac{c}{2n} t \sin nt + a \cos nt + \frac{u}{n} \sin nt.$$

Here we have resonance in the exact sense, the amplitude of the forced oscillation increasing indefinitely with t. In practice the differential equation is usually a sufficiently accurate representation of the motion only for reasonably small values of x, so the solution is not reliable for large values of t.

(iii) The solution of the differential equation

$$\ddot{x} + n^2 x = c \cos qt \tag{3}$$

consists in each case of the sum of a complementary function and a particular integral. The complementary function is a solution of the equation obtained by omitting the second member, and is of the form

$$A \cos nt + B \sin nt.$$

It represents, as we saw, a free oscillation. The particular integral, representing the forced oscillation, is a solution of the complete equation.

We can usually find the particular integral in this kind of work most expeditiously by using the complex variable. As an illustration, let us find a particular integral of equation (3) for the case $q = n$. The equation is

$$\ddot{x}+n^2x = c\cos nt. \tag{4}$$

Let us take for the particular integral the real part of ye^{int}, where y is a function of t, not necessarily real, to be determined. We anticipate that we can find a particular integral for which y has a simple form, and this expectation is fulfilled. We have

$$(D^2+n^2)ye^{int} = ce^{int},$$

where we have written D for the operator d/dt, and using a well-known property of this operator we obtain

$$\{(D+in)^2+n^2\}y = c.$$

Thus the differential equation satisfied by y is

$$(D^2+2inD)y = c,$$

and our task is reduced to the finding of a particular integral of this equation. A very simple answer is obvious, namely,

$$y = \frac{c}{2in}t,$$

and this gives as a particular integral of the equation (4) the real part of

$$\frac{c}{2in}te^{int}, \quad \text{i.e.} \quad \frac{c}{2n}t\sin nt.$$

(iv) The solution of

$$\ddot{x}+n^2x = \psi(t),$$

if $x = a$ and $v = u$ at $t = 0$, is

$$x = \frac{1}{n}\int_0^t \psi(\theta)\sin n(t-\theta)\,d\theta+a\cos nt+\frac{u}{n}\sin nt.$$

Again the first term may be spoken of as the forced oscillation. The preceding results, when ψ has the form $c\cos qt$, are simple special cases of this result.

(v) *Damped harmonic motion.* The equation of motion, when the particle suffers, in addition to the harmonic attraction, a resistance proportional to the speed, is

$$\ddot{x}+2k\dot{x}+n^2x = 0,$$

where k is a positive constant. There are three types of solution, according as (a) $k>n$, (b) $k = n$, (c) $k<n$.

(a) *Heavy damping*, $k>n$. The solution can be expressed in a great variety of forms. Perhaps the simplest is

$$x = \frac{1}{\alpha-\beta}\{(\alpha a+u)\,e^{-\beta t}-(\beta a+u)e^{-\alpha t}\},$$

where α, β $(\alpha>\beta>0)$ are roots of the quadratic equation

$$m^2-2km+n^2 = 0;$$

$$\alpha = k+\lambda, \quad \beta = k-\lambda, \quad \lambda^2 = k^2-n^2 \quad (\lambda>0).$$

(We can also express this in the form $\alpha = n\xi$, $\beta = n/\xi$, where ξ is the number greater than unity defined by $k = \frac{n}{2}\left(\xi+\frac{1}{\xi}\right)$). Thus $\alpha>n>\beta>0$. If $a>0$ and $u>0$, x rises to a maximum, and then tends rapidly to zero as t tends to infinity (Fig. 7·6*b*).

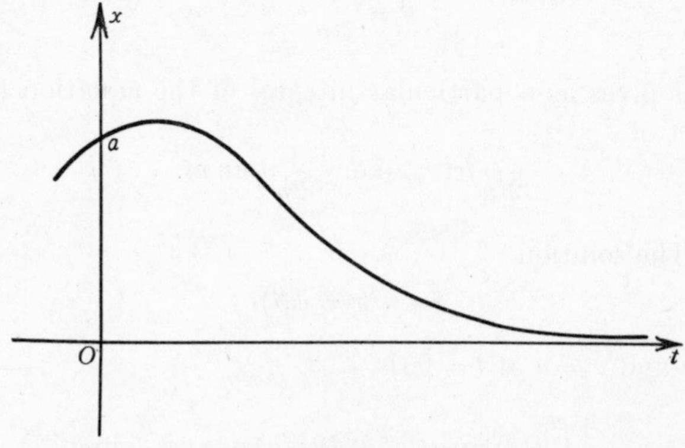

Fig. 7·6*b*

(b) $k = n$. Here $x = \{a+(na+u)t\}e^{-nt}$. Again, if a and u are positive, x rises to a maximum, and then tends to zero as t tends to infinity.

(c) *Light damping*, $k<n$. This is perhaps the most important case. The solution is

$$x = e^{-kt}\left(a\cos\mu t + \frac{ka+u}{\mu}\sin\mu t\right),$$

where $\mu^2 = n^2 - k^2$. The terms in the bracket represent a harmonic oscillation of the type $A\cos\mu(t-t_0)$, and we can think of the motion as a harmonic motion, of period $2\pi/\mu$, with a decreasing amplitude which tends to zero like e^{-kt} (Fig. 7·6c).

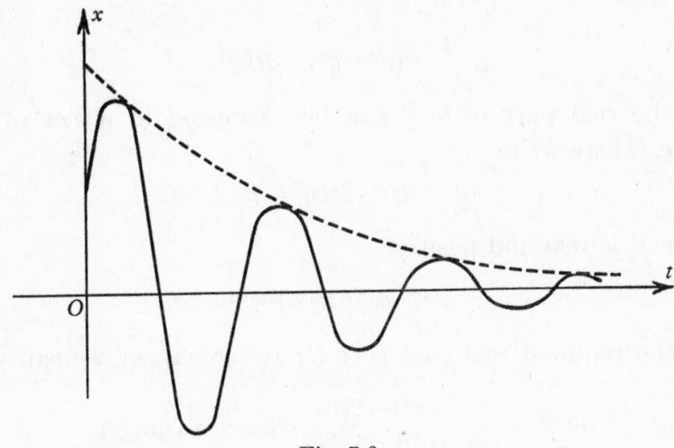

Fig. 7·6c

(vi) Damped harmonic motion with an applied force $\psi(t)$. We have

$$\ddot{x} + 2k\dot{x} + n^2 x = \psi(t),$$

and we will suppose, for definiteness, $n>k$. The solution is obtained by adding to the free oscillation just found the forced oscillation

$$\frac{1}{\mu}\int_0^t e^{-k(t-\theta)}\sin\mu(t-\theta)\,\psi(\theta)\,d\theta.$$

The case of most interest is that in which $\psi(t)$ has the form $c\cos qt$. We can obtain the solution by the method just explained, but if we require only the forced oscillation we can find it more simply as follows. We observe that the free oscillation contains the damping factor e^{-kt}, which tends rapidly to zero. The forced oscillation is a harmonic oscillation with the period of the applied force. Thus the free oscillation soon becomes negligible—it is said to be 'damped out'. We can find the forced oscillation very simply by choosing the constants A and B so that

$$x = A\cos qt + B\sin qt$$

satisfies the differential equation

$$\ddot{x} + 2k\dot{x} + n^2x = c\cos qt. \tag{5}$$

Or we can use a complex variable, as in (iii) above, and find a complex number a such that the real part of ae^{iqt} satisfies the equation (5). Using this latter method we find

$$a = \frac{c}{(n^2-q^2)+2ikq},$$

and the real part of ae^{iqt} can be expressed in either of two forms. If we write

$$(n^2-q^2)+2ikq = Re^{i\alpha},$$

where R is real and positive,

$$ae^{iqt} = (c/R)\,e^{i(qt-\alpha)},$$

and the required real part is $(c/R)\cos(qt-\alpha)$; or we can write

$$ae^{iqt} = \frac{(n^2-q^2)-2ikq}{(n^2-q^2)^2+4k^2q^2}\,c(\cos qt + i\sin qt)$$

and the real part is

$$\frac{c}{(n^2-q^2)^2+4k^2q^2}\{(n^2-q^2)\cos qt + 2kq\sin qt\}.$$

The two forms we have obtained are of course equivalent. We notice that the forced oscillation is independent of the initial conditions.

Example. *A particle hangs by a vertical spring which is such that if the particle is displaced vertically* in vacuo *it will describe a harmonic oscillation with period $2\pi/n$. The motion of the particle is resisted by a force equal to $2k$ times the velocity per unit mass, where $n > k > 0$, and the upper end of the spring is caused to move vertically so that its downward displacement at time t is $\alpha\sin pt$. If the particle starts from rest at its equilibrium position when $t = 0$, show that the displacement of the particle at time t is*

$$\frac{n^2\alpha}{R}\left\{\sin(pt-\gamma)+e^{-kt}\left(\sin\gamma\cos rt - \frac{p\cos\gamma - k\sin\gamma}{r}\sin rt\right)\right\},$$

where $\quad R\sin\gamma = 2kp,\quad R\cos\gamma = n^2-p^2,\quad r^2 = n^2-k^2.$

In the first case, when the upper end of the spring is fixed, and there is no resistance, the equation of motion is

$$m\ddot{x} = -\kappa x,$$

where x is measured downwards from the equilibrium position, and the tension is k times the stretch. Thus $\kappa = mn^2$.

In the problem proposed, the extra stretch of the spring (beyond the stretch when the particle hangs in equilibrium) is $x - \alpha \sin pt$, and the equation of motion is

$$m\ddot{x} = -mn^2(x - \alpha \sin pt) - 2km\dot{x},$$

giving

$$\ddot{x} + 2k\dot{x} + n^2 x = n^2 \alpha \sin pt.$$

We need the solution of this equation for which $x = \dot{x} = 0$ at $t = 0$. A particular solution is the imaginary part of Ae^{ipt}, where

$$A = \frac{n^2 \alpha}{-p^2 + 2ikp + n^2} = \frac{n^2 \alpha}{Re^{i\gamma}},$$

giving the solution $\dfrac{n^2 \alpha}{R} \sin(pt - \gamma)$; adding the complementary function we obtain

$$x = \frac{n^2 \alpha}{R}\{\sin(pt - \gamma) + e^{-kt}(B \cos rt + C \sin rt)\},$$

and we have to choose B and C to satisfy the conditions $x = \dot{x} = 0$ at $t = 0$. These conditions give

$$B = \sin \gamma, \quad \text{and} \quad rC = -(p \cos \gamma - k \sin \gamma).$$

We have now completed our survey of the particular problems of rectilinear motion as enumerated in § 4·2.

7·7 The three forms of the equation of energy.

We have already discussed, in § 5·3, the classical form of the energy equation for the rectilinear motion of a particle in a field of force, namely,

$$T + V = \text{constant}, \tag{1}$$

or

$$T + \Omega = \text{constant}. \tag{2}$$

We consider some cognate results.

The equation of motion

$$mf = X$$

leads to the equation

$$mv\frac{dv}{dt} = Xv,$$

and the first member of this is equal to dT/dt. It follows that

$$\frac{dT}{dt} = Xv = R, \tag{3}$$

where R is the power, or rate of working, of the resultant force X which acts on the particle. The rate of working by this resultant force is equal to the rate of increase of the kinetic energy. Equation (3) is the first or *primitive form* of the equation of energy, and holds generally, in whatever way X is defined. If R can be expressed as a function of the time, the increase in the kinetic energy in the interval from t_1 to t_2 is

$$T(t_2) - T(t_1) = \int_{t_1}^{t_2} R dt. \tag{4}$$

In particular, if R is constant, the change in T during this interval is $R(t_2 - t_1)$.

Next, if $X = X(x)$, we have that the change in T while the particle moves from $x = x_1$ to $x = x_2$ is

$$\begin{aligned} T(x_2) - T(x_1) &= \int_{x_1}^{x_2} X dx \\ &= V(x_1) - V(x_2), \end{aligned}$$

and this is of course equivalent to the second or *classical form* of the equation of energy

$$T + V = C. \tag{1}$$

Finally, suppose the force acting on the particle is the resultant of two forces X' and X'' in the direction Ox, where X' is a function of x, say $X' = -dV/dx$. If, as in § 5·3, equation (1), we form the function $\Omega(t)$ by substituting in $V(x)$ the position of the particle at time t, we have $X'v = -d\Omega/dx$. Now

$$m\frac{dv}{dt} = X' + X'',$$

whence
$$mv\frac{dv}{dt} = X'v + X''v,$$

i.e.
$$\frac{dT}{dt} = -\frac{d\Omega}{dt} + X''v.$$

Thus
$$\frac{d}{dt}(T + \Omega) = X''v. \tag{5}$$

The rate of increase of the total energy $(T+\Omega)$ of the particle is equal to the rate of working of the remaining forces (i.e. the forces not derived from the potential V). It is clear that this third or *general form* (equation (5)) embraces both the primitive form and the classical form.

There is another aspect of the general form of the equation of energy

$$\frac{d}{dt}(T+\Omega) = X''v \tag{5}$$

which is sometimes important. Instead of thinking of the particle as acted on by two kinds of forces, one derived from the conservative field and one extraneous, we can think of the particle-and-field together as a single physical entity whose total energy is $(T+\Omega)$. This notion of the total energy goes back to the classical form of the equation of energy, and in the classical case, where there are no extraneous forces, the total energy remains constant; in the general case it does not remain constant, and its rate of increase is equal to the power of the extraneous forces.

EXAMPLES VII

1. A particle moving in a straight line is subject to a resistance which produces the retardation kv^3, where v is the velocity and k is a constant. Show that v and t (the time) are given in terms of s (the distance) by the equations

$$v = \frac{u}{1+ksu}, \quad t = \frac{s}{u} + \tfrac{1}{2}ks^2,$$

where u is the initial velocity.

As a result of certain experiments with a rifle, it was estimated that the bullet left the muzzle with a velocity of 2400 ft./sec. and that the velocity was reduced to 2350 ft./sec. when 100 yd. had been traversed. Assuming that the air-resistance varied as v^3, and neglecting gravity, calculate the time of traversing a range of 1000 yd.

2. A particle moves in a straight line under a retardation kv^{m+1}, where v is the velocity at time t. Show that, if u is the velocity at $t=0$,

$$kt = \frac{1}{m}\left(\frac{1}{v^m} - \frac{1}{u^m}\right),$$

and obtain a corresponding formula for the space in terms of v.

A bullet fired with a horizontal velocity of 2500 ft./sec. is travelling with a velocity of 1600 ft./sec. at the end of one second. Assuming that $m=\frac{1}{2}$, calculate k and the space traversed in the first second, neglecting the effect of gravity.

3. A particle moves along a straight line against a resistance equal to $av+bv^3$ per unit mass, a and b being positive. If the initial velocity of projection is V_1, show that the velocity will be reduced to V_2 in a time

$$\frac{1}{2a}\log_e\frac{V_1^2(a+bV_2^2)}{V_2^2(a+bV_1^2)}$$

after travelling a distance $$\frac{1}{c}\tan^{-1}\frac{(V_1-V_2)\,c}{a+bV_1V_2},$$
where $c^2=ab$.

Show that the particle will not come to rest in any finite time, and that however great its initial velocity, it will not travel a distance greater than $\pi/2c$.

4. A ball is thrown vertically upwards in a medium which offers a resistance kv per unit mass when the speed is v. If v_0 is the speed of projection, prove that the ball returns to the starting-point after a time t_1, where

$$(g+kv_0)\,(1+e^{-kt_1})=gkt_1.$$

If $t_1=t_0(1-\lambda)$, where t_0 is the time after which the ball would return to the starting-point if there were no resistance, prove that, if kt_0 is small, the value of k is approximately $3g\lambda/v_0$.

5. A particle is projected vertically upwards with velocity u in a medium offering a resistance kv^2 per unit mass. Prove that it returns to the starting-point after a time

$$\frac{1}{\sqrt{(kg)}}\,\{\alpha+\log\,(\sec\alpha+\tan\alpha)\},$$

where $\tan\alpha=u\sqrt{(k/g)}$.

If k is small, and the particle returns to the starting-point after a time $(2u/g)\,(1-\lambda)$, prove that an approximation to the value of k is $4g\lambda/u^2$.

6. A particle is projected vertically upwards with velocity u and moves in a medium offering resistance kv^4 per unit mass, where v is the velocity. Prove that the greatest height reached above the point of projection is $\dfrac{\alpha}{2\sqrt{(kg)}}$, where $g\tan^2\alpha=ku^4$.

If the particle has velocity w downwards on regaining the point of projection, show that $w^2=\sqrt{\dfrac{g}{k}}\tanh\alpha$.

7. A particle falls from rest under gravity in a medium offering a resistance $av+bv^2$ per unit mass, where a and b are positive constants. Find the velocity acquired and the distance fallen in time t.

8. In starting a train the pull of the engine on the rails is at first constant, and equal to P; and after the speed attains a certain value u, the engine works at a constant rate $R=Pu$. Prove that when the engine has attained a speed v greater than u, the time t and the distance x from the start are given by

$$t=\frac{1}{2}\frac{M}{R}\,(v^2+u^2),\quad x=\frac{1}{3}\frac{M}{R}\,(v^3+\tfrac{1}{2}u^3),$$

where M is the mass of the engine and train together.

Calculate the time occupied in attaining a speed of 45 m.p.h. when the total mass is 300 tons, the engine has 420 h.p. and can exert a pull equal to 12 tons weight.

(The acceleration of gravity may be taken as 32 in foot-second units.)

9. A train of total mass M lb. runs on a horizontal track, the frictional resistance being negligible. The engine can exert a maximum tractive force of P poundals, but cannot work at a rate exceeding Pu ft.-poundals/sec. Show that, if the train starts from rest, it cannot attain a speed of $2u$ ft./sec. in less than $5Mu/(2P)$ sec., and that it cannot attain this speed in a run of less than $17Mu^2/(6P)$ ft.

10. If a motor-car weighing 10 cwt. travels at a uniform speed of 25 m.p.h. up a hill which is 800 ft. high and a mile long, find the horse-power exerted by the engine in overcoming the force of gravity. If the engine is actually working at 15 h.p., find in lb. weight the resistance to motion.

11. A truck runs down an incline of 1 in 100; the resistance to motion is proportional to the square of the speed, and the terminal velocity is 40 m.p.h. Prove that the truck, starting from rest, acquires a velocity of 20 m.p.h. in a distance of about 1550 ft.

12. The resistance to the motion of a train, for speeds between 20 and 30 m.p.h., may be taken as $\frac{1}{400}V^2+9$ in lb. wt. per ton, where V is the velocity in miles per hour. Sketch a curve showing how the horse-power per ton, necessary to overcome the resistance, increases with the speed as the speed rises from 20 to 30 m.p.h., the train being on the level.

Steam is shut off when the speed is 30 m.p.h. and the train slows down under the given resistance. In what time will the speed fall to 20 m.p.h.?

13. A train is drawn by an engine which exerts a constant pull at all speeds. Find the equation of motion at any time, assuming that the total resistance to the motion of the train varies as the square of the velocity.

In the above case, the mass of the engine and train combined is 300 tons, the maximum speed on the level is 60 m.p.h., and the horse-power then developed is 1500. Show that, when climbing a slope of 1 in 100, the maximum speed is nearly 32 m.p.h.

14. The external resistance to the motion of a bicycle and rider may be supposed to consist of two parts, a constant force and a force varying as the square of the speed. A rider observes that his speed when free-wheeling down a hill of slope 1 in 50 is sensibly constant when it reaches 10 ft./sec., and that on a slope of 1 in 25 it becomes constant at 20 ft./sec. The mass of the bicycle and rider is 200 lb. Find the power expended by the rider in maintaining a steady speed of 15 ft./sec. on the level, assuming that when the rider is propelling the bicycle 10 per cent of the work he does is lost in internal friction in the pedalling gear.

15. A cyclist works at the constant rate of P horse-power. When there is no wind, he can ride at 22 ft./sec. on level ground, and at 11 ft./sec. up a hill making an angle $\sin^{-1}\frac{1}{20}$ with the horizon. The total mass of man and cycle is 180 lb. The resistance of the air is kv^2 lb. wt., where the velocity of the man relative to the air is v ft./sec.; the other frictional forces are negligible. Find P, and show that the speed of the cyclist when riding on level ground against a wind of 22 ft./sec. is between 10 and 10·5 ft./sec.

16. The horse-power required to propel a steamer of 10,000 tons displacement at a steady speed of 20 knots is 15,000. If the resistance is proportional to the square of the speed, and the engines exert a constant propeller thrust at all speeds, find the acceleration when the speed is 15 knots.

Show that the time taken from rest to acquire a speed of 15 knots is about $1\frac{1}{2}$ min., given $\log_e 7 = 1{\cdot}946$, one knot = 100 ft./min.

17. A car of mass m moves on a level road. The engine works at a constant rate R and there is a constant frictional resistance, the maximum velocity attainable being w. Prove that the distance in which the car, starting from rest, acquires a velocity V is

$$\frac{mw^3}{R}\left(\log\frac{w}{w-V}-\frac{V}{w}-\frac{1}{2}\frac{V^2}{w^2}\right).$$

If the mass of the car is 1 ton, the rate of working is 10 h.p., and the maximum velocity is 60 m.p.h., find, correct to the nearest foot, the distance in which the car acquires from rest a velocity of 30 m.p.h.

18. A locomotive drawing a total weight of 264 tons on the level is exerting a tractive force of 20,000 lb. wt. at the speed of 15 m.p.h. It works at constant horse-power until its speed is 60 m.p.h., when it is just able to overcome the resistances to motion, which may be taken to vary as the square of the velocity. Show that it reaches the speed of 45 m.p.h. from the speed of 15 m.p.h. in a distance of approximately 5080 ft.

19. Find the average horse-power of an engine required to pump out a dock 400 ft. long, 90 ft. wide, and 20 ft. deep, in 4 hours, if the water is delivered at a uniform rate through a pipe of section 1 sq.ft. at a level of 40 ft. above the bottom of the dock, the efficiency of the pump being 80 per cent and account being taken of the energy of the water as it leaves the pipe. (A cubic foot of sea-water weighs 64 lb.)

20. A stream of water, 1 sq.ft. in section, flowing at the rate of 16 ft./sec., enters a turbine. At what rate, in horse-power, does the water deliver energy? (A cubic foot of water weighs 62·5 lb.)

What fraction of this energy is used when the water-power drives a shaft, at a speed of 100 r.p.m., with a couple of which the moment is 140 with the pound weight and foot as units?

21. Let it be assumed that the effective rate of working of the engine of a motor-car is represented by the expression $hn(n_1-n)$, where n is the number of revolutions per second and h, n_1 are certain constants; also that the velocity v is equal to kn, where k is a constant. If there is a fixed resistance to the motion of the car equal to λ times its weight w, find the speed with which the car will ascend an incline of angle α. If the value of k is capable of adjustment, show that to secure the best speed k must be equal to

$$hn_1/2w(\lambda+\sin\alpha),$$

the best speed being $$hn_1^2/4w(\lambda+\sin\alpha).$$

If the maximum power of the engine is 20 h.p., the weight of the car is 1 ton, and the best speed up a slope of 1 in 10 is 20 m.p.h., show that the best speed on the level is nearly 50 m.p.h., and find the best speed up a slope of 1 in 5.

*22. In Example 1 of § 7·5, prove that the time from u to v is

$$\frac{mc^2}{6K}\left\{\log\frac{(c^3-v^3)}{(c^3-u^3)}\frac{(c-u)^3}{(c-v)^3}-2\sqrt{3}(\beta-\alpha)\right\},$$

where $$\tan\beta=\frac{1}{\sqrt{3}}\left(\frac{2v}{c}+1\right),\quad \tan\alpha=\frac{1}{\sqrt{3}}\left(\frac{2u}{c}+1\right).$$

Find the time for the numerical case in Example 2.

23. A particle moves in a straight line with an acceleration directed towards a fixed point in the line and equal to μ times the distance of the particle from the point. Show that the motion is periodic with period $2\pi/\sqrt{\mu}$.

A body is attached to one end of an inextensible string and the other end moves in a vertical line with simple harmonic motion of amplitude a, making n complete oscillations per second. Show that the string will not remain tight during the motion unless $n^2 < g/4\pi^2 a$.

24. A particle of unit mass, moving in a straight line, is at a distance x from a fixed point O of the line at time t, and is acted on by forces $4\pi n_1^2 x$ towards O and $k\cos 2\pi nt$ away from O. If the particle is at rest at O when $t=0$, find an expression for x at time t. (n_1 is not equal to n.)

Two bodies, A of mass 2 lb. and B of mass 1 lb., are on a smooth horizontal table and are connected by a spring. B is connected by a second spring to a fixed frame on the opposite side of B from A. The springs are in line, and all the motions to be considered take place parallel to the line of the springs. When B is fixed A oscillates with a frequency of 5 per second. When A is fixed, B oscillates with a frequency of 30 per second about the point B_0. Find an expression for the displacement of B towards the fixed frame at time t if A is made to execute the oscillation in which its displacement is $a\cos 40\pi t$ towards the fixed frame, and B is at rest at B_0 when $t=0$.

25. A particle of mass m lies in a rough fixed horizontal tube, and is attached to one end of a light spiral spring. The other end of the spring is attached to a fixed point of the tube and the spring lies along the tube. When the spring is extended a distance x the tension is mn^2x. Initially the spring is compressed by an amount c. If the coefficient of friction between m and the tube is $n^2c/6g$, determine the subsequent motion, and show that m eventually remains at rest with the spring neither extended nor compressed. Find the time during which the particle is in motion.

26. A particle moves on a straight line, in a medium whose resistance is proportional to the velocity, under an attractive force to a fixed point of the line, proportional to the distance. Prove that if σ is the period, and a, b, c are the coordinates of the extremities of three consecutive semi-vibrations, then the coordinate of the position of equilibrium is

$$\frac{ac-b^2}{a+c-2b},$$

and the time of vibration if there were no resistance would be

$$\sigma\left[1+\frac{1}{\pi^2}\left(\log\frac{a-b}{c-b}\right)^2\right]^{-\frac{1}{2}}.$$

27. A man of mass m stands on an escalator of inclination α which ascends with uniform velocity. He walks up the escalator, and finally comes to rest again relative to the escalator. Show that if a is the distance travelled by the escalator, and b the distance travelled by the man relative to the escalator, then of the total work done $mga \sin \alpha$ is supplied by the engine and $mgb \sin \alpha$ by the man. (The man is to be considered as a particle, and the escalator as a continuous gradient without steps. The acceleration of the man relative to the escalator is assumed to be a continuous function of the time.)

Consider next the more general problem in which the velocity of the escalator at any instant is u, and of the man relative to the escalator v, u and v being functions of the time having continuous derivatives. Show that of the total work done on the man in any interval of time t_1 to t_2 the amount supplied by the engine is

$$mga \sin \alpha + \tfrac{1}{2}m(u_2^2 - u_1^2) + m\int_{t_1}^{t_2} u \frac{dv}{dt}\, dt,$$

and by the man $$mgb \sin \alpha + \tfrac{1}{2}m(v_2^2 - v_1^2) + m\int_{t_1}^{t_2} v \frac{du}{dt}\, dt,$$

where a is the distance travelled by the escalator, and b by the man relative to the escalator, in the given interval.

CHAPTER VIII

MOTION OF TWO PARTICLES ON A STRAIGHT LINE; IMPULSES

8·1 Two particles moving on a straight line. We now consider briefly the theory of the motion of a pair of particles on a straight line. This is the simplest case of a *system* of particles, and will serve as an introduction to the general case where there are ν particles moving in space. We denote the masses of the particles P_1, P_2 by m_1, m_2, and their positions on the line at time t by x_1, x_2, supposing in the first place that P_2 is to the right of $P_1, x_2 > x_1$.

Newton's Third Law of Motion states that *action and reaction are equal and opposite*. This means that the force exerted by P_1 on P_2 is equal and opposite to the force exerted by P_2 on P_1. We speak of these forces as *internal forces*. The characteristic property of the internal forces is that implied in the third law of motion, namely, that the internal forces consist of pairs of equal and opposite collinear forces. The internal forces may be, for example, a gravitational attraction between the particles, or the tension or thrust in a light spring joining the particles, or the tension or thrust in a light rod to whose ends the particles are attached. We will denote the internal forces by $-R$ on m_1 and by R on m_2. In the gravitational case, where the internal forces are a mutual *attraction*, $R < 0$ (supposing $x_2 > x_1$).

In addition to the internal forces there may be other forces, called *external forces*, acting on the particles. We will denote these by X_1 and X_2.

As a specific example, if the particles are joined by a light spring, and move in a vertical line, the tension or thrust in the spring provides the internal forces, the weights are the external forces. (We are here supposing that the gravitational attraction of the particles on one another is for practical purposes too small to be sensible.) In this case the external forces are proportional to the masses; taking Ox upwards we have $X_1 = -m_1 g$, $X_2 = -m_2 g$.

The equations of motion for the particles are therefore

$$m_1 \ddot{x}_1 = X_1 - R, \tag{1}$$

$$m_2 \ddot{x}_2 = X_2 + R. \tag{2}$$

We deduce from these equations two classical theorems. The first concerns the motion of the centre of gravity G of the particles, and is a special case of a result valid for a general system. The second concerns the motion of P_2 relative to P_1.

(i) Adding the two equations we get

$$m_1 \ddot{x}_1 + m_2 \ddot{x}_2 = X_1 + X_2. \tag{3}$$

If we denote the position of the centre of gravity G by ξ, we have

$$m_1 x_1 + m_2 x_2 = M\xi, \tag{4}$$

where M is the mass of the whole system, $M = m_1 + m_2$. The equation (3) therefore gives

$$M\ddot{\xi} = X_1 + X_2. \tag{5}$$

The motion of G is given by Newton's second law for a single particle, if we think of all the mass as concentrated at G, and all the external forces as acting at G. In certain cases (for example if $X_1 + X_2$ is constant, or if $X_1 + X_2$ can be expressed as a function of ξ) the motion of G can be determined from this equation by the theory for a single particle.

An important special case is that in which $X_1 + X_2 = 0$; this occurs, in particular, when no external forces act on the particles, $X_1 = X_2 = 0$. If $X_1 + X_2 = 0$, we have

$$m_1 \ddot{x}_1 + m_2 \ddot{x}_2 = 0,$$

whence $$m_1 \dot{x}_1 + m_2 \dot{x}_2 = \text{constant},$$

and this implies $$\dot{\xi} = \text{constant}.$$

This is a simple special case of the famous theorem of *the conservation of momentum*. If $X_1 + X_2 = 0$, the momentum of the system $m_1 \dot{x}_1 + m_2 \dot{x}_2$ remains constant. And this is equivalent to saying that the velocity of G is constant.

We shall find that this theorem can readily be extended to a more general system. (The general case of the theorem will be found in § 24·1.)

(ii) To discuss the motion of P_2 relative to P_1 write

$$s = x_2 - x_1, \tag{6}$$

and we have
$$\ddot{s} = \frac{X_2}{m_2} - \frac{X_1}{m_1} + \frac{R}{\mu}, \tag{7}$$

where $\dfrac{1}{\mu} = \dfrac{1}{m_1} + \dfrac{1}{m_2}$. Notice that μ is positive and less than the smaller of m_1 and m_2.

There is one special case of equation (7) which is of outstanding importance. It occurs if $X_1 = X_2 = 0$ (i.e. there are no external forces) or if $X_1/m_1 = X_2/m_2$ (i.e. the external forces are proportional to the masses). If one of these conditions is fulfilled

$$\mu\ddot{s} = R, \tag{8}$$

and *the relative motion* (i.e. the motion of P_2 relative to P_1) *is the same as if P_1 were fixed and the mass of P_2 reduced to μ.* In this case the relative motion can be studied in isolation, without reference to the individual motions of the separate particles.

The condition $X_1/m_1 = X_2/m_2$ is approximately satisfied in any problem where the variation in the external force per unit mass is small in a distance of the scale of the problem. When this holds the relative motion is given approximately by equation (8) and, to a first approximation, it is independent of the absolute motions.

Another way of seeing the genesis of equation (8) is by marking the accelerations in a diagram. The actual accelerations of P_1 and P_2 are as indicated in Fig. 8·1*a*; if therefore $X_1 = X_2 = 0$, or if $X_1/m_1 = X_2/m_2$, the acceleration of P_2 relative to P_1 is as indicated in Fig. 8·1*b*, and this leads to $\ddot{s} = R/\mu$ as before.

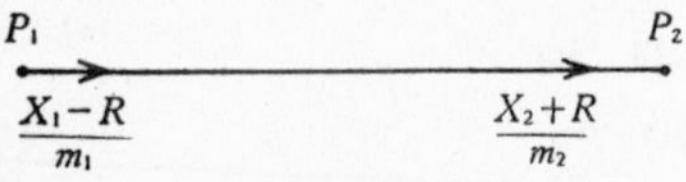

P_1 P_2

$\dfrac{X_1 - R}{m_1}$ $\dfrac{X_2 + R}{m_2}$

Fig. 8·1*a*

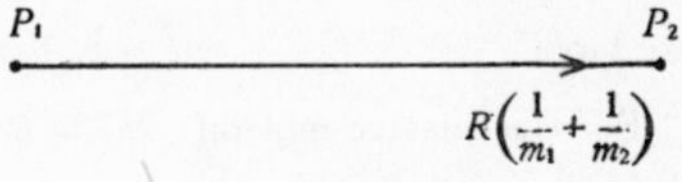

P_1 P_2

$R\left(\dfrac{1}{m_1} + \dfrac{1}{m_2}\right)$

Fig. 8·1*b*

If we can find ξ and s as functions of t we can easily deduce the values of x_1 and x_2 as functions of t, since (from equations (4) and (6))

$$x_1 = \xi - \frac{m_2}{M} s, \tag{9}$$

$$x_2 = \xi + \frac{m_1}{M} s. \tag{10}$$

We frequently use α_1, α_2 to denote the coordinates of P_1, P_2 *relative to G*, so that

$$x_1 = \xi + \alpha_1,$$

$$x_2 = \xi + \alpha_2.$$

The characteristic property of these relative coordinates is that described by the equation

$$m_1\alpha_1 + m_2\alpha_2 = 0.$$

Example 1. *Two particles, whose masses are m_1 and m_2, are initially at rest at a distance b apart. They move towards each other under their mutual gravitation. To find how long before they collide.*

Here $$R = -\gamma m_1 m_2/s^2,$$

and the equation for the relative motion is

$$\mu\ddot{s} = -\gamma m_1 m_2/s^2,$$

i.e. $$\ddot{s} = -k/s^2,$$

where $k = \gamma(m_1+m_2)$. (We may notice that this result can be arrived at very simply by marking the accelerations in a diagram as in Fig. 8·1*a*. The accelerations of the individual particles, directed towards one another, are $\gamma m_2/s^2$ and $\gamma m_1/s^2$, so the relative acceleration of m_2 towards m_1 is $\gamma(m_1+m_2)/s^2$.) The time to the collision is therefore given by § 6·12 (i), namely, $\pi\sqrt{(b^3/8k)}$.

Example 2. *Two particles m_1, m_2 are attached to the ends of a light spring. The natural length of the spring is a, and its tension is k times its extension. Initially the particles are at rest with m_2 at a height a vertically above m_1. At the instant $t = 0$ the particle m_2 is projected vertically upwards with velocity u, where $u < na$, and $n^2 = k/\mu$. To discuss the subsequent motion.*

Measuring upwards we have (from equation (5) above)

$$\ddot{\xi} = -g,$$

and since initially $$M\dot{\xi} = m_2 u$$

we have $$\xi = \xi_0 + (m_2/M)ut - \tfrac{1}{2}gt^2.$$

For the relative motion $$\mu\ddot{s} = R = -k(s-a),$$

$$\ddot{s} + n^2(s-a) = 0,$$

and since $s = a$, $\dot{s} = u$, at $t = 0$,

$$s - a = \frac{u}{n}\sin nt.$$

We have now found the values of ξ and of s at time t, and from these we can deduce the values of x_1 and of x_2 at time t by means of the equations (9) and (10) above.

In this problem the mutual gravitation of the particles has been neglected in comparison with the other forces; these other forces are the weights (the attraction of the earth on the particles) and the tension or thrust in the spring.

Example 3. *Atwood's machine. Two particles, whose masses are M and m ($M>m$) are attached to the ends of a light perfectly flexible inelastic string that passes over a smooth pulley or peg (Fig. 8·1c). To find the motion, the particles being initially at rest with the two parts of the string taut and vertical.*

The particles move vertically. The tension T in the string at any instant is the same all along the string, because the string is supposed to be of negligible mass. If f is the downward acceleration of M at time t, then f is the upward acceleration of m, and we have the equations of motion

$$Mg-T=Mf,$$

$$T-mg=mf,$$

whence $f=\dfrac{M-m}{M+m}g, \quad T=\dfrac{2Mm}{M+m}g.$

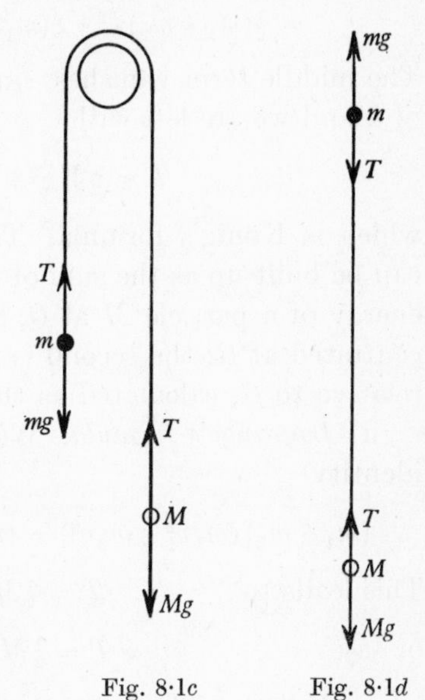

Fig. 8·1c Fig. 8·1d

Thus f remains constant (until m reaches the peg), and the motion of either particle is given by § 4·4.

This example does not properly come under the present heading (strictly speaking it belongs the theory of systems, which is discussed in Chapter XXIII), but it is convenient to include it here. In fact the formula for f is most easily remembered if we think of the string as straightened out (Fig. 8·1*d*), and then we have a problem of two particles moving on a straight line. The external forces are then Mg and $-mg$, and the internal force is the tension in the string. The equation (5) above now gives

$$(M+m)f=(M-m)g,$$

as though a mass $(M+m)$ were moving under the action of a force $(M-m)g$.

If $M-m$ is small, the machine gives us a practicable method of finding g; g itself is too large to measure easily, but f can be measured with reasonable accuracy.

8·2 Kinetic energy.

We have already spoken of the kinetic energy T of a particle. The kinetic energy T of a system is defined as the sum of the kinetic energies of the separate particles. There are two classical formulae for the kinetic energy of a pair of particles moving on a straight line; both of these can be extended to the case of a general system of particles moving in space.

(i) *König's formula.* We have, using the notation of § 8·1,

$$T = \tfrac{1}{2}(m_1 \dot{x}_1^2 + m_2 \dot{x}_2^2)$$
$$= \tfrac{1}{2}m_1(\dot{\xi} + \dot{\alpha}_1)^2 + \tfrac{1}{2}m_2(\dot{\xi} + \dot{\alpha}_2)^2$$
$$= \tfrac{1}{2}(m_1 + m_2)\dot{\xi}^2 + \dot{\xi}(m_1 \dot{\alpha}_1 + m_2 \dot{\alpha}_2) + \tfrac{1}{2}(m_1 \dot{\alpha}_1^2 + m_2 \dot{\alpha}_2^2).$$

The middle term vanishes, since $m_1 \alpha_1 + m_2 \alpha_2 = 0$ for all values of t, and we are left with

$$T = \tfrac{1}{2}M\dot{\xi}^2 + \tfrac{1}{2}(m_1 \dot{\alpha}_1^2 + m_2 \dot{\alpha}_2^2),$$

which is König's formula. The kinetic energy of the system can be built up as the sum of two terms: the first is the kinetic energy of a particle M at G, as though all the mass were concentrated at G; the second is the kinetic energy of the motion relative to G, calculated as though G were fixed.

(ii) *Lagrange's formula.* We start from the simple algebraic identity

$$(m_1 + m_2)(m_1 v_1^2 + m_2 v_2^2) = (m_1 v_1 + m_2 v_2)^2 + m_1 m_2 (v_2 - v_1)^2.$$

This leads to $$T = \tfrac{1}{2}M\dot{\xi}^2 + \tfrac{1}{2}\mu \dot{s}^2,$$

or say $$T = \tfrac{1}{2}MU^2 + \tfrac{1}{2}\mu W^2,$$

where U is the velocity of G, and W the velocity of one particle relative to the other. This is Lagrange's formula for the kinetic energy.

Both the formulae we have found are important; the first (König's) is the more generally useful.

Example. *A body of mass $m_1 + m_2$ is split into two parts of masses m_1 and m_2 by an internal explosion which generates kinetic energy E. Show that if after the explosion the parts move in the same line as before, their relative speed is*

$$\sqrt{\{2E(m_1 + m_2)/m_1 m_2\}}.$$

We interpret this as meaning that the kinetic energy after the explosion exceeds that before the explosion by E. The kinetic energy after the explosion is (Lagrange's formula)

$$\tfrac{1}{2}MU^2 + \tfrac{1}{2}\mu W^2,$$

and the first term measures the kinetic energy before the explosion, so $\tfrac{1}{2}\mu W^2 = E$. This is equivalent to the result stated.

8·3 The equation of energy.

We now consider the extension of the equation of energy, already discussed for the motion of a single particle, to a system of two particles.

Let us suppose, to fix the ideas, that the external forces arise from a field of force, in which the force on a given particle depends only on its position, and the forces on different particles are proportional to their masses. Then there exists a potential per unit mass V_0, with the property that the force on a particle of mass m is $-m\dfrac{dV_0}{dx}$. (For example, if the field is the gravitational field of a fixed mass M at the origin, then the force on a particle of mass m is $-\gamma Mm/x^2$, and $V_0 = -\gamma M/x$, for $x > 0$. We observe, with reference to this example, that the description of a force as external or internal is not absolute, but depends on what particular collection of particles we isolate as the system to be discussed. If our system consists of two particles P_1 and P_2 moving in the field of the fixed mass M, the attraction of this fixed mass on either particle is an external force. If we took as our system three particles moving under their mutual attractions, all the attractions between the pairs of particles would be internal forces.)

Next we suppose that the internal force, the mutual repulsion R, is a function only of the distance s between the particles, $R = R(s)$. We introduce a function $V_1(s)$ defined by the equation

$$V_1 = -\int^s R(\theta)\, d\theta,$$

so that

$$\frac{dV_1}{ds} = -R.$$

We shall find that the total potential energy $V(x_1, x_2)$,

$$V = m_1 V_0(x_1) + m_2 V_0(x_2) + V_1(s), \tag{1}$$

plays a role similar to that of the potential energy in the equation of energy for a single particle moving in a field of force. If we denote by $\Omega(t)$ the value of V when we substitute for x_1 and x_2 their values at time t in the actual motion, we have

$$\begin{aligned}\frac{d\Omega}{dt} &= m_1 \dot{x}_1 \left(\frac{dV_0}{dx}\right)_{x=x_1} + m_2 \dot{x}_2 \left(\frac{dV_0}{dx}\right)_{x=x_2} + \frac{dV_1}{ds}\dot{s} \\ &= -X_1 \dot{x}_1 - X_2 \dot{x}_2 - R(\dot{x}_2 - \dot{x}_1) \\ &= -(X_1 - R)\,\dot{x}_1 - (X_2 + R)\,\dot{x}_2,\end{aligned}$$

and in virtue of the equations of motion for the individual particles this leads to

$$\frac{d\Omega}{dt} = -m_1\dot{x}_1\ddot{x}_1 - m_2\dot{x}_2\ddot{x}_2 = -\frac{dT}{dt}.$$

Thus

$$T + \Omega = C, \tag{2}$$

as in the simpler problem of a single particle moving in a field of force.

The important thing to observe here is that there is a term in V (and in Ω) contributed by the mutual attraction or repulsion in addition to the terms arising from the external field. (The analogous theory for a plane system is discussed in § 23·8.) We may remark again, as in the case of a single particle already discussed, that there is no objection to writing the equation of energy in the form

$$T + V = C \tag{3}$$

when once the significance of V is understood; namely, that it is here the value for a particular configuration (the configuration of the system at a certain instant t) in contrast to its original significance as the total potential energy for arbitrary values of x_1 and x_2.

We shall find later (§ 23·9) that the theorem of the conservation of energy, which we proved for a single particle in § 5·3, and which we have proved here for a pair of particles moving on a straight line, is capable of wide extension; the two theorems mentioned are special cases of a general and far-reaching result.

8·4 Impulses. Consider first a single particle moving on a straight line. We have the equation of motion

$$m\ddot{x} = X. \tag{1}$$

If we can express X as a function of t this leads to

$$mv_2 - mv_1 = \int_{t_1}^{t_2} X\,dt, \tag{2}$$

where v_1 is the velocity at the instant t_1, and v_2 is the velocity at the instant t_2. The vector $m\mathbf{v}$ is called the momentum of the particle. In our present problem the motion is rectilinear, and the momentum is effectively a scalar quantity mv. Our equation (2) shows that the integral $\int_{t_1}^{t_2} X\,dt$ is equal to the change in momentum that takes place during the time-interval from t_1 to t_2. (Cf. § 4·8.)

If $(t_2 - t_1)$ is small, and X is large during part of this interval, we may get a sensible change of momentum in a short time. When the interval is so short as to be negligible for ordinary purposes, but the value of X is so large that the integral has a finite value, we speak of an *impulse*. Effectively we have an instantaneous change of momentum. We take $B = \int_{t_1}^{t_2} X\,dt$ as the measure of the impulse, and the fundamental equation (2) becomes

$$mv_2 - mv_1 = B.$$

Familiar examples are provided by a bat striking a cricket ball, and by the collisions of billiard balls.

The essentially new feature is the sudden change of velocity; the velocity, considered as a function of t or of x, is a *discontinuous* function. The theory of impulses is concerned with the study of such discontinuities. It is one of the simplest branches of mechanics, because, in a pure impulse problem, we have no change of time or configuration to consider, but only the instantaneous change in velocity.

If the particle is initially at rest and an impulse B is applied we have

$$mv = B.$$

The impulse is equal to the momentum set up when the particle starts from rest. The relation between an impulse and the momentum it sets up is similar to the relation between a force and the kineton it sets up.

When an impulse acts there is a sudden change in the kinetic energy. Thus, if a particle moves in a field of force, and at some point an impulse is applied, there is a sudden change in the kinetic energy. The energy is no longer constant during an interval including the instant at which the impulse acts. It remains constant until the impulse

$$T + V = C_0,$$

and again after the impulse

$$T + V = C_1,$$

but $C_0 \neq C_1$.

8·5 Collisions. We now consider the impact of smooth solid spheres moving, without rotation, in the same line. The spheres are to be thought of as uniform solids, nearly rigid, but not quite rigid, because there will be a slight distortion during the small interval of contact. But we suppose the distortion to be only temporary, the spheres regaining their original forms immediately they are separated. Throughout the small interval of contact the forces on the two spheres are at each instant equal and opposite. Thus the impulses on the two spheres are equal and opposite, and the momentum gained by one is precisely equal to the momentum lost by the other. The total momentum is conserved. Thus, if u_1, u_2 are the velocities of the centres immediately before the impact, and v_1, v_2 immediately after, we have

$$m_1v_1 + m_2v_2 = m_1u_1 + m_2u_2. \tag{1}$$

This implies that the velocity U of G does not change at the impulse, and each member of equation (1) is equal to $(m_1+m_2)U$.

The law of conservation of momentum, equation (1), was given by Newton as an empirical result, but, as we have seen, it is already implied in the third law of motion.

In order to determine v_1 and v_2 (when u_1 and u_2 are given) we need another relation. This is

$$\frac{v_2 - v_1}{u_2 - u_1} = -e, \tag{2}$$

where e is a constant for the two spheres, the *coefficient of restitution*; the velocity of separation $(v_2 - v_1)$ is e times the velocity of approach $(u_1 - u_2)$. This law was also given, as an empirical result, by Newton. In practice $0 < e < 1$, but theoretically we often deal with one of the limiting cases, (i) $e = 0$, when the bodies are said to be *inelastic*, (ii) $e = 1$, when they are said to be *perfectly elastic*.

The equations (1) and (2) suffice to determine v_1 and v_2, the velocities after the impact,

$$v_1 = u_2 + \frac{m_1 - em_2}{m_1 + m_2}(u_1 - u_2),$$

$$v_2 = u_1 + \frac{em_1 - m_2}{m_1 + m_2}(u_1 - u_2).$$

In particular in the 'billiard-ball problem', where the masses are equal and the spheres are perfectly elastic ($m_1 = m_2$, $e = 1$), we have

$$v_1 = u_2, \quad v_2 = u_1.$$

Thus, in the billiard-ball problem, *the effect of the collision is to interchange the velocities.*

Finally, returning to the general case, we have

$$m_1(v_1 - U) + m_2(v_2 - U) = 0,$$

$$m_1(u_1 - U) + m_2(u_2 - U) = 0,$$

so that

$$\frac{v_2 - U}{u_2 - U} = \frac{v_1 - U}{u_1 - U},$$

and each fraction is equal to

$$\frac{v_2 - v_1}{u_2 - u_1} = -e.$$

Thus for either sphere the velocity of separation from G is e times the velocity of approach to G.

Example 1. *Three small spheres A, B, C, whose masses are $8m, m, 7m$, are at rest in line, and $AB = BC = a$. The middle sphere B is projected towards C with velocity U. Assuming all the spheres to be perfectly elastic, discuss the subsequent motion, and illustrate by a graph showing the positions of the spheres at any time after the start.*

Let us first find the velocities after the various impacts; then we can calculate the times at which the collisions occur, and the positions of the spheres at each collision.

If v, w are velocities of B, C after the first impact

$$U = v + 7w = w - v,$$

giving

$$\frac{v}{-3} = \frac{w}{1} = \frac{U}{4}.$$

If u, v now denote velocities of A, B after the second impact

$$-\tfrac{3}{4}U = 8u + v = u - v,$$

giving

$$\frac{u}{-2} = \frac{v}{7} = \frac{U}{12}.$$

If v, w now denote velocities of B, C after the third impact

$$v + 7w = \tfrac{7}{12}U + \tfrac{7}{4}U,$$

$$w - v = \tfrac{7}{12}U - \tfrac{1}{4}U,$$

giving

$$v = 0, \quad w = \tfrac{1}{3}U,$$

and there are no further collisions.

Thus in the four stages of the motion the velocities of the spheres (without reference to position) are as shown in Fig. 8·5*a*.

We now calculate times and positions. The time to the first impact is a/U, and the positions at the first impact—taking the initial positions as $(-a, 0, a)$—are $(-a, a, a)$.

The time from the first impact to the second is $2a/(\frac{3}{4}U) = 8a/3U$, and the position of C at the second impact is $a+\frac{1}{4}U(8a/3U) = \frac{5}{3}a$. Thus the positions at the second impact are $(-a, -a, \frac{5}{3}a)$.

The time from the second impact to the third is, by considering the motion of B relative to C, $\frac{8}{3}a/(\frac{1}{3}U) = 8a/U$. The positions at the third impact are,

for A
$$-a-\tfrac{1}{6}U\left(\frac{8a}{U}\right) = -\tfrac{7}{3}a,$$

for B
$$-a+\tfrac{7}{12}U\left(\frac{8a}{U}\right) = \tfrac{11}{3}a,$$

for C (as a check of the arithmetic) $\frac{5}{3}a+\frac{1}{4}U\left(\frac{8a}{U}\right) = \frac{11}{3}a.$

Thus the impacts take place at times

$$3t_0, \quad 11t_0, \quad 35t_0$$

after the start, where $t_0 = a/(3U)$, and the positions of the spheres at these times have been found, so we can now draw the graph (Fig 8·5*b*).

Example 2. *Two buckets of given depth are suspended by a fine inelastic string placed over a smooth pulley; at the centre of the base of one of the buckets a frog is sitting; at an instant of instantaneous rest of the buckets the frog leaps vertically upwards so as just to arrive at the level of the rim of the bucket; find the ratio of the absolute length of the frog's vertical ascent in space to the depth of the bucket, and show that the time which elapses before the frog again arrives at the base of the bucket is independent of the frog's mass.* (*From Walton's problems*, 1858.)

Let m be the frog's mass, M_1 that of his bucket, M_2 of the other, h the depth of a bucket; let u be the frog's velocity upwards, and v the velocity of M_1 downwards, immediately after the jump. Then

$$mu = (M_1+M_2)v.$$

(To prove this formally we notice that the downward impulse on M_1 is mu, so if T is the impulsive tension in the string

$$mu-T = M_1 v, \quad T = M_2 v,$$

and the result follows; but the result is evident without bringing in the impulsive tension if we think of the string as straightened out, as in Example 3 of § 8·1.) During the jump the frog has uniform acceleration g downwards, and M_1 has uniform acceleration f upwards, where

$$f = \frac{M_2-M_1}{M_2+M_1}g.$$

Of course f may be negative, but it is numerically less than g, so $(f+g)$ is positive always.

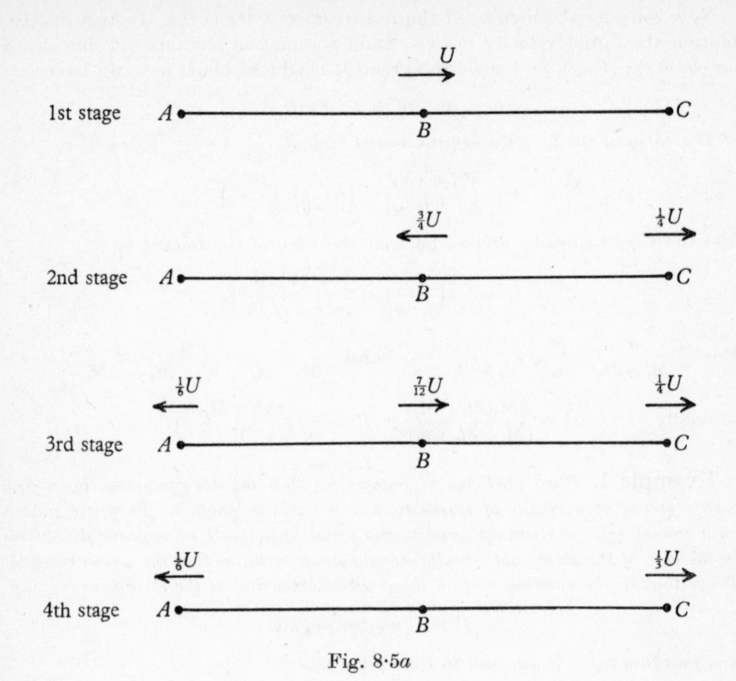

Fig. 8·5*a*

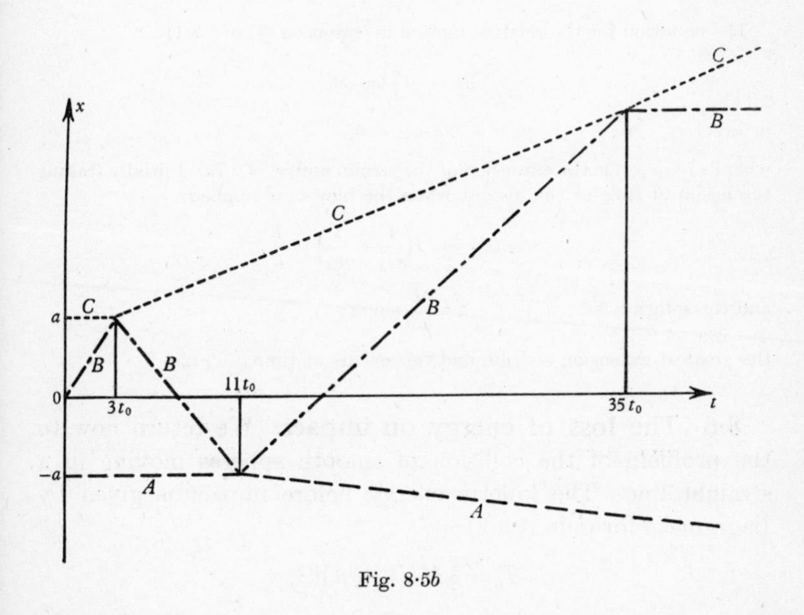

Fig. 8·5*b*

Now consider the motion of the frog *relative to the bucket*. In this relative motion the initial velocity is $(u+v)$ and the motion has uniform downward acceleration $(f+g)$, and since the greatest height attained is h, we have

$$(u+v)^2 = 2(f+g)h.$$

The ratio of the frog's absolute ascent to h is

$$\lambda = \frac{u^2}{2g}\bigg/\frac{(u+v)^2}{2(f+g)} = \left(\frac{u}{u+v}\right)^2\left(\frac{f+g}{g}\right),$$

and the time before he arrives back at the base of the bucket is

$$t_1 = 2\sqrt{\left(\frac{2h}{f+g}\right)} = 2\sqrt{\left(\frac{2h}{g}\cdot\frac{g}{f+g}\right)}.$$

Now $$\frac{u}{M_1+M_2} = \frac{v}{m} = \frac{u+v}{M_1+M_2+m}, \quad \text{and} \quad \frac{f}{M_2-M_1} = \frac{g}{M_2+M_1} = \frac{f+g}{2M_2},$$

so finally $$\lambda = \frac{2M_2(M_1+M_2)}{(M_1+M_2+m)^2}, \quad t_1 = 2\sqrt{\left(\frac{(M_1+M_2)h}{M_2 g}\right)}.$$

Example 3. *Two particles of masses m_1 and m_2 are connected by a fine elastic string of modulus of elasticity λ and natural length l. They are placed on a smooth table at distance l apart, and equal impulses I in opposite directions in the line of the string act simultaneously upon them, so that the string extends. Prove that in the ensuing motion the greatest extension of the string is*

$$I\{(m_1+m_2)l/m_1 m_2 \lambda\}^{\frac{1}{2}},$$

and that this value is attained in time $\pi/2n$, where

$$n^2 = (m_1+m_2)\lambda/m_1 m_2 l.$$

The equation for the relative motion is (equation (7) of § 8·1)

$$\mu\ddot{x} = -\frac{\lambda}{l}(x-l),$$

or say $$\ddot{z}+n^2 z = 0,$$

where $z\ (= x-l)$ is the extension of the string, and $n^2 = \lambda/l\mu$. Initially (taking the origin of time at the instant when the blows are applied)

$$z = 0, \quad \dot{z} = I\left(\frac{1}{m_1}+\frac{1}{m_2}\right) = \frac{I}{\mu},$$

and the solution is $$z = \frac{I}{\mu n}\sin nt;$$

the greatest extension is $I/\mu n$, and this occurs at time $t = \pi/2n$.

8·6 The loss of energy on impact.

We return now to the problem of the collision of smooth spheres moving in a straight line. The kinetic energy before impact is given by Lagrange's formula (§ 8·2)

$$T_0 = \tfrac{1}{2}MU^2 + \tfrac{1}{2}\mu W^2,$$

and the kinetic energy after impact is

$$T = \tfrac{1}{2}MU^2 + \tfrac{1}{2}\mu(eW)^2,$$

since U is unchanged by the collision, and W is changed to $-eW$. Thus

$$T_0 - T = \tfrac{1}{2}\mu(1-e^2)W^2, \tag{1}$$

and there is a loss of energy, unless $e = 1$. If the spheres are moving in a field of force, and subject also to a mutual attraction which is a function of their distance apart, the energy is conserved during any interval not including the collision, i.e. an equation

$$T + V = C_0$$

holds for a period preceding the collision, and an equation

$$T + V = C_1$$

holds for a period after the collision. But $C_0 \neq C_1$, in fact

$$C_1 = C_0 - \tfrac{1}{2}\mu(1-e^2)\,W^2.$$

(Compare the case of a single particle already mentioned in § 8·4.)

As we have remarked already, although from the point of view of theoretical mechanics there is a loss of energy in this problem, from the wider point of view of general physics the energy is not lost, but only converted into other forms. Elastic vibrations are set up in the bodies, and these are dissipated by friction, the energy being converted into heat; and some of the energy may be imparted to the surrounding air in waves of sound. It is perhaps of interest to examine more closely the mechanism of the collision in the light of the general theory of energy. In order to put the argument into a simple form, we use a simplified model; we think of the spheres as rigid bodies provided with spring buffers—like a railway engine—which are compressed during the collision. The buffers are to be thought of as light springs, and we examine what happens during the short interval of contact.

The equation for the relative motion is

$$\mu\ddot{s} = X,$$

where X is the thrust in the springs. If we write v for $\dot{s}$ this can be written

$$\frac{d}{ds}\left(\tfrac{1}{2}\mu v^2\right) = X.$$

Suppose that during compression s decreases from a to b; initially $v = -W$, where W is the velocity of approach, and, at maximum compression, $v = 0$, so we have

$$\tfrac{1}{2}\mu W^2 = \int_b^a X\,ds.$$

Thus the work done in compressing the springs is $\frac{1}{2}\mu W^2$, and we regard this as energy stored up in the springs. Similarly, the work done by the springs during the expansion is $\frac{1}{2}\mu W'^2$, where W' is the velocity of separation.

Now if all the energy stored up during the compression is given out again during the expansion we have

$$W' = W,$$

and the impact is perfectly elastic. But if only part of the stored energy is given up, say a fraction e^2, then

$$W' = eW,$$

and a fraction $(1-e^2)$ of the relative kinetic energy (i.e. of the term $\frac{1}{2}\mu W^2$ in Lagrange's formula) appears to be lost. This 'lost' energy, the difference between the mechanical energy stored during compression and the energy given out during expansion, has been converted into some other form. And this effects the reconciliation between our result and the general principle of the conservation of energy.

Nevertheless we must remember that, from the point of view of dynamics, there *is* a loss of energy on impact, just as there is when a body moves in a resisting medium, and in other cases of friction.

Example. *Pile driver. The pile, of mass M, is driven into the ground by dropping a hammer, called a* monkey, *on to it. The monkey falls through a distance h, and the impact is inelastic. The motion is opposed by a constant resisting force R. To find the depth z through which the pile is driven by one blow.*

The velocity of the monkey just before impact is u, and the velocity of the monkey and pile together just after impact is v, where

$$u^2 = 2gh, \quad v = \frac{m}{M+m}u.$$

The retardation (negative acceleration) is

$$f = \frac{R-(M+m)g}{M+m}$$

and $$z = \frac{v^2}{2f} = \frac{m}{M+m}\cdot\frac{mgh}{R-(M+m)g}.$$

Alternatively from the energy: the loss of potential energy is

$$mg(h+z)+Mgz,$$

and this is equal to the work done against the resistance, plus the loss on impact,

$$Rz + \frac{1}{2}\frac{Mm}{M+m}2gh.$$

On equating these we recover the same result. (The resistance R is called into play during the motion, and ceases when motion ceases. After the motion ceases the upward force supplied by the ground is simply $(M+m)g$).

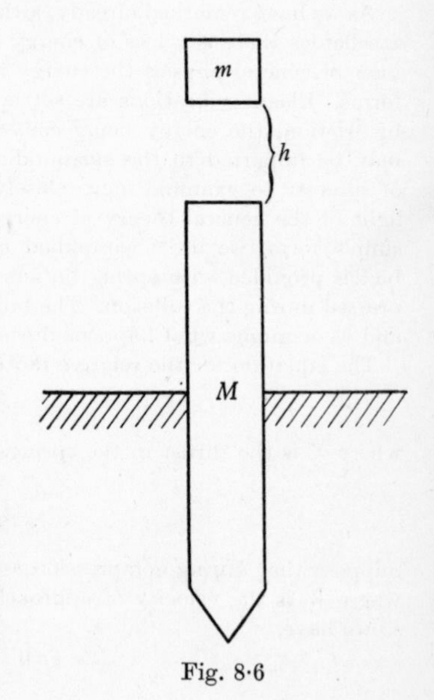

Fig. 8·6

8·7 The bouncing ball. When a particle impinges normally on a fixed wall the velocity of separation is $e\times$ (velocity of approach). This is merely Newton's law of impact. The result may be regarded as a limiting case of our general formulae when $u_2 = 0$ and $m_2 \to \infty$.

Example. *A ball is dropped from a height h on to a level floor. To find the time during which the ball is in motion, and the total distance travelled by it.*

If t_0 is the time taken to reach the floor, and v_0 the velocity on arrival, we have

$$v_0 = \sqrt{(2gh)}, \quad t_0 = v_0/g.$$

Now let v_1 be the velocity upwards just after the first bounce, and t_1 the time from the first bounce to the second. Then

$$v_1 = ev_0, \quad t_1 = 2v_1/g = 2et_0.$$

The downward velocity before the second bounce is v_1, so if v_2 is the upward velocity just after the second bounce, and t_2 the time from the second bounce to the third, we have

$$v_2 = ev_1 = e^2v_0, \quad t_2 = 2v_2/g = 2e^2t_0.$$

And generally, if v_n is the upward velocity just after the nth bounce, and t_n the time from the nth impact to the $(n+1)$th, we have, by induction,

$$v_n = e^nv_0, \quad t_n = 2e^nt_0.$$

The total time of motion is

$$t_0(1+2e+2e^2+2e^3+\ldots) = t_0\frac{1+e}{1-e} = \frac{1+e}{1-e}\sqrt{\left(\frac{2h}{g}\right)}.$$

After this time the ball is at rest on the floor. The distance travelled is

$$h+2\left(\frac{v_1^2}{2g}+\frac{v_2^2}{2g}+\ldots\right) = h(1+2e^2+2e^4+\ldots) = \frac{1+e^2}{1-e^2}h.$$

It is interesting to verify that the total loss of energy amounts to mgh; the loss is

$$\tfrac{1}{2}m(1-e^2)(v_0^2+v_1^2+v_2^2+\ldots) = \tfrac{1}{2}m(1-e^2)(1+e^2+e^4+\ldots)v_0^2 = \tfrac{1}{2}mv_0^2,$$

and this is equal to mgh as we wished to prove.

EXAMPLES VIII

1. A bullet of 0·1 lb. weight is fired with a speed of 2200 ft./sec. into the middle of a block of wood of 30 lb. weight, which is at rest but free to move. Find the speed of the block and bullet afterwards, and the loss of kinetic energy in ft. lb. What becomes of this energy?

2. Two marbles, A and B, lie on a smooth horizontal circular groove at opposite ends of a diameter. A is projected along the groove and after time t impinges on B; show that the second impact will occur after a further time $2t/e$, where e is the coefficient of restitution.

3. A smooth rectangular plank of mass M fits accurately in a smooth horizontal groove along which it is free to slide. The plank has a deep rectangular groove with vertical sides cut in its upper face, the edges of the groove being parallel to the edges of the plank, and in this groove a ball of mass m fits closely and runs freely. With the plank at rest the ball is projected with velocity V along the groove. The coefficient of restitution between the ball and the ends of the groove is e. Find an expression for the velocities of the plank and the ball just after the nth impact. Show that the velocities of the plank and the ball both tend to the same limit.

4. Two elastic spheres collide directly, one being initially at rest. Determine their velocities after impact.

A large cube, with its edge of length a, moves with velocity v in a direction perpendicular to one face through a medium formed of small spheres, each of mass m and of coefficient of elasticity e. The spheres are initially at rest and their mutual collisions can be neglected. Prove that when the number (n) of spheres per unit volume is large enough, the cube is practically subjected to a retarding force

$$(1+e)nma^2v^2.$$

5. A locomotive of mass M can exert a pull P. It starts into motion from rest a train of n trucks, each of mass m. The couplings are loose and inelastic so that each truck moves forward a distance a before jerking the next truck into motion. A similar coupling connects the engine to the first truck. Show that the velocity v_n of the train when the last truck has just been started into motion is given by

$$(M+nm)^2v_n^2 = Pan\{2M+(n-1)m\}.$$

6. A number n of imperfectly elastic spheres are suspended by long threads so that their centres are in the same horizontal line and their surfaces very nearly touching. They are all made of the same material (coefficient of restitution e), and their masses are $M, 2M, \ldots, nM$. If a velocity u in the line of centres is given to the smallest sphere, prove that the largest will be set in motion with velocity

$$V = u(1+e)^{n-1}\frac{(n-1)!}{1.3.5\ldots(2n-1)}.$$

Prove also that, if the velocity u is given to the largest sphere, the smallest will be set in motion with velocity nV.

7. The masses of three spheres A, B, C are $7m$, $7m$, m; their coefficient of restitution is unity. Their centres are in a straight line and C lies between A and B. Initially A and B are at rest and C is given a velocity along the line of centres in the direction of A. Show that it strikes A twice and B once, and that the final velocities of A, B, C are proportional to 21, 12, 1.

8. Two particles of masses m_1, m_2 are connected by a spring of such strength that when m_1 is held fixed, m_2 makes n complete vibrations per second. Prove that, if m_2 is held, m_1 will make $n\sqrt{(m_2/m_1)}$ vibrations per second, and that if both are free they will make $n\sqrt{\{(m_1+m_2)/m_1\}}$ vibrations per second. The vibrations in all cases are in the line of the spring.

9. Two particles of masses m_1 and m_2 are joined by a light elastic thread of natural length a and modulus λ. Initially they are lying on a smooth table at distance a apart. If one particle is given a velocity V in a direction directly away from the other, how long will it be before they collide?

10. Two particles, of masses m and M, lying on a smooth horizontal plane, are connected by a light elastic rod, of length l and modulus of elasticity λ. The particle m is struck by a blow B in the direction of the rod. Determine the subsequent motion, on the assumption that the mass of the rod may be neglected, and that consequently the stress in it is the same at all points; and show that, if its length at any time t is $l-x$, then x is given by the equation
$$\frac{Mm\dot{x}^2}{M+m}+\frac{\lambda x^2}{l}=\frac{MB^2}{m(M+m)}.$$

11. A projectile, of mass m, is fired horizontally from a gun, of mass M, which is free to recoil. The length of the barrel is l, and the force P exerted by the propellant has the constant value P_0 until the projectile has travelled a length $\frac{3}{4}l$ of the barrel, after which it steadily decreases from P_0 to zero so that P is proportional to the distance from the muzzle. Show that the square of the velocity of the projectile just after leaving the muzzle is
$$\frac{7MlP_0}{4m(M+m)}.$$

12. A short train consists of an engine of mass m_2 coupled to a single coach of mass m_1 whose bearings are smooth. Between the engine and the coach there are two pairs of spring buffers of negligible mass, one pair being on the engine and the other on the coach. The coupling is such that, with the train at rest and the coupling taut, the buffers on the engine are just in contact with the corresponding buffers on the coach but none of the buffers is compressed. Each buffer has a compliance C, compliance being the ratio of compression to force producing compression. When the train is in steady motion with uniform velocity along a straight track, brakes are applied, but only to the engine. The braking force is such that it would produce a retardation f, if there were no coach. Prove that, after application of the brakes, the separation between engine and coach oscillates with period $2\pi\sqrt{(\mu C)}$. Prove further that, if this oscillation is damped out, the effect of the retardation is to reduce the separation between engine and coach by μCf.

13. A light vertical spiral spring is suspended by one end and a weight A of 10 lb. is attached to the other. Another weight B of 20 lb. is hung from A by an inextensible string. The weight B is initially supported, and under the influence of A alone, the spring elongates 0·01 ft. B is then let go and acquires a velocity of 16 ft./sec. just before the string becomes taut. Calculate the maximum elongation produced in the spring.

14. Each of three particles A, B, C has mass m, and A is joined to B, and B to C by similar light springs of natural length a. The particles move in a straight line under no forces save the tensions of the springs. Show that if the lengths of AB, BC respectively at time t are denoted by $a+x$, $a+y$ respectively, then
$$\ddot{u}+n^2u=0, \quad \ddot{v}+3n^2v=0,$$
where $u=x+y$, $v=x-y$, and amn^2 is the tension required to double the length of either spring.

Hence determine the length of AB at any time if the system, originally at rest with the springs unstretched, is set in motion by an impulse I on the particle C in the direction AC.

15. Two particles A and B are attached to the ends of a light inextensible string passing over a smooth fixed pulley. In the ensuing motion A suffers no collision, but B strikes the ground, the coefficient of elasticity being e. Prove that, if $e < \frac{1}{2}$, B ceases to rebound before the string is again taut.

16. Masses P, Q of 4 and 5 kg. respectively are attached to the ends of the string of an Atwood's machine; the mass of the wheel is negligible. A rider R, of 2 kg., carried by P, can be removed by a ring through which P passes freely. The system is released from rest when P is at a point A above the ring. After P has passed through the ring, P comes (instantaneously) to rest at B, where $AB = h$. The time of travel from A to B is T and the velocity when R strikes the ring is u. Show that

$$T = \sqrt{(40h/g)}, \quad u = 2h/T.$$

After coming to rest at B, P ascends, picks up the rider, and comes (instantaneously) to rest at C. Show that $BC = 9h/11$, and explain why BC is less than h.

17. Two masses $M+m$ and M are connected by a light inextensible string which passes over a light pulley which is free to turn about a horizontal axis. A rider of mass $2m$ is placed on the smaller mass and the system is released from rest. When the velocity has attained the value V the rider is removed by a fixed inelastic ring which allows the mass M to pass through. The system continues to oscillate, the rider being picked up and deposited alternately. Show that the system will finally come to rest after a time

$$\frac{(2M+3m)(4M+3m)}{m^2}\frac{V}{g}.$$

18. Two particles of masses M and m are connected by a fine inextensible string passing over a fixed smooth pulley, and the motion of the heavier particle, M, is limited by a fixed horizontal inelastic plane, on which it can impinge. The system starts from rest with M at a given height above the plane; show that the successive heights of M at which it comes to instantaneous rest form a geometrical progression of ratio $\{m/(M+m)\}^2$, and that the whole time during which the system is in motion is three times the interval from the beginning of the motion to the first impact on the plane.

19. A long light inelastic string passes over a light frictionless pulley and carries a bucket of mass M at one end and a counterpoise of mass M at the other. The bucket and counterpoise are in equilibrium when a small elastic ball of mass m is dropped into the bucket so that it hits the horizontal bottom of the bucket with velocity u. Show that the ball ceases to bounce after a time

$$t_0 = \frac{2eu}{g(1-e)},$$

where e is the coefficient of restitution between the ball and the bucket.

By considering the fact that the total momentum of the system in the direction of the string is unaltered by impulsive tensions in the string, or otherwise, prove that the velocity of the bucket after time t_0 is

$$\frac{1+e}{1-e}\frac{mu}{2M+m}.$$

20. An inelastic pile of mass m is driven into the ground by a weight of mass M which falls on it. Show that at each blow a fraction $m/(M+m)$ of the kinetic energy of the moving weight is wasted.

When the clear fall of the weight in each stroke is 10 ft., and $M = 7m$, it is found that 15 strokes are necessary to drive the pile 7 in. Show that if the weight is doubled, 7 strokes only will be necessary.

21. A block of mass m falls vertically with velocity V on a pile of mass M. There is a coefficient of elasticity e, and the resistance of the earth to the movement of M is a constant force R. Show that the pile will come to rest before the second impact of the block upon it if

$$R > \frac{2Mg(M+m)e}{2Me-m(1-e)},$$

and that the distance it is driven in by the first impact is

$$\frac{m^2(1+e)^2 V^2}{2(m+M)^2\left(\dfrac{R}{M}-g\right)}.$$

22. A steel ball is released from rest and falls upon a fixed steel anvil and rebounds, the coefficient of restitution being 0·9. The lowest point of the ball is initially at a distance of 1 ft. above the anvil, and the gravitational acceleration is 32 ft.sec.$^{-2}$. Find the position and velocity of the ball half a second after its release.

Show that the ball finally comes to rest on the anvil 4·75 sec. after its release and that the total distance travelled by the ball is $\frac{181}{19}$ ft.

23. A horizontal rod of mass M is movable along its length, and its motion is controlled by a light spring which exerts a restoring force Ex when the rod is displaced through a distance x. A spider of mass m stands on the rod, and everything is initially at rest. The spider then runs a distance a along the rod, and then stops, his velocity relative to the rod being constant and equal to u. Show that the total energy of the system after the run is

$$\frac{2m^2u^2}{M+m}\sin^2\left[\frac{a}{2u}\sqrt{\left(\frac{E}{M+m}\right)}\right],$$

and find the amplitude of the final motion.

24. A particle of unit mass moves on the axis Ox under the action of a force $-kv^2-n^2x$, where k and n are positive constants. It is projected from the origin with speed $n/(k\sqrt{2})$. Prove that it moves with constant retardation.

CHAPTER IX

VARIABLE MASS

9·1 Variable mass, mass picked up from rest. So far we have supposed that the particle whose motion we are considering has constant mass. We consider now the rectilinear motion of a body that accumulates additional matter as it moves, and we suppose in the first instance that the added mass is picked up from rest; we may consider, for example, a raindrop gathering moisture as it falls through a cloud. To deal with problems of this type we shall need to introduce a new principle, whose validity must ultimately be tested experimentally.

It should be noticed that for most of the problems discussed in this chapter the traditional name 'variable mass' is something of a misnomer. The mass of the body varies as it moves because matter is acquired or ejected, not because the mass of any portion of matter actually changes. It is not until we reach § 9·4, where we consider the motion of an electron, that we encounter particles whose mass actually varies.

The problem is best approached through the theory of impulses. We suppose first that the increase of mass occurs by discrete amounts picked up from rest. Between two accretions, occurring at the instants t_1 and t_2, the mass remains constant, and we have

$$mv_2 - mv_1 = \int_{t_1}^{t_2} X\,dt,$$

where the force X acting on the body is supposed expressed as a function of t. The integral $\int_{t_1}^{t_2} X\,dt$ measures the increase in momentum in the interval (t_1, t_2). If we write p for the momentum mv we have

$$p(t_2) - p(t_1) = \int_{t_1}^{t_2} X\,dt.$$

Now when an increase of mass occurs *there is no discontinuity of momentum*, since the added mass has no momentum before the impact. It follows that $\int_{t_0}^{T} X dt$ *always* measures the increase of momentum in the interval (t_0, T), whether in this interval accretions of mass occur or not.

It is natural to suppose that this will be true in general when the increase of mass is continuous instead of by jumps; this is the new principle mentioned above. Thus we have, when the mass changes continuously,

$$p(T) - p(t_0) = \int_{t_0}^{T} X\,dt,$$

which implies as the equation of motion

$$X = \frac{dp}{dt} \tag{1}$$

for the case in which the added mass is picked up from rest. The rate of change of momentum in the second member replaces the kineton that appears in the ordinary case of constant mass.

For a raindrop falling through a stationary cloud we have, if v is the downward velocity,

$$mg = mf + v\frac{dm}{dt},$$

so that

$$f = g - \frac{1}{m}\frac{dm}{dt}v,$$

and, in comparison with the case of constant mass, the variation of mass produces a retardation somewhat similar to that in a resisting medium.

Notice that problems with variable mass, like problems with impulses, will in general involve dissipation of energy.

Example 1. *An engine of mass M picks up water from a trough between the rails.* If the mass of water in the tank at time t is m we have

$$\frac{d}{dt}\{(M+m)\,v\} = X - F,$$

where X is the driving force, and F the resistance due to friction. The equation can be written

$$(M+m)f + v\frac{dm}{dt} = X - F.$$

Example 2. *A raindrop of mass $3kV/g$ is falling vertically with speed $2V$ when it overtakes a cloud which is falling vertically with constant speed V. During its passage through the cloud the mass of the raindrop increases by condensation from the cloud at a constant rate k, and the force of resistance is k times the relative speed. Find the differential equation relating the mass of the drop when in the cloud to the relative velocity, and hence express the relative velocity in terms of the mass. Find the displacement relative to the cloud at any time.*

Take the cloud as base (it is still a Newtonian base, since it moves uniformly and without rotation relative to the earth, which is here supposed a Newtonian base) and denote the velocity of the drop relative to the cloud, measured downwards, by w. If m is the (variable) mass of the drop, the equation of motion is

$$\frac{d}{dt}(mw) = mg - kw,$$

and since $\dfrac{d}{dt} \equiv k\dfrac{d}{dm}$, this can be written

$$\frac{d}{dm}(mw) + w = \frac{g}{k}m,$$

i.e.

$$m\frac{dw}{dm} + 2w = \frac{g}{k}m.$$

It is obvious that m is an integrating factor, and if we multiply by m and integrate we get

$$m^2 w = \frac{g}{3k}m^3;$$

the constant of integration is zero, since $m = \dfrac{3k}{g}w$ initially. Thus the relation connecting m and w is

$$w = \frac{g}{3k}m.$$

Since

$$m = \left(\frac{3V}{g} + t\right)k$$

this gives us, for the displacement z relative to the cloud,

$$\dot{z} = w = V + \frac{g}{3}t,$$

$$z = Vt + \frac{g}{6}t^2,$$

measuring z from the top of the cloud, and taking $t = 0$ at the instant when the drop enters the cloud.

9·2 Variable mass, the general problem.

We consider now a problem of a more general type. The motion is still entirely on a straight line, but we suppose that the added mass is not necessarily at rest before it is picked up, and that matter is ejected as well as absorbed. The new theory will enable us to deal not only with problems where the matter picked up is itself in motion, but also with problems such as that of the motion of a rocket, or of an engine with a leaking tank. We again begin by considering the problem of discrete changes of mass, and then derive the equation of motion for continuous changes by a limiting process.

Suppose then that in an interval of time (t_0, T) changes of mass occur at the instants $t_1, t_2, \ldots, t_{n-1}$, where $t_0 < t_1 < t_2 < \ldots < t_{n-1} < t_n = T$. At the instant t_r a mass δm_1^r is picked up, its velocity immediately before the impact being U_1^r, and that at the same instant a mass δl_2^r is ejected, its velocity immediately after ejection being U_2^r. All these velocities are in the same straight line. The momentum p of the body is now discontinuous at t_r, indeed,

$$p(t_r+0)-p(t_r-0) = U_1^r \delta m_1^r - U_2^r \delta l_2^r.$$

Considering the interval (t_{r-1}, t_r), in which the mass remains constant, we have

$$p(t_r-0)-p(t_{r-1}+0) = \int_{t_{r-1}}^{t_r} X\,dt,$$

and summing over the n intervals we obtain

$$\begin{aligned} p(T)-p(t_0) &= \int_{t_0}^{T} X\,dt + \sum_{r=1}^{n-1} \{p(t_r+0)-p(t_r-0)\} \\ &= \int_{t_0}^{T} X\,dt + \sum_{r=1}^{n-1} (U_1^r \delta m_1^r - U_2^r \delta l_2^r) \\ &= \int_{t_0}^{T} X\,dt + \sum_{r=1}^{n-1} (U_1^r \delta m_1^r + U_2^r \delta m_2^r), \end{aligned}$$

where we have written $-\delta m_2^r$ for δl_2^r.

We now make an assumption analogous to that made above—that we may replace the summation by an integration when the change is continuous. We thus obtain, for the general case of continuous change of mass,

$$p(T)-p(t_0) = \int X\,dt + \int U_1 dm_1 + \int U_2 dm_2,$$

where we think of the mass m as a sum $m_0+m_1+m_2$, m_1 varying by accretion, and m_2 by ejection. This gives as the general equation of motion

$$\frac{dp}{dt} = X + U_1 \frac{dm_1}{dt} + U_2 \frac{dm_2}{dt},$$

where U_1 is the velocity of the mass absorbed immediately before absorption, and U_2 is the velocity of the mass ejected immediately after ejection. Now

$$\frac{dp}{dt} = m\frac{dv}{dt} + v\left(\frac{dm_1}{dt} + \frac{dm_2}{dt}\right),$$

and the equation of motion takes the final form

$$X = m\frac{dv}{dt} + (v-U_1)\frac{dm_1}{dt} + (v-U_2)\frac{dm_2}{dt}. \tag{1}$$

Example 1. *Consider again, this time as an application of equation* (1), *the problem of the raindrop* (*Example* 2 *of* § 9·1), *this time using fixed axes.*

If v is the velocity of the drop downwards at time t we have, since $U_1 = V$,

$$m\frac{dv}{dt} + (v-V)\frac{dm}{dt} = mg - k(v-V),$$

and if we write w for $v-V$ this becomes

$$\frac{d}{dt}(mw) = mg - kw,$$

as before.

Example 2. *An engine with a leaking tank.* Here $U_2 = v$, and the equation of motion is

$$m\frac{dv}{dt} = X.$$

Example 3. *A machine-gun of mass M stands on a horizontal plane and contains shot of mass M'. The shot is fired, at a rate of mass k per unit of time, with velocity u relative to the ground. If the coefficient of sliding friction between the gun and the plane is μ, show that the velocity of the gun backward by the time the mass M' is fired is*

$$\frac{M'}{M}u - \frac{(M+M')^2 - M^2}{2kM}\mu g.$$

If v is the backward velocity of the gun at time t after the start we have, since $dm_2/dt = -k$,

$$m\frac{dv}{dt} - (v+u)k = -\mu mg,$$

where $m = M + M' - kt$. Thus

$$(M+M'-kt)\frac{dv}{dt} - (v+u)k = -\mu g(M+M'-kt),$$

whence, on integrating,

$$(M+M'-kt)\,v = ukt - \mu g\{(M+M')\,t - \tfrac{1}{2}kt^2\}.$$

The value of v when $kt = M'$ is V, where

$$V = \frac{M'}{M}u - \frac{(M+M')^2 - M^2}{2kM}\mu g.$$

9·3 Alternative method. Another approach to problems of this type is worthy of notice. Consider first a system of bodies in motion on a straight line, and suppose that occasional collisions occur, which may involve interchanges of mass between one body and another. Now in the interval of time between two collisions the increase in the momentum of the system is $\int X dt$, where X is the sum of the forces on all the bodies. Moreover, at a collision there is no change in the momentum of the system, so that, just as in § 9·1, $\int X dt$ measures the change of momentum of the system in any interval, whether collisions occur in the interval or not. If we now assume as before that the same is true in the limiting case of a continuous interchange of mass, we obtain the equation

$$p(T) - p(t_0) = \int_{t_0}^{T} X\,dt, \tag{1}$$

whence also

$$X = \frac{dp}{dt}, \tag{2}$$

X being the total force on the system, and p the total momentum.

Example 1. *As an illustration of this line of thought consider again the general problem of* § 9·2.

The mass of the body at time t is $(m_0 + m_1 + m_2)$, and at a later time $t + \delta t$ it is $(m_0 + m_1 + m_2 + \delta m_1 + \delta m_2)$, where of course $\delta m_1 \geqslant 0$ and $\delta m_2 \leqslant 0$.

The mass with which we are concerned is $(m_0+m_1+m_2+\delta m_1)$, the momentum at the instant t is

$$(m_0+m_1+m_2)v+U_1\,\delta m_1,$$

and at $t+\delta t$ it is

$$(m_0+m_1+m_2+\delta m_1+\delta m_2)\,(v+\delta v)-U_2\,\delta m_2.$$

Thus the change of momentum in the interval δt is

$$(m_0+m_1+m_2)\,\delta v+(v-U_1)\,\delta m_1+(v-U_2)\,\delta m_2+O(\delta t^2),$$

and this differs from $X\,\delta t$ by $O(\delta t^2)$. Letting $\delta t\to 0$ we recover the equation (1) of § 9·2, viz.

$$X=m\frac{dv}{dt}+(v-U_1)\frac{dm_1}{dt}+(v-U_2)\frac{dm_2}{dt}.$$

Example 2. *Consider again as an illustration of equation* (2) *the machine-gun problem, Example* 3 *of* § 9·2.

Here
$$p=ktu-(M+M'-kt)v,$$

where v is the backward velocity of the gun, and

$$X=\mu(M+M'-kt)g.$$

Thus, from equation (2), we have

$$\frac{d}{dt}\{ktu-(M+M'-kt)v\}=\mu(M+M'-kt)g.$$

Hence, since $v=0$ at $t=0$, we have

$$ktu-(M+M'-kt)v=\mu g\{(M+M')t-\tfrac{1}{2}kt^2\},$$

as before.

9·4 Particle whose mass is a function of its speed.

An example of a different type is provided by an electron, whose mass is not constant, but a function of the speed v,

$$m=\frac{m_0}{\sqrt{\left(1-\frac{v^2}{c^2}\right)}},\tag{1}$$

where c is a positive constant, the velocity of light. Here the variation of mass is not effected by interchange with other bodies as in the previous examples, but is inherent in the principle of relativity, and the appropriate equation of motion on a straight line cannot be derived by the kind of reasoning previously used. Nevertheless the equation (1) of § 9·1

$$X=\frac{d}{dt}(mv)\tag{2}$$

is still valid, the results based on it being verified in the experimental work.

Consider then the motion of a particle whose mass is a function of its velocity, say $m=\phi(v)$, and let us suppose, for definiteness, that the particle moves in a field of force, $X=X(x)$. The equation of motion is

$$\frac{d}{dt}(v\phi)=X,\tag{3}$$

where we now write ϕ rather than m for the mass. (This notation will help us to remember, not only that the mass is not constant, but also that it is a given function of v.) The equation (3) can be written

$$v\frac{d}{dx}(v\phi) = X, \tag{4}$$

and on integration by parts we have

$$[v^2\phi]_{v_1}^{v_2} - \int_{v_1}^{v_2} v\phi\,dv = \int_{x_1}^{x_2} X\,dx = V(x_1) - V(x_2), \tag{5}$$

where V is the potential of the field; of course V is such that the force derived from it is the actual force, not the force per unit mass.

We write
$$T = v^2\phi(v) - \int_0^v \theta\phi(\theta)\,d\theta, \tag{6}$$
and equation (5) takes the form

$$T+V = C. \tag{7}$$

If $\phi(v)$ is constant the formula (6) falls back on the familiar form $\frac{1}{2}mv^2$. The new 'equation of energy' (7) is the (x, v) relation for a mass which is a function of v moving in a field of force.

9·5 Motion of an electron.

Consider now more particularly the case of the electron, where

$$\phi(v) = \frac{m_0}{\sqrt{(1-v^2/c^2)}}.$$

We will consider first the simple case of a uniform field F, and then a field of more general type.

(i) The uniform field; for example, if the electron carries a charge $-e$ and moves in a uniform electric field $-E$, then $F = eE$. Here

$$\frac{d}{dt}\left[\frac{m_0 v}{\sqrt{(1-v^2/c^2)}}\right] = F,$$

or say
$$\frac{d}{dt}\left[\frac{v/c}{\sqrt{(1-v^2/c^2)}}\right] = \frac{F}{m_0 c} = k.$$

If the electron starts from rest at $t = 0$ we have

$$\frac{v/c}{\sqrt{(1-v^2/c^2)}} = kt,$$

whence
$$\frac{v^2}{c^2} = 1 - \frac{1}{1+k^2t^2}. \tag{1}$$

We notice that v/c is always less than 1, and $v/c \to 1$ as $t \to \infty$; the velocity approaches but never reaches the velocity of light.

The (t, x) relation is readily found. We have

$$\frac{dx}{dt} = \frac{ckt}{\sqrt{(1+k^2t^2)}},$$

and, if $x = 0$ at $t = 0$, this leads to

$$x = \frac{c}{k}[\sqrt{(1+k^2t^2)} - 1], \tag{2}$$

and x approximates to ct as $t \to \infty$. If we write

$$kt = \sinh \theta,$$

the relations (1) and (2) take the simpler forms

$$v = c \tanh \theta,$$

$$kx = c(\cosh \theta - 1).$$

We can now easily obtain the (x, v) relation. We can either eliminate t between equations (1) and (2), or quote the equation of energy, equation (7) of § 9·4. Either method gives us immediately

$$\frac{v^2}{c^2} = 1 - \left(\frac{c}{c+kx}\right)^2.$$

We notice that $$m - m_0 = m_0\left[\frac{1}{\sqrt{(1-v^2/c^2)}} - 1\right] = \frac{m_0 k}{c} x,$$

and the increase in mass is proportional to the distance travelled. This is a special case of a more general law which we shall find in a moment (equation (3) below).

(ii) The general field. We have, from equation (6) of § 9·4,

$$T = \frac{m_0 v^2}{\sqrt{(1-v^2/c^2)}} - m_0 c^2 \int_0^v \frac{\theta/c^2}{\sqrt{(1-\theta^2/c^2)}}\, d\theta = m_0 c^2\left[\frac{1}{\sqrt{(1-v^2/c^2)}} - 1\right] = c^2(m - m_0),$$

and the 'kinetic energy' T is $c^2 \times$ (increase in mass). Suppose the electron starts from rest at $x = 0$ at the instant $t = 0$; we may suppose without loss of generality that $V(0) = 0$. Then the energy equation, equation (7) of § 9·4, gives us

$$c^2(m - m_0) + V = 0. \tag{3}$$

(For the particular case of a uniform field we recover the result $(m - m_0) \propto x$.)

Equation (3) can also be written in the form

$$\frac{v^2}{c^2} = 1 - \left(1 - \frac{V}{m_0 c^2}\right)^{-2},$$

and this is the (x, v) relation for the problem.

EXAMPLES IX

1. A jet of water, moving at a speed of 64 ft./sec., impinges normally, without appreciable rebound, on a vertical door. If the force exerted on the door is 250 lb. weight, find in square inches the cross-sectional area of the jet.

If this water is being pumped from a pond whose surface is 20 ft. below the jet, find the horse-power at which the pump is working.

(Take g to be 32 ft./sec./sec., and assume that the mass of a cubic foot of water is $62\frac{1}{2}$ lb. Frictional losses are to be neglected.)

2. A particle whose mass at time t is $m_0(1+\alpha t)$ is projected vertically upwards at time $t = 0$ with velocity u, the added mass being picked up from rest. Show that it rises to a height

$$\frac{g + 2\alpha u}{4\alpha^2} \log\left(1 + \frac{2\alpha u}{g}\right) - \frac{u}{2\alpha}.$$

3. A raindrop falls from rest through an atmosphere containing water vapour at rest. The mass of the raindrop is initially m and increases by condensation uniformly with the time in such a way that after a given time T it is equal to $2m$. The motion is opposed by a frictional resistance equal to $\lambda m/T$ times the velocity, where λ is a positive constant. Show that after time T the velocity is

$$\frac{gT}{2+\lambda}\{2-2^{-(1+\ )}\}.$$

4. A raindrop falls through a stationary cloud, its mass m increasing by accretion uniformly with the distance fallen, $m = m_0(1+kx)$. The motion is opposed by a resisting force $m_0 k\lambda v^2$ proportional to the square of the speed v. If $v = 0$ when $x = 0$ prove that

$$v^2 = \frac{2g}{(3+2\lambda)k}\left\{1+kx-\frac{1}{(1+kx)^{2+2\lambda}}\right\}.$$

5. A small raindrop falling through a stationary cloud acquires moisture by condensation from the cloud. When the mass of the raindrop is m, the rate of increase of mass per unit time is km. The raindrop starts from rest. Neglecting resistance to motion, find the relation between the velocity v and the distance fallen x; and prove that if k is small the velocity is given approximately by

$$v^2 = 2gx\left[1-\frac{2}{3}k\sqrt{\left(\frac{2x}{g}\right)}\right].$$

6. A train of mass M is moving with velocity V when it begins to pick up water from rest at a uniform rate. The power is constant and equal to H. If after time t a mass m of water has been picked up, find the velocity and show that the loss of energy is

$$\frac{m(Ht+MV^2)}{2(m+M)}.$$

7. Reconsider Example 3 of § 9·2 on the assumption that the velocity of the shot *relative to the gun* is constant and equal to u.

8. Two buckets of water each of total mass M are suspended at the ends of a cord passing over a smooth pulley and are initially at rest. Water begins to leak from a small hole in the side of one of the buckets at a steady slow rate of m units of mass per second. Establish the equations of motion, and prove that the velocity V of the bucket when a mass M' of water has escaped is given by

$$V = \frac{2Mg}{m}\log\frac{2M}{2M-M'}-\frac{gM'}{m}.$$

9. An engine contains a quantity of fuel which is being steadily consumed at the rate of m units of mass per unit time, the products of combustion being ejected with the speed of the engine. There is a constant propulsive force $k\alpha^2$, and a resistance kv^2, where v is the velocity at time t. If at $t = 0$ the mass is M and the velocity is zero, prove that

$$v = \alpha\left\{\frac{M^\lambda-(M-mt)^\lambda}{M^\lambda+(M-mt)^\lambda}\right\},$$

where $\lambda = 2k\alpha/m$.

10. A uniform chain of length l and weight wl is suspended by one end and the other end is at a height h above a smooth inelastic table. Prove that if the upper end is let go the pressure on the table as the coil is formed increases from $2hw$ to $(2h+3l)w$.

11. A uniform perfectly flexible chain is coiled at the edge of a table with one end just hanging over. Prove that, if a length x of the chain has fallen over the edge at time t after the start, then $6x = gt^2$. How much energy has been dissipated in this time?

12. A uniform chain, 30 cm. long, having a mass of 1 g./cm., lies partly in a straight line along a rough horizontal table, perpendicular to the edge. The portion hanging over the edge is just sufficient to cause the chain to commence to slip. The coefficient of friction with the table being $\frac{1}{2}$, find the velocity of the chain and its tension at the edge of the table when x cm. have slipped off, the part of the chain off the table being constrained by a smooth wall to remain vertical.

13. A uniform fine chain of length $3l/2$ and mass $3ml/2$ hangs over a small smooth peg at a height l above a horizontal table. The chain is released from rest in the position in which it hangs in two vertical straight pieces with one end just touching the table. Show that when the other end is leaving the peg the force on the table is

$$mgl(4\log\tfrac{3}{2}-\tfrac{1}{2}).$$

14. Two scale-pans of equal mass m are connected by a light string passing freely over a smooth peg and hang in equilibrium under gravity. A uniform flexible chain of length l and total mass m is held by one end over one of the scale-pans, so that the lower end is just in contact with it, and is then released. Show that, if the impacts of the elements of the chain with the scale-pan are inelastic, the chain will fall into the scale-pan in time $\sqrt{(5l/2g)}$, and find the speed of the scale-pan at this instant.

15. An electron moves on a straight line under a harmonic attraction μm_0 times the distance from the centre of attraction. Prove that the period of an oscillation of amplitude a is

$$\frac{4}{c}\int_0^a \frac{\alpha^2-x^2}{\sqrt{[(a^2-x^2)(\beta^2-x^2)]}}\,dx,$$

where $\alpha^2 = a^2+(2c^2/\mu)$, $\beta^2 = a^2+(4c^2/\mu)$.

CHAPTER X

MOTION OF A PARTICLE IN A PLANE

10·1 The fundamental vectors. When we studied the motion of a particle on a straight line the position of the particle at time t was defined by the scalar quantity whose measure is x. When we turn to problems of motion in a plane this scalar quantity is replaced by the vector quantity whose vector is $\mathbf{r}$, the position vector of the particle relative to a Newtonian base in the plane. We fix a definite sense Oz in the normal to the plane as the *positive normal* (in the diagrams the positive normal is conventionally drawn from the paper towards the reader), and we speak of a right-handed rotation about the positive normal as a positive (anticlockwise) rotation. We take axes Ox, Oy fixed in the base. Usually we take rectangular axes, and the positive directions in the axes are so chosen that a line OL rotating in the plane suffers a positive rotation through one right angle when it moves from Ox to Oy. (In the diagrams we conventionally draw Ox to the right and Oy upwards.) When we use oblique axes we conventionally choose the positive sense in Oy, when the positive sense in Ox has been fixed, so that a positive rotation through an angle less than π carries OL from Ox to Oy. The position vector $\mathbf{r}$ represents the displacement OP of the particle relative to the base. We denote the (constant) mass of the particle by m.

From the *velocity*, the vector quantity whose vector is $\dot{\mathbf{r}}$ or $\mathbf{v}$, we form the *momentum*, whose vector is $m\mathbf{v}$. Momentum is a vector quantity of a different kind from velocity in the same direction as velocity. From the *acceleration*, the vector quantity whose vector is $\ddot{\mathbf{r}}$ or $\dot{\mathbf{v}}$ or $\mathbf{f}$, we form the *kineton*, whose vector is $m\mathbf{f}$. Kineton is a vector quantity of a different kind from acceleration in the same direction as acceleration. All these vectors lie in the plane.

Consider now more particularly the vector $\dot{\mathbf{r}}$ or $\mathbf{v}$. We have already seen that, if the particle moves along a smooth rectifiable curve, and we denote by $\mathbf{T}$ a unit vector along the tangent to the curve, drawn in the direction of increasing s, then $\mathbf{v} = \dot{s}\mathbf{T}$. (Cf. § 2·9.)

Suppose now that we have any vector function of t, say $\mathbf{G}(t)$, and we represent this vector by the position vector OA in a vector diagram. Then the velocity of A in this diagram represents the derivative $\dot{\mathbf{G}}$ of $\mathbf{G}$; for the velocity of A is the rate of change of the position vector OA, and this position vector now represents $\mathbf{G}$. In particular, if OA represents the *velocity* of a particle P then the velocity of A in the vector diagram represents the *acceleration* of P. Such a diagram, in which OA is a velocity vector, is called a *hodograph*.

Next let OA in a vector diagram represent a unit vector $\mathbf{E}$. Then the path of A is a circle. We denote the polar angle of OA, measured positively from Ox, by θ, and then the length of the arc A_0A is θ (Fig. 10·1a). Then $\dot{\mathbf{E}}$ is the velocity of A, which is of magnitude $\dot{\theta}$ along the tangent to the circle. If we denote by $\mathbf{E}'$ the unit vector obtained from $\mathbf{E}$ by a positive rotation through one right angle we have $\dot{\mathbf{E}} = \dot{\theta}\mathbf{E}'$. We notice also, for future reference, that $\mathbf{E}'' = -\mathbf{E}$, and if we write $\mathbf{F}$ for $\mathbf{E}'$ we have the important formulae

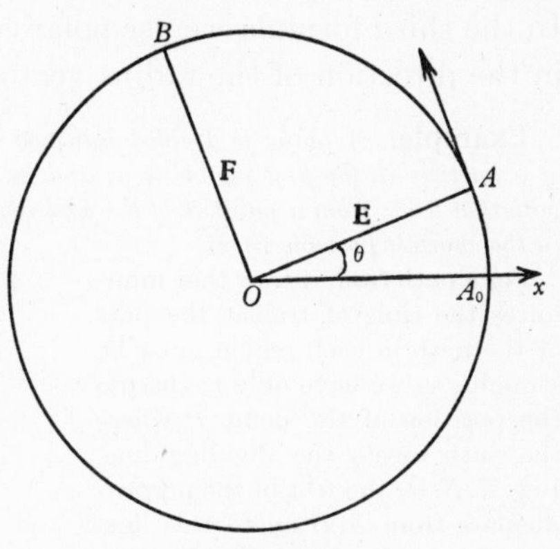

Fig. 10·1a

$$\dot{\mathbf{E}} = \dot{\theta}\mathbf{F}, \quad \dot{\mathbf{F}} = -\dot{\theta}\mathbf{E}. \tag{1}$$

We now determine the radial and transverse components of velocity when the position of the particle P is defined by the polar coordinates (r, θ). We have

$$\mathbf{r} = r\mathbf{E},$$

where $\mathbf{E}$ is a unit vector in the direction OP. Then

$$\mathbf{v} = \dot{\mathbf{r}} = \dot{r}\mathbf{E} + r\dot{\mathbf{E}} = \dot{r}\mathbf{E} + r\dot{\theta}\mathbf{F},$$

and the radial and transverse components of velocity are $\dot{r}$ and $r\dot{\theta}$.

Thus, collecting together the results, we have three formulae for velocity, and all of these will turn out to be important; they are

$$\mathbf{v} = \dot{x}\mathbf{i} + \dot{y}\mathbf{j} \tag{2}$$

$$= \dot{s}\mathbf{T} \tag{3}$$

$$= \dot{r}\mathbf{E} + r\dot{\theta}\mathbf{F}. \tag{4}$$

Here $\mathbf{i}, \mathbf{j}$ are unit vectors along the axes of coordinates, and the first formula is valid whether the axes are rectangular or not. In the second formula $\mathbf{T}$ is a unit vector in the direction of the tangent to the path of the particle in the sense of increasing s. In the third formula we use polar coordinates; $\mathbf{E}$ is a unit vector in the direction of the radius vector OP, and $\mathbf{F} = \mathbf{E}'$.

Example. *A plane is divided into two regions by a straight line; the speed of a particle in the first region is u_1 and in the second region it is u_2. Find the path that leads from a point A of the first region to a point B of the second region in the shortest possible time.*

For a path from A to B that minimizes the time of transit the part of the path in each region must be straight, so we have only to discuss the position of the point P where the path meets the dividing line. Let M, N be the feet of the perpendiculars from A, B on to this line, and denote the distance MN by b and the distances AP, PB by r_1, r_2. The time of transit from A to B is

$$t = \mu_1 r_1 + \mu_2 r_2,$$

where $\mu_1 = 1/u_1$ and $\mu_2 = 1/u_2$. We will therefore study the behaviour of this function t as P moves on the dividing line.

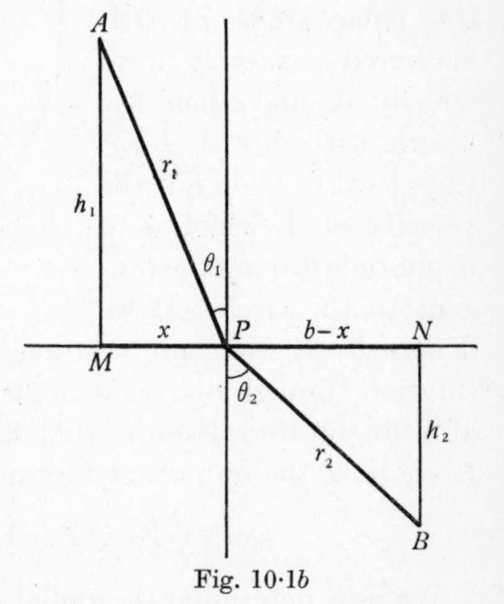

Fig. 10·1b

If M and N coincide it is evident that the straight line AB gives a minimum value to t. Suppose then that M and N are distinct points. Take M as origin and MN as the positive sense on the line, and denote MP by x. We shall see that as x increases from $-\infty$ to $+\infty$ the time t has a unique minimum when P is at a point P_0 between M and N. There is a unique minimizing path AP_0B.

As x increases from $-\infty$ to 0, both r_1 and r_2 steadily decrease, so t steadily decreases; and similarly as x increases from b to $+\infty$, t steadily increases. Therefore t, which is a continuous function of x, attains its least value in the range $0 \leqslant x \leqslant b$. We have

$$\frac{dt}{dx} = \mu_1 \sin\theta_1 - \mu_2 \sin\theta_2,$$

where we have used the formulae

$$\frac{dr_1}{dx} = \frac{x}{r_1} = \sin\theta_1, \quad \frac{dr_2}{dx} = -\frac{b-x}{r_2} = -\sin\theta_2. \quad \text{(See Fig. 10·1}b\text{.)}$$

Now as x increases from 0 to b, $\sin\theta_1/\sin\theta_2$ steadily increases from 0 to ∞, and passes just once through the value μ_2/μ_1 when $x = x_0$. For $0<x<x_0$, $dt/dx<0$, and for $x_0<x<b$, $dt/dx>0$, so there is a unique minimum of t at $x = x_0$. The minimizing path is the path AP_0B, where P_0 is given by the equation

$$\mu_1 \sin\theta_1 = \mu_2 \sin\theta_2.$$

Alternatively we can show that the stationary value of t is a minimum by noticing that

$$\begin{aligned}\frac{d^2t}{dx^2} &= \frac{d}{dx}\left(\mu_1\frac{x}{r_1} - \mu_2\frac{b-x}{r_2}\right)\\ &= \mu_1\left(\frac{1}{r_1} - \frac{x^2}{r_1^3}\right) - \mu_2\left(-\frac{1}{r_2} + \frac{(b-x)^2}{r_2^3}\right)\\ &= \frac{\mu_1 h_1^2}{r_1^3} + \frac{\mu_2 h_2^2}{r_2^3},\end{aligned}$$

and $d^2t/dx^2>0$ at $x = x_0$.

The result has an important significance in geometrical optics. It exhibits a relation between Fermat's principle of least time and Snell's law of refraction.

10·2 Acceleration.

We derive very simply the formulae for acceleration corresponding to the formulae for velocity found in § 10·1.

Using Cartesian axes, whether rectangular or not, we have

$$\mathbf{f} = \ddot{x}\mathbf{i} + \ddot{y}\mathbf{j},$$

so $\ddot{x}\mathbf{i}$ and $\ddot{y}\mathbf{j}$ are the components in the directions of the axes.

Next, in terms of the path, we have

$$\begin{aligned}\mathbf{v} &= \dot{s}\mathbf{T},\\ \mathbf{f} &= \ddot{s}\mathbf{T} + \dot{s}\dot{\mathbf{T}}\\ &= \ddot{s}\mathbf{T} + \dot{s}\dot{\psi}\mathbf{T}'.\end{aligned}$$

The components of acceleration along tangent and normal are $\ddot{s}$, $\dot{s}\dot{\psi}$, where ψ is the inclination of the tangent to Ox. Suppose, to fix the ideas, that $\dot{s}>0$. Then the normal component $\dot{s}\dot{\psi}$ has the sign of $\dot{\psi}$; if $\dot{\psi}>0$, $\mathbf{T}'$ is along the inward normal to the curve (i.e. towards the centre of curvature), and if $\dot{\psi}<0$, $\mathbf{T}'$ is along the outward normal. We include all cases very simply by saying that the normal component of acceleration is $|\dot{s}\dot{\psi}|$ along the inward normal to the curve.

There are equivalent forms of these formulae which are also very important. If we write v for $\dot{s}$ we can write for the tangential component $v\dfrac{dv}{ds}$ or $\dfrac{d}{ds}(\tfrac{1}{2}v^2)$, just as for the rectilinear case. Observe again that the second of these formulae is more general than the first, since it is valid for a value of s for which $v = 0$ and dv/ds does not exist. For the normal component we can write $|\dot{s}|/\rho$ in place of $|\dot{\psi}|$, where ρ is the radius of curvature, and the normal component is therefore v^2/ρ along the inward normal.

In polars
$$\mathbf{v} = \dot{r}\mathbf{E} + r\dot{\theta}\mathbf{F},$$
$$\begin{aligned}\mathbf{f} &= \ddot{r}\mathbf{E} + \dot{r}\dot{\mathbf{E}} + (r\ddot{\theta} + \dot{r}\dot{\theta})\mathbf{F} + r\dot{\theta}\dot{\mathbf{F}}\\ &= \ddot{r}\mathbf{E} + \dot{r}\dot{\theta}\mathbf{F} + (r\ddot{\theta} + \dot{r}\dot{\theta})\mathbf{F} - r\dot{\theta}^2\mathbf{E}\\ &= (\ddot{r} - r\dot{\theta}^2)\mathbf{E} + (r\ddot{\theta} + 2\dot{r}\dot{\theta})\mathbf{F}.\end{aligned}$$

The radial and transverse components are $\ddot{r} - r\dot{\theta}^2$ and $r\ddot{\theta} + 2\dot{r}\dot{\theta}$. These important formulae are of wide applicability.

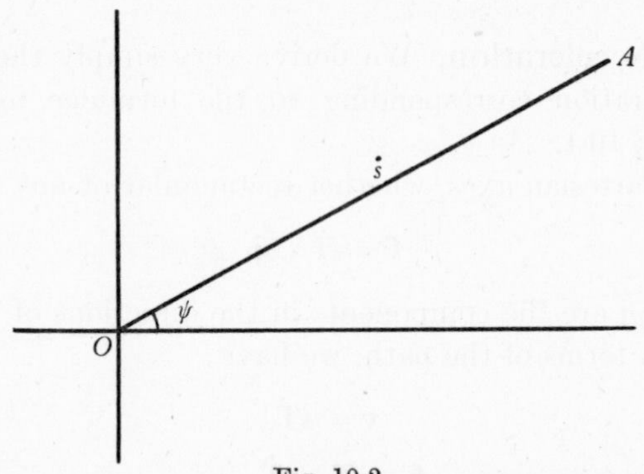

Fig. 10·2

We can also find the formulae for acceleration by means of the hodograph. Take for example the formulae in terms of the path. We have to take a vector diagram in which OA represents velocity, and then the velocity of A represents acceleration. Now the polar coordinates of A in the vector diagram are $(\dot{s}, \psi)$, so, from the formulae for velocity in polar coordinates the tangential and normal components of velocity of A are $\ddot{s}, \dot{s}\dot{\psi}$. We thus recover the formulae already found for the tangential and normal components of acceleration.

Collecting up our results for the acceleration we have

$$\mathbf{f} = \ddot{x}\mathbf{i} + \ddot{y}\mathbf{j} \tag{1}$$

$$= \ddot{s}\mathbf{T} + \dot{s}\dot{\psi}\mathbf{T}' \tag{2}$$

$$= \frac{d}{ds}(\tfrac{1}{2}v^2)\,\mathbf{T} + \frac{v^2}{\rho}\mathbf{N} \tag{3}$$

$$= (\ddot{r} - r\dot{\theta}^2)\,\mathbf{E} + (r\ddot{\theta} + 2\dot{r}\dot{\theta})\,\mathbf{F}. \tag{4}$$

The first refers to Cartesian axes, whether rectangular or not. In the second $\mathbf{T}$ is a unit vector in the direction of the tangent to the path of the particle in the sense of increasing s. In the third $\mathbf{N}$ is a unit vector along the inward normal. In the fourth we use polar coordinates; $\mathbf{E}$ is a unit vector in the direction of the radius vector, and $\mathbf{F} = \mathbf{E}'$.

10·3 Motion in a circle. If the particle moves in a circle of radius a the components of acceleration are $(a\ddot{\theta}, a\dot{\theta}^2)$ along the tangent and along the inward normal. This we derive immediately either from the second or the fourth of the formulae of the preceding paragraph. In particular, for uniform motion in a circle ($\dot{\theta}$ = constant = ω) the acceleration is towards the centre, and its constant magnitude can be expressed as $a\omega^2$ or as v^2/a.

10·4 Acceleration when v is given in terms of r. The reader may be tempted to ask, what is the analogue in plane motion of the formula $v\frac{dv}{dx}$ for the acceleration in rectilinear motion? This formula was applicable when we thought of v as a function of position rather than of time. When we consider instead motion in a plane, v is replaced by the vector $\mathbf{v}$, and x by the vector $\mathbf{r}$. If we use Cartesian axes, denoting the components of $\mathbf{r}$ by (x, y), and the components of $\mathbf{v}$ by (u, v), the components of acceleration are $\left(u\frac{\partial u}{\partial x} + v\frac{\partial u}{\partial y},\ u\frac{\partial v}{\partial x} + v\frac{\partial v}{\partial y}\right)$. But these formulae are somewhat misleading, since u, for example, considered as a function of x and y, is not defined for all (x, y) in some domain, but only for points on the path of the particle. It is more natural to think of u and v as functions of the single variable s, the arc-length of the path, and then the components of acceleration are

$$\left(w\frac{du}{ds},\ w\frac{dv}{ds}\right),\ \text{where } w = \dot{s}.$$

10·5 Relative motion. Suppose we have two points A and B moving in the plane. If $\mathbf{r}_1$ is the position vector of A, and $\mathbf{r}_2$ is the position vector of B, we say that $\mathbf{s} = \mathbf{r}_2 - \mathbf{r}_1$ is the position vector of B relative to A. We call $\dot{\mathbf{s}}$ the velocity of B relative to A, and $\ddot{\mathbf{s}}$ the acceleration of B relative to A.

Suppose we take rectangular axes Ox, Oy in the plane. If the components of $\mathbf{r}_1$ are (x_1, y_1), and the components of $\mathbf{r}_2$ are (x_2, y_2), the components of $\mathbf{s}$ are $(x_2 - x_1, y_2 - y_1)$. Thus the components of $\mathbf{s}$ are the coordinates of B measured in axes with A as origin and parallel to the original axes Ox, Oy; and we get the same vector $\mathbf{s}$ if we change the origin O *provided we do not change the orientation of the axes.* But we get a different vector to describe the position of B relative to A if we change the orientation of the axes. It is particularly important to observe that the relative position vector is not independent of the base of reference if changes of orientation are allowed.

Since

$$\mathbf{r}_2 = \mathbf{r}_1 + \mathbf{s},$$

we have

$$\dot{\mathbf{r}}_2 = \dot{\mathbf{r}}_1 + \dot{\mathbf{s}},$$

and

$$\ddot{\mathbf{r}}_2 = \ddot{\mathbf{r}}_1 + \ddot{\mathbf{s}}.$$

We have here two rules which, in spite of their extreme simplicity, are of immense value in the applications. (i) The velocity of B is the velocity of A, plus the velocity of B relative to A. (ii) The acceleration of B is the acceleration of A, plus the acceleration of B relative to A. The word 'plus' in the enunciation of these rules refers of course to the vector law of addition.

Consider, as an example of the use of these rules, the *linkage* shown in Fig. 10·5*a*. It consists of two rods OA, AB, whose lengths are a and b; one end of the first rod is pivoted at the fixed point O, and the rods are hinged together at A. The system moves in a plane. We denote the inclinations of OA and of AB to Ox at time t by θ and ϕ.

Now consider the motion of B. To find its velocity we notice that the velocity of A is $a\dot{\theta}$ perpendicular to OA. Next we need the velocity of B relative to A. Now the motion of B relative to A is motion in a circle of radius b, and the velocity of B relative to A is $b\dot{\phi}$ perpendicular to AB. Thus the velocity of B is the vector sum of $a\dot{\theta}$ perpendicular to OA and $b\dot{\phi}$ perpendicular to AB as illustrated in Fig. 10·5*a*.

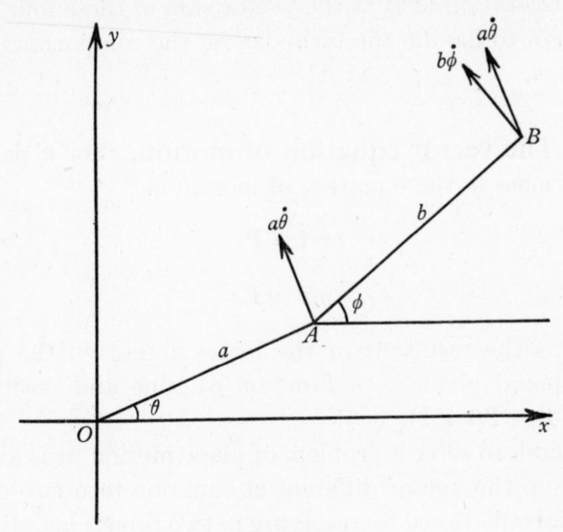

Fig. 10·5*a*

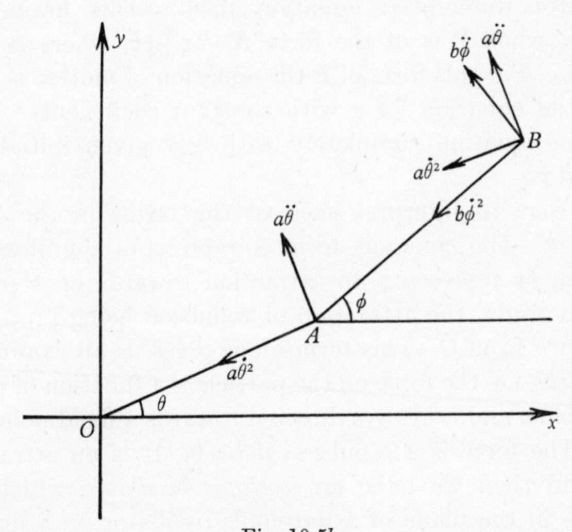

Fig. 10·5*b*

Similarly for the acceleration of B. The acceleration of A has components $(a\ddot{\theta}, a\dot{\theta}^2)$, and the relative acceleration has components $(b\ddot{\phi}, b\dot{\phi}^2)$, as shown in Fig. 10·5*b*. The acceleration of

B is the vector sum of these four vectors. In the applications the simplest way to hold this result in mind is usually to think of the acceleration of B as the vector sum of these four vectors, rather than to handle the formulae for the components parallel to the axes.

10·6 The vector equation of motion.

For a particle of constant mass m the equation of motion is

$$m\mathbf{f} = \mathbf{P},$$

or

$$m\ddot{\mathbf{r}} = \mathbf{P},$$

where $\mathbf{P}$ is the resultant of the forces acting on the particle. $\mathbf{P}$ is supposed given as a function of time and position and velocity, $\mathbf{P} = \mathbf{P}(t, \mathbf{r}, \dot{\mathbf{r}})$.

In general, to solve a problem of plane motion, it is expedient to break up the vector differential equation into two ordinary differential equations, by resolving in two directions. But there are a few special cases in which we can with comfort integrate the vector differential equation itself. This happens, for example, when $\mathbf{P}$ is of the form $\mathbf{A} + b\mathbf{r} + c\dot{\mathbf{r}}$, where $\mathbf{A}, b, c$ are constants. For this form of $\mathbf{P}$ the equation of motion is a linear differential equation for $\mathbf{r}$ with constant coefficients. We can solve the equation completely with any given initial values for $\mathbf{r}$ and $\dot{\mathbf{r}}$.

It is easy to interpret each of the terms in the formula $\mathbf{A} + b\mathbf{r} + c\dot{\mathbf{r}}$. The constant term $\mathbf{A}$ represents a uniform field. The term $b\mathbf{r}$ represents an attraction towards or a repulsion from the origin, the attraction or repulsion being proportional to distance from O. This term alone gives us an example of a *central field*, i.e. the force on the particle is a function of position in the plane, and is always directed towards a fixed point of the plane. The force is a repulsion if $b > 0$. It is an attraction if $b < 0$, and then we have an *isotropic oscillator*, which is the analogue in the plane of a harmonic oscillator on a line. The term $c\dot{\mathbf{r}}$, with $c < 0$, represents a resistance proportional to velocity.

We will consider more particularly the two classical problems of the uniform field and the isotropic oscillator.

10·7 The uniform field (1). For the problem of the uniform field we write $\mathbf{A} = m\mathbf{g}$, and the equation of motion is

$$\ddot{\mathbf{r}} = \mathbf{g}.$$

Hence
$$\dot{\mathbf{r}} = \mathbf{u} + \mathbf{g}t,$$
and
$$\mathbf{r} = \mathbf{a} + \mathbf{u}t + \tfrac{1}{2}\mathbf{g}t^2,$$

where $\mathbf{a}, \mathbf{u}$ are the values of $\mathbf{r}, \dot{\mathbf{r}}$ at $t = 0$. The solution is the general form of that found previously (§ 4·4) for motion on a straight line. (We write $\mathbf{g}t, \mathbf{u}t, \mathbf{g}t^2$ instead of the more natural $t\mathbf{g}, t\mathbf{u}, t^2\mathbf{g}$ to preserve the analogy with the straight-line problem.)

The path is a parabola with its axis parallel to the field.

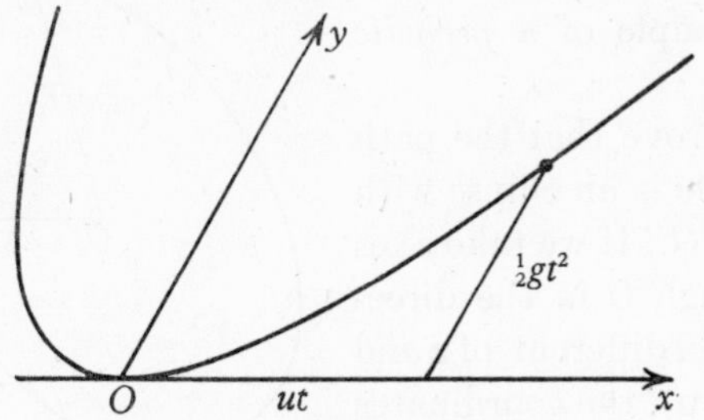

Fig. 10·7

A simple way of proving this is to consider the components of $\mathbf{r} - \mathbf{a}$ in the directions, assumed different, of $\mathbf{u}$ and of $\mathbf{g}$. If we take axes Ox, Oy parallel to $\mathbf{u}$ (or to $-\mathbf{u}$) and to $\mathbf{g}$, with the point of projection O as origin (Fig. 10·7), the vector $\mathbf{r} - \mathbf{a}$ has components $(\pm ut, \tfrac{1}{2}gt^2)$, where $u = |\mathbf{u}|$ and $g = |\mathbf{g}|$. The path is the parabola

$$x^2 = \frac{2u^2}{g} y.$$

This theory gives us an approximation to the motion in any field of force if the scale of the problem is sufficiently small, i.e. if the part of the path considered is so small that the change in the field is insensible.

10·8 The isotropic oscillator (1). We next consider the isotropic oscillator. Here $b < 0$, and we write $b = -mn^2$. The equation of motion takes the form

$$\ddot{\mathbf{r}} + n^2\mathbf{r} = 0,$$

and we have the analogue in a plane of the harmonic oscillator on a line. The solution is

$$\mathbf{r} = \mathbf{a}\cos nt + \frac{1}{n}\mathbf{u}\sin nt, \qquad (1)$$

where $\mathbf{a}$ is the value of $\mathbf{r}$, and $\mathbf{u}$ the value of $\dot{\mathbf{r}}$, at $t = 0$. The solution is again the general form of that found already (§ 5·5) for the rectilinear problem.

The values of $\mathbf{r}$ and of $\dot{\mathbf{r}}$ at $t = t_0$ are recovered at $t = t_0 + \sigma$, where $\sigma = 2\pi/n$, and the motion that occurs in the interval t_0 to $t_0 + \sigma$ is repeated in the interval $t_0 + \sigma$ to $t_0 + 2\sigma$. The motion is periodic with period σ, and we have an example of a *periodic orbit*.

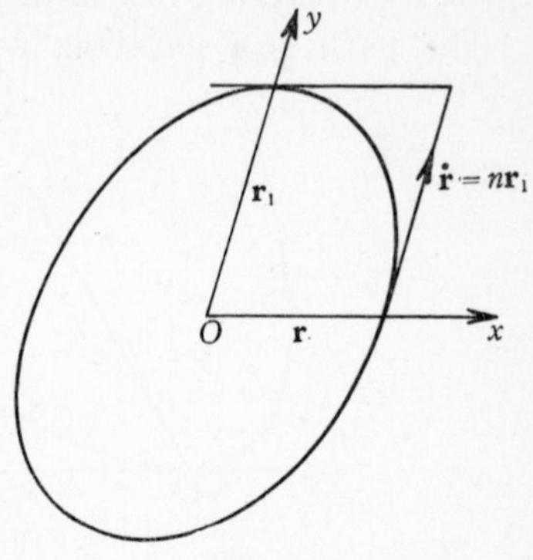

Fig. 10·8

We now prove that the path of the particle is an ellipse with its centre at O. If we take axes Ox, Oy through O in the directions, assumed different, of $\mathbf{a}$ and of $\mathbf{u}$ (or of $-\mathbf{u}$), the coordinates of P at time t are

$$x = a\cos nt, \quad y = \pm b\sin nt,$$

where $a = |\mathbf{a}|$, $b = |\mathbf{u}|/n$. The path or orbit is the curve

$$\frac{x^2}{a^2} + \frac{y^2}{b^2} = 1, \qquad (2)$$

which represents an ellipse for which the radii parallel to $\mathbf{a}$ and to $\mathbf{u}$ are conjugate.

Returning now to the vector formula (1) for $\mathbf{r}$ we see that the velocity at time t is

$$\dot{\mathbf{r}} = n\left(-\mathbf{a}\sin nt + \frac{1}{n}\mathbf{u}\cos nt\right) = n\mathbf{r}_1,$$

where $\mathbf{r}_1$ is the value of $\mathbf{r}$ at time $t + \pi/2n$, a quarter-period after the instant t. Since $\dot{\mathbf{r}}$ is in the direction of the tangent, $\mathbf{r}_1$ is the radius conjugate to $\mathbf{r}$. It takes a quarter-period to reach the end of the conjugate radius: the velocity at any point of the ellipse is in the direction of the conjugate radius, and n times this radius in magnitude. Thus the hodograph is a similar ellipse.

If we refer the orbit to the principal axes of the ellipse we have

$$x = a_0 \cos nt, \quad y = b_0 \sin nt,$$

where a_0, b_0 are the principal semi-axes, and we have taken $t = 0$ when the particle is at $(a_0, 0)$. Thus nt is the eccentric angle. The motion on the ellipse is determined by the simple rule that the corresponding point on the auxiliary circle moves uniformly with angular velocity n.

Example. *A particle of unit mass moves under the action of a constant force* $(n^2+k^2)\mathbf{b}$, *together with an attractive force of magnitude* $(n^2+k^2)r$ *towards the origin* $(n>0)$, *in a medium which offers a retardation* $2k$ *times the speed, and* $\mathbf{r} = \mathbf{a}$, $\mathbf{v} = \mathbf{u}$, at $t = 0$. *To find the position at any subsequent time.*

The vector equation of motion is

$$\ddot{\mathbf{r}} + 2k\dot{\mathbf{r}} + (n^2+k^2)\mathbf{r} = (n^2+k^2)\mathbf{b},$$

and the solution is of the form

$$\mathbf{r} = \mathbf{b} + e^{-kt}(\mathbf{A}\cos nt + \mathbf{B}\sin nt).$$

We have only to adjust the values of $\mathbf{A}$ and $\mathbf{B}$ to make $\mathbf{r} = \mathbf{a}$ and $\dot{\mathbf{r}} = \mathbf{u}$ at $t = 0$, giving

$$\mathbf{r} = \mathbf{b} + \frac{1}{n}e^{-kt}\{n(\mathbf{a}-\mathbf{b})\cos nt + [\mathbf{u}+k(\mathbf{a}-\mathbf{b})]\sin nt\}.$$

10·9 Cartesian axes.

If we take Cartesian axes Ox, Oy, and denote by $\mathbf{i}, \mathbf{j}$ unit vectors in these directions, we have

$$\mathbf{r} = x\mathbf{i} + y\mathbf{j},$$

where (x, y) are the Cartesian coordinates of the particle, and

$$\ddot{\mathbf{r}} = \ddot{x}\mathbf{i} + \ddot{y}\mathbf{j}.$$

Also we can resolve $\mathbf{P}$ into components $X\mathbf{i}$, $Y\mathbf{j}$, parallel to the axes,

$$\mathbf{P} = X\mathbf{i} + Y\mathbf{j}.$$

The vector equation of motion,

$$\mathbf{P} = m\ddot{\mathbf{r}},$$

is equivalent to the two differential equations

$$X = m\ddot{x}, \quad Y = m\ddot{y}.$$

The axes are not necessarily rectangular.

In the particular problem considered in § 10·6, where

$$\mathbf{P} = \mathbf{A} + b\mathbf{r} + c\dot{\mathbf{r}},$$

we have $\quad X = A_1 + bx + c\dot{x}, \quad Y = A_2 + by + c\dot{y},$

where $A_1\mathbf{i}, A_2\mathbf{j}$ are the components of $\mathbf{A}$. The equations of motion are therefore

$$m\ddot{x} = A_1 + bx + c\dot{x}, \quad m\ddot{y} = A_2 + by + c\dot{y}.$$

(We can simplify the equations a little if desired by taking the axis Oy parallel to $\mathbf{A}$; we then get rid of the term A_1 in the first equation.)

The problem therefore possesses the peculiar feature that the x-motion and the y-motion can be studied independently. The simultaneous differential equations for x and y as functions of t have separated out into two independent equations, one involving the variables x and t only, the other the variables y and t only. This is true whether the axes are rectangular or not.

The most important case, however, is that in which the axes are rectangular. It is usually simplest to work with rectangular axes, unless there is some specific reason for choosing oblique axes. Such a reason did in fact emerge in the problems of the uniform field and of the isotropic oscillator, where our choice of axes depended on the initial conditions. But usually such oblique axes are to be regarded as a temporary expedient, to be discarded later on in favour of rectangular axes. This preference for rectangular axes holds for the general problem of plane motion, not merely for problems in which the x-motion and the y-motion are independent.

It will be interesting and useful to reconsider the two classical problems dealt with in §§ 10·7 and 10·8, the uniform field and the isotropic oscillator, working with rectangular axes throughout.

10·10 The uniform field (2). For the uniform field we conventionally take Ox perpendicular to the field, Oy opposite to the direction of the field, so that $\mathbf{P} = -mg\mathbf{j}$. If the field is that of gravity in a region near a point on the earth's surface, this means that Ox is horizontal and Oy vertically upwards.

The equations of motion are

$$\ddot{x} = 0, \quad \ddot{y} = -g,$$

and the position at time t is given by

$$x = x_0 + u_0 t, \quad y = y_0 + v_0 t - \tfrac{1}{2}gt^2,$$

where (x_0, y_0) is the point of projection, and (u_0, v_0) the velocity of projection. Geometrically, these equations define the path in terms of the parameter t. The elimination of t leads to the equation

$$\left(x - x_0 - \frac{u_0 v_0}{g}\right)^2 = \frac{2u_0^2}{g}\left(y_0 + \frac{v_0^2}{2g} - y\right),$$

representing a parabola with its axis vertical and vertex upwards.

10·11 The isotropic oscillator (2). For the isotropic oscillator we have

$$\ddot{x} + n^2 x = 0, \quad \ddot{y} + n^2 y = 0,$$

and the solution is

$$\left.\begin{aligned} x &= x_0 \cos nt + \frac{u_0}{n}\sin nt, \\ y &= y_0 \cos nt + \frac{v_0}{n}\sin nt. \end{aligned}\right\}$$

The elimination of t leads to the equation of an ellipse with O as centre,

$$(xv_0 - yu_0)^2 + n^2(xy_0 - yx_0)^2 = (x_0 v_0 - y_0 u_0)^2.$$

In practice we can simplify the result by taking the axis Ox, say, through the point of projection, giving $y_0 = 0$; we can simplify it still further, when once we have shown that the orbit is an ellipse, by taking Ox, Oy along the principal axes of the ellipse.

10·12 The anisotropic oscillator, general theory of small oscillations. Another important problem in which the x-motion and the y-motion are independent is that of the anisotropic oscillator, for which the equations of motion, referred to rectangular axes, are

$$\ddot{x} + p^2 x = 0, \quad \ddot{y} + q^2 y = 0, \tag{1}$$

where p and q are positive unequal numbers. The solutions of the equations are

$$\left.\begin{aligned} x &= x_0 \cos pt + \frac{u_0}{p} \sin pt, \\ y &= y_0 \cos qt + \frac{v_0}{q} \sin qt. \end{aligned}\right\}$$

The curves which are defined parametrically by these equations are called Lissajous' figures.

The x-motion is a periodic motion with period $2\pi/p$, and the y-motion is a periodic motion with period $2\pi/q$. The motion in the plane is therefore periodic, with period σ, if there are positive integers r, s such that

$$\sigma = r\frac{2\pi}{p} = s\frac{2\pi}{q}.$$

The motion is periodic if and only if p/q is rational. If $p/q = r/s$, where r and s are positive integers with no common factor, the period is $2\pi(r/p) = 2\pi(s/q)$.

The most striking and important feature of this problem is that the presence or absence of the property of periodicity does not depend on the circumstances of projection. Many dynamical systems are capable both of periodic and of non-periodic motions, according to the circumstances of projection. But for the anisotropic oscillator the periodicity or non-periodicity is inherent in the system, and is beyond our control.

Consider more particularly the case when p/q is irrational. We can write the solution, without loss of generality, in the form

$$x = A \cos pt, \quad y = B \cos (qt + \alpha),$$

where A and B are positive constants, and the origin of time has been chosen suitably. The orbit is inscribed in a rectangle whose sides are of lengths $2A$ and $2B$. The position and velocity at any instant t_0 never recur, and the orbit fills the rectangle more and more thickly as time goes on.

The equations (1) of this paragraph are of far-reaching importance, because, by using the so-called *normal coordinates*, the equations of motion for the small oscillations of any vibrating system with two degrees of freedom can be reduced to this form.

The anisotropic oscillator thus serves as a model of vibrating systems in general. (A concrete example of normal coordinates for a vibrating system will be found in § 23·14, Example 3.)

10·13 Uniform motion in a circle. If the particle moves in a circle of radius a with uniform angular velocity ω, the tangential component of acceleration is zero, and the acceleration at each instant is directed towards the centre of the circle, and is of magnitude $a\omega^2$. Thus the force required to maintain the motion is a force $ma\omega^2$, or mv^2/a, along the inward radius (§ 10·3).

Suppose we have a central field of force, i.e. the force on the particle depends only on its position, and is directed always towards a fixed point O of the plane. Suppose also that the force is an attraction, and is a function only of r, the distance from O; say the attraction is $mf(r)$. Then the particle can move in a circle of any radius a about O as centre, the velocity in the circle being $\sqrt{\{af(a)\}}$. However, although such a motion is theoretically possible, it could not always be achieved in practice, because, as we shall see, the circular orbit may be unstable. This means that an accidental disturbance, however small, would lead to a motion far removed from the original uniform motion in a circle. We have already met with this notion of an unstable motion in the simpler problem of motion on a straight line (§ 6·8).

A classical example of uniform circular motion is provided by the conical pendulum. A particle is attached to one end of a light string, the other end being fixed, and the particle moves in a horizontal circle with the string inclined at a constant angle α to the downward vertical. The tension T in the string and the weight mg must have as their resultant $mOP\omega^2$ along PO, the notation being that shown in the figure (Fig. 10·13a). Thus we have

$$\frac{T}{AP} = \frac{mg}{AO} = \frac{mOP\omega^2}{OP},$$

whence
$$\omega^2 = \frac{g}{AO} = \frac{g}{h},$$

where h is the depth of the plane of motion below the point of support A. The period is

$$\frac{2\pi}{\omega} = 2\pi\sqrt{\left(\frac{h}{g}\right)}.$$

The result is valid whether the string is inelastic or elastic. The motion is stable; we can feel sure of this in a general way, without formal proof, from our everyday experience.

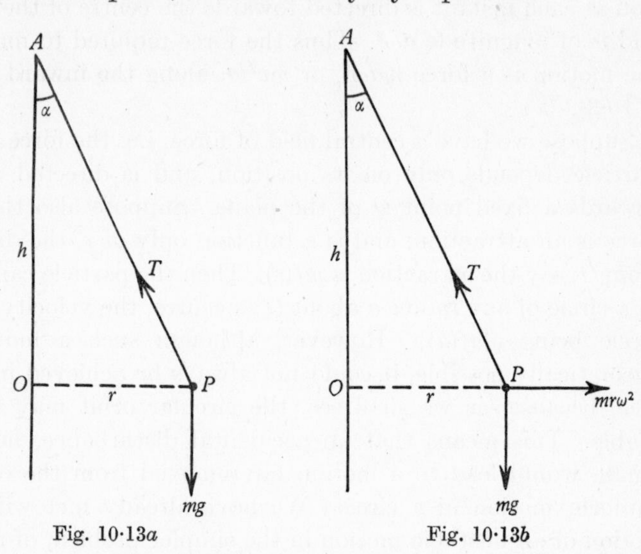

Fig. 10·13a

Fig. 10·13b

We can look at this and similar problems in another way. Instead of saying that the resultant of T in PA and mg downwards is $mr\omega^2$ *inwards*, we can say that three forces T in PA, mg downwards, and $mr\omega^2$ *outwards* are in equilibrium. We thus reduce the dynamical problem to a statical one. Of course the remark is a truism, in the sense that any problem of particle dynamics can be reduced to a statical one in the same way. The forces acting on the particle and the reversed kineton form a system in equilibrium. But this device is of particular interest only in a few special cases. More generally, if we deal with a dynamical system instead of a single particle, the device is most valuable when the system remains rigid, i.e. when there exists a (moving) frame of reference in which the system is actually at rest. In our simple problem the system is at rest in a frame of reference rotating, with uniform angular velocity ω, about the vertical through A.

The force $mr\omega^2$ which is applied along the outward radius to produce the equivalent statical problem (Fig. 10·13*b*) is called the *centrifugal force*. We must remember, if we use this device, that when the centrifugal force has been added to the actual forces, we have a statical problem to deal with, and we must think of the system as being at rest. From the triangle of forces, or by resolving perpendicular to the string, we recover the result $\omega^2 = g/h$.

10·14 The rotation of the earth. When a heavy body is suspended from the ceiling by a fine thread, and is at rest relative to the room, the line of the thread is not in the direction of the earth's field. If the earth were not rotating, the tension in the thread would be equal and opposite to the gravitational force on the body. But the earth *is* rotating, and the body, instead of being at rest, is actually moving in a circle. As we have seen, a simple way to take account of the rotation is to introduce the notion of the centrifugal force; then the tension in the thread balances the resultant of weight and centrifugal force.

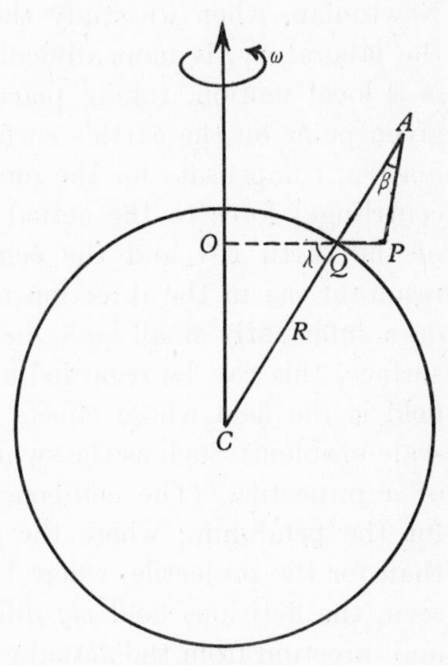

Fig. 10·14

On the assumption of a spherical earth, it is easy to find the deviation of the plumb-line from the radial line. We consider a place Q in north latitude λ. The resultant of the gravitational force mG, represented in Fig. 10·14 by AQ, and the centrifugal force represented by QP, is represented by AP, which is in the direction of the thread. We call this resultant mg, because this is in fact the weight of the body, as measured by the tension in the thread. If R is the earth's radius, the centrifugal force is $mR\omega^2 \cos\lambda$, and the small angle QAP, which we denote by β, is given by the equation

$$\sin\beta = \frac{PQ}{PA}\sin\lambda = \frac{R\omega^2 \cos\lambda \sin\lambda}{g},$$

or, to a good approximation,

$$\beta = R(\omega^2/g)\cos\lambda \sin\lambda.$$

For London, in latitude $51\frac{1}{2}°$, the value of β is about $6'$.

In this problem the theory of the centrifugal force is only applicable when the body is at rest relative to the earth. The problem of how to compensate for the fact that the base is not Newtonian, when we study the *motion* of a body relative to the laboratory, is more difficult. Nevertheless, if the motion is a local motion, taking place within a small distance of a given point on the earth's surface, we can, as a first approximation, compensate for the motion of the base by adding the centrifugal force to the actual forces. The gravitational pull of the earth mG and the centrifugal force together give a resultant mg in the direction of the plumb-line. For motion on a sufficiently small scale, near a given point on the earth's surface, this can be regarded as a uniform field. Indeed, this field is the field whose effects we observe in ordinary small-scale problems, such as the swing of a pendulum, or the motion of a projectile. (The compensation is of course more exact for the pendulum, where the scale of motion is very small, than for the projectile, where the scale is larger.) As we have seen, the fictitious field mg differs only slightly in magnitude and direction from the actual gravitational field mG.

10·15 Motion in a field of force, reversibility. When the force is a function of position in the plane

$$\mathbf{P} = \mathbf{P}(\mathbf{r}),$$

the particle is said to move in a field of force. The motion is determined by the equation

$$m\ddot{\mathbf{r}} = \mathbf{P}(\mathbf{r}).$$

The motion is reversible, i.e. if at any instant the velocity is reversed (as, for example, by normal impact with a perfectly elastic wall) the particle retraces its path, the velocity at any point on the return journey being equal and opposite to the velocity at that point in the original journey. Explicitly, if $\mathbf{r} = \mathbf{F}(t)$ in the interval $(0, t_0)$, and the velocity is reversed at the instant t_0, then in the interval $(t_0, 2t_0)$ we have $\mathbf{r} = \mathbf{F}(2t_0 - t)$. The proof is similar to that in § 5·4, the vector $\mathbf{r}$ replacing the scalar x.

EXAMPLES X

1. A particle moves in a plane, its coordinates at time t referred to rectangular axes in the plane being (x, y). The component acceleration in the direction Oy has the constant value $-g$, and the component acceleration in the direction Ox is $k(y-h)$, where k and h are positive constants. The particle is projected from the origin in the direction Oy with speed $\sqrt{(3gh)}$. Discuss the motion, and prove that the path of the particle is a parabola.

2. If a high throw is made with a diabolo spool the vertical resistance may be neglected, but the spin and the vertical motion together account for a horizontal drifting force which may be taken as proportional to the vertical velocity. Show that if the spool is thrown so as to rise to a height h and return to the point of projection, the spool is at its greatest distance c from the vertical through that point when at a height $2h/3$; and find the equation of the trajectory in the form

$$4h^3x^2 = 27c^2y^2(h-y).$$

3. Find the formulae for acceleration in polar coordinates by means of the hodograph.

4. The velocities of a particle along and perpendicular to the radius vector from a fixed origin are λr^2 and $\mu\theta^2$; find the polar equation of the path of the particle and also the component accelerations in terms of r and θ.

5. A particle, initially at rest at the origin, moves in a plane with an acceleration that is the resultant of a constant radial acceleration f outwards and a variable acceleration equal to 2ω times the velocity in a direction perpendicular to the velocity, where ω is constant. Show that its angular velocity about the origin is equal to ω, and find the polar equation of its path, taking the direction of f at the origin as the initial line.

6. A bead can slide freely on a straight wire AB of length l which is rotated in a horizontal plane with constant angular velocity ω about its end A. Initially the bead is projected along the wire with velocity V from A. When the bead leaves the wire, what is the angle between the line of the wire and the direction of motion of the bead?

7. A fine smooth tube rotates about a point O of itself with uniform angular velocity ω in a horizontal plane. The tube contains a particle of mass m attached to O by a fine elastic thread of natural length a and modulus of elasticity $2ma\omega^2$. Initially the thread is just taut and the particle is at rest relative to the tube. Show that the particle makes one complete oscillation in the tube in each rotation, and that the greatest horizontal pressure of the tube on the particle is $2ma\omega^2$.

Make a drawing of the path of the particle in the plane.

8. A narrow straight tube of length $2a$ is rotating in a plane about one end with constant angular velocity ω. Inside the tube is a small bead of mass m which is instantaneously at relative rest at the mid-point. The coefficient of friction between the bead and the tube is $\frac{3}{4}$. Obtain an expression for the reaction between the particle and the tube in terms of m, a, ω, and the time that has elapsed since the bead was at the mid-point, neglecting the effect of gravity.

Show that the particle reaches one end of the tube after time $\dfrac{2\log_e 2{\cdot}5}{\omega}$ approximately.

9. Two particles of masses m_1 and m_2 are connected by a light string passing through a small smooth ring, and one is held at rest with the string tight, the vertical lengths of string being a and b. They are then simultaneously projected horizontally with velocities u and v respectively. Prove that the initial tension in the string is

$$\mu\left(\frac{u^2}{a}+\frac{v^2}{b}+2g\right),$$

where

$$\frac{1}{\mu}=\frac{1}{m_1}+\frac{1}{m_2}.$$

10. The angular velocity of the chain-wheel of a bicycle is one nth of the angular velocity of a road-wheel, the radii of these wheels being r and a respectively. At a certain instant the velocity of the bicycle is v and its acceleration is f. Find the acceleration at any point on the circumference of the chain-wheel, and show that there is one such point with zero acceleration if and only if

$$\frac{v^4}{f^2}=a^2n^2\left(\frac{a^2n^2}{r^2}-1\right).$$

11. A and B are two fixed points on the circumference of a circle and the distances from A and B of any other point P on the circumference are r and s respectively. If u and v are the components of P's velocity, as it moves round the circumference, along $\overrightarrow{AP}$ and $\overrightarrow{BP}$ respectively, prove that, α being the angle APB,

$$u\sin^2\alpha=\dot{r}-\dot{s}\cos\alpha,\quad v\sin^2\alpha=\dot{s}-\dot{r}\cos\alpha,$$

and deduce that $ur+vs=0$.

12. A particle of unit mass moves under the action of a force F which is constant in magnitude and direction, and is also subject to a force of magnitude G times its speed, in a direction normal to its path. The particle is released from rest. Show that if $\psi=0$ is the direction of F, the intrinsic equation of its path is

$$s=\frac{4F}{G^2}(1-\cos\psi).$$

Show that the normal acceleration is equal and opposite to the normal component of F, and that the direction of motion changes at a constant rate $\frac{1}{2}G$.

13. A Mississippi steamer takes m minutes to go a mile upstream and n minutes to go a mile downstream. Prove that it takes $\sqrt{(mn)}$ minutes to go a mile across the stream.

14. A man walks with uniform velocity across a road of breadth h between two buses of breadth b travelling one behind the other with speed V; the distance between the back of the first bus and the front of the second is a. What is the least speed at which he must walk, and how long will he take at this speed to cross the road?

15. A boat which can travel at $4\frac{1}{2}$ m.p.h. in still water has to make a trip from a point A to a point B, one mile away from A, and back to A. On account of the tide, the shortest times in which the outward and return journeys can be made are respectively 16 min. and 12 min. By accurate drawing find the speed of the tidal current and the angle which its direction makes with the line AB. Verify the results by calculation.

16. A torpedo which drives itself through the water at a constant speed V, its axis preserving in all circumstances an invariable direction, is fired from O at an enemy ship P whose constant speed through the water is v. At the moment of firing the direction of motion of the enemy ship makes an angle ϕ with OP. (When the ship is going straight away $\phi = 0$.) Prove that in order to allow for the motion of the enemy ship, the axis of the torpedo must be aimed ahead of the ship by an angle δ, where

$$\sin \delta = \frac{v}{V} \sin \phi,$$

so that δ is independent of the range OP.

Prove that this result is unaffected by a uniform tidal current.

Prove further that, at a time t after firing, the locus of the positions relative to the ground of torpedoes, fired in all horizontal directions from a fixed point, is a circle whose radius is independent of the tidal current, and find the centre of this circle.

17. An aeroplane has a speed of v m.p.h., and a range of action (out and home) of R miles in calm weather. Prove that in a north wind of w m.p.h. its range of action is

$$\frac{R(v^2 - w^2)}{v(v^2 - w^2 \sin^2 \phi)^{\frac{1}{2}}}$$

in a direction whose true bearing is ϕ. If $R = 200$ miles, $v = 80$ m.p.h. and $w = 30$ m.p.h., find the direction in which its range is a maximum, and the value of the maximum range.

18. Two particles A, B, travel in the same sense in coplanar circular paths of radii a and b respectively with a common centre O, the speeds being inversely proportional to the square root of their distances from O. Prove that their relative velocity will be in the direction AB when the angle AOB is $\cos^{-1} \sqrt{(ab)}/[a+b-\sqrt{(ab)}]$.

Show that whatever values a and b may have, there is always a real angle AOB having this value.

19. A light rod AB, carrying a heavy particle at B, is connected by a smooth hinge at A to a second light rod AO. OA, AB are initially in the same straight line, rotating with uniform angular velocity about a fixed centre O. AB is now displaced through a small angle, relatively to OA, in the plane of rotation. Show that, if the angular velocity of OA be maintained constant, AB will oscillate relatively to OA with a period $T\sqrt{(l/a)}$, where T is the time of revolution of OA about O, and a and l are the lengths of OA and AB respectively.

20. A light rod PQ, of length b, has a massive particle attached at Q. The rod rests on a smooth table when the end P is seized and moved off in a horizontal circle of radius a with constant velocity $a\omega$, the initial position of the rod being outside the circle and in line with the centre O. In the subsequent motion the angle which PQ makes with OP produced is denoted by ϕ; show that

$$b^2\dot{\phi}^2 = (a^2+b^2+2ab\cos\phi)\,\omega^2.$$

Show further that if $a = b$ then $\phi \to \pi$ as $t \to \infty$.

21. The speed of an aeroplane relative to the air is V, and the time it takes to traverse a certain circuit in still air is t_0. Show that the time taken to traverse the same circuit in a wind of velocity kV $(k<1)$ lies between t_0/k' and t_0/k'^2, where $k'^2 = 1-k^2$.

22. Atwood's machine consists of two particles, of mass m and m' respectively, attached to the ends of a light string which passes over a pulley. Find the acceleration of the system and the tension of the string, neglecting the inertia of the pulley and the friction of the axle.

Further, find the circumstances of the motion, and the tension of the string, if the machine be placed in a railway truck which is allowed to run freely down a slope of given inclination.

23. A particle describes an elliptic orbit under a force to the centre proportional to distance. Prove that the angular velocity about a focus is inversely proportional to the distance from that focus.

24. A horizontal bar AB of length a is made to rotate with a constant angular velocity ω about a vertical axis through the end B. If a particle is attached to A by a string of length l, the string makes an angle θ with the vertical when the motion is steady. Prove that

$$l\cos\theta + a\cot\theta = g/\omega^2.$$

25. A particle moves in a circle of radius a under an attraction μr^n per unit mass to the centre. For what value of n is (i) the speed, (ii) the period independent of r?

26. If the earth is suddenly stopped in its orbit, how long will it take to fall into the sun? (Assume the earth to move in a circle about the sun, which is supposed fixed.)

27. A heavy particle is attached by a light string to a fixed point O, and moves so as to describe a circle in a horizontal plane with uniform angular velocity ω. Show that this plane is at a distance g/ω^2 below the point O.

An elastic string of unstretched length l is extended by an amount λ_1 when it supports a mass m at rest, and is extended by an amount λ_2 when it is rotating as above carrying a particle of the same mass m. Show that $g\lambda_2/\lambda_1 = \omega^2(l+\lambda_2)$.

28. Show that a point moving in a plane curve with velocity v has acceleration v^2/ρ towards the centre of curvature, where ρ is the radius of curvature.

The point A is vertically above B, and $AB = l$. The ends of a string ACB of length $2l$ are fixed at A, B. A bead C, of mass m, which can slide freely on the string, describes a horizontal circle with angular velocity ω about AB. The plane in which C moves is at depth y below A. Show that

$$y = \tfrac{1}{2}l + 4g/(3\omega^2),$$

and find the tension in the string in terms of ω.

29. A car takes a banked corner of a racing track at a speed V, the lateral gradient α being designed to reduce the tendency to side-slip to zero for a lower speed U. Prove that (supposing that the car can be treated as a particle) the coefficient of friction necessary to prevent side-slip for the greater speed V must be at least

$$\frac{(V^2-U^2)\sin\alpha\cos\alpha}{V^2\sin^2\alpha+U^2\cos^2\alpha}.$$

30. A particle hangs by an inelastic string of length a from a fixed point, and a second particle of the same mass hangs from the first by an equal string. The whole moves with uniform angular velocity ω about the vertical through the point of suspension, the strings making constant angles α and β with the vertical. Show that

$$\tan\alpha = k(\sin\alpha+\tfrac{1}{2}\sin\beta), \quad \tan\beta = k(\sin\alpha+\sin\beta),$$

where $k = a\omega^2/g$.

Hence show that if α and β are small such a steady motion is only possible if k has one of the values $(2\pm\sqrt{2})$, and that $\beta/\alpha = \pm\sqrt{2}$.

31. A particle of mass m_1 is attached to the end B of a string AB and a particle m_2 to the end C of a string BC, the end A being attached to a fixed point. AB, BC lie in a vertical plane which is rotating with angular velocity ω, the strings AB, BC making angles α and β respectively with the downward vertical. Show that, whatever the angular velocity,

$$\frac{\tan\alpha}{\tan\beta} = \frac{\bar{x}}{x_2},$$

where $\bar{x}$ is the distance of the centre of gravity of m_1 and m_2 from the vertical line through A, and x_2 is the distance of m_2 from the same line.

Hence show that, if α and β have the same sign, β is greater than α and that, if α and β have opposite signs, B and C are on opposite sides of the vertical through A. Show also that if the string BC meets the vertical through A in P and from P a line be drawn parallel to AB to meet the horizontal through C in K, then K lies on the vertical through the centre of gravity.

32. Two light rods AB, BC, each of length a, are freely jointed at B, and particles of masses m_1, m_2, m_3 are attached at A, B, C respectively. The

system is placed on a rough horizontal turn-table, the particles alone making contact with it, so that A, B, C are at distances $a, 2a, 3a$ respectively from the centre of rotation. Prove that, if the table rotates with constant angular velocity ω and

$$a\omega^2(m_1+2m_2+3m_3) < \mu g(m_1+m_2+m_3),$$

where μ is the coefficient of friction at each contact, the system can remain upon the table without slipping.

33. An elastic ring of mass M, natural length $2\pi a$, and modulus of elasticity λ, is placed upon a rough, horizontal, steadily rotating turn-table; the coefficient of friction is μ. Find the range of values of ω, the angular velocity of the turn-table, for which the ring can remain on the turn-table, without slipping, in the form of a circle of radius $2a$ with its centre at the centre of rotation.

CHAPTER XI

THE CONSERVATIVE FIELD; CONSTRAINED MOTION

11·1 The conservative field. When the force acting on the particle is a function of position in the plane,

$$\mathbf{P} = \mathbf{P}(\mathbf{r}),$$

we speak of a *field of force*. We shall sometimes use rectangular Cartesian coordinates (x, y), and then we shall denote the components of $\mathbf{P}$ at (x, y) by (X, Y), where

$$X = X(x, y), \quad Y = Y(x, y).$$

Strictly speaking, of course, the components are $X\mathbf{i}$ and $Y\mathbf{j}$, where as usual $\mathbf{i}, \mathbf{j}$ are unit vectors in the directions of the axes, but the looser form of expression is convenient and unambiguous. We have

$$X = (\mathbf{P}.\mathbf{i}), \quad Y = (\mathbf{P}.\mathbf{j}),$$

and

$$\mathbf{P} = (\mathbf{P}.\mathbf{i})\mathbf{i} + (\mathbf{P}.\mathbf{j})\mathbf{j}.$$

On other occasions, particularly in the theory of central orbits, we shall use polar coordinates (r, θ), and then we shall denote the radial and transverse components (projections) of $\mathbf{P}$ by (R, S), where

$$R = R(r, \theta), \quad S = S(r, \theta).$$

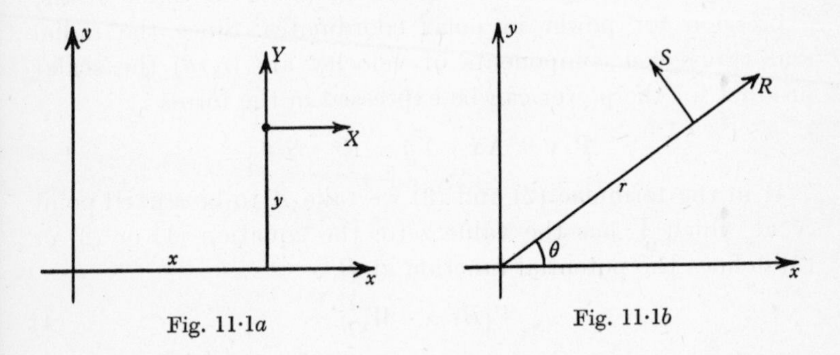

Fig. 11·1*a*

Fig. 11·1*b*

Again, the components, strictly speaking, are $R\mathbf{E}$ and $S\mathbf{F}$, where $\mathbf{E}$ is a unit vector in the direction of the radius vector, and $\mathbf{F} = \mathbf{E}'$. We have

$$R = (\mathbf{P}.\mathbf{E}), \quad S = (\mathbf{P}.\mathbf{F}),$$

and

$$\mathbf{P} = (\mathbf{P}.\mathbf{E})\mathbf{E} + (\mathbf{P}.\mathbf{F})\mathbf{F}.$$

The equations of motion are

$$m\ddot{\mathbf{r}} = \mathbf{P};$$

or

$$m\ddot{x} = X, \quad m\ddot{y} = Y;$$

or

$$m(\ddot{r} - r\dot{\theta}^2) = R, \quad m(r\ddot{\theta} + 2\dot{r}\dot{\theta}) = S.$$

In the particular case when the field $\mathbf{P}$ is the gradient of a uniform function of position, $-V(\mathbf{r})$,

$$\mathbf{P} = -\operatorname{grad} V, \tag{1}$$

we call the field *conservative* (§ 3·10). The scalar function V is called the *potential* or *potential function* of the field. We now turn to the study of motion in a conservative field.

The central property of the conservative field is that the work done by the field when the particle moves from A to B is a function only of the termini and is independent of the path. As we have noticed already (equation (2) of § 3·10),

$$W_{AB} = \int_A^B X\,dx + Y\,dy = V(A) - V(B). \tag{2}$$

If we use polar, instead of Cartesian, coordinates, the equation takes the form

$$W_{AB} = \int_A^B R\,dr + Sr\,d\theta = V(A) - V(B). \tag{3}$$

One way of finding the integrand in (3) is to think of the expression for power in polar coordinates. Since the radial and transverse components of velocity are $(\dot{r}, r\dot{\theta})$ the scalar product for the power can be expressed in the forms

$$\mathbf{P}.\mathbf{v} = X\dot{x} + Y\dot{y} = R\dot{r} + Sr\dot{\theta}.$$

If in the formulae (2) and (3) we take A to be a fixed point N at which V has the value zero, the equation (1) or (2) or (3) defines the potential function at B,

$$V(B) = -W_{NB}. \tag{4}$$

Since N can be chosen arbitrarily there is an additive constant in the expression for V to which we can give any convenient value.

When the potential function V is known we can at once determine the field. If V is expressed as a function of (x, y) we have

$$X = -\frac{\partial V}{\partial x}, \quad Y = -\frac{\partial V}{\partial y}, \tag{5}$$

and if V is expressed as a function of (r, θ),

$$R = -\frac{\partial V}{\partial r}, \quad S = -\frac{1}{r}\frac{\partial V}{\partial \theta}. \tag{6}$$

(To find the last of these we can use equation (3), taking the path from (r, θ) to $(r, \theta+\delta\theta)$ to be the circle $r =$ constant.)

Conversely, if we know the value of $\mathbf{P}(\mathbf{r})$ we can test if the field is conservative by calculating the work done by the field when the particle moves from a fixed point N to a point P. If W_{NP} is independent of the path, depending only on the position of P, the field is conservative, and

$$V(P) = -W_{NP}.$$

Another way of stating the same test for a conservative field is as follows: the field is conservative if the work done by the field when the particle moves round an arbitrary closed curve is zero.

The curves $V =$ constant are called the *equipotentials* or *level curves*; the field at any point is normal to the level curve through that point. A curve drawn so that the tangent at each point is in the direction of the field at that point is called a *line of force*. The lines of force are the integral curves of the differential equation

$$\frac{dx}{X} = \frac{dy}{Y}. \tag{7}$$

In general one line of force passes through a given point of the plane, though an exception to this rule may occur at a point at which $\mathbf{P} = 0$, or at which $\mathbf{P}$ is not defined. Since at any point the field is normal to the level curve through that point, *the lines of force are the orthogonal trajectories of the level curves.*

We have used the symbol V to denote the potential function, whether it is expressed in terms of (x, y) or in terms of (r, θ). This notation is convenient, and usually it will not give rise to any confusion, because we do not often need the expression in terms of (x, y) and the expression in terms of (r, θ) in the same problem. But occasionally both expressions do appear in the same problem, and then we need different symbols for the two functions, say $V(x, y) = \Psi(r, \theta)$.

11·2 Examples of conservative fields.

(i) Consider the uniform field

$$X = 0, \quad Y = -mg.$$

Here
$$W_{OP} = \int_O^P (X\,dx + Y\,dy) = -mgy,$$

so the field is conservative, and

$$V = mgy.$$

The level curves are the lines y = constant, and the lines of force are the lines x = constant. Of course we can approach this from the other end; given $V = mgy$ the field is conservative, and, by equations (5) of § 11·1, $X = 0$, $Y = -mg$. Indeed, it is clear that the field is uniform when V is any linear function of x and y.

(ii) Another simple example is that of the anisotropic oscillator already mentioned in § 10·12. Here

$$V = \tfrac{1}{2}m(p^2x^2 + q^2y^2),$$

giving
$$X = -mp^2x, \quad Y = -mq^2y.$$

The level curves are the ellipses

$$\frac{x^2}{q^2} + \frac{y^2}{p^2} = \text{constant}.$$

(iii) Now consider a conservative *central field*. A central field is one in which the force at any point P is in the line PO, or OP, where O is a fixed point of the plane. If we use polar coordinates, with O as origin, we have, in any central field, $R = R(r, \theta)$, $S = 0$. The lines of force are the straight lines through O. Now if the central field is conservative, the level curves are the orthogonal trajectories of these lines, i.e. the level curves are the circles with O as centre. Therefore V is a function of r, and it follows that R is a function of r, $R = R(r)$. We have

$$R = -\frac{dV}{dr},$$

$$V = -\int_a^r R(\xi)\, d\xi,$$

the lower limit a being any convenient fixed number. If $R < 0$ at all points of the field we have a field of attraction, and in that case V increases steadily with r.

(iv) An important special case of a central field of attraction is the isotropic oscillator

$$R = -mn^2 r, \quad V = \tfrac{1}{2}mn^2 r^2;$$

here the origin is taken as the point at which $V = 0$.

(v) A still more important example is the gravitational attraction of a fixed particle, of mass M, at O,

$$R = -\gamma Mm/r^2, \quad V = -\gamma Mm/r.$$

In this case V tends to a limit as r tends to infinity, and (as in § 5·2) we have chosen the arbitrary constant in V so that this limit is zero.

11·3 The potential function is uniform. It is important to observe that the uniform (one-valued) character of V is an important part of the idea of a conservative field. For example, if we have the field

$$X = -ky/r^2, \quad Y = kx/r^2,$$

or, expressed more simply,

$$R = 0, \quad S = k/r,$$

the field is the gradient of a function of position $-k\theta$, and in a sense a potential function exists,

$$V = -k\theta.$$

But this potential function is not uniform, and the field does not possess the characteristic property of the conservative field, that the work done by the field when the particle traverses a closed curve is zero. If the particle travels round a circle about the origin, or indeed round any simple closed curve about the origin, the work done by the field is $2\pi k$.

11·4 Free motion in a conservative field; the equation of energy. Suppose a particle moves under the action of a conservative field of force whose potential function is V. If we denote by $\Omega = \Omega(t)$ the value of V at the point occupied by the particle at time t we have

$$\frac{d\Omega}{dt} = \frac{\partial V}{\partial x}\dot{x} + \frac{\partial V}{\partial y}\dot{y} = -(X\dot{x} + Y\dot{y}). \tag{1}$$

(The formula $-d\Omega/dt$ for the power we have already encountered in § 3·10.) Now the equations of motion are

$$X = m\ddot{x}, \quad Y = m\ddot{y}, \tag{2}$$

giving

$$X\dot{x} + Y\dot{y} = m\dot{x}\ddot{x} + m\dot{y}\ddot{y} = dT/dt, \tag{3}$$

where as usual T denotes the kinetic energy $\frac{1}{2}m(\dot{x}^2 + \dot{y}^2)$. From the equations (1) and (3) we find

$$\frac{d}{dt}(T + \Omega) = 0,$$

whence

$$T + \Omega = C. \tag{4}$$

As we have remarked already in a simpler case (§ 5·3), we can safely write the equation in the classical form

$$T + V = C \tag{5}$$

when once the special significance of V in this equation has been understood. Here V denotes *the potential energy of the particle*, i.e. the value of the potential function at the point occupied by the particle.

Equation (4) or (5) is the *equation of energy*. It dominates the whole theory of motion in a conservative field, just as it did in the simpler case of rectilinear motion in a field of force.

The level curve $V = C$ is called the *energy level* for the problem under discussion. We denote it by λ. Since $T \geqslant 0$ it is clear that the motion takes place in the region of the plane $V \leqslant C$, a region bounded by the curve λ. If the particle reaches λ its velocity when it does so is zero.

11·5 Constrained motion in a conservative field; the equation of energy. Suppose that a bead is free to slide on a fixed smooth wire, in the form of a simple curve, and that the bead is acted on by a conservative field of force. In this problem an extra force acts on the particle in addition to the field. This extra force is the reaction of the wire on the bead, and it is always normal to the wire, because the wire is smooth. The extra force therefore has the property that it does no work, because it is always perpendicular to the direction of motion. Such a force is called a *force of constraint*.

Now the intrusion of this extra force does not invalidate the equation of energy. If we denote the components of the force of constraint by (X', Y') we have

$$X'\dot{x} + Y'\dot{y} = 0$$

at every instant during the motion. We have now

$$m\ddot{x} = X + X', \quad m\ddot{y} = Y + Y',$$

and therefore

$$m(\dot{x}\ddot{x} + \dot{y}\ddot{y}) = (X + X')\dot{x} + (Y + Y')\dot{y}$$
$$= X\dot{x} + Y\dot{y},$$

which leads to

$$T + \Omega = C$$

as before.

Notice that the problem in question has only one degree of freedom, i.e. we can specify the configuration of the system at time t by the value at that instant of just one parameter. We could take as this parameter the arc-length s to the point on the curve occupied by the particle. The problem is solved when we have expressed s as a function of t.

11·6 Constrained motion, another approach. Now if we think of the value of V at points on the curve as a function of s, the tangential component of force acting on the particle is $-dV/ds$, and the equation of motion is

$$mf_t = -\frac{dV}{ds},$$

where f_t is the tangential component of acceleration. As we have seen (equation (3) of § 10·2)

$$f_t = \frac{d}{ds}(\tfrac{1}{2}v^2).$$

Substituting this formula for f_t in the equation of motion, we have

$$m\frac{d}{ds}(\tfrac{1}{2}v^2) = -\frac{dV}{ds},$$

whence $$\tfrac{1}{2}mv^2 + V(s) = C,$$

which is again the equation of energy. We can write this equation in the form $$\tfrac{1}{2}m\dot{s}^2 = C - V(s);$$

now this is vitally important, because it is a differential equation of a type which we have already studied in detail in Chapter VI. As we saw, we have only to consider the graph of $-V(s)$, and the types of motion that can take place with different values of C can be read off.

In particular, a point a of the curve at which $V(s)$ is a minimum is a position of stable equilibrium. If the bead is set in motion from a point sufficiently near to a with a sufficiently small speed the motion will be a libration motion in the neighbourhood of a. For small values of $C - V(a)$ the motion is approximately a harmonic motion with period $2\pi/n$, where $mn^2 = V''(a)$.

Other examples of constrained motion are provided by a particle sliding in a fine-bored smooth fixed tube, and by a particle fixed to one end of a light rod, the other end of which is freely attached to a fixed point; in the last case the tension or thrust in the rod is a force of constraint.

Example. *A bead of mass m slides on a smooth straight wire and is attracted to two centres of force A and B. The attraction to A is a constant force $m\mu_1$ and the attraction to B is a constant force $m\mu_2$. Prove that there is a position of stable equilibrium.*

The potential function is $m(\mu_1 r_1 + \mu_2 r_2)$, where r_1 and r_2 are distances from A and B, and we have to examine the behaviour of this function on the line MN (Fig. 10·1b). As we have seen, there is a minimum of $\mu_1 r_1 + \mu_2 r_2$ at a point P_0 such that

$$\mu_1 \sin\theta_1 = \mu_2 \sin\theta_2, \tag{1}$$

where the notation is the same as in § 10·1; and P_0 is therefore a position of stable equilibrium. The period of oscillation when the equilibrium is slightly disturbed is $2\pi/n$, where

$$n^2 = \frac{\mu_1 h_1^2}{r_1^3} + \frac{\mu_2 h_2^2}{r_2^3}.$$

It is clear, by resolving along the wire, that if a position of equilibrium exists it must satisfy the equation (1). If then, returning to the example in § 10·1, we are content to assume the existence of a unique minimizing path, the present problem determines the path without formal calculation of the minimum. We thus obtain a 'physical' deduction of Snell's law from Fermat's principle.

11·7 Unilateral constraint. A new feature appears when we consider a particle sliding on a fixed smooth surface, or a particle tied to a fixed point by a light inelastic string. These are examples of *unilateral* or *one-sided constraints*. The reaction of the smooth surface can only act outwards, away from the surface; the tension can only be positive. We must be prepared for the possibility that the particle will leave the surface, or that the string will slacken. In such cases the usual attack is to begin by ignoring the one-sided character of the constraint, and to study the motion on the assumption that the particle does remain on the curve. Then we calculate the value of the force of constraint required to maintain this motion. At an instant when the sign of this constraint changes from the permissible to the impossible the particle leaves the curve, and the motion that immediately follows is free motion in the field. In some cases the one-sided constraint may come into play again later on.

To begin with, however, we will confine our attention to complete two-sided constraints. We will take as the typical problem of constrained motion a bead sliding on a smooth fixed wire. Usually we shall be content to consider the simple case where the field is uniform; we can picture the problem as that of a heavy bead sliding on a fixed smooth wire in a vertical plane.

11·8 Lines of quickest descent. The simplest problem of constrained motion is that in which the wire is straight, and the field is uniform. Here the matter is very simple, and we can write down the complete solution without any appeal to the general theory. If the field is that of gravity and the wire makes an acute angle α with the downward vertical, the particle moves with uniform acceleration $g\cos\alpha$. If the particle starts from rest at $s = 0$ at the instant $t = 0$, we have, measuring s downwards,

$$s = \tfrac{1}{2}gt^2\cos\alpha.$$

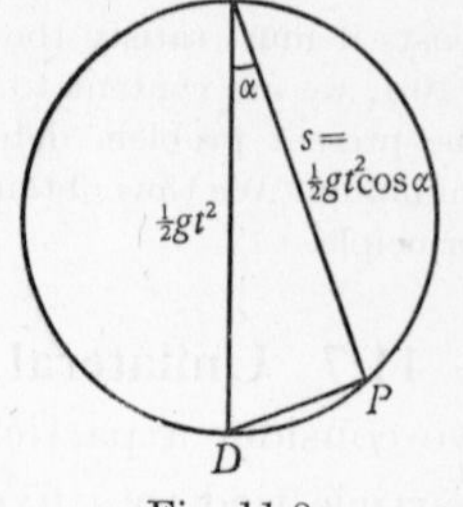

Fig. 11·8a

Suppose now we have a number of smooth straight wires through a point O, and particles sliding down the wires, all starting from rest at O at $t = 0$. The particles at time t all lie on a circle, the circle of diameter $\frac{1}{2}gt^2$ with O as highest point.

Consider, then, how to find the line of quickest descent from a point O to a given curve Γ, i.e. how to find the point Q on Γ so that the time taken by a bead to slide down a smooth straight wire from O to Q is a minimum. We must consider the family of circles with O as highest point. The chord of quickest descent from O to Γ is a chord of the smallest circle of the family which has a point in common with Γ. Usually this circle touches Γ. We think of a circle, with O as highest point, *growing* from the point-circle with radius zero, the radius steadily increasing until the critical instant when the circle makes contact with Γ at Q. At this instant the process of growth is stopped, and we have found the chord of quickest descent OQ.

The simplest problem is that in which Γ is a straight line, and the solution is obvious. If the horizontal line through O meets Γ in A, the point Q is the point on Γ below A such that $AQ = AO$ (Fig. 11·8b). The method fails when Γ is a horizontal line above O, when there is clearly no solution.

If Γ is a circle, and OQ meets Γ again in B, the tangents at O and at B are parallel, and B is the lowest point of Γ. Thus to find the line of quickest descent from a point to a circle we take the join of the point to the lowest point of the circle (Fig. 11·8*c*). There is no solution if the lowest point of Γ is higher than *O*.

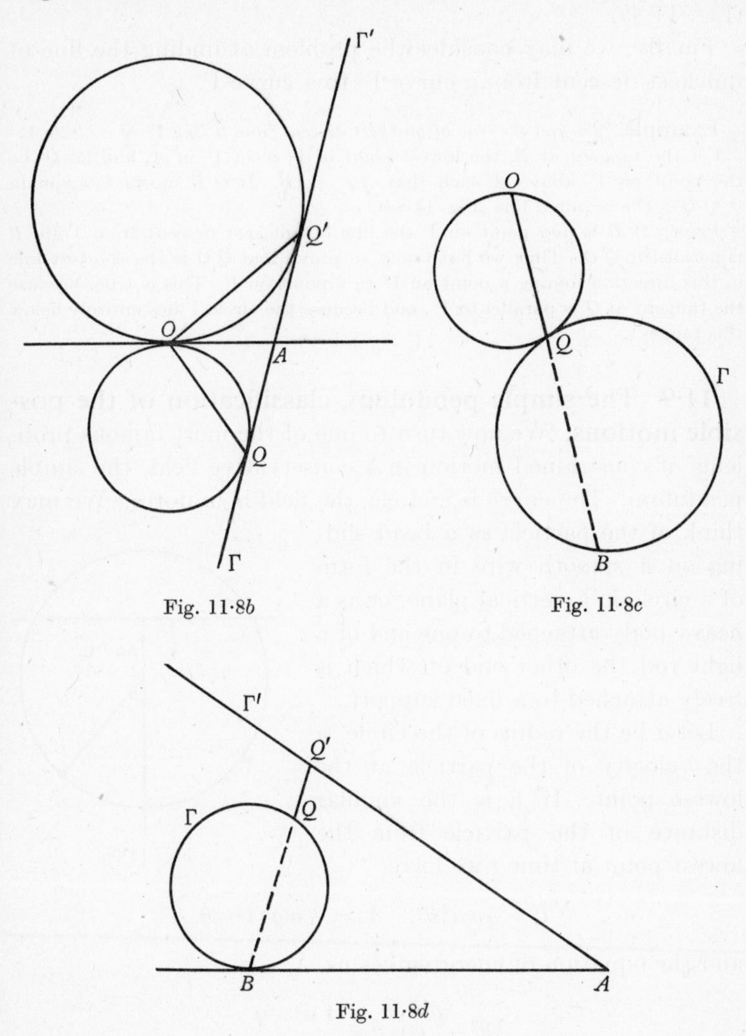

Fig. 11·8*b*

Fig. 11·8*c*

Fig. 11·8*d*

Again, we can find in a similar way the line of quickest descent *from* a curve Γ′ to a point *O*. This time we construct the family of circles having *O* as lowest point, and find the smallest of these circles which has a point *Q*′ in common with Γ′.

For example, if Γ' is a straight line the line of quickest descent is got by drawing the horizontal line through O to meet Γ' in A, and then Q' is the point on Γ' above A such that $AQ' = AO$ (Fig. 11·8*b*).

Finally, we may consider the problem of finding the line of quickest descent from a curve Γ' to a curve Γ.

Example. *To find the line of quickest descent from a line* Γ' *to a circle* Γ.

Let the tangent at B, the lowest point of Γ, meet Γ' in A, and let Q' be the point on Γ' above A such that $AQ' = AB$. If $Q'B$ meets Γ again in Q, $Q'Q$ is the required line (Fig. 11·8*d*).

Proof. If R is *any* point on Γ the line of quickest descent from Γ' to R is parallel to $Q'Q$. Thus we have only to prove that $Q'Q$ is the shortest line in this direction joining a point on Γ' to a point on Γ. This is true, because the tangent at Q is parallel to Γ', and because the circle Γ lies entirely below this tangent.

11·9 The simple pendulum, classification of the possible motions. We now turn to one of the most famous problems of constrained motion in a conservative field, the simple pendulum. The curve is a circle, the field is uniform. We may think of the particle as a bead sliding on a smooth wire in the form of a circle in a vertical plane, or as a heavy body attached to one end of a light rod the other end of which is freely attached to a fixed support.

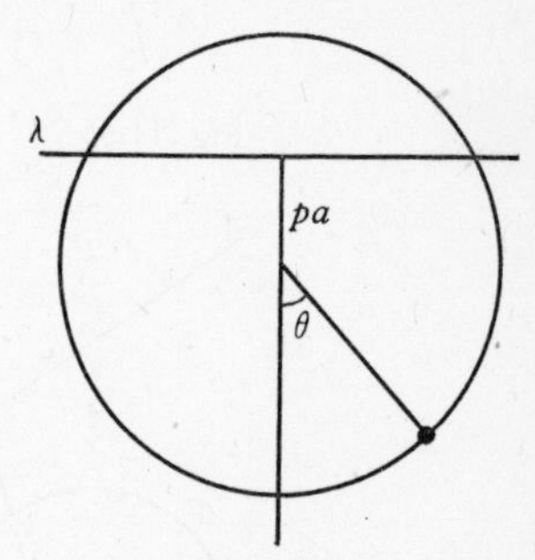

Fig. 11·9*a*

Let a be the radius of the circle, u the velocity of the particle at the lowest point. If θ is the angular distance of the particle from the lowest point at time t we have

$$T = \tfrac{1}{2}ma^2\dot{\theta}^2, \quad V = -mga\cos\theta,$$

and the equation of energy gives us

$$\tfrac{1}{2}\dot{\theta}^2 - \frac{g}{a}\cos\theta = \frac{1}{2}\frac{u^2}{a^2} - \frac{g}{a}.$$

If we write $$u^2 = 2ga(1+p) \quad (p \geqslant -1)$$

and $$\frac{g}{a} = n^2,$$

the equation takes the form

$$\tfrac{1}{2}\dot{\theta}^2 = n^2(\cos\theta + p). \qquad (1)$$

The horizontal line at height pa above the centre of the circle is the energy level λ for the problem (Fig. 11·9*a*).

The differential equation (1) is of the type that we have already studied in Chapter VI; we have only to look at the graph of

$$y = \cos\theta + p,$$

shown in Fig. 11·9*b*, and we see that there are four distinct cases arising from different values of p. (We assume throughout

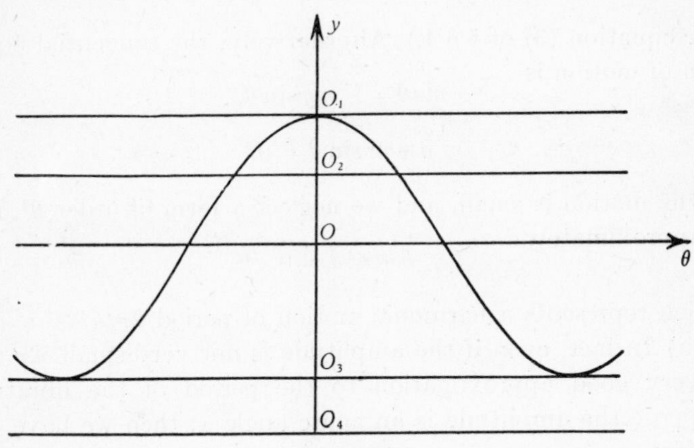

Fig. 11·9*b*

$p \geqslant -1$, since if $p < -1$ the energy level λ is lower than the lowest point of the circle, and there is no corresponding physical problem.) The four cases are these:

(i) $p = -1$ (origin at O_1); the particle rests in stable equilibrium at the lowest point $\theta = 0$.

(ii) $-1 < p < 1$ (origin at O_2); a periodic oscillation bounded by the points where λ cuts the circle; this is the typical 'pendulum motion'.

(iii) $p = 1$ (origin at O_3); the particle rests in unstable equilibrium at the highest point of the circle, $\theta = \pi$ (or $-\pi$), or there is a limitation motion in which $\theta \to \pi$ (or $-\pi$) as $t \to \infty$.

(iv) $p > 1$ (origin at O_4); λ lies above the highest point of the circle; if $\dot{\theta} > 0$ initially, θ increases indefinitely with t.

11·10 The simple pendulum, detailed study of the motion. It will be worth while, especially in view of the importance of this problem in the history of dynamics, to consider some of these cases in rather more detail.

(i) Suppose p is a little greater than -1; the particle is slightly disturbed from stable equilibrium at the lowest point. Then, as we saw in § 11·6, the motion is approximately a harmonic motion about $\theta = 0$ as centre. The period is $2\pi/n$; this follows either from the rule in § 11·6, or from the approximate equation of energy

$$\dot{\theta}^2 + n^2\theta^2 = \text{constant} \quad (n^2 = g/a).$$

(Cf. equation (3) of § 6·4.) Alternatively, the tangential equation of motion is

$$ma\ddot{\theta} = -mg\sin\theta,$$

i.e.

$$\ddot{\theta} + n^2\sin\theta = 0. \tag{1}$$

If the motion is small, and we neglect a term of order θ^3, this is approximately

$$\ddot{\theta} + n^2\theta = 0,$$

which represents a harmonic motion of period $2\pi/n$.

(ii) In fact, even if the amplitude is not very small, $2\pi/n$ is a very good approximation to the period of the libration. Suppose the amplitude is an acute angle α; then we have, for $0 \leqslant \theta \leqslant \alpha$,

$$\begin{aligned}\tfrac{1}{2}\dot{\theta}^2 &= n^2(\cos\theta - \cos\alpha)\\ &= n^2\int_\theta^\alpha \sin x\,dx\\ &= n^2\int_\theta^\alpha \frac{\sin x}{x}\,x\,dx.\end{aligned}$$

Take for definiteness a quarter-period in which θ and $\dot{\theta}$ are positive. As x increases from 0 to α, $\sin x/x$ decreases steadily from 1 to $\sin\alpha/\alpha$, so for a given θ,

$$\tfrac{1}{2}\dot{\theta}^2 < n^2\int_\theta^\alpha x\,dx = \tfrac{1}{2}n^2(\alpha^2 - \theta^2)$$

and

$$\tfrac{1}{2}\dot{\theta}^2 > n^2\frac{\sin\alpha}{\alpha}\int_\theta^\alpha x\,dx = \frac{\sin\alpha}{\alpha}\tfrac{1}{2}n^2(\alpha^2 - \theta^2).$$

The quarter period

$$\tfrac{1}{4}\sigma = \int_0^\alpha \frac{d\theta}{\dot\theta} > \frac{1}{n}\int_0^\alpha \frac{d\theta}{\sqrt{(\alpha^2-\theta^2)}} = \frac{\pi}{2n}$$

and $$\tfrac{1}{4}\sigma < \sqrt{\left(\frac{\alpha}{\sin\alpha}\right)}\frac{1}{n}\int_0^\alpha \frac{d\theta}{\sqrt{(\alpha^2-\theta^2)}} = \sqrt{\left(\frac{\alpha}{\sin\alpha}\right)}\frac{\pi}{2n}.$$

Thus the period σ is greater than the classical value $2\pi/n$, but less than $\sqrt{\left(\frac{\alpha}{\sin\alpha}\right)}\frac{2\pi}{n}$. Even if we take so large an amplitude as $30°$, $\sqrt{\left(\frac{\alpha}{\sin\alpha}\right)} = 1{\cdot}023\ldots$, and the classical value is only wrong by less than 2·4 per cent. If the amplitude is 5°, $\sqrt{\left(\frac{\alpha}{\sin\alpha}\right)} = 1{\cdot}0007\ldots$, and the classical value differs from the true period by less than one part in 1,000.

(iii) We can easily express the period of the oscillation, whether α is small or not, as a definite integral. We have

$$\sigma = \frac{4}{n}\int_0^\alpha \frac{d\theta}{\sqrt{\{2(\cos\theta-\cos\alpha)\}}}$$

$$= \frac{4}{n}\int_0^{\theta=\alpha} \frac{d(\frac{1}{2}\theta)}{\sqrt{(\sin^2\frac{1}{2}\alpha-\sin^2\frac{1}{2}\theta)}}.$$

If we use the substitution

$$\sin\tfrac{1}{2}\theta = k\sin\phi,$$

where $$k = \sin\tfrac{1}{2}\alpha$$

we obtain the classical formula for the period

$$\sigma = \frac{4}{n}\int_0^{\frac{1}{2}\pi} \frac{d\phi}{\sqrt{(1-k^2\sin^2\phi)}}.$$

We have already met integrals of this form in §§ 6·8 and 6·9.

(iv) Next, consider the limitation motion, case (iii) of § 11·9; suppose the particle is projected from the lowest point of the circle at $t = 0$ with velocity $2\sqrt{(ga)}$, which gives $p = 1$. Then

$$\tfrac{1}{2}\dot\theta^2 = n^2(1+\cos\theta),$$

whence $$\tfrac{1}{2}\dot\theta = n\cos\tfrac{1}{2}\theta.$$

The relation between time and position is

$$nt = \int_0^{\frac{1}{2}\theta} \sec x\,dx = \log(\sec\tfrac{1}{2}\theta + \tan\tfrac{1}{2}\theta).$$

We can write this in a simpler form. Since

$$\sec\tfrac{1}{2}\theta + \tan\tfrac{1}{2}\theta = e^{nt},$$

we have

$$\sec\tfrac{1}{2}\theta - \tan\tfrac{1}{2}\theta = e^{-nt},$$

whence

$$\tan\tfrac{1}{2}\theta = \sinh nt,$$

$$\sec\tfrac{1}{2}\theta = \cosh nt,$$

and

$$\sin\tfrac{1}{2}\theta = \tanh nt.$$

The last of these is perhaps the simplest form of the relation connecting t and θ. We see again that $\theta \to \pi$ as $t \to \infty$, as we know already from the general theory.

If we write $\theta = \pi - \phi$ we have

$$\cot\tfrac{1}{2}\phi = \sinh nt;$$

when t is large and ϕ is small this is approximately

$$\phi = 4e^{-nt}.$$

(v) Finally, in case (iv) of § 11·9, $\dot{\theta}$ remains positive, and θ increases always with t. But here a new and important phenomenon makes its appearance. Although θ always increases, the values θ and $\theta + 2\pi$ represent the same point on the circle; θ is said to be a *cyclic coordinate*. If the particle starts from the lowest point with velocity u ($> 2\sqrt{(ga)}$) it returns, after a certain interval σ, to this same point with the same velocity u. Then the same motion is repeated in the interval $\sigma < t < 2\sigma$, and so on in successive intervals $r\sigma < t < (r+1)\sigma$. So we again have a *periodic motion* though the values of θ are not repeated in successive periods, unless indeed we replace the value of θ at time t by the remainder after subtracting an integral multiple of 2π. We can if we wish bring this periodic motion also into the general category of libration motions by working with another coordinate instead of θ, say with $z = \cos\theta$; the z-motion is a libration between the values $+1$ and -1. But this procedure would be rather perverse; indeed, our more usual procedure is the very opposite, to replace the oscillating variable z in the libration motion by a variable θ which continually increases (§ 6·5).

The value of σ is easily found. Since

$$(\tfrac{1}{2}\dot{\theta})^2 = \tfrac{1}{2}n^2(\cos\theta + p)$$
$$= n^2\left(\frac{1+p}{2} - \sin^2\frac{\theta}{2}\right),$$

where $p > 1$, we have

$$(\tfrac{1}{2}\dot{\theta})^2 = \frac{n^2}{k^2}\left(1 - k^2\sin^2\frac{\theta}{2}\right),$$

where $k^2 = 2/(1+p)$. Hence

$$\sigma = \frac{2k}{n}\int_0^{\frac{1}{2}\pi} \frac{d\phi}{\sqrt{(1-k^2\sin^2\phi)}}.$$

11·11 The simple pendulum, unilateral constraint. Suppose now that we have a simple pendulum in which a particle is tied to a fixed point by a light *string* of length a. The theory we have discussed already remains valid so long as the tension T in the string is positive. But so soon as we require a negative tension to keep the particle moving in the circle the former theory breaks down. Then the particle leaves the circle and moves freely under gravity until the string tightens again.

Suppose, as before, that we project the particle from the lowest point with velocity u, where

$$u^2 = 2ga(1+p).$$

Then, assuming for the moment that the particle does remain on the circle,

$$\tfrac{1}{2}\dot{\theta}^2 = n^2(\cos\theta + p),$$

and from the radial equation of motion

$$T - mg\cos\theta = ma\dot{\theta}^2 = 2mg(\cos\theta + p),$$

giving

$$T = 3mg(\cos\theta + \tfrac{2}{3}p).$$

If, during the motion, the requisite value of T becomes negative, the assumption that the particle remains on the circle is falsified.

Now if $-1 \leqslant p < 0$ there is no danger, because $\cos\theta + p \geqslant 0$ during the motion, and *a fortiori* $\cos\theta + \frac{2}{3}p > 0$, so $T > 0$. If $p = 0$, both $\dot\theta^2$ and T are proportional to $\cos\theta$, and T just vanishes at the end of the range ($\theta = \frac{1}{2}\pi$) but is positive elsewhere. Suppose, then, $p > 0$, and write $\cos\theta = -z$, so that z is proportional to the height above the centre of the circle. We have

$$\dot\theta^2 \propto p - z,$$

and

$$T \propto \tfrac{2}{3}p - z.$$

If $0 < p < 1$, the motion takes place in the range $-1 \leqslant z \leqslant p$ ($z = p$ is on the energy level), and $T < 0$ in the range $\frac{2}{3}p < z \leqslant p$. If $1 < p < \frac{3}{2}$, the motion takes place in the range $-1 \leqslant z \leqslant 1$, and $T < 0$ in the range $\frac{2}{3}p < z \leqslant 1$. If $p > \frac{3}{2}$, T remains positive always.

Thus the particle leaves the circle (when $z = \frac{2}{3}p$) if $0 < p < \frac{3}{2}$, i.e. if

$$2ga < u^2 < 5ga.$$

Example. *A particle slides on a solid circular cylinder, which is fixed with its axis horizontal. The surface is smooth, and the particle is let go from rest at an angular distance α from the highest generator. To find where the particle leaves the cylinder.*

The motion takes place in a vertical plane perpendicular to the axis. If we denote the angular distance of the particle from the highest point of the circle by θ, and if N is the reaction of the cylinder on the particle (along the outward normal), we have, so long as the particle remains on the cylinder,

$$N - mg\cos\theta = -ma\dot\theta^2.$$

If we substitute in this equation for $\dot\theta^2$ from the energy equation

$$\tfrac{1}{2}ma^2\dot\theta^2 = mga(\cos\alpha - \cos\theta),$$

we find

$$N = mg(3\cos\theta - 2\cos\alpha).$$

Thus N remains positive so long as $\cos\theta > \frac{2}{3}\cos\alpha$. The particle leaves the cylinder when its height above the axis is two-thirds of its original height.

11·12 Other problems of small oscillations; the cycloid.

We return now to ordinary (two-sided) constraints. The small oscillation of the simple pendulum is a phenomenon which recurs in any problem of motion on a smooth curve if the curve has a point O where the tangent is horizontal and concavity upwards. We shall suppose that the radius of curvature at O is finite.

We take as axes the horizontal and vertical lines through O (Fig. 11·12a) and take the positive sense to the right for Ox and upwards for Oy. We denote the arc-length measured from O to a point P of the curve by s, and the inclination of the tangent at P to the horizontal by ψ. Taking $V = 0$ at O we

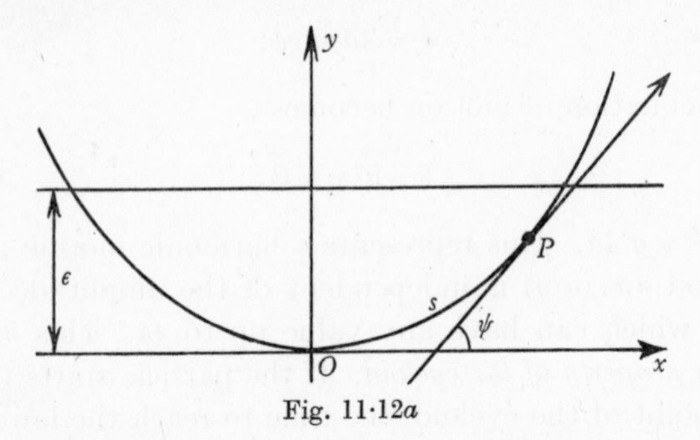

Fig. 11·12a

have $V = mgy$, and it is clear that the function $V(s)$, defining V at points on the curve, has a minimum at O. If the energy constant has a sufficiently small value $mg\epsilon$ the motion is a libration bounded by the points where the horizontal line $y = \epsilon$ cuts the curve. For small values of ϵ the motion is approximately a harmonic motion with the period $2\pi/n$, where $mn^2 = V''(0)$. Now

$$\frac{dV}{ds} = mg\frac{dy}{ds} = mg\sin\psi,$$

$$\frac{d^2V}{ds^2} = mg\cos\psi\frac{d\psi}{ds} = mg\frac{\cos\psi}{\rho},$$

so $n^2 = g/\rho_0$, where ρ_0 is the radius of curvature at O. Of course this is the result we anticipate from the theory of the simple pendulum. An alternative method of proof comes from the tangential equation of motion (cf. § 11·10 (i)). If $s = s(t)$ is the position of the particle on the curve at time t we have

$$\ddot{s} = -g\sin\psi.$$

In the motion we are considering s remains small for all time, and when s is small $\sin\psi$ is approximately equal to s/ρ_0. We thus obtain as a first approximation to the equation of motion

$$\ddot{s} + \frac{g}{\rho_0}s = 0,$$

and this represents a harmonic motion in s about the position of stable equilibrium with period $2\pi\sqrt{(\rho_0/g)}$.

There is one curve for which the motion in s is *accurately* harmonic, even when the amplitude is not small, and that is the curve for which $s \propto \sin\psi$. This curve is the *cycloid*. We write

$$s = 4a \sin\psi,$$

and the equation of motion becomes

$$\ddot{s} + n^2 s = 0,$$

where $n^2 = g/4a$. This represents a harmonic motion in s, and the period $4\pi\sqrt{(a/g)}$ is independent of the amplitude b of the motion, which can have any value up to $4a$. This is *the isochronous property of the cycloid*. If the particle starts from rest at any point of the cycloid, the time to reach the lowest point is always the same, the quarter-period $\pi\sqrt{(a/g)}$.

The cycloid is the path of a point P on the rim of a wheel that rolls without skidding on a straight line; we have met the curve already (§ 6·12). In the figure (Fig. 11·12b) the wheel, of radius a, rolls on the under side of the horizontal line AB. Since there is no skidding, when the wheel has turned through an angle θ, its centre has moved through a distance $a\theta$. We take

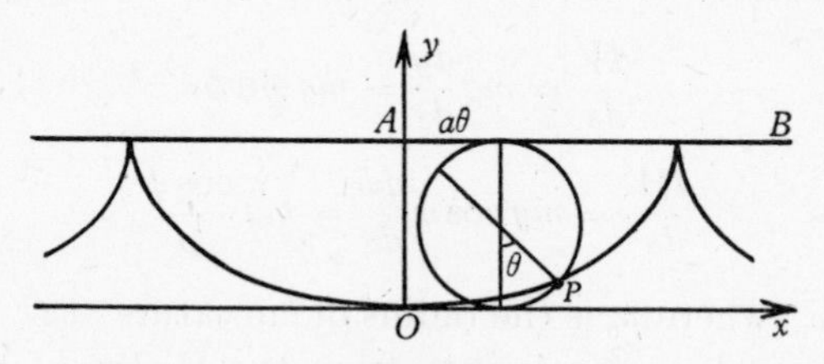

Fig. 11·12b

as origin the lowest position of P, and, with the axes shown in the figure, we have, considering first the range $0 \leqslant \theta \leqslant \pi$,

$$\begin{cases} x = a(\theta + \sin\theta), \\ y = a(1 - \cos\theta). \end{cases}$$

The coordinates of a point on the curve are thus expressed in terms of the parameter θ. Also

$$\frac{dy}{dx} = \frac{\sin\theta}{1+\cos\theta} = \tan\frac{\theta}{2},$$

and

$$\psi = \tfrac{1}{2}\theta.$$

To find the intrinsic equation we have

$$ds = \sec\psi\,dx$$
$$= a\sec\psi(1+\cos\theta)\,d\theta$$
$$= 4a\cos\psi\,d\psi,$$

since $\theta = 2\psi$. Thus
$$s = 4a\sin\psi,$$

and this is the equation already quoted. The path of P as the wheel continues to roll consists of a series of inverted arches, with cusps on AB. The equation we have found represents, in the range $-\frac{1}{2}\pi \leqslant \psi \leqslant \frac{1}{2}\pi$, a single arch between two cusps.

A particularly interesting case occurs when the amplitude has its maximum value $4a$; the particle falls from rest at a cusp. Then in the subsequent motion

$$s = 4a\cos nt,$$

so that
$$\cos nt = \sin\psi.$$

It is easy to see from this equation how ψ varies with t. For the first half-period, from $t = 0$ to $t = \pi/n$,

$$\psi = \tfrac{1}{2}\pi - nt,$$

and for the second half-period, from $t = \pi/n$ to $t = 2\pi/n$,

$$\psi = -\tfrac{3}{2}\pi + nt.$$

The angular velocity of the tangent during the motion is equal always either to n or to $-n$.

11·13 Some general theorems on constrained motion. The general form of some theorems, of which special cases have already appeared in the problem of the simple pendulum, may be of interest at this point.

As we have seen, if a particle moves in a plane conservative field of force, either freely or on a smooth guiding curve, there is an equation of energy

$$\tfrac{1}{2}mv^2 = C - V.$$

The constant C is fixed so soon as we know the origin from which the particle is projected and the speed of projection.

The particle cannot leave the region defined by the inequality $V \leqslant C$. This region is bounded by the energy level λ, $V = C$. It can, with a suitable guiding curve, reach any point A of this region in a finite time. (An exception may occur if there is a point of equilibrium on λ, i.e. a point at which $\mathbf{P} = \mathbf{0}$, but we will suppose, as is usually true, that no such point exists.) No doubt of the truth of this theorem arises when A is an interior point of the region. Suppose, therefore, that A lies on λ. Then, if we provide a guiding curve which cuts λ at a finite angle at A, the particle reaches A in a finite time.

To prove this we have only to show that $C-V$, considered as a function of the arc-length s of the guiding curve, has a *simple* zero at A. If we take axes as in the figure, with A as origin, and the tangent to λ as axis of x, we have $\partial V/\partial x = 0$ at A; but $\partial V/\partial y \neq 0$ at A, since by hypothesis A is not a

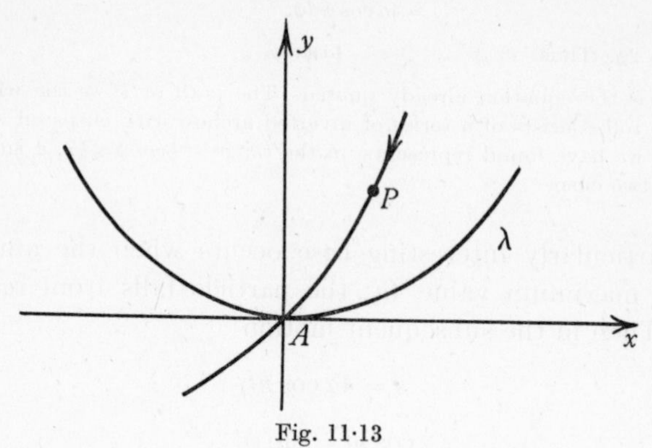

Fig. 11·13

point of equilibrium. (Actually, of course, in the case shown in Fig 11·13 $\partial V/\partial y < 0$ at A, since the particle approaches A from the side $y > 0$.) Now if we consider $C-V$ as a function of s, we have

$$C-V=0$$

at A, but $d(C-V)/ds \neq 0$. For

$$\frac{dV}{ds}=\frac{\partial V}{\partial x}\frac{dx}{ds}+\frac{\partial V}{\partial y}\frac{dy}{ds}=\frac{\partial V}{\partial y}\frac{dy}{ds}.$$

Now $\partial V/\partial y \neq 0$ at A, and $dy/ds \neq 0$, since the guiding curve cuts λ at a finite angle. Thus $dV/ds \neq 0$, and $C-V$ has merely a simple zero at A. We know, therefore, from the theory of the differential equation

$$\tfrac{1}{2}m\dot{s}^2 = C-V,$$

that the particle reaches A in a finite time.

If, however, the guiding curve *touches* λ at A, the particle does not reach A in a finite time. In this case $dy/ds = 0$ at A, so both $C-V$ and $d(C-V)/ds$ vanish at A. Thus $C-V$ has a multiple zero, and the motion on the guiding curve is a limitation motion.

Examples of both cases occur in the theory of the simple pendulum. In (ii), § 11·9, the guiding curve (the circle) cuts λ at a finite angle, and the particle reaches λ in a finite time. In (iii) the guiding curve touches λ (at the highest point of the circle) and the particle does not reach λ in a finite time.

11·14 The brachistochrone. Another very famous problem of constrained motion is that of the *brachistochrone*. Two points O and A are given, and a wire, in the form of a simple curve in a vertical plane, joins O to A. What is the shape of this curve if a bead, sliding on the wire and starting from rest at O, reaches A in the shortest possible time? (Of course A must not be at a higher level than O, or there is no solution.)

A complete discussion of the brachistochrone problem is a little beyond the scope of this book, but it is easy to produce a plausible argument which does in fact lead to the correct result. The essence of the argument is that, in virtue of the energy equation, the speed depends only on the depth below O. Let us therefore start with the simpler problem in which the plane is divided into strips by a number of horizontal lines, the velocity v in each strip being constant and prescribed in advance. The shortest-time path from one point of the plane to another will consist of a number of straight segments (Fig. 11·14a), and the directions of these segments will satisfy the condition

$$\frac{\sin\psi}{v} = \text{constant}$$

where ψ denotes the inclination of the segment to the vertical (§ 10·1, example).

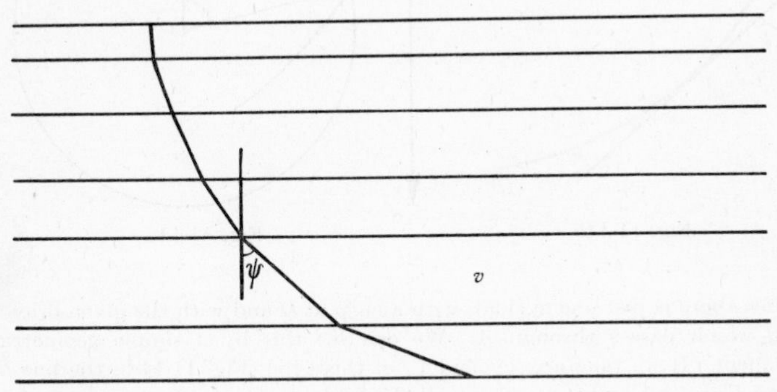

Fig. 11·14a

If now the strips are very narrow, the prescribed velocity in a strip being $\sqrt{(2g\xi)}$, where ξ is the average depth of the strip below O, we have a close approximation to the problem proposed. We can think of the brachistochrone problem as a limiting case when the width of the strips tends to zero. This suggests that in the shortest-time path from O to A, if it exists, the tangent at depth x below O will make an angle ψ with the vertical such that $\sin\psi \propto \sqrt{x}$.

Let us assume that this is true, and write

$$x = 2a\sin^2\psi,$$

where a is a positive constant. We should expect on general common-sense grounds that $\psi = 0$ at $x = 0$; the particle gets a good start on its journey by moving vertically downwards from O.

To find y in terms of ψ, Oy being horizontal, we have

$$dy = \tan\psi\, dx = 4a\sin^2\psi\, d\psi,$$

$$y = a(2\psi - \sin 2\psi).$$

Thus we have arrived at the curve whose parametric expression in terms of ψ is

$$\begin{cases} x = a(1-\cos 2\psi), \\ y = a(2\psi - \sin 2\psi), \end{cases}$$

and this is a cycloid (Fig. 11·14b), with the line of cusps horizontal, one cusp being at O. Fig. 11·14*c* shows how the formulae given arise from the definition of a cycloid as the path of a point on the rim of a rolling wheel.

We are thus led to suppose that the shortest-time path from O to A will be an arc of a cycloid, and this result is correct; but the argument given here is clearly incomplete, especially if the cycloidal arc joining O to A contains a point at which the tangent is horizontal.

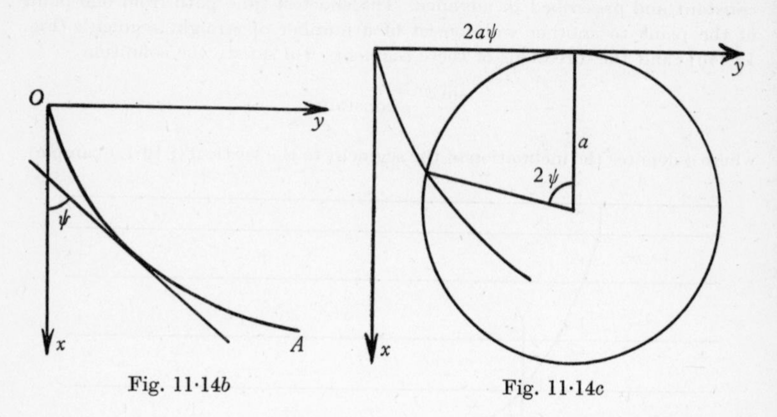

Fig. 11·14*b* Fig. 11·14*c*

Now there is just one cycloid, with a cusp at O and with the given orientation, which passes through A. We can see this by a simple geometrical argument. If we take any cycloid Γ_0 of this type (Fig. 11·14*d*), the line OA meets Γ_0 in one point A_0. Since all the cycloids are similar curves we have only to enlarge Γ_0 in the ratio $OA : OA_0$ and we have a unique cycloid joining O to A.

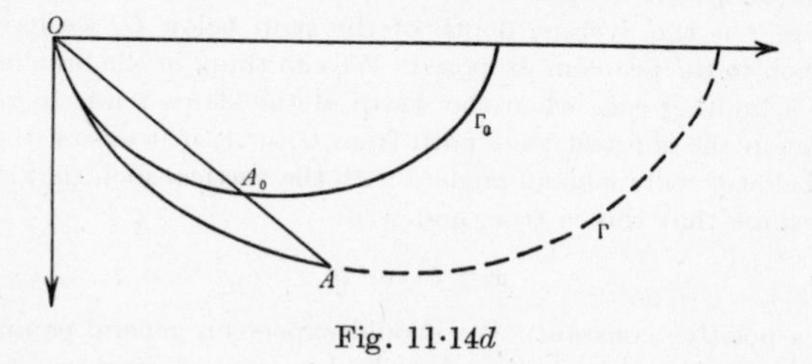

Fig. 11·14*d*

The actual value of t_0, the time from O to A, is now easily found. We have

$$t_0 = \int \frac{ds}{v} = \frac{1}{\sqrt{(2g)}} \int_0^a \sqrt{\left(\frac{1+y'^2}{x}\right)}\, dx = \sqrt{\left(\frac{4a}{g}\right)} \int_0^\beta d\psi,$$

where β is the value of ψ at A. Thus

$$t_0 = \sqrt{\left(\frac{4a}{g}\right)}\, \beta = \frac{\beta}{n}$$

if we write $n^2 = g/4a$ as before. Of course we could have anticipated this result from § 11·12.

In particular, for two points O, A at the same level, and at distance $2b$ apart, we have $b = \pi a$, and the minimum time from O to A is

$$\frac{\pi}{n} = 2\pi\sqrt{\left(\frac{a}{g}\right)} = 2\sqrt{\left(\frac{\pi b}{g}\right)} = 3{\cdot}54\ldots\sqrt{\left(\frac{b}{g}\right)}.$$

It may be of interest to compare this result with the least time from O to A by a path consisting of two vertical pieces and a horizontal piece as

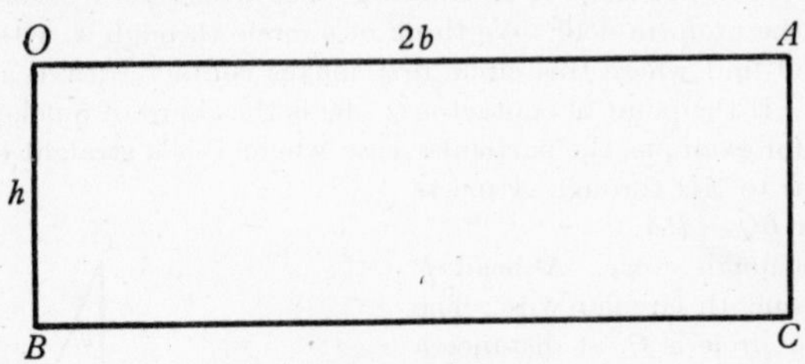

Fig. 11·14e

shown (Fig. 11·14e). The corners at B and C are rounded, so that there is no discontinuity of velocity at B and C. If the depth of BC below OA is h, the time from O to A is

$$2\sqrt{\left(\frac{2h}{g}\right)} + \frac{2b}{\sqrt{(2gh)}} = 2\sqrt{\left(\frac{b}{g}\right)}\left[\sqrt{\left(\frac{2h}{b}\right)} + \sqrt{\left(\frac{b}{2h}\right)}\right],$$

and the minimum value of this, which occurs when $h = \frac{1}{2}b$, is $4\sqrt{(b/g)}$.

11·15 Harmonic field.

So far in our discussion of constrained motion we have dealt almost entirely with the uniform field of gravity, though the theory we have developed is of course applicable also to other conservative fields. We take, as an example of a field other than the uniform field, the harmonic field, a central attraction mn^2r towards a fixed point O of the plane. We will consider for this field some of the problems we have already discussed for the uniform field.

(i) *Lines of quickest descent.* A bead P slides on a smooth straight wire, starting at $t = 0$ from rest at a point A at distance b from O. The wire makes an angle θ with AO. We denote the foot of the perpendicular from O on the wire by M, and the distance MP at time t by x. Then with the notation of Fig. 11·15a,

$$m\ddot{x} = -mn^2r\cos\phi = -mn^2x,$$

whence

$$\ddot{x} + n^2x = 0.$$

The solution of this, with $x = b\cos\theta$ and $\dot{x} = 0$ at $t = 0$, is

$$x = b\cos\theta\cos nt,$$

so

$$AP = b(1-\cos nt)\cos\theta.$$

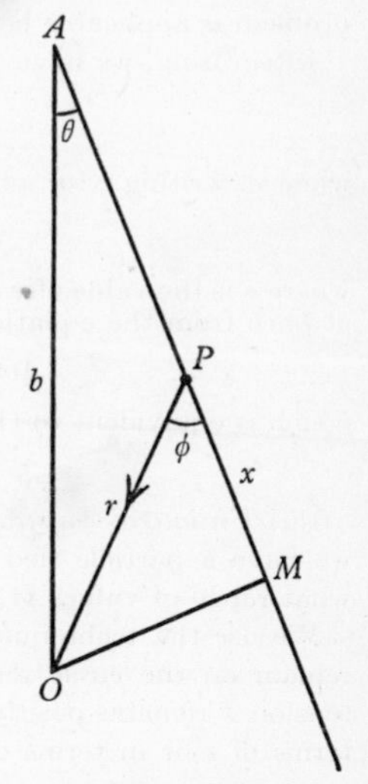

Fig. 11·15a

If therefore we think of a number of wires through A, all in a given plane through A and O, and beads starting from rest at A at $t=0$, the beads at time t all lie on the circle through A whose centre lies on AO, and whose diameter is $b(1-\cos nt)$. This diameter grows from zero to $2b$ in time π/n, and it never exceeds the value $2b$.

We can now easily determine the line of quickest descent from A to a given curve Γ; of course Γ must have a point inside or on the circle through A with O as centre, or there is no solution. The technique is now almost the same as for the uniform field. We think of a circle through A with its centre C in AO, and find where this circle first makes contact with Γ as C moves from A to O. If the point of contact is Q, AQ is the chord of quickest descent.

Consider, for example, the particular case where Γ is a straight line. If the perpendicular to AO through A meets Γ in B, then $BQ = BA$.

(ii) *Motion on a circle.* A bead P slides on a smooth circular wire. The centre of the circle is C, at distance h from O, and the radius of the circle is a. We will suppose for definiteness $h>a$. Denoting the angle PCO at time t by θ we have, with the notation of Fig. 11·15b,

$$ma\ddot{\theta} = -mn^2 r\sin\phi = -mn^2 h\sin\theta,$$

whence $$\ddot{\theta}+n^2\frac{h}{a}\sin\theta = 0.$$

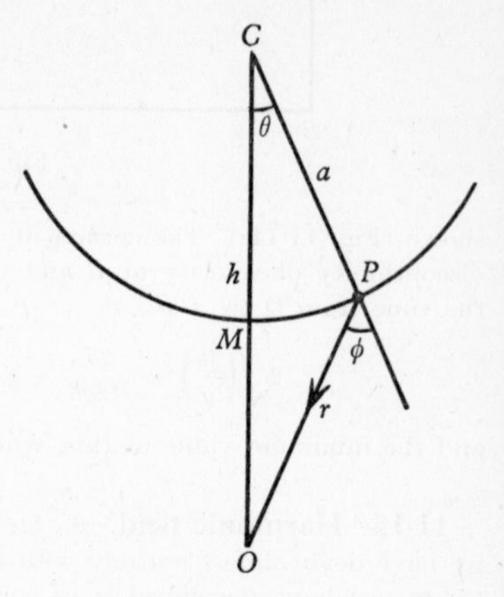

Fig. 11·15b

This is similar to the equation of motion for a simple pendulum under gravity (§ 11·9). The classification of the possible motions already given for that problem is applicable here.

Integrating, we have

$$\tfrac{1}{2}\dot{\theta}^2-n^2\frac{h}{a}\cos\theta = \text{constant}, \tag{1}$$

whence, writing v for $a\dot{\theta}$, the velocity of P on the circle, we have

$$v^2 = 2n^2ah\cos\theta+u^2-2n^2ah,$$

where u is the value of v at M. Of course the same result can be written down at once from the equation of energy

$$\tfrac{1}{2}m(v^2+n^2r^2) = \tfrac{1}{2}m\{u^2+n^2(h-a)^2\},$$

which is equivalent to (1), since

$$r^2 = a^2+h^2-2ah\cos\theta.$$

(iii) *Unilateral constraint.* If instead of a bead sliding on a smooth wire we have a particle tied to C by a light inextensible string of length a, for what range of values of u will the string remain taut?

We use the technique explained in § 11·7, i.e. we suppose the particle to remain on the circle and find for what range of values of u the requisite tension T remains positive. To study the variation of T we can express T in terms of r or in terms of θ; perhaps r is simpler. We have

$$T-mn^2r\cos\phi = mv^2/a,$$

and remembering that $$v^2 = u^2+n^2\{(h-a)^2-r^2\},$$
$$h^2 = a^2+r^2+2ar\cos\phi,$$
we find $$2aT/m = 2u^2+n^2(3h^2-4ha+a^2-3r^2).$$
We see that T is least when r is greatest.

Now $v^2>0$ at the point of the circle farthest from O if
$$u^2>4n^2ha,$$
and in this case the particle moves right round the circle. The tension remains positive throughout if $T>0$ for $r=h+a$, i.e. if
$$u^2>n^2a(5h+a).$$
If $$u^2<4n^2ha$$
the value of r varies from its least value $h-a$ to a greatest value given by
$$r^2 = (u/n)^2+(h-a)^2.$$
$T>0$ throughout if it is positive for the greatest value of r, i.e. if
$$u^2<2n^2a(h-a).$$
Thus the string will slacken, and the particle will leave the circle, if u^2 lies between
$$2n^2a(h-a) \quad \text{and} \quad n^2a(5h+a).$$
For the lower of these limits the maximum value of r, where T just vanishes, is $\sqrt{(h^2-a^2)}$; at this point OP is a tangent to the circle.

EXAMPLES XI

1. A centre of force is said to be of strength μ if it exerts a radial force μr on a particle of unit mass at distance r from it, μ being a constant. A particle moves in a plane under the influence of two such centres of force in the plane of strengths μ_1, μ_2 respectively. Show that, if $\mu_1+\mu_2$ does not vanish, the particle moves as if under the influence of a single centre of force of strength $\mu_1+\mu_2$ situated at a certain point in the plane, and that if $\mu_1+\mu_2$ vanishes the particle moves in a constant field of force.

Generalize the result to the case of n centres of force of strengths $\mu_1, \mu_2, \ldots, \mu_n$ distributed arbitrarily in a plane, obtaining the condition that the field becomes uniform in a specified direction.

Obtain the corresponding condition for the case of n centres of force, where the sth centre exerts not a radial but a *tranverse* force $\mu_s r_s$ on a particle of unit mass at distance r_s from it ($s = 1, 2, \ldots, n$).

2. A bead slides on a smooth straight wire under gravity. Show how to find the chord of quickest descent from rest at a point O to a curve Γ in a vertical plane through O.

Consider the case in which Γ is a parabola, of latus rectum $4a$, with its axis vertical and vertex downwards, and O is a point on the axis at a height c above the vertex. Prove that the time t of quickest descent from O to the parabola is given by
$$\tfrac{1}{2}gt^2 = 4\sqrt{a}(\sqrt{c}-\sqrt{a})$$
if $c>4a$, and by
$$\tfrac{1}{2}gt^2 = c$$
if $c<4a$.

3. A bead starting from rest slides on a smooth straight wire in a vertical plane. Find the chord of quickest descent from one circle to another.

4. The acceleration g of gravity at a point on the earth's surface at sea-level is given approximately by the formula $g = g_0(1-a\cos 2\lambda)$, where g_0 and a are constants and λ is the latitude of the point. At the equator $g = 32{\cdot}091$ ft./sec./sec., and at the North Pole $g = 32{\cdot}252$ ft./sec./sec. approximately. Discuss the variation of the period of oscillation of a simple pendulum with λ, and show that the change in this period due to a small change $\delta\lambda$ in λ is approximately equal to that due to a percentage shortening of the pendulum of amount $\frac{1}{2}\delta\lambda\sin 2\lambda$.

5. A heavy particle of weight W, attached to a fixed point by a light inextensible string, describes a circle in a vertical plane. The tension in the string has the values mW and nW, respectively, when the particle is at the highest and lowest points in its path. Show that $n = m+6$.

6. A heavy particle of weight W, attached to a fixed point by a light inelastic string, describes a circle in a vertical plane. The velocities at the highest and lowest points are V and nV respectively. Show that when the inclination of the string to the downward vertical is θ the tension in the string is

$$\left\{\frac{2(n^2+1)}{n^2-1}+3\cos\theta\right\}W.$$

Hence find the limits within which n must lie if the string cannot support a tension greater than $9W$.

7. A particle is moving under gravity on a smooth curve in a vertical plane. Obtain equations determining the velocity and pressure on the curve at any point.

If the curve is a vertical circle of radius a and the velocity at the top is $\sqrt{(ga)}$, find at what point of the curve the horizontal component of the pressure on the curve is a maximum, and determine its value there.

8. A particle is free to move on a smooth vertical circle of radius a. It is projected from the lowest point with velocity just sufficient to carry it to the highest point. Show that after a time

$$\log(\sqrt{5}+\sqrt{6})\sqrt{(a/g)},$$

the reaction between the particle and the wire is zero.

9. On a smooth plane inclined at an angle α to the horizontal a particle is lying at rest attached to a fixed point above the plane by an inextensible string making an acute angle β with the plane. Prove that it is possible to project the particle so that it describes a complete circle on the plane if $\cot\alpha \geqslant 6\tan\beta$.

10. A box of mass M rests on a rough horizontal table, and from the centre of the lid of the box there hangs a pendulum of length l whose bob has mass m. The pendulum vibrates through a right angle on either side of the vertical. Assuming that the box does not tilt, show that it does not slide on the table if the coefficient of friction between box and table is greater than

$$\frac{3m}{2\sqrt{\{M(M+3m)\}}}.$$

11. A bullet of small mass m was fired into the bob of a ballistic pendulum of mass M, in which it remained embedded. The pendulum was subject to a small frictional force $2k \times$ (velocity) per unit mass, and the observed time of a small oscillation was T. If x was the horizontal displacement of the pendulum from its equilibrium position, show that

$$\frac{d^2x}{dt^2} + 2k\frac{dx}{dt} + \left(\frac{4\pi^2}{T^2} + k^2\right)x = 0.$$

If x_1 was the horizontal departure of the pendulum from its equilibrium position, in the direction in which the bullet was moving, on the first swing after the impact, and x_n that on the nth in the same direction, show that the velocity of the bullet at impact was

$$\frac{2\pi}{T}.\frac{M+m}{m}.x_1.\left(\frac{x_1}{x_n}\right)^{\frac{1}{4(n-1)}},$$

approximately, provided that $2\pi/kT$ is large.

12. Two small spheres A and B of equal mass m, the coefficient of restitution between which is e, are suspended in contact by two equal vertical strings so that the line of centres is horizontal. The sphere A is drawn aside through a small distance and allowed to fall back and collide with the other, its velocity on impact being u. Show that all subsequent impacts occur when the spheres are in the same position as for the first impact, and that the velocity of the sphere A immediately after the third impact is $\frac{1}{2}u(1-e^3)$. Show that the kinetic energy of the system tends to the value $\frac{1}{4}mu^2$.

13. The bob P, of mass m, of a simple pendulum OP, of length l, is projected horizontally from its lowest position A with velocity u. Find the tension in the string when the angle POA is θ. Show that the string will never slacken, if u^2 does not lie between $2gl$ and $5gl$.

Two equal pendulums OP, CQ, of length l, are suspended from two points O, C in a horizontal line, such that when the bobs are hanging at rest they are just in contact. The bob P is projected horizontally with velocity $\sqrt{(gl)}$ from the point at height l vertically above O, and strikes the bob Q which was previously hanging at rest. Show that the string of Q will become slack before Q reaches its highest point, if the coefficient of restitution lie between $\sqrt{(1\cdot 6)}-1$ and unity.

14. A bead slides on a smooth parabolic wire held in a vertical plane with axis vertical and vertex downwards. Prove that the square of the normal reaction at any point is inversely proportional to the cube of the height of the point above the directrix.

15. A particle is constrained to move on a smooth curve under gravity. Find a general formula for the pressure on the curve, and show that if the curve is a parabola with axis vertical and vertex upwards the pressure on the curve is numerically

$$m(u_0^2-gl)/\rho,$$

where u_0 is the velocity at the vertex, $2l$ the latus rectum and ρ the radius of curvature at any point.

What is the meaning of the special case $u_0^2 = gl$?

16. A particle placed close to the vertex of a smooth cycloid whose axis is vertical and vertex upward is allowed to run down the curve under gravity. Show that it leaves the curve when it is moving in a direction making an angle of 45° with the horizontal.

17. A particle is projected horizontally from the lowest point of a smooth elliptic arc, whose major axis $2a$ is vertical, and moves under gravity along the concave side. Prove that it will leave the curve if the velocity of projection lies between $\sqrt{(2ga)}$ and $\sqrt{\{ga(5-e^2)\}}$, where e is the eccentricity of the ellipse.

18. A particle oscillates on a smooth cycloid from rest at a cusp, the axis being vertical and the vertex downwards. Show that (i) the hodograph is a pair of circles touching each other, (ii) the resultant acceleration of the particle is equal to g, (iii) the pressure on the curve is $2mg\cos\psi$, where ψ is the inclination of the tangent to the horizontal, (iv) the projection of the particle on the axis has simple harmonic motion.

19. Obtain a general formula for the pressure exerted by a smooth wire in a vertical plane on a particle constrained to move on the wire under gravity.

If the wire is in the form of the cycloid

$$x = a(\theta + \sin\theta), \quad y = a(1-\cos\theta)$$

with the axis of x horizontal and the axis of y vertically upwards, and the velocity at the origin is $\sqrt{(2ag)}$, calculate the pressure when the particle is at the point θ.

20. A smooth rigid wire bent in the form of a cardioid with polar equation $r = a(1+\cos\theta)$ is held fixed with axis vertical and cusp above the vertex. A bead is projected horizontally at the lowest point and slides up the wire until it comes to rest at the level of the cusp. Find the velocity of projection and prove that the pressure of the bead on the wire vanishes when $r = 6a/5$.

21. A light inextensible string AB of length l has the end A attached to a point of the surface of a fixed cylinder, whose cross-section is a simple closed oval curve whose intrinsic equation is $s = f(\psi)$. Both ψ and s vanish at A, and s increases always with ψ. A particle of mass m is attached to B, and is acted on by a constant force mc at right-angles to the string, so that the string wraps itself round the cylinder, the whole motion being in a plane at right-angles to the generators. Find the relation connecting the time with the inclination ψ of the straight part of the string to the tangent at A, and show that the tension of the string is

$$m\frac{u^2 + 2c\{l\psi - F(\psi)\}}{l - f(\psi)},$$

where u is the velocity of the particle when $\psi = 0$, and $F(\psi) = \int_0^\psi f(x)\,dx$.

22. A particle is attached to the end of a light inextensible string and hangs at rest with the upper end of the string attached to a point on the upper half of a fixed horizontal circular cylinder of radius a, so that the straight part of the string is of length l and is a vertical tangent to the cylinder. The particle is projected horizontally with a small velocity V at right angles to the axis of the cylinder, so that the string begins to wrap round the

cylinder. Prove that when the straight part of the string has turned through an angle θ the velocity v of the particle is given by the equation

$$v^2 = 2ga(\sin\theta - \theta\cos\theta) - 2gl(1 - \cos\theta - \tfrac{1}{2}\alpha^2),$$

where $\alpha^2 = V^2/gl$. Hence show that to a first approximation the motion is an oscillation of amplitude α, but that to a second approximation the extreme displacements on either side of the vertical have the unequal values

$$\alpha\left(1 \pm \frac{a\alpha^2}{3l}\right).$$

*23. A particle is attached to the end of a light inextensible string and hangs at rest with the upper end of the string attached to a point on the upper half of a fixed horizontal circular cylinder of radius a, so that the straight part of the string is of length $a\pi$ and is a vertical tangent to the cylinder. The particle is then projected horizontally with velocity V at right angles to the axis of the cylinder so that the string begins to wrap round the cylinder. Show that when the length of the straight part of the string is $a\phi$, the velocity v of the particle and the tension T in the string are given by

$$\frac{v^2}{ga} = 2\sin\phi - 2\phi\cos\phi - \left(2\pi - \frac{V^2}{ga}\right),$$

$$\frac{T\phi}{mg} = 2\sin\phi - 3\phi\cos\phi - \left(2\pi - \frac{V^2}{ga}\right).$$

Hence show that the string will become slack during the motion if

$$2\pi - 2 < V^2/ga < 2\pi + \lambda,$$

where $\lambda = 3\beta\cos\beta - 2\sin\beta$, and β is the acute angle defined by $\beta\tan\beta = \frac{1}{3}$.

24. The points O and A are at the same level and at a distance $2b$ apart. A wire consisting of two straight pieces, OB, BA joins O to A, the point B being vertically below the mid-point of OA. A bead starting from rest at O slides on the wire from O to A, the path being supposed rounded off at B so that there is no loss of energy. Find the shortest time for the journey from O to A. (The depth of B below OA is adjustable.)

Solve the same problem with a path $OBCA$ consisting of three straight pieces, OB and CA being equally inclined to the vertical and BC horizontal; the corners are supposed rounded off as before.

Compare the results with the genuinely shortest time from O to A by the cycloidal path (§ 11·14).

25. A particle moves on a smooth horizontal table and is attached to one end of a light rod of length a. The other end of the rod is smoothly pivoted at a fixed point C of the table. The particle is attracted towards another fixed point O in the table by a force proportional to its distance from O. The distance OC is $2a$ and when the particle is at the mid-point A of OC the force is equal to its weight. If the particle is initially projected from A with velocity V, find the condition for the particle to complete a full circle. Assuming this condition to be satisfied, find the value of V which makes the force in the rod zero when the particle is at its greatest distance from O.

26. A particle moves in a smooth tube in the form of a catenary, being attracted to the directrix with a force proportional to the distance from the directrix. Prove that the period of oscillation is independent of the amplitude.

CHAPTER XII

FREE MOTION IN A CONSERVATIVE FIELD, ACCESSIBILITY AND THE THEORY OF PROJECTILES

12·1 The conservative field, accessibility. Suppose a particle moves in a conservative field of force, and is projected from a point K with speed w_0. Then the energy constant of the motion, C, is given by

$$C = V(K) + \tfrac{1}{2}mw_0^2.$$

C is determined when K and w_0 are given, and its value is independent of the direction of projection. By projecting from K in different directions we obtain a single infinity of orbits all with the same energy constant C. All these orbits lie in the region $V \leqslant C$. This region is bounded by the energy level λ, the equipotential $V = C$.

But in general it is not possible to reach all points of this region by projection from K with the given velocity. In general the single infinity of orbits has an envelope E which divides the region $V \leqslant C$ into two parts; points in one part are accessible from K, and points in the other are not.

What is probably the simplest example of this theory is provided by the uniform field. We have already given a brief discussion of motion in a uniform field; we now consider this 'theory of projectiles' in somewhat greater detail.

12·2 The theory of projectiles, the fundamental theorem. Without loss of generality we can take the point of projection as origin of coordinates. We take the axis Ox perpendicular to the field, and Oy in a direction opposite to the field (as in § 10·10). If the field is that of gravity, Oy is vertically upwards. Then $\mathbf{P}$ has components $(0, -mg)$, and $V = mgy$. We consider the orbits got by projection from O, with speed w_0 at the instant $t = 0$. If we denote the components of the velocity of projection by u_0, v_0 we have

$$u_0^2 + v_0^2 = w_0^2.$$

We denote the speed at time t by w, its components by u, v. The energy equation is

$$\tfrac{1}{2}mw^2 + mgy = \tfrac{1}{2}mw_0^2,$$

and if we write $w_0^2 = 2gh$, this becomes

$$w^2 = 2g(h-y).$$

The energy level for the problem is the horizontal line λ, whose equation is $y = h$. The speed at any point P of the path is the speed which would be acquired by falling from rest on the energy level, either freely or along a smooth guiding curve, to P.

The fundamental theorem in the theory of motion in a uniform field is that *the path is a parabola whose directrix is the energy level.* That the path is a parabola we have proved already, but it may be worth while to repeat the argument here for the sake of completeness. The equations of motion are

$$\ddot{x} = 0, \quad \ddot{y} = -g,$$

whence $$u = \dot{x} = u_0, \quad v = \dot{y} = v_0 - gt,$$

and $$x = u_0 t, \quad y = v_0 t - \tfrac{1}{2}gt^2,$$

since $x_0 = y_0 = 0$. Eliminating t gives us the equation of the path

$$\left(x - \frac{u_0 v_0}{g}\right)^2 = 4a\left(\frac{v_0^2}{2g} - y\right), \tag{1}$$

where $a = u_0^2/2g$. This is a parabola with its axis vertical and vertex upwards. The height of the vertex (above $y = 0$) is $v_0^2/2g$, and the height of the directrix is $(v_0^2/2g) + a = w_0^2/2g = h$. The line λ is the directrix of the parabola, and this is the result we wished to prove.

Another proof of the fundamental theorem, that the directrix is the energy level, may be of interest. We take it as known that the orbit is a parabola with axis vertical and vertex upwards. With the notation of the figure (Fig. 12·2a) the time from O to A is v_0/g, and the distance ON is therefore $u_0 v_0/g$. Now the tangent at O, which is in the direction of projection, bisects the angle SOM,

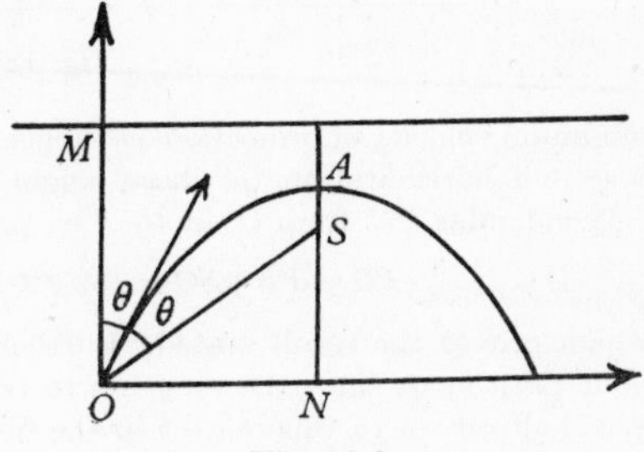

Fig. 12·2a

where S is the focus and M is the point on the directrix vertically above O.

Hence
$$ON = \frac{u_0 v_0}{g} = OS \sin 2\theta = 2OS\frac{u_0 v_0}{w_0^2}.$$

Therefore
$$OM = OS = w_0^2/2g.$$

Example 1. *It is required to throw a ball from a point P to a point Q. Prove that if the velocity of projection is a minimum, the focus of the parabolic path lies on PQ.*

This follows at once from the fundamental theorem. Since the velocity of projection is a minimum, the energy level λ, which is the directrix of the parabola, must be as low as possible. This means that $PM+QN$ must be a minimum, where PM and QN are the perpendiculars from P and Q on to the directrix. If S is the focus

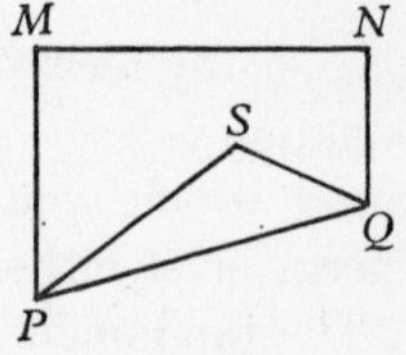

Fig. 12·2b

$$PM+QN = PS+SQ,$$

so $PS+SQ$ must be as small as possible, which implies that S lies on PQ. In fact, if PQ is of length l, and Q is at a height b above P,

$$PS = \tfrac{1}{2}(l+b), \quad SQ = \tfrac{1}{2}(l-b).$$

Another point should be noticed before leaving this problem. The point Q lies on the parabola E which has P as focus and λ (the energy level for the

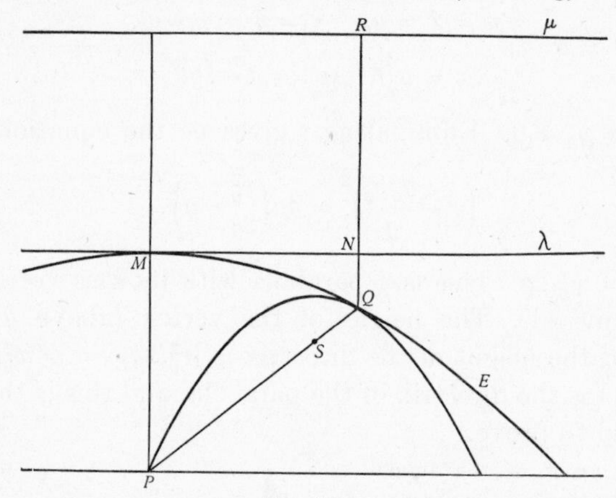

Fig. 12·2c

minimum velocity of projection) as tangent at the vertex M. To prove this draw a second horizontal line μ whose height P above is twice that of λ. Let the perpendicular QN from Q on to λ be produced to meet μ in R. Then

$$PQ = PS+SQ = PM+QN = QN+NR = QR,$$

which proves the result stated. Further, the parabola E touches the parabolic path at Q, since the tangents to both parabolas bisect the angle PQR. We shall return to this matter in the next paragraph, when the significance of the parabola E will be made clear.

Example 2. *It is required to throw a ball from some point O on a given horizontal plane to pass successively just over the tops P, Q of two given vertical poles PU, QV. Prove that the velocity of projection will be a minimum when the focus of the parabolic trajectory lies on PQ.*

If the poles are 20 ft. apart, and their tops at heights 33 ft. and 48 ft. above the plane, and if the velocity of projection is a minimum, find (i) the time that elapses while the ball travels from P to Q, (ii) the horizontal distance of the point of projection O from the nearer pole PU.

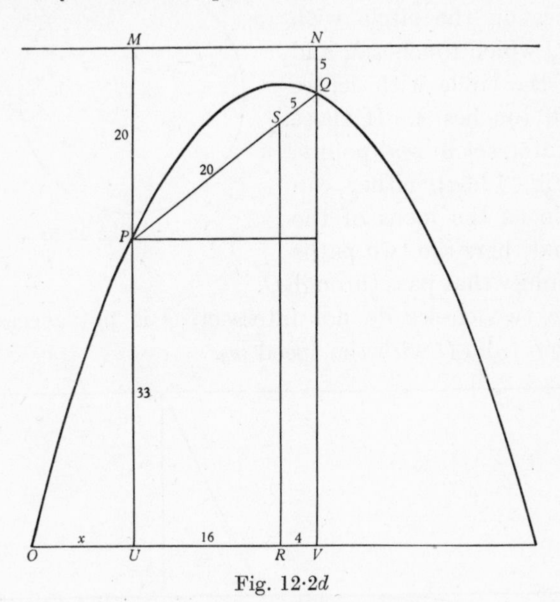

Fig. 12·2*d*

As in the preceding example, the directrix of the parabolic path must be as low as possible, from which it follows as before that S lies on PQ. If PM, QN are the perpendiculars from P and Q on to the directrix

$$PM = PS = 20, \quad SQ = QN = 5,$$

the distances being in feet throughout. The depth of S below the directrix is

$$2a = 5+(15/5) = 8, \quad a = 4,$$

and the horizontal component of velocity, which is the velocity at the vertex of the parabola, is u, where

$$u^2 = 2ga = 256, \quad u = 16,$$

where we have taken $g = 32$. The time from P to Q is therefore $20/16 = 5/4$ seconds.

The height of S above the plane is $53-8 = 45$, and the length OS is equal to the distance of O from the directrix, which is 53. Then

$$OR^2 = OS^2 - RS^2,$$

$$(x+16)^2 = 53^2 - 45^2 = 28^2, \quad OU = x = 12.$$

12·3 The enveloping parabola (1). Suppose now that we wish to project the particle from O with the given speed w_0 so as to pass through a given point Q of the vertical plane in which the motion takes place. The possible paths are the directrix. If one of these parabolas passes through Q, its focus lies on the circle with centre O which touches λ, and also on the circle with centre Q which touches λ. If these circles intersect in real points, S, H (Fig. 12·3*a*), either can be taken as the focus of the path, and there are two paths of the family that pass through Q.

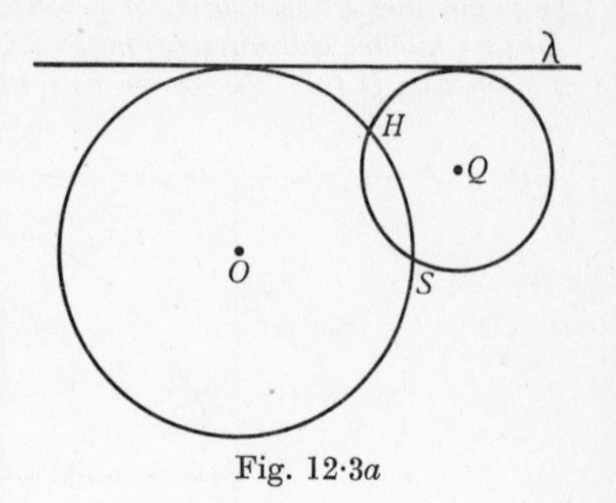

Fig. 12·3*a*

If the two circles do not intersect, Q is not accessible by projection from O with the speed w_0.

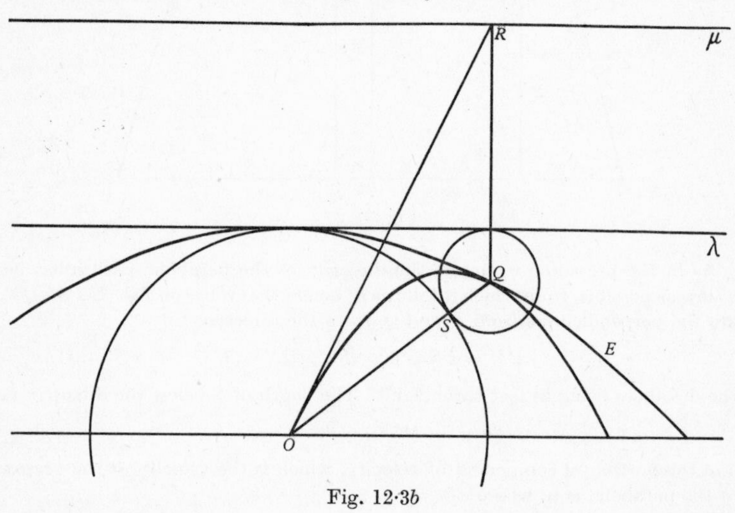

Fig. 12·3*b*

In the critical case the circles *touch*; Q is just accessible from O, and lies on the boundary of the region which is accessible from O. If we take a second line μ, at height $2h$ above O (Fig. 12·3*b*), and if R is the point on μ vertically above Q, we see that $OQ = QR$. The locus of Q is the parabola E having O as focus, μ as directrix, and λ as tangent at the vertex. There is just one path from O to Q, and its focus is the point S on OQ where the circles touch. Now the path touches E at Q, because the tangent to the path and the tangent to E at Q both bisect the angle OQR. Thus E is the envelope of the paths obtained by projection from O with the given speed w_0: it is called the 'enveloping parabola'. We have already encountered this curve in Example 1 of § 12·2.

Since OQ is a focal chord of the path, the directions of motion at O and at Q are at right angles. This is a special case of a more general theorem which we shall prove later (§ 16·6). Notice that the direction of projection from O is OR, so the time t_1 from O to a point Q which is just accessible from O is given by $OQ = QR = \frac{1}{2}gt_1^2$; if $OQ = r, t_1 = \sqrt{(2r/g)}$.

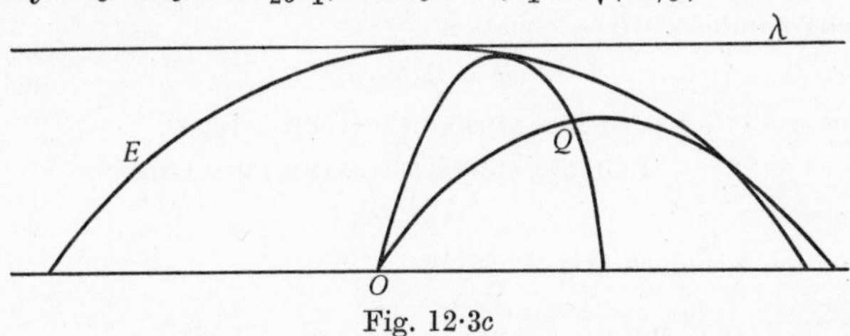

Fig. 12·3c

In this problem the region $V \leqslant C$ is the part of the plane on or below the line λ. The region accessible from O with the given energy is the part of the plane on or below E. If Q lies below E there are two paths with the given energy joining O to Q (Fig. 12·3c).

Example 1. *The speed of projection from O being given, to find the maximum range on a line through O which is inclined at an angle α to Ox.*

We have only to find where this line cuts E. The maximum range up the plane is R_1, where (Fig. 12·3d)

$$R_1(1+\sin\alpha) = 2h,$$

and the maximum range down the plane is R_2, where

$$R_2(1-\sin\alpha) = 2h.$$

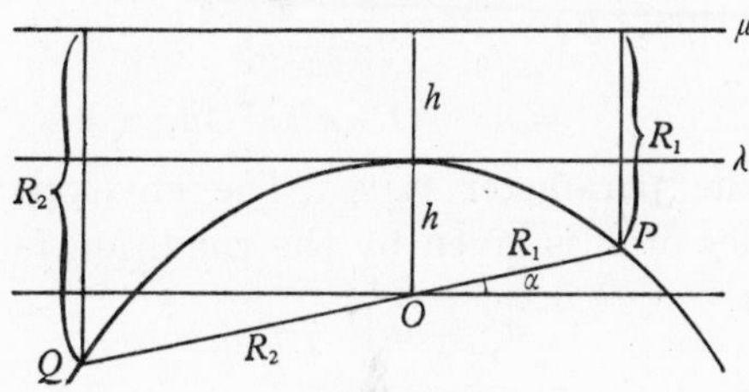

Fig. 12·3d

The trajectory which gives the maximum range OP up the plane has its focus on OP, so the direction of projection from O bisects the angle between OP and the upward verticle.

We can interpret the first result (say) in another way. Suppose we ask, what is the minimum velocity of projection from O so that the projectile may pass through P? If the velocity of projection is a minimum, P lies on the enveloping parabola for projection from O with this velocity, so

$$2h = R_1(1+\sin\alpha)$$

as before. The minimum velocity of projection is

$$\sqrt{(2gh)} = \sqrt{\{gR_1(1+\sin\alpha)\}}.$$

Example 2. *A gun fires a shell with a muzzle velocity of* 1040 *ft./sec. Neglecting the resistance of the air, what is the greatest horizontal distance at which an aeroplane at* 2500 *ft. can be hit and what gun elevation is required? Show that the shell would then take approximately* 44·2 *sec. to reach the aeroplane.*

We use feet and seconds as units, and adopt the conventional approximation $g = 32$. The point at which the shell hits the aeroplane lies on the enveloping parabola, whose equation is

$$x^2 = 4h(h-y).$$

In this case $\quad h = w_0^2/2g = (1040/8)^2 = (130)^2 = 16900,$

and $\quad x^2 = 4\times 130^2\times(130^2-50^2) = (2\times 130\times 120)^2,$

$$x = 31200.$$

The elevation α is given (Fig. 12·3*b*) by

$$\tan\alpha = \frac{2h}{x} = \frac{169}{156} = \frac{13}{12}, \quad \alpha = 47^\circ\, 17'.$$

The time to reach the aeroplane is (§ 12·3)

$$\sqrt{\left(\frac{2r}{g}\right)} = \sqrt{\left(\frac{2(2h-y)}{g}\right)} = \sqrt{\left(\frac{31300}{16}\right)} = \frac{10}{4}\sqrt{(313)} = 44{\cdot}2.$$

12·4 The enveloping parabola (2).

There are many other ways of proving the envelope theorem. (i) The equation of the path, taking the point of projection O as origin, is given by equation (1) of § 12·2, which can be written in the form

$$y = \frac{v_0}{u_0}x - \frac{g}{2u_0^2}x^2.$$

We can write this $\quad y = \dfrac{v_0}{u_0}x - \dfrac{g}{2w_0^2}\dfrac{u_0^2+v_0^2}{u_0^2}x$

$$= \kappa x - (1+\kappa^2)\,x^2/4h,$$

where κ is the parameter v_0/u_0. The envelope of this for different values of κ is given by the condition for equal roots in κ, which is

$$x^2 = 4h(h-y).$$

This is the equation of the enveloping parabola.

(ii) If we think of particles projected simultaneously at $t = 0$ in different directions, all with the same velocity of projection w_0, the particles at time t lie on the circle (Fig. 12·4).

$$x^2+(y+\tfrac{1}{2}gt^2)^2 = w_0^2t^2.$$

We can write this in the form

$$r^2-2(2h-y)p+p^2 = 0,$$

where $r^2 = x^2+y^2$ and $p = \tfrac{1}{2}gt^2$. The envelope of this family of circles for different values of p is the enveloping parabola

$$r = 2h-y.$$

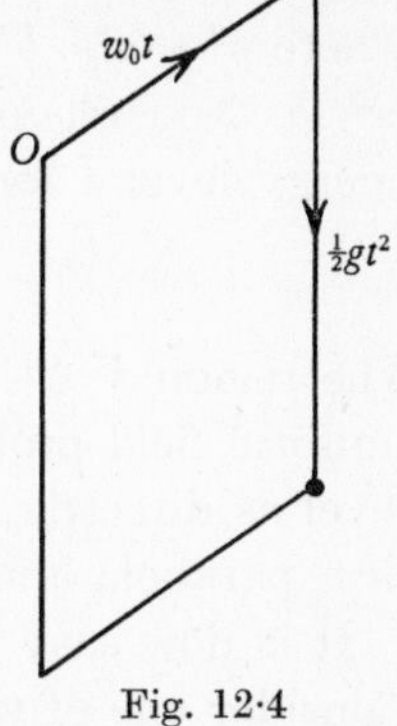

Fig. 12·4

(iii) The relation between the two preceding methods is worth noticing. If we project at $t = 0$ at an inclination α to the horizontal the position of the particle at time t is given by formulae of the type

$$x = \phi(\alpha, t), \quad y = \psi(\alpha, t). \tag{1}$$

If we eliminate t between these equations we get the parabola corresponding to the angle α, and the envelope of these parabolas is the envelope E. If we eliminate α we get the circle corresponding to the instant t, and the envelope of these circles is again the enveloping parabola E.

The result is general. If we start with the formulae (1), the envelope of the α-curves got by eliminating t is the same as the envelope of the t-curves got by eliminating α. This common envelope is the curve obtained by eliminating α, t from the equations (1) and the equation

$$\frac{\partial(\phi, \psi)}{\partial(\alpha, t)} = 0. \tag{2}$$

12·5 The isotropic oscillator, the enveloping ellipse.

As another illustration of the theory of accessibility we find the region accessible from a point K when the field is that of the isotropic oscillator, an attraction mn^2r towards O. Here $V = \tfrac{1}{2}mn^2r^2$. We denote the distance OK by a, the velocity of projection by nb, so that $b = w_0/n$. For a particular orbit, if we take oblique axes Ox, Oy, with Ox along OK, and Oy parallel to the direction of projection, the equation of the orbit is

$$\frac{x^2}{a^2}+\frac{y^2}{b^2} = 1,$$

an ellipse with the axes along conjugate radii. Moreover, a and b are the lengths of conjugate radii, and by a familiar property of the ellipse, the radius of the director circle is ρ, where $\rho^2 = a^2+b^2$. Thus all the ellipses have the same director circle, the circle of radius ρ with O as centre. This circle is the energy level λ for the problem, since

$$C = \tfrac{1}{2}mn^2(a^2+b^2) = \tfrac{1}{2}mn^2\rho^2.$$

The region $V \leqslant C$ is this circle and its interior. (Compare the uniform field problem, where all the parabolas had the energy level as directrix. We recall an analogy between the directrix of a parabola and the director circle of an ellipse.)

It is now easy to find the envelope of the family of ellipses. Consider one of the family, and let Q be a point on the ellipse where the tangent is perpendicular to the direction of projection, i.e. to the tangent at K. We denote the other end of the diameter through K by H; the tangent at Q meets the tangents at K and H in M and N; these points M, N lie on the director circle (Fig. 12·5a).

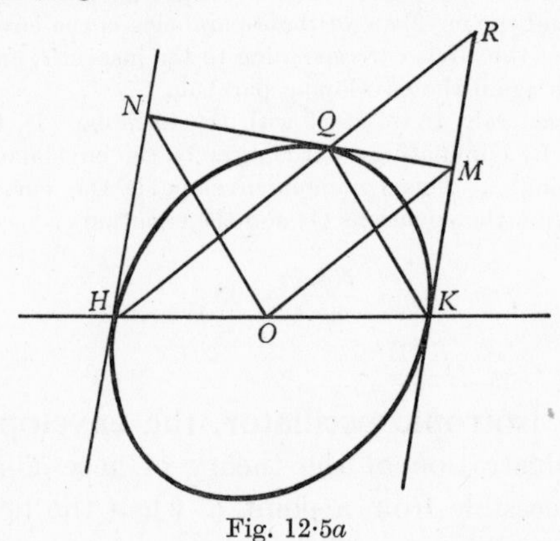

Fig. 12·5a

Now HQ is parallel to the diameter conjugate to KQ, i.e. to OM; and similarly KQ is parallel to ON. Since the triangle OMN is isosceles, we see that KQ, HQ are equally inclined to MN.

If HQ meets KM in R, R is the image of K in MN, and

$$HQ+QK = HQ+QR = HR = 2OM = 2\rho.$$

Thus the locus of Q is an ellipse E, whose foci are H and K, and whose major axis is of length 2ρ. Moreover, the tangent to E at Q is MN, since HQ, KQ are equally inclined to MN. Thus the orbit touches E at Q, and E is the envelope of the orbits. The region of accessibility consists of E and its interior (Fig. 12·5*b*).

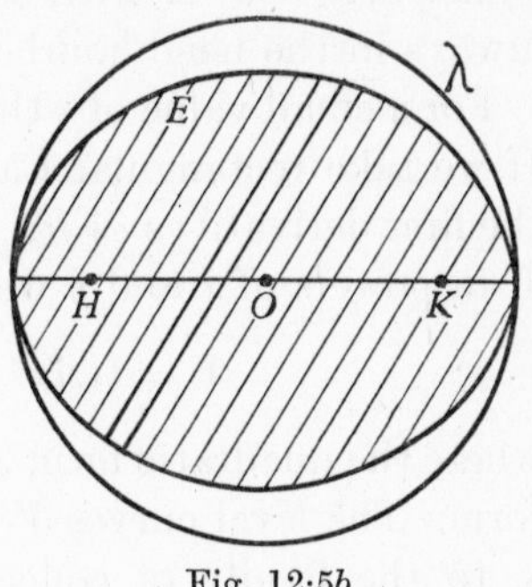

Fig. 12·5*b*

12·6 Orbits in which the particle comes to rest. If a particle moving in a plane conservative field reaches the energy level λ at a point A, the orbit at A is perpendicular to λ; and after reaching A the particle retraces its path. The proof is simple. By the conservation of energy the particle comes to rest at A; indeed, the theorem is essentially about motions in which the particle starts from rest, or comes to rest at some instant. The fact that the particle retraces its path is a special case of the theorem of reversibility (§ 10·15). Also, in the motion from A, the acceleration is initially in the direction of the field, and this direction is perpendicular to λ; and this initial acceleration defines the direction of the tangent to the path. Therefore the path meets λ normally at A.

12·7 Positions of equilibrium in the field. If V has a minimum value at a point A of the field, $X(= -\partial V/\partial x)$ and $Y(= -\partial V/\partial y)$ vanish at A, and A is a point of equilibrium or neutral point. The level curves in the neighbourhood of A are closed oval curves enclosing A. If we take $V = 0$ at A (as we may do without loss of generality) the level curve $V = \epsilon$, for a small positive value of ϵ, consists (at least in part) of a simple closed curve λ enclosing A, and the value of V at any point inside λ is less than ϵ. If therefore the particle moves in the neighbourhood of A with energy ϵ its path lies wholly in the region consisting of λ and its interior. This follows from the energy equation, since

$$V = \epsilon - T \leqslant \epsilon.$$

The particle can rest in equilibrium at A, and the equilibrium is stable. This means that if the particle starts from a point sufficiently near A with a sufficiently small speed, its path lies always in the neighbourhood of A.

For a small value of ϵ the curve λ is approximately an ellipse. If we take rectangular Cartesian coordinates with A as origin the first derivatives $\partial V/\partial x$, $\partial V/\partial y$ vanish at A, and if we expand V in powers of x and y we obtain

$$V = \tfrac{1}{2}(Ax^2 + 2Hxy + By^2) + O(r^3),$$

where the quadratic form $Ax^2 + 2Hxy + By^2$ is a positive-definite form. The level curves $V = \epsilon$ approximate, for small values of ϵ, to the family of concentric similar and similarly situated ellipses

$$Ax^2 + 2Hxy + By^2 = 2\epsilon.$$

If we take new rectangular axes coinciding with the axes of the ellipse we obtain the form

$$V = \tfrac{1}{2}m(p^2x^2 + q^2y^2) + O(r^3).$$

Now we know that, for sufficiently small values of the energy, x and y remain small throughout the motion, so we obtain as an approximation to the equations of motion for the small oscillations

$$\ddot{x} + p^2x = 0, \quad \ddot{y} + q^2y = 0.$$

These are the equations for the anisotropic oscillator, already discussed in § 10·12. The new coordinates are normal coordinates for the vibrating system.

If V is a maximum at A we have a position of unstable equilibrium. Theoretically the particle can rest in equilibrium at A, but after a small disturbance, however slight, the particle does not remain in the neighbourhood of A.

Sometimes there is a point of equilibrium A where V is stationary but neither a maximum nor a minimum; this is called a *saddle point*. The level curve $V = V(A)$ has a double point at A. A simple example occurs in the gravitational field of two fixed attracting particles, each of mass M, at $(-a, 0)$ and $(a, 0)$. For this problem

$$V = -\gamma Mm\left(\frac{1}{r_1} + \frac{1}{r_2}\right),$$

where r_1 and r_2 are distances from the two particles. In this case the origin is a saddle point. Explicitly, if we expand in powers of x and y, we find

$$V = -\frac{\gamma Mm}{a}\left\{2+\frac{1}{a^2}(2x^2-y^2)+O\left(\frac{r^4}{a^4}\right)\right\}.$$

If the particle were constrained to move on the line $x = 0$ the equilibrium would be stable (cf. Example 1 of § 6·9), but for motion on $y = 0$, or for a general plane motion in the neighbourhood of A, it is unstable.

EXAMPLES XII

1. Show that a particle moving under gravity describes a parabola, and find the position of the focus S relative to the vertex A, the velocity at A being u.

Prove that, when the particle is at P, where $SP = r$, the components of its velocity along and perpendicular to SP are respectively equal to the vertical and horizontal components of its velocity.

Show that the component in the direction SP of the acceleration of the particle is $g-u^2/r$.

2. A body is projected from a given point with velocity u, so as to pass through another point which is at a horizontal distance b from the point of projection and at a height h above it. Find an equation to determine the angles at which the body may be projected.

A shot has a range b on the horizontal plane when the angle of elevation is θ and just reaches the base of a vertical target of height $2a$, where $a = b\tan\alpha$; show that with the same initial velocity and elevation $\alpha+\theta$ it will strike the target at a depth $a\sin^2\theta\sec^2(\alpha+\theta)$ below the centre.

3. A projectile is fired from a point O with velocity $\sqrt{(2gh)}$ and hits a target at the point $P\ (x, y)$, the axes being horizontal and *downward* vertical lines through O. Show that in general there are two possible directions of projection.

Show that the two directions are at right angles if $x^2 = 2hy$, and that if this condition is fulfilled one of the possible directions of projection bisects the angle POx.

If $x = h$, $y = \frac{1}{2}h$, find to the nearest minute the inclinations to the horizontal of the two possible directions of projection.

4. A particle is projected at angle of elevation α to strike a smooth vertical wall and after rebounding it passes through the point of projection. If ϕ is the angle of inclination to the horizontal at which the particle rebounds from the wall and e the coefficient of restitution, show that

$$e(1+e)\tan\phi = (1-e)\tan\alpha.$$

5. A shell fired with velocity V at elevation θ hits an airship at height H, which is moving horizontally away from the gun with velocity v. Show that if

$$(2V\cos\theta - v)(V^2\sin^2\theta - 2gH)^{\frac{1}{2}} = vV\sin\theta,$$

the shell might also have hit the airship if the latter had remained stationary in the position it occupied when the gun was actually fired.

6. A particle is projected *in vacuo* with a given velocity in a given direction; find the time of flight and the horizontal range.

A juggler keeps three equal balls, each of radius c, in the air by throwing them up successively with one hand from the same point at equal intervals of time. Show that the distance through which the hand moves horizontally cannot be less than $6c$, if the velocity of the hand be supposed infinitely great, nor less than $5c$, if the hand be supposed to move with uniform horizontal velocity, and not to rest at the points of turning.

7. A ship is rolling with a period of 10 sec. A man at the masthead 100 ft. above the deck is swung to and fro 25 ft. on either side of the vertical with a motion which is approximately horizontal and simple harmonic. The man weighs 200 lb. and his horizontal hold failing at 50 lb. weight, he is thrown off the mast. The width of the deck being 80 ft., prove that he falls clear of the ship.

8. It is required to throw a ball from some point on a given horizontal plane to pass just over the tops of two given vertical poles; prove that the velocity of projection will be a minimum when the focus of the parabolic trajectory is collinear with the tops of the poles.

If the poles are 24 ft. apart and their tops at heights 57 ft. and 64 ft. above the point of projection, and if the velocity of projection is a minimum, find (i) the time which elapses while the ball passes from the top of the first pole to the top of the second, (ii) the horizontal distance of the point of projection from the first pole.

9. A projectile is shot from the ground and grazes the tops of two chimneys at heights of 36 and 64 ft. respectively which stand 96 ft. horizontally apart. Show that the initial velocity must be at least 80 ft./sec., and that, if the time from chimney to chimney is 5 sec., the initial velocity is 100 ft./sec.

10. Two particles are projected under gravity from a point O with the same initial velocity in the same vertical plane through O at angles of elevation α, β. If the trajectories meet at the point (h, k) referred to horizontal and upward vertical axes at O, prove that

$$k = h\tan(\alpha+\beta-\tfrac{1}{2}\pi).$$

By considering the limiting case of this result as $\alpha \to \beta$, prove that to attain maximum range along a given straight line through O with a given initial velocity, the direction of projection must bisect the angle between the line and the vertical through O.

11. A particle is projected in a given vertical plane from an origin O, with velocity $\sqrt{(2gh)}$. It passes through the point (x, y) at time t after projection the axes being horizontal and vertically upwards. Prove that

$$p^2 - 2(2h-y)p + r^2 = 0,$$

where $p = \frac{1}{2}gt^2$, and $r^2 = x^2+y^2$.

Show that the points of the plane which are accessible by projection from O with the given velocity lie on or under the parabola having O as focus and a horizontal line at height h above O as tangent at the vertex; and that the time taken to reach a point *on* this parabola is $\sqrt{(2r/g)}$.

12. A particle is projected at time $t = 0$ in a fixed vertical plane from a given point S with given velocity $\sqrt{(2ga)}$, of which the *upward* vertical component is v. Show that at time $t = 2a/v$ the particle is on a fixed parabola (independent of v), that its path touches the parabola, and that its direction of motion is then perpendicular to its direction of projection.

13. A fort and a ship are both armed with guns which give their projectiles a muzzle velocity $\sqrt{(2gk)}$, and the guns in the fort are at a height h above the guns in the ship. If d_1 and d_2 are the greatest (horizontal) ranges at which the fort and ship respectively can engage, prove that

$$\frac{d_1}{d_2} = \sqrt{\left(\frac{k+h}{k-h}\right)}.$$

14. A particle is projected from a given point A so as to pass through a given point B, where the distance AB is l and the line AB makes an angle θ with the horizontal. Prove that the least possible velocity of projection is

$$\sqrt{\{gl(1+\sin\theta)\}}.$$

Suppose now that the particle instead of moving freely in space is projected along the surface of a smooth plane inclined at an angle ϕ to the horizontal. Prove that the least velocity of projection for the particle to pass through B is

$$\sqrt{\{gl(\sin\phi+\sin\theta)\}}.$$

15. AB is a horizontal straight line of length $2a$; the middle point of AB is the centre of a sphere of radius $b(<a)$. If a particle projected from A with velocity v just clears the sphere and passes through B, prove that

$$v^2 = \frac{g(2a^2-b^2)}{\sqrt{(a^2-b^2)}} \quad \text{or} \quad \frac{g(a^2+4b^2)}{2b},$$

according as a is less or greater than $b\sqrt{2}$.

16. A shell explodes on the surface of horizontal ground. Earth is scattered in all directions with varying velocities up to a certain maximum and falls over a circular area of radius $2a$. Prove that the interval during which earth falls upon a spot on the ground at a distance a from the place of explosion may be as long as $2\sqrt{(a/g)}$.

17. An aeroplane is flying horizontally at height k with velocity U. An anti-aircraft gun is situated on the ground at a distance h from the vertical plane in which the aeroplane is flying. The gun can fire shells with velocity V. Prove that the aeroplane is within range of the gun for a time

$$\frac{2}{gU}(V^4-2V^2gk-g^2h^2)^{\frac{1}{2}},$$

assuming that $$g^2h^2+2V^2gk < V^4.$$

18. Find the least velocity u with which a particle must be projected from a point O on the ground so that it may pass over a wall of height h at a distance a (measured horizontally in a direction perpendicular to the wall) from O. Find also the velocity u' with which it then reaches the wall.

If the particle is projected from O with velocity $v > u$, prove that the length of the top of the wall within range is

$$2\sqrt{\{(v^2-u^2)(v^2+u'^2)\}}/g.$$

19. Particles of mud are thrown off the tyres of a car travelling along a level road at v ft./sec. from points on the tyre where the tyre makes an angle of 45° or less with the road. A car following at the same pace has a wind-screen h ft. above the ground. Prove that, neglecting air effect, the wind-screen will be splashed unless the following car is more than

$$\frac{v^2}{2g}+\frac{v}{2g}\sqrt{(v^2-4gh)}$$

feet behind. The height of the mud particles above the road at the moment of projection may be neglected.

20. A railway truck carrying a gun is moving with velocity U on a horizontal track directly towards a target, which is not necessarily at ground-level. When the truck is at a horizontal distance l from the target the gun is fired with velocity V relative to the truck. If α is the angle at which the gun should be elevated in order to hit the target, and α_0 is the corresponding elevation when the truck is not moving, show that, when U/V is small,

$$\alpha_0-\alpha=\frac{\mu\sec\alpha_0-\sin\alpha_0}{1-\mu\tan\alpha_0}\cdot\frac{U}{V}$$

approximately, where $\mu = gl/V^2$.

21. A gun is fired from a point O at a car which is moving horizontally with uniform speed U directly away from O; at the instant when the gun is fired the distance of the car from O is b. Prove that, if the car is to be hit, the horizontal component u of the velocity of projection of the shell, and the vertical component v, must satisfy the relation

$$v(u-U) = \alpha^2,$$

where $\alpha^2 = \frac{1}{2}gb$.

Prove that, if the car is to be hit, and if the velocity of projection is as small as possible, then $u = k^3\alpha$, $v = k\alpha$, where k is the positive root of the equation

$$k^4-(U/\alpha)k-1 = 0.$$

22. A gun is to be fired at a moving target, which starts from the gun itself, and moves directly away from it along a level road. Prove that if the speed of projection is V and the speed of the moving target is $7V/20$, and if the target is to be hit when it is at a distance $2h\sin\theta\,(0<\theta<\frac{1}{2}\pi)$ from the gun (where $V^2 = 2gh$), the gun must be fired at one of two instants, the later of which is at an interval of time

$$\left(\frac{20}{7}\sin\theta-2\sin\frac{\theta}{2}\right)\sqrt{\left(\frac{2h}{g}\right)}$$

after the target leaves the gun.

Prove that this has a maximum value $\frac{54}{35}\sqrt{\left(\frac{2h}{g}\right)}$ when $\sin\theta = \frac{24}{25}$.

23. A particle hangs from a fixed point by a light inextensible string of length a. It is projected with horizontal velocity u. Find conditions for the string to become slack during the subsequent motion, and prove that in this case the string is slack for a time $\sqrt{(8a\sin\phi\sin 2\phi/g)}$, where $3\cos\phi = (u^2/ga)-2$.

24. A particle is attached to one end of a light inelastic string of length a, the other end of which is tied to a fixed peg O. The particle is projected horizontally with velocity u from the point at depth a vertically below O. Show that if $2ga < u^2 < 5ga$ the string will not remain taut during the subsequent motion.

Show (i) that if $u^2 = (2+\sqrt{3})ga$ the particle will subsequently strike the peg, and (ii) that if $u^2 = (7/2)ga$ the string will slacken when it has turned through an angle $2\pi/3$, and that when it again tightens the particle will be at the point from which it was originally projected.

25. A particle attached by a light inextensible inelastic string of length a to a fixed point O is projected from the point at a height a vertically above O with horizontal velocity $\frac{1}{2}\sqrt{(ga)}$. Find where the string becomes taut again, and determine whether the string will subsequently slacken.

26. A particle is projected along the outside surface of a smooth sphere of radius a from the highest point with velocity $\frac{1}{2}\sqrt{(ga)}$. Prove that it strikes a horizontal plane through the centre of the sphere at a distance

$$(q\sqrt{3q}+7\sqrt{7})\,a/64$$

from the centre.

*27. A particle is projected in a given vertical plane from a point O, the horizontal and vertical components of velocity being u and v respectively. If u, v are connected by a relation

$$\alpha u^2 + v^2 = 2gh,$$

where α, h are positive constants, show that the envelope of the trajectories is a parabola whose vertex is at a height h above O and whose latus rectum is $4h/\alpha$. Show also that to reach the point Q on the envelope the elevation of the direction of projection must be $\tan^{-1}(2h/x)$, where x denotes the horizontal projection of OQ.

A gun is mounted on a truck and can fire a shell of mass m in a vertical plane parallel to the rails, the mass of the gun and truck together being M. Find the envelope of the trajectories (i) on the assumption that the velocity of the shell relative to the gun is constant and equal to $\sqrt{(2gh)}$, (ii) on the assumption that the total kinetic energy immediately after the shell leaves the gun is constant and equal to mgh.

*28. An insect of mass m stands on a small plate of mass M which is lying on a rough table. The insect can leap with velocity $\sqrt{(ga)}$ relative to the plate. Prove that the greatest distance along the table he can reach by leaping is

$$a \quad \text{if} \quad \mu > 1,$$

$$\frac{2\mu}{1+\mu^2}a \quad \text{if} \quad 1 > \mu > \mu_0,$$

$$\frac{\mu m + \sqrt{(M^2+\mu^2 m^2)}}{M+m}a \quad \text{if} \quad \mu_0 > \mu,$$

where μ is the coefficient of friction between the plate and the table, and

$$\mu_0^2 = \frac{M}{M+2m}.$$

*29. An insect of mass m stands on a rod, also of mass m, which is lying on a rough table. The coefficient of friction between the rod and the table is μ, both for finite forces and for impulses. The insect leaps from the rod, his velocity relative to the rod on leaping having the prescribed value w, and alights again on the rod. Prove that the greatest distance along the rod that he can reach with one leap is

$$\frac{w^2}{g} \quad \text{if} \quad \mu > 1,$$

$$\frac{2\mu}{1+\mu^2}\frac{w^2}{g} \quad \text{if} \quad 1 > \mu > \frac{1}{\sqrt{3}},$$

$$\frac{1+9\mu^2}{8\mu}\frac{w^2}{g} \quad \text{if} \quad \frac{1}{\sqrt{3}} > \mu > \frac{1}{\sqrt{15}},$$

$$\{\sqrt{(1+\mu^2)}-\mu\}\frac{w^2}{g} \quad \text{if} \quad \frac{1}{\sqrt{15}} > \mu.$$

*30. A particle of mass m moves in a plane, and is attracted towards a fixed origin O in the plane with a force mn^2r, where r denotes distance from O. It is projected from the point $(c, 0)$, the axes being rectangular, with velocity nb and in a direction inclined at an angle θ to the axis Ox. Show that the path of the particle is the ellipse

$$b^2(x \sin\theta - y\cos\theta)^2 + c^2y^2 = b^2c^2\sin^2\theta.$$

Show further that the points of the plane which are accessible by projection from the given point with the given velocity lie within the ellipse E whose equation is

$$\frac{x^2}{b^2+c^2}+\frac{y^2}{b^2} = 1.$$

Show also that if a number of particles are projected simultaneously from the given point with the given velocity their positions at any instant lie on a circle, and that the envelope of these circles is again the ellipse E.

31. A particle moves in a plane field of force, the value of V at the point $P(x, y)$ being given by

$$V = \frac{k}{r} - \frac{4k}{r'},$$

where k is a positive constant and r and r' are the distances of P from the points $(a, 0)$ and $(2a, 0)$ respectively. Show that the origin is a point of equilibrium for a unit mass which is free to move in the plane and discuss the stability or instability of this point of equilibrium.

CHAPTER XIII

CENTRAL ORBITS

13·1 Motion in a central field. We consider now the motion of a particle in a central field. The force on the particle is a function of position, and the force at a point P is always in the line OP, where O is a fixed point. If the particle is projected from a point K, it is clear that the motion takes place in the plane defined by OK and the direction of projection, so the problem is one of motion in a plane. We have already studied one such problem—the isotropic oscillator (§§ 10·8 and 10·11).

The fundamental theorem is that of *the conservation of angular momentum* about O (§ 13·2). If the field is conservative (and we shall be mainly concerned with fields which are conservative) we have also *the equation of energy*. If the field is conservative the potential function and the force are functions of r, the distance from O (§ 11·2 (iii)); and conversely, if the force is a function of r, the field is conservative.

These two principles, the conservation of angular momentum and the conservation of energy, form the basis of the theory of motion in a central conservative field.

13·2 The conservation of angular momentum. (i) We take rectangular axes, with O as origin. Since the acceleration is in the direction of the radius vector we have

$$x\ddot{y}-y\ddot{x} = 0,$$

whence $$x\dot{y}-y\dot{x} = \text{constant} = h, \text{ say.} \tag{1}$$

The scalar quantity $$m(x\dot{y}-y\dot{x})$$

is called the angular momentum about O, and equation (1) shows that it remains constant during the motion.

The expression $(x\dot{y}-y\dot{x})$ which occurs in the definition of the angular momentum can also be written as a determinant

$$\begin{vmatrix} x & y \\ \dot{x} & \dot{y} \end{vmatrix}.$$

This determinant is called the Wronskian of the functions $x(t), y(t)$. Determinants of this form are important in the theory of linear differential equations.

(ii) If we now introduce polar coordinates (r, θ), where $x = r\cos\theta$, $y = r\sin\theta$, we have

$$x\dot{y} - y\dot{x} = r\cos\theta(\dot{r}\sin\theta + r\cos\theta\dot{\theta}) - r\sin\theta(\dot{r}\cos\theta - r\sin\theta\dot{\theta}) = r^2\dot{\theta},$$

whence, from (1), $$r^2\dot{\theta} = h. \tag{2}$$

It is easy to prove this result directly from the equations of motion in polar coordinates. Since the transverse component S of force is zero, the transverse equation of motion gives us

$$r\ddot{\theta} + 2\dot{r}\dot{\theta} = 0,$$

i.e. $$\frac{1}{r}\frac{d}{dt}(r^2\dot{\theta}) = 0,$$

whence $$r^2\dot{\theta} = \text{constant}$$

as before.

The polar form, equation (2), leads at once to an important observation. So long as r remains positive, $\dot{\theta}$ has always the same sign throughout the motion; this is certainly true when, as most frequently happens, r has a positive lower bound. In problems where r actually attains the value zero in a finite time (i.e. the particle falls into the centre of attraction) our interest usually ceases at the instant of this catastrophe—indeed, the physical conditions may become obscure at this point. Thus we may assert as a general rule that $\dot{\theta}$ retains the same sign throughout the motion. We conventionally take this sign, without loss of generality, to be positive. In a central orbit, θ increases always with t.

(iii) We need still a third form of the equation. If we denote the velocity of the particle by v, and the inclination of the direction of motion to Ox by ψ, we have (Fig. 13·2)

$$\dot{x} = v\cos\psi, \quad \dot{y} = v\sin\psi,$$

and $$x\dot{y} - y\dot{x} = (x\sin\psi - y\cos\psi)v = pv,$$

where p is the perpendicular distance of O from the tangent to the orbit. Thus we obtain the third form

$$pv = h. \tag{3}$$

Notice that in general p does not vanish during the motion. Our formula (3) is of course analogous to the similar formula for the moment of a force about a point.

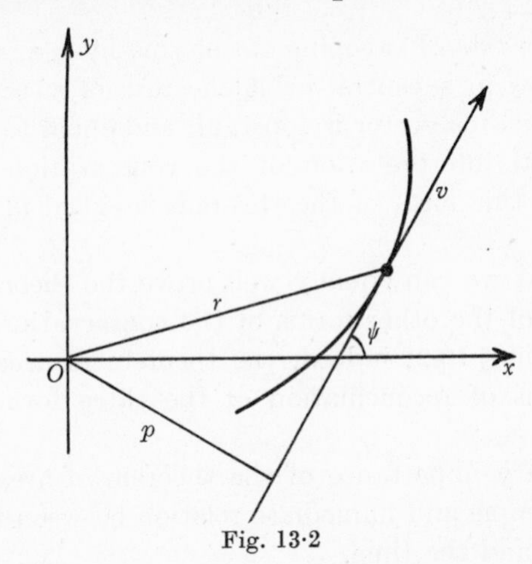

Fig. 13·2

Another way of establishing the third formula pv for the angular momentum about O is to notice that the value at any instant of the angular momentum about O is independent of the orientation of the axes. This is evident from the second formula $r^2\dot{\theta}$. The third formula follows from the first formula by referring to axes having Oy in the direction in which the particle is moving at the instant considered.

13·3 The converse theorem. The converse of the theorem of § 13·2 is true. If the angular momentum about O is conserved, we have a problem of a central attraction to O. One simple way of seeing this is to notice that the equation

$$r^2\dot{\theta} = \text{constant}$$

implies

$$r\ddot{\theta} + 2\dot{r}\dot{\theta} = 0,$$

whence $S = 0$.

13·4 The theorem of areas. If we have a curve $r = f(\theta)$ the area of the sector bounded by the radii $\theta = \theta_1$, $\theta = \theta_2$ $(\theta_1 < \theta_2)$ and the curve is

$$\Delta_{12} = \frac{1}{2}\int_{\theta_1}^{\theta_2} r^2 d\theta.$$

Thus the area swept out by the radius vector as the particle moves in its orbit is

$$\Delta_{12} = \frac{1}{2}\int_{t_1}^{t_2} r^2\dot{\theta}\,dt.$$

Therefore the rate of sweeping out of area by the radius vector is $\frac{1}{2}r^2\dot{\theta}$. Thus in a central orbit the rate of sweeping out of area by the radius vector is constant, and equal to $\frac{1}{2}h$. This is a geometrical interpretation of the conservation of angular momentum; this form of the theorem is called *the theorem of areas.*

Notice that we can equally well prove the theorem of areas from either of the other forms of the conservation of angular momentum in § 13·2; indeed, the theorem of areas affords a simple means of reconciliation of the three forms with one another.

The primary importance of the theorem of areas is that it gives us a simple and immediate relation between the position in the orbit and the time.

13·5 Further remarks on the conservation of angular momentum; vector product. The angular momentum $m(x\dot{y}-y\dot{x})$ can be regarded as the scalar product $\mathbf{r}'.\mathbf{p}$, where $\mathbf{r}'$ is obtained from $\mathbf{r}$ by turning through a right angle in the positive sense (as in § 10·1), and $\mathbf{p}$ is the momentum $m\mathbf{v}$. The components of $\mathbf{r}'$ are $(-y, x)$, and the components of $\mathbf{p}$ are $(m\dot{x}, m\dot{y})$.

But this interpretation of the angular momentum as a scalar product is to some extent misleading. It is adequate for the study of plane motion, but not for the general theory of motion in space. A deeper examination reveals that the angular momentum is properly a vector quantity; but in plane motion this vector quantity is always normal to the plane of motion, and, as we have seen in § 1·8, a family of vector quantities all parallel to a line (in this case the normal to the plane) effectively reduce to scalar quantities.

The same point arises in statics, in the theory of a plane system of forces acting on a rigid body. If we have a number of couples acting in the plane, they are equivalent to a single couple, whose moment is the algebraic sum of the moments of the given couples. Couples in one plane can therefore be thought of as scalar quantities. But in the more general theory of forces in space we prove that couples are vector quantities. The direction of the vector is normal to the plane of the couple, the modulus of the vector is the moment of the couple, and the sense is such that the moment about it is right-handed. Couples in a plane can masquerade as scalar quantities, because their vectors are all parallel.

We now define the vector product $\mathbf{P}\times\mathbf{Q}$ of a vector $\mathbf{P}$ and a vector $\mathbf{Q}$. It will be convenient to think of the vectors as represented in the diagram by segments starting from the same point. If the vectors are parallel, their vector product is defined to be the null vector. If they are not parallel, the vector product $\mathbf{P}\times\mathbf{Q}$ is a vector of directed magnitude $|\mathbf{P}|\,|\mathbf{Q}|\sin\alpha$, where α

is the angle between the directions of **P** and **Q**. Its direction is perpendicular to the plane of **P** and **Q**, and its sense is such that a right-handed rotation about it through an angle α carries **P** into coincidence with **Q**. (It is clear that it does not matter which of the two possible values of α we choose; if $0<\alpha<\pi, \sin\alpha>0$, so that the sense of the vector product is such that a right-handed rotation about it through an angle less than π carries **P** into coincidence with **Q**.)

The vector product is not commutative; indeed, it is evident that

$$\mathbf{Q}\times\mathbf{P} = -\mathbf{P}\times\mathbf{Q}.$$

If we introduce rectangular Cartesian axes, and denote the components of **P** by X, Y, Z, and the components of **Q** by L, M, N, the components of $\mathbf{P}\times\mathbf{Q}$ are

$$YN-ZM, \quad ZL-XN, \quad XM-YL.$$

The angular momentum, or moment of momentum, about O is the vector quantity $\mathbf{r}\times\mathbf{p}$. In the case under discussion, where the motion takes place in the plane $z=0$, the components of $\mathbf{r}$ are $(x, y, 0)$, the components of $\mathbf{p}$ are $(m\dot{x}, m\dot{y}, 0)$. Thus the components of the angular momentum are

$$0, 0, m(x\dot{y}-y\dot{x}).$$

The angular momentum is therefore always in the direction Oz; because of this we can safely treat the angular momentum in the plane problem as a scalar quantity.

The general theory of the vector product will not be required in this book, and we shall therefore not develop the theory in detail. It is of prime importance in three-dimensional mechanics. In three-dimensional statics the vector character of the vector product is roughly equivalent to the theorem already mentioned, that couples acting on a rigid body are vector quantities.

13·6 Apses. A point A on a central orbit at which the particle is moving at right angles to OA is called an *apse*; the line OA is called an apse-line, and the distance OA an apsidal distance. At an apse $\dot{r}=0$, so at an apse r has a stationary value.

Let us assume now that the field is conservative. Then *the orbit is symmetrical about each apse-line.* For if at an apse A the velocity is reversed, the particle retraces its path (§ 10·15), whereas, on account of the symmetry of the field about OA, this reversed path is the mirror image in OA of the original path.

We shall find that in most cases there are two apsidal distances, r_1 and r_2, which are the minimum and maximum values of r $(0<r_1<r_2)$. An apse at which $r=r_1$ is called a *perihelion*, and an apse at which $r=r_2$ is called an *aphelion.* The value of r oscillates continually from r_1 to r_2 and back again. The orbit lies in the annulus between the circles $r=r_1$ and $r=r_2$, and touches these circles alternately.

The angle α between two successive apse-lines (one perihelion and one aphelion) is called the *apsidal angle*. Since there is symmetry about each apse-line, we can infer the whole orbit from the part of it between one apse-line and the next. If the field is one of attraction towards O, the orbit is everywhere concave to O.

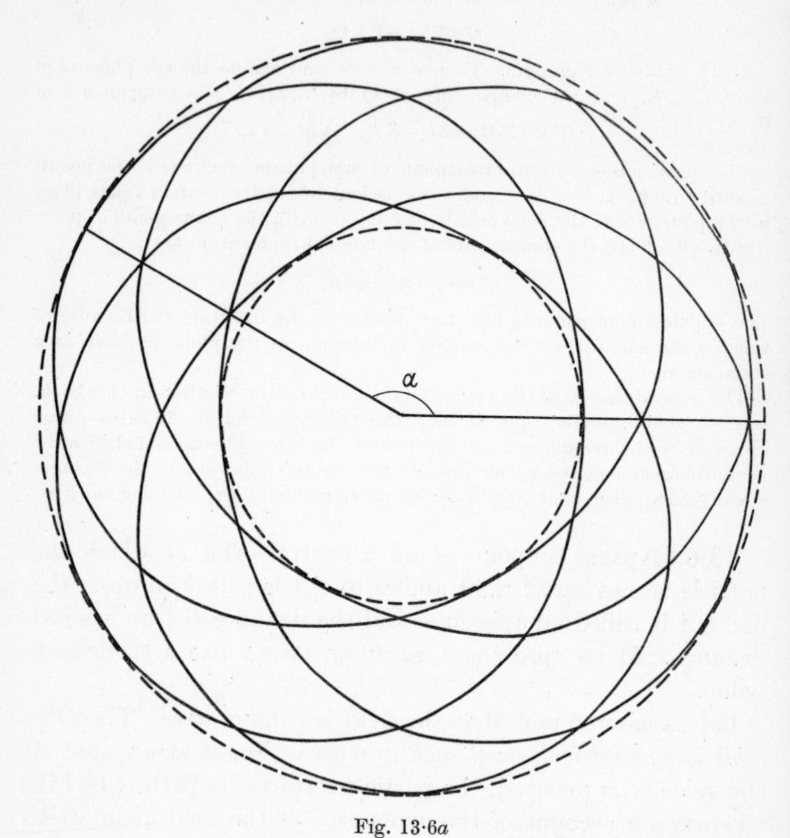

Fig. 13·6a

If α is a rational multiple of π, the orbit is periodic. Suppose $\alpha/\pi = p/q$, where p and q are positive integers with no common factor, and consider the tracing out of the orbit, starting say from a perihelion. After q librations of r, from r_1 to r_2 and back, θ has increased by $2q\alpha = 2p\pi$; the radius vector has made p complete rotations, and the orbit has returned to the same perihelion point from which it started. After this the same curve is traversed again, and so on. The figures illustrate two simple cases; in Fig. 13·6*a*, $\alpha/\pi = \frac{5}{6}$, and in Fig. 13·6*b*, $\alpha/\pi = \frac{7}{6}$.

If α/π is irrational the orbit never closes, but fills the annulus more and more thickly as time goes on.

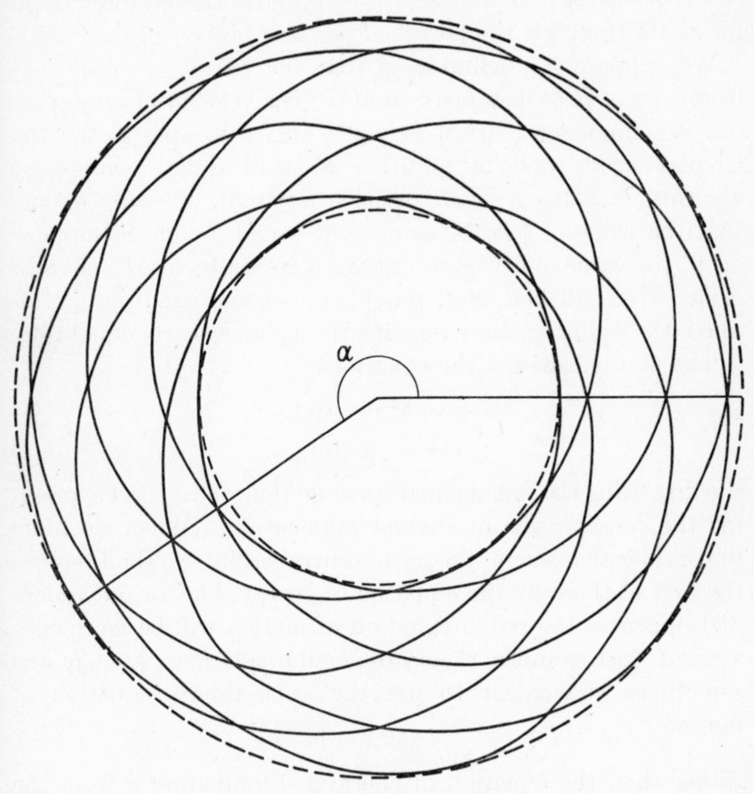

Fig. 13·6*b*

Consider by way of illustration the problem of the isotropic oscillator already discussed in §§ 10·8 and 10·11. In this case the orbit is an ellipse with its centre at O. The apsidal distances are the semi-axes of the ellipse, and the apsidal angle is a right angle. The orbit is a periodic orbit in which $\alpha/\pi = \frac{1}{2}$.

13·7 The nearly circular orbit. Perhaps the simplest illustration of the ideas of the preceding paragraph is given by the nearly circular orbit. We saw that, provided the field is an attraction $mf(r)$, we can have a circular orbit of any radius a, the speed of the particle on the circle being $\sqrt{\{af(a)\}}$. In certain cases we can have a nearly circular orbit in which r executes a small oscillation about the value a. We suppose that $f(r)$ is positive, and possesses a continuous second derivative, for $r > 0$. If $V(r)$ is the potential per unit mass, $f(r) = dV/dr$.

We suppose for definiteness that the particle is projected from a point K at distance a from O. The velocity of projection has a component $\sqrt{\{af(a)\}}$ at right angles to OK, as for the circular orbit, and in addition a small radial component $\lambda\sqrt{\{af(a)\}}$, λ being a small positive constant. It will be seen that the value of h is the same as it would be for the circular orbit, the value of C is greater than it would be for the circular orbit. We shall find that, provided a certain condition is fulfilled, the radius vector r remains throughout nearly equal to a.

The motion satisfies the equations

$$\ddot{r} - r\dot{\theta}^2 = -f(r), \tag{1}$$

$$r^2\dot{\theta} = \sqrt{\{a^3 f(a)\}}, \tag{2}$$

the first being the radial equation of motion, the second expressing the conservation of angular momentum. (When we come to consider the general theory of central orbits, we shall replace the first of these by the equation of energy, but for our immediate purpose the radial equation of motion will be more convenient; just as in the theory of small oscillations, § 6·9, it was sometimes convenient to use the approximate equation of motion

$$\ddot{x} + n^2 x = 0$$

rather than the equation of energy.) Eliminating $\dot{\theta}$ from the equations (1) and (2) we obtain

$$\ddot{r} - \frac{a^3}{r^3} f(a) = -f(r). \tag{3}$$

If we substitute $r = a + x$ and expand in powers of x we get

$$\ddot{x} - \left(1 - 3\frac{x}{a}\right) f(a) = -f(a) - xf'(a) + O(x^2),$$

whence, on rearranging,

$$\ddot{x} + \left\{3\frac{f(a)}{a} + f'(a)\right\} x = O(x^2). \tag{4}$$

This equation is of the type already studied in § 6·8. The essential condition that we need is that the coefficient of x in the first member should be positive,

$$3\frac{f(a)}{a}+f'(a)>0. \tag{5}$$

We assume this condition to be fulfilled. Then, provided the initial disturbance is not too large, x is a periodic function of t such that x oscillates through a small interval, say from β to $-\gamma$, about $x = 0$. To the first order of approximation, $\beta = \gamma$, and the mean of the apsidal distances is a.

If the condition (5) is fulfilled we say that the circular orbit of radius a is *stable*. This merely means that the disturbed orbit lies in a narrow annulus about the undisturbed orbit; every point of the disturbed orbit is near to some point of the undisturbed orbit. It does *not* mean that if two particles start together from the same point at $t = 0$, one moving in the original and one in the disturbed orbit, they will remain near together for all time (cf. § 14·12).

The proof we have given is not quite complete because, for the sake of simplicity, we have considered a particular kind of disturbance, namely, one in which C is increased but h is unaltered. For a more general disturbance the condition (5) is still sufficient to ensure stability. The most striking difference that appears if we change h as well as C is that the mean of the apsidal distances in the first approximation to the disturbed orbit is no longer exactly equal to a.

In the problem before us the disturbance is small, and we obtain a good approximation to the solution of equation (4) by omitting the term $O(x^2)$, which leads to

$$\ddot{x}+n^2x = 0,$$

where

$$n^2 = 3\frac{f(a)}{a}+f'(a).$$

The solution, with $x = 0$ and $\dot{x} = \lambda\sqrt{\{af(a)\}}$ at $t = 0$, is

$$x = \frac{\lambda}{n}\sqrt{\{af(a)\}}\sin nt = \frac{\lambda a}{p}\sin nt,$$

where

$$p^2 = 3+\frac{af'(a)}{f(a)},$$

and
$$\frac{n^2}{p^2} = \frac{f(a)}{a} = \omega^2,$$

where ω is the angular velocity in the undisturbed circular orbit. The value of r at time t in the disturbed orbit is approximately

$$r = a\left(1+\frac{\lambda}{p}\sin nt\right), \tag{6}$$

the approximation being the better the smaller the value of λ. Now

$$\dot{\theta} = \frac{a^2\omega}{r^2} = \omega\left(1-\frac{2\lambda}{p}\sin nt\right)$$

approximately, giving as the approximate value of θ at time t

$$\theta = \omega t - \frac{2\lambda}{p^2}(1-\cos nt),$$

taking $\theta = 0$ at $t = 0$. Finally, the polar equation of the orbit, correct to order λ, is found by writing θ for ωt in equation (6), giving

$$r = a\left(1+\frac{\lambda}{p}\sin p\theta\right). \tag{7}$$

We see from equation (6) or equation (7) that the apsidal distances are $a(1 \pm \lambda/p)$, and from equation (7) that the apsidal angle is π/p; these statements refer, of course, to the approximate solution, not to the exact solution.

13·8 The inverse *k*th power. An important special case is that in which $f(r) = \mu/r^k$, where $\mu > 0$. The condition for stability of the circular orbit (the inequality (5) of § 13·7) is $k < 3$; and if this is true $p^2 = 3-k$. The condition covers all values of a.

Two cases of particular interest are $k = -1$, the isotropic oscillator, and $k = 2$, the law of gravity. For $k = -1$, $p = 2$, the apsidal angle in the disturbed orbit is a right angle. Actually this is exactly, not merely approximately, true; in fact, the disturbed orbit is an ellipse with its centre at O. For $k = 2, p = 1$, the apsidal angle in the disturbed orbit is π. Again, this is exactly true; the disturbed orbit is an ellipse with a focus at O, as we shall prove in the next chapter.

13·9 The maximum-h theorem. A particle moves in a central attractive field; $f(r) > 0$ for all positive r, $V(r)$ increases always with r. The orbits with energy C all lie inside the circle $V = C$ if such a circle exists. It will not exist if V tends to a finite limit V_∞ as $r \to \infty$, and $C > V_\infty$. (In practice we choose the arbitrary constant in V to make $V_\infty = 0$.)

Of all orbits with energy C the orbit with the greatest angular momentum, if such an orbit exists, is circular. To prove this let γ be an orbit with energy C and with angular momentum h as large as possible. Then γ is a circle with O as centre. For if not there is a point on γ at which the direction of motion is not at right angles to the radius vector. If at this point we change the direction to that at right angles to the radius vector, without altering the speed, we obtain a new orbit with energy C and a larger value of h, which is impossible.

We consider this question a little further. Suppose the particle projected from an apse at distance r from O, the energy C being fixed, so that the velocity of projection u is given by

$$u^2 = 2(C - V),$$

then

$$h^2 = r^2u^2 = 2r^2(C - V),$$

and

$$\frac{d}{dr}(h^2) = 2r\{2(C - V) - rV'\},$$

which vanishes if r is the radius of a circular orbit with energy C, $u^2 = rV'(r)$. Proceeding

$$\frac{d^2}{dr^2}(h^2) = 4(C - V) - 8rV' - 2r^2V'',$$

and for a circular orbit

$$\frac{d^2}{dr^2}(h^2) = -2r(3V' + rV'')$$

$$= -2r^2\left(\frac{3f(r)}{r} + f'(r)\right).$$

The second member is negative for a stable circular orbit, and this gives a maximum value of h. An unstable circular orbit corresponds to a minimum value of h.

13·10 The missing constants. In the theory of central orbits we are concerned with the motion of a particle in a plane. The solution of the problem therefore involves *four* independent constants—say the values of two coordinates and of two velocity-components at $t = 0$. On the other hand, since the theory is based on the principles of the conservation of energy and momentum, only *two* constants, C and h, seem at first sight to appear conspicuously in the discussion. What is the explanation of this discrepancy?

The answer is simple. Suppose the solution of a particular problem, with given values of C and h, is

$$r = \xi(t), \quad \theta = \eta(t). \tag{1}$$

Then another solution of the equations of motion, with the same values of C and h, is

$$r = \xi(t - t_0), \quad \theta - \theta_0 = \eta(t - t_0), \tag{2}$$

the values of the constants t_0 and θ_0 being arbitrary. The new orbit is the old orbit turned through an angle θ_0, and with each point reached at a time t_0 later than the corresponding point in the original orbit. The solution (1) with the given values of C and h gives rise to a double infinity of orbits, not merely to one single orbit.

The value of t_0 in the description of a particular motion depends on the instant we choose as the zero of time, the value of θ_0 depends on the particular direction we choose as the zero of θ. The two new constants merely depend upon our choice of zeros of scales of measurement, and for this reason their roles in the theory are of less importance than those of C and h.

13·11 Examples.

We now consider some simple problems as applications of the general principles established in this chapter.

Example 1. *A particle moves in an ellipse, of eccentricity e, under an attraction to a centre of force in a focus S of the ellipse. For what fraction of the period does the distance of the particle from S exceed the semi-axis major of the ellipse?*

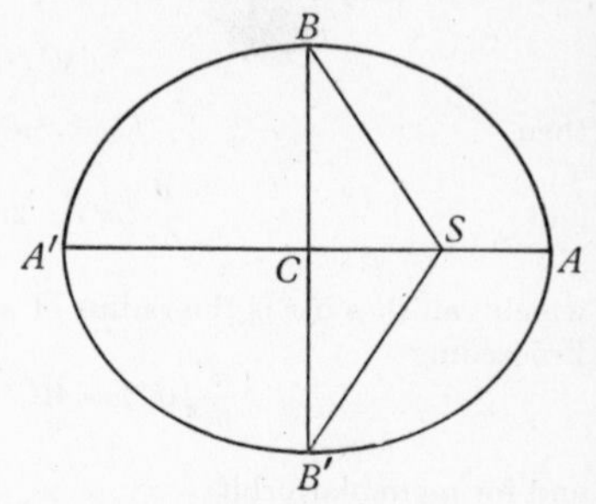

Fig. 13·11

The answer comes immediately from the theorem of areas (§ 13·4). Let a, b be the lengths of the semi-axes as usual, B and B' the ends of the minor axis. Now, with the notation of Fig. 13·11, $SB = a$, and $SP > a$ so long as P lies on the arc $BA'B'$. The fraction of the period for which $SP > a$ is therefore

$$\frac{\text{time to traverse } BA'B'}{\text{period}} = \frac{\text{area } SBA'B'S}{\Delta} = \frac{\frac{1}{2}\Delta + \delta}{\Delta},$$

where Δ is the area of the ellipse, δ the area of the triangle SBB'. Thus the required fraction is

$$\frac{1}{2} + \frac{\delta}{\Delta} = \frac{1}{2} + \frac{eab}{\pi ab} = \frac{1}{2} + \frac{e}{\pi}.$$

We shall see in the next chapter that the attraction in this problem varies as the inverse square of the distance from S.

Example 2. *The acceleration of a particle in the gravitational field of a star is μ/r^2 towards the centre of the star, where μ is a constant and r is the distance from the centre of the star. A particle starts at a great distance with velocity V, the length of the perpendicular from the centre of the star on the tangent to the initial path of the particle being p. Show that the least distance of the particle from the centre of the star is λ, where*

$$\lambda V^2 = (\mu^2 + p^2 V^4)^{\frac{1}{2}} - \mu.$$

If $\mu = 1$, $V = 2$, in the system of units chosen, and if the radius of the star is 0·005, prove that the particle will strike the star if p is less than 0·05.

When the particle is at its least distance from the centre of the star it is travelling at right angles to the radius vector, since $\dot{r} = 0$, so if its velocity at this point is U we have, from the fundamental theorem of the conservation of angular momentum (§ 13·2),

$$pV = \lambda U.$$

Also the potential function is $-\mu m/r$, so from the equation of energy

$$\tfrac{1}{2}V^2 = \tfrac{1}{2}U^2 - \mu/\lambda.$$

Eliminating U we obtain $\quad \lambda^2 V^4 + 2\mu\lambda V^2 = p^2 V^4,$

whence $\quad (\lambda V^2 + \mu)^2 = \mu^2 + p^2 V^4,$

and $\quad \lambda V^2 = (\mu^2 + p^2 V^4)^{\frac{1}{2}} - \mu.$

The particle will strike the star if λ is less than the radius a, i.e. if

$$(\mu^2 + p^2 V^4)^{\frac{1}{2}} - \mu < aV^2,$$

$$\mu^2 + p^2 V^4 < (\mu + aV^2)^2,$$

$$p^2 < a^2\left(1 + \frac{2\mu}{aV^2}\right).$$

With the given numbers this condition is $p^2 < 101a^2$, and is certainly satisfied if $p < 10a = 0{\cdot}05$.

We shall see in the next chapter that the path of the particle in this problem is a branch of a hyperbola.

Example 3. *A particle moves in a field of attraction $m\mu/r^3$, the orbit being a circle of radius a; to find the new orbit if a radial impulse is applied.* (*See* §§ 13·7, 13·8; *here $k = 3$, and the circular orbit is unstable.*)

The velocity in the circle is β/a, and $h = \beta$, where $\beta^2 = \mu$. The equations (1) and (2) of § 13·7 are

$$\ddot{r} - r\dot{\theta}^2 = -\beta^2/r^3, \quad r^2\dot{\theta} = \beta,$$

and the elimination of $\dot{\theta}$ gives $\ddot{r} = 0$. If the outward normal velocity at the impulse is $\lambda\,\beta/a$ (where now λ need not be small) we have

$$r = a + (\lambda\beta/a)t,$$

taking $t = 0$ when the impulse is applied. To find the orbit

$$\frac{dr}{d\theta} = \frac{\dot{r}}{\dot{\theta}} = \frac{\lambda r^2}{a},$$

whence $\dfrac{1}{r} = \dfrac{\lambda}{a}\left(\dfrac{1}{\lambda} - \theta\right)$, taking $\theta = 0$ at the point where the impulse is applied. If at the impulse the direction of motion is turned through an angle α, the orbit is such that $r \to \infty$ as $\theta \to \cot\alpha$.

EXAMPLES XIII

1. Prove that, in the orbit described by a small body P under the action of a force which is always directed towards a fixed point S, the velocity varies inversely as the perpendicular from S upon the tangent at P. Hence show that, if the tangent at P meet any line through S in T, the component of the velocity of P in a direction perpendicular to ST varies inversely as ST.

In an elliptic orbit described under the action of a force to a focus, find the two points of the orbit at which the component of the velocity in any direction LM and the component in the opposite direction ML have maximum values; and show that the sum of the two maximum values is the same for all directions of LM.

2. The eccentricity of the earth's orbit round the sun is $1/60$; prove that the earth's distance from the sun exceeds the length of the semi-axis major of the orbit during about 2 days more than half the year.

3. A planet moves in an elliptic orbit with the centre of attraction, the sun, in one focus. If the greatest and least velocities of the planet are v_1 and v_2, find the eccentricity of the ellipse.

4. A particle moves in a circular orbit of radius a under an attraction to a point inside the circle. The greatest and least velocities of the particle in its orbit are v_1 and v_2. Prove that the period is

$$\frac{\pi a(v_1+v_2)}{v_1 v_2}.$$

5. A particle describes a circle of radius a steadily under a central attraction $\phi(r)/r^3$ to the centre of the circle, where $\phi'(a)>0$. Prove that the period of a small oscillation about the steady motion is $2\pi\sqrt{\{a^3/\phi'(a)\}}$.

6. A weight can slide along the spoke of a horizontal wheel but is connected to the centre of the wheel by means of a spring. When the wheel is fixed the period of a small oscillation of the weight is $2\pi/n$; show that when the wheel is made to rotate with constant angular velocity ω the period of oscillation is

$$2\pi/\sqrt{(n^2-\omega^2)}.$$

If the wheel is a light frame whose mass may be neglected, and is started to rotate freely with angular velocity Ω, show that if $\Omega = 6n/5\sqrt{11}$ the greatest stretch of the spring is 20 per cent of its original length.

7. A particle of mass m is projected in any direction with velocity V from a point at distance R from a centre of force whose attraction at distance r is μmr. Prove that the orbit is in general an ellipse, and that the mean kinetic energy (with respect to time) in a period is

$$\tfrac{1}{4}m(V^2+\mu R^2).$$

8. A particle is *repelled* from a centre of force O with a force $n^2 r$ per unit mass, where r denotes distance from O. It is projected from a point distant c from O with velocity nc. Prove that the orbit is a branch of a rectangular hyperbola.

*9. Prove the statement made in § 13·7, that the condition (5) is sufficient to ensure the stability of the circular orbit for a general small disturbance, not merely for a disturbance in which h does not change.

CHAPTER XIV

THE INVERSE-SQUARE LAW

14·1 Kepler's laws. The most famous of all problems of central orbits—perhaps, indeed, the most famous problem of particle dynamics—is that in which the force is an attraction varying inversely as the square of the distance from the centre of force.

The reason for the importance of this problem is its relation to the law of gravitation. This law, discovered by Newton, asserts that any two particles of matter, whose masses are M and m, attract one another when at distance r apart with a force $\gamma Mm/r^2$, where γ is a universal constant. The problem proposed arises from the law of gravitation when we study the motion of m under the attraction of M, if M is held fixed, or if the mass M is so large that the effect on its motion of the attraction of m is insensible.

It might seem natural to start our investigation with the study of the motion of a particle in the field $f(r) = \mu/r^2$. Instead, we choose as an introduction to this part of the subject a line of approach roughly parallel to the historical development. We start from the observed facts, and, assuming the validity of Newton's laws of motion, we deduce the law of gravitation.

Newton deduced the law of gravitation from Kepler's laws, which describe the motion of the planets about the sun in the solar system. Kepler's laws are not accurately obeyed, but they are a valuable first approximation. We shall shortly be able to infer what degree of accuracy they represent. For the present, then, let us consider a solar system in which Kepler's laws are exactly true, the sun being supposed fixed, and the sun and planets being treated as particles. How does the law of gravitation emerge from Kepler's laws?

Kepler's laws are:

1. Each planet moves in an ellipse with the sun in one focus.

2. The radius vector from sun to planet sweeps out equal areas in equal times.

3. The squares of the periods are proportional to the cubes of the mean distances (semi-axes major).

The first and second of these refer to the motion of a single planet; the third expresses a relation between the motions of different planets. Laws (1) and (2) were given in 1609, law (3) in 1619.

14·2 The law of gravitation. Consider the motion of a planet P in its elliptic orbit, the sun S being at a focus of the ellipse (Fig. 14·2). The second of Kepler's laws shows that the

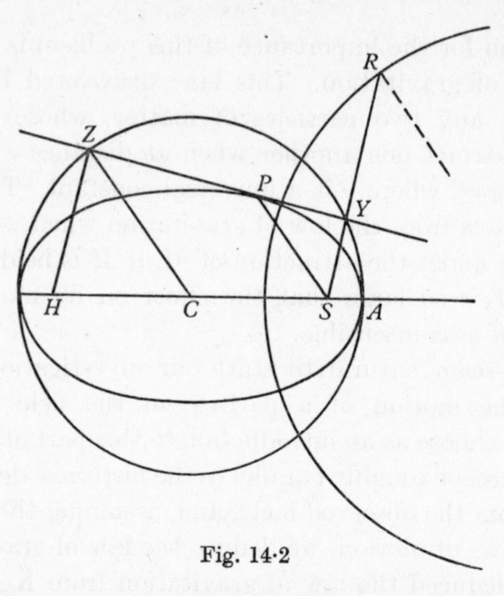

Fig. 14·2

force on the planet is an attraction to S (§§ 13·3 and 13·4). If we denote the velocity of the planet at some instant by v, and the length of the perpendicular SY on the tangent at P to the ellipse by p, we have

$$pv = h. \tag{1}$$

Produce SY to R, the length SR being numerically equal to v. If HZ is the perpendicular on to the tangent at P from the other focus H, SR is parallel and proportional to HZ; it is clear that these lines are parallel, and for the ratio of the lengths we have

$$\frac{SR}{HZ} = \frac{SR.SY}{HZ.SY} = \frac{h}{b^2},$$

where we have quoted a well-known property of the ellipse $SY.HZ = b^2$. (We denote the lengths of the major and minor semi-axes of the ellipse as usual by a and b.) We recall that, if C is the centre of the ellipse, CZ is parallel to SP. We shall need polar coordinates, r denoting the length SP, and θ the angle ASP measured from the perihelion A.

Now if SR were turned through a right angle in the positive sense the path of R would be a hodograph of the motion of P (§ 10·1), and the velocity of R would represent the acceleration of P. We anticipate therefore that the velocity of R is perpendicular to SP; and this is easily verified. For the velocity of R is parallel to that of Z, and the locus of Z (and the locus of Y also) is the auxiliary circle of the ellipse. The velocity of Z is perpendicular to CZ, and therefore the velocity of R is perpendicular to CZ, and therefore it is perpendicular to SP.

The acceleration of P is in the direction PS, its magnitude is measured by the speed of R. Now the speed of R is (h/b^2) times the speed of Z, i.e. the speed of R is $(h/b^2)a\dot{\theta}$. Since $r^2\dot{\theta} = h$, the speed of R is μ/r^2, where $\mu = h^2a/b^2$. Thus the acceleration of P is in the direction PS and of magnitude μ/r^2. This is the fundamental result we have to establish.

But we can go further. The area swept out in unit time by the radius vector is $\frac{1}{2}h$. Therefore the periodic time σ for the elliptic orbit is $2\Delta/h$, where Δ is the area enclosed by the ellipse. Thus

$$\sigma = 2\pi ab/h. \tag{2}$$

Now

$$\mu = h^2a/b^2, \tag{3}$$

and substituting for h from (2) in (3) we have

$$\mu = 4\pi^2a^3/\sigma^2.$$

By Kepler's third law a^3/σ^2 has the same value for all the planets, i.e. the coefficient μ is the same for all the planets. The acceleration at distance r from the sun is the same for all the planets, and therefore the force on each planet is proportional to its mass.

It needs now no great leap of imagination to arrive at the complete law of gravitation. We have found that the attraction on a planet of mass m is proportional to m/r^2. If there is a law of universal gravitation it is natural to expect that the masses of the two particles will appear symmetrically in the formula for the attraction. We thus find the general law $\gamma Mm/r^2$. The factor μ in the planetary problem just considered is γM, where M is the mass of the sun.

We can now see the conditions in which Kepler's laws would be exactly fulfilled. Consider a system in which the mass M of the sun is very large in comparison with the mass m of a planet. The attraction of the planet on the sun may be too small in relation to the sun's mass to effect sensibly the sun's motion, and we may suppose the sun to remain at rest. We must suppose also that the attraction of the planets on one another is also negligible—this of course involves the assumption that no two of the planets come dangerously near to one another during the motion. Then we have an approximation to the problem of a fixed sun, with planets moving under a central attraction $\mu m/r^2$. More precisely, Kepler's laws represent the limiting case obtained when M tends to infinity, γ tends to zero, and the product γM tends to a finite limit.

The value of γ in c.g.s. units, found experimentally, is $6{\cdot}658 \times 10^{-8}$.

14·3 Motion in a central gravitational field.

We now turn to the direct problem—the determination of the motion of a particle in a central field of force varying inversely as the square of the distance. We denote by $f(r)$, $V(r)$ the attraction and potential per unit mass, as in § 13·7, so that

$$f(r) = \mu/r^2,$$
$$V(r) = -\mu/r.$$

We suppose in the first instance that the field is one of attraction towards O, so that $\mu > 0$. The velocity from infinity in this field (i.e. the velocity that would be acquired by a particle falling, either freely or along a smooth guiding curve, from rest at a great distance) is $\sqrt{(2\mu/r)}$. This result follows immediately from the equation of energy, as in the simpler case considered in § 6·12.

The orbit is a conic having one focus at the centre of force. Many proofs of this famous theorem have been given; for example, we can deduce it from the general theory to be dealt with in § 15·1. We begin, however, by giving two simple proofs, both of which take advantage of the special features of the particular problem. We may suppose that the direction of projection is not along the radius vector through the centre of force; this problem has already been dealt with in § 6·12. We assume therefore that $h > 0$.

14·4 The orbit is a conic, first proof. We take rectangular axes through the centre of force O, the axis Ox through the point of projection $(c, 0)$. The equations of motion can be written

$$\ddot{x} = -\frac{\mu}{r^2}\cos\theta, \quad \ddot{y} = -\frac{\mu}{r^2}\sin\theta, \tag{1}$$

where (x, y) are Cartesian coordinates, and (r, θ) polar coordinates, as usual. Now from the conservation of angular momentum we have

$$r^2\dot{\theta} = h,$$

and using this result the equations (1) lead to the equations

$$\ddot{x} = -\frac{\mu}{h}\dot{\theta}\cos\theta, \quad \ddot{y} = -\frac{\mu}{h}\dot{\theta}\sin\theta.$$

Integration of each equation gives us

$$\dot{x} = \frac{\mu}{h}(-\sin\theta + A), \quad \dot{y} = \frac{\mu}{h}(\cos\theta + B), \tag{2}$$

where A, B are constants. Now

$$x\dot{y} - y\dot{x} = h$$

and substituting for $\dot{x}$ and $\dot{y}$ from (2) we have

$$\frac{\mu}{h}x(\cos\theta + B) - \frac{\mu}{h}y(-\sin\theta + A) = h.$$

Since $x\cos\theta + y\sin\theta = r$ this equation may be written

$$r = -Bx + Ay + (h^2/\mu). \tag{3}$$

This shows that the orbit is the locus of a point that moves so that its distance from O is proportional to its distance from the line

$$-Bx + Ay + (h^2/\mu) = 0. \tag{4}$$

The orbit is a conic, with O as focus and the line (4) as directrix; this is the result we set out to establish.

14·5 Size and shape of the orbit. The length of the *latus rectum* of the conic (equation (3) of § 14·4), i.e. the chord through the focus O parallel to the directrix, is $2h^2/\mu$. If we denote the length of the *latus rectum* by $2l$ we have

$$h^2 = \mu l. \tag{1}$$

We next determine the eccentricity. If the components of the velocity of projection from $(c, 0)$ are (u_0, v_0), we have

$$h = cv_0, \quad A = hu_0/\mu, \quad B = (hv_0/\mu)-1. \tag{2}$$

Thus, if e denotes the eccentricity of the conic, we have

$$\begin{aligned} e^2 &= A^2+B^2 \\ &= 1+\frac{h^2}{\mu^2}\left\{(u_0^2+v_0^2)-\frac{2\mu}{c}\right\}. \end{aligned} \tag{3}$$

Now $\mu > 0$, so three cases arise. If the velocity of projection is less than the velocity from infinity, $e < 1$, and the orbit is an ellipse. If the velocity of projection is equal to the velocity from infinity, $e = 1$, and the orbit is a parabola. If the velocity of projection is greater than the velocity from infinity, $e > 1$, and the orbit is the branch of a hyperbola nearer to and concave to the focus O.

We may express these results in another way. If C is the total energy (per unit mass) in the motion

$$C = \tfrac{1}{2}(u_0^2+v_0^2)-\frac{\mu}{c}, \tag{4}$$

and $e \lesseqqgtr 1$ according as $C \lesseqqgtr 0$.

Now in virtue of (3) and (1) we can rewrite equation (4) as

$$C = \frac{\mu^2}{2h^2}(e^2-1) = \frac{\mu}{2l}(e^2-1). \tag{5}$$

If we now denote the semi-axis major of the conic by a as usual, we have for the ellipse $(e < 1)$

$$l = a(1-e^2),$$

whence $$C = -\frac{\mu}{2a}. \tag{6}$$

Similarly for the hyperbola $(e > 1)$

$$l = a(e^2 - 1)$$

and
$$C = \frac{\mu}{2a}. \tag{7}$$

The equations (1) and (6) (or (7)) are the classical formulae expressing the size and shape of the orbit in terms of the constants of angular momentum and of energy. The values of h and C are determined by the circumstances of projection, and the equations express a and e in terms of h and C.

14·6 Further discussion of the orbit. A number of famous and important theorems can be deduced from the results established in the preceding paragraphs.

From the equations (2) of § 14·4 we see that the velocity of the particle is the vector sum of two components of constant magnitude; one of these components has a fixed direction (parallel to the directrix), the other is of magnitude μ/h in a direction perpendicular to the radius vector OP.

From equation (6) of § 14·5 we see that all elliptic orbits with the same energy C (<0) have major axes of the same length $2a$, where $C = -\mu/2a$; they do not all have the same eccentricity. The energy level λ for these orbits is the circle $r = 2a$.

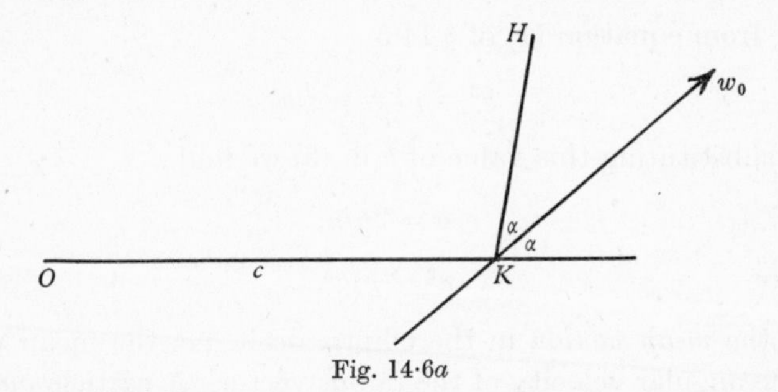

Fig. 14·6a

It is easy to construct the orbit geometrically. Suppose the particle is projected from a point K at distance c from the centre of force O, the velocity of projection being $w_0(<\sqrt{(2\mu/c)})$. Then a is determined from the equation

$$\tfrac{1}{2}w_0^2 - \frac{\mu}{c} = C = -\frac{\mu}{2a}, \tag{1}$$

and we can easily find the second focus H; OK and KH are equally inclined to the direction of projection (Fig. 14·6a) and $OK+KH = 2a$. The orbit can now be constructed from the well-known property of the ellipse as the locus of a point P such that

$$OP+PH = 2a.$$

There is a simple relation connecting the speed w of the particle at a point of the elliptic orbit with the distance of this point from the focus O. The energy equation gives us

$$\tfrac{1}{2}w^2 - \frac{\mu}{r} = C = -\frac{\mu}{2a}.$$

So
$$w^2 = \mu\left(\frac{2}{r} - \frac{1}{a}\right). \tag{2}$$

This is a well-known result, of which a special case has already been encountered in equation (3) of § 6·12.

We have already found (in § 14·2) a classical formula for the period of the elliptic orbit, but we repeat the argument here for the sake of completeness. The period depends on C only, not on h, so σ can be expressed in terms of a only. Explicitly we have (as in equation (2) of § 14·2)

$$\sigma = 2\pi ab/h. \tag{3}$$

Now from equation (1) of § 14·5

$$h^2 = \mu l = \mu b^2/a, \tag{4}$$

and substituting this value of h in (3) we find

$$\sigma = 2\pi/n, \tag{5}$$

where
$$n^2 = \mu/a^3. \tag{6}$$

n is the *mean motion* in the elliptic orbit, i.e. the mean value of the angular velocity of the radius vector. A particle moving in a circle with constant angular velocity n would have the same period as the elliptic orbit.

Now consider a hyperbolic orbit. There is no energy level (since $C > 0$). The result analogous to (2) is

$$w^2 = \mu\left(\frac{2}{r} + \frac{1}{a}\right).$$

We have found the eccentricity of the orbit and hence we can determine the angle between the asymptotes; with the notation of Fig. 14·6*b* we have

$$\cos\phi = 1/e. \tag{7}$$

But there is a more useful formula to determine the angle ϕ. Let us for the moment take as the point of projection the apse, the vertex A of the hyperbola, and then the first of the equations (2) of § 14·4 gives

$$\dot{x} = -\frac{\mu}{h}\sin\theta. \tag{8}$$

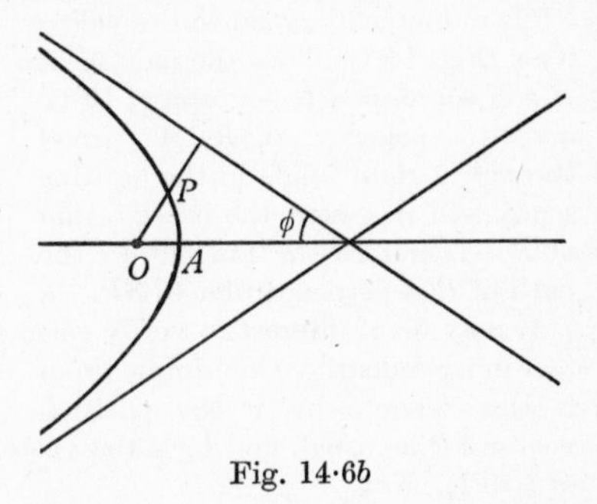

Fig. 14·6*b*

The motion ultimately approximates to uniform rectilinear motion with speed W say along an asymptote, and equation (8) leads to

$$W\cos\phi = \frac{\mu}{h}\sin\phi,$$

whence
$$\tan\phi = \frac{hW}{\mu} = \frac{PW^2}{\mu}, \tag{9}$$

where P denotes the *impact parameter*, i.e. the perpendicular distance of O from the asymptote. The result is important in physics. The motion starts in a region remote from O as a nearly uniform rectilinear motion, and it finally approximates again to uniform rectilinear motion; the angle through which the direction of motion has been turned by the interaction is $\pi - 2\phi$.

So far we have supposed $\mu > 0$, and the field is one of attraction towards O. It may be of interest to mention at this point another case. If the field is one of repulsion (as between two charged particles, with charges of the same sign) $\mu < 0$, and the orbit is always the branch of a hyperbola remote from and convex to the focus O. If the repulsion is k/r^2 per unit mass, $k = -\mu$, we easily prove by reasoning precisely similar to that already given

$$h^2 = kl, \quad C = k/2a, \quad \tan\phi = PW^2/k. \tag{10}$$

14·7 The orbit is a conic, second proof. *A particle moves under an attraction μ/r^2 per unit mass towards a centre of attraction S. To determine the motion.*

We draw the perpendicular SY on the tangent to the path at P, and produce SY to R such that the length SR is numerically equal to the velocity v (Fig. 14·7). Thus the motion of R will serve as a hodograph (§ 10·1), and the velocity of R, if turned through a right angle in the positive sense, will represent the acceleration of P. Therefore the tangent to the path of R is perpendicular to SP.

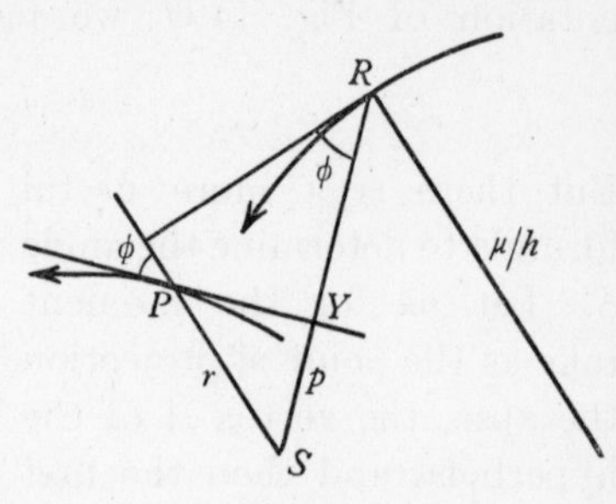

Fig. 14·7

It may be of interest to verify this fact independently. One simple proof is this: denote by $\mathbf{r}$ the position vector SP as usual, and by $\mathbf{s}$ the vector SR (so that $\mathbf{s}' = \mathbf{v}$, in the notation of § 10·1). Now

$$\mathbf{r}.\mathbf{s} = h,$$

whence

$$\dot{\mathbf{r}}.\mathbf{s}+\mathbf{r}.\dot{\mathbf{s}} = 0.$$

But $\dot{\mathbf{r}}.\mathbf{s} = 0$, and therefore

$$\mathbf{r}.\dot{\mathbf{s}} = 0,$$

and the velocity of R is perpendicular to SP.

Let W denote the speed of R, and ψ the angle that the tangent to the path of R makes with a fixed line, so that we have

$$W = \mu/r^2,$$

and

$$\psi = \theta+\frac{\pi}{2},$$

where (r, θ) are polar coordinates of P. Then

$$\frac{W}{\dot{\psi}} = \frac{\mu}{r^2\dot{\theta}} = \frac{\mu}{h},$$

and the path of R is a circle, of radius μ/h. By inversion, the path of Y is a circle, since

$$SY.SR = pv = h.$$

Therefore the envelope of YP is a conic with S as one focus, and this is the orbit of P.

The orbit is an ellipse if S lies *inside* the Y-circle, i.e. if S lies inside the R-circle. This is so if

$$v < 2(\mu/h)\sin\phi,$$

and, since $h = rv\sin\phi$, this implies

$$v^2 < 2\mu/r.$$

The orbit is an ellipse if the velocity at a point of the orbit is less than the velocity from infinity. The equation of energy is

$$\tfrac{1}{2}v^2-\frac{\mu}{r} = C,$$

and the orbit is an ellipse if $C < O$.

We notice again a fact already mentioned (in § 14·6) that the velocity in the elliptic orbit is the vector sum of two components, both of constant magnitude, and one constant also in direction. This follows at once from the fact that the hodograph is a circle.

14·8 The apsidal quadratic. The constants C and h are determined by the initial conditions, the position and velocity at the instant $t = 0$; we suppose $C < 0$, so that the orbit is an ellipse. We have seen already, in § 14·5, how to determine the major axis and eccentricity of the ellipse in terms of C and h. We now exhibit an alternative and perhaps simpler method of achieving this end; the new method involves the *apsidal quadratic*.

The equations of energy and of angular momentum are

$$\tfrac{1}{2}v^2 - \frac{\mu}{r} = C, \tag{1}$$

$$pv = h. \tag{2}$$

Eliminating v we obtain the equation

$$\frac{h^2}{2p^2} - \frac{\mu}{r} = C. \tag{3}$$

Now at an apse $p = r$, so writing r for p in equation (3) we obtain a quadratic equation

$$2Cr^2 + 2\mu r - h^2 = 0 \tag{4}$$

whose roots are the apsidal distances.

Equation (4) is the *apsidal quadratic*, and its roots are $a(1-e)$ and $a(1+e)$. Thus, since the sum of the roots is $-\mu/C$, we have

$$-\frac{\mu}{C} = 2a, \tag{5}$$

and, since the product of the roots is $-h^2/2C$, we have

$$-\frac{h^2}{2C} = a^2(1-e^2). \tag{6}$$

If we write as before $\quad a(1-e^2) = l,$

so that l denotes the length of the *semi-latus rectum* of the ellipse, the equations (5) and (6) give us

$$C = -\frac{\mu}{2a},$$

$$h^2 = \mu l.$$

These are the classical formulae already found in § 14·5 connecting the constants of energy and of angular momentum with the size and shape of the elliptic orbit. A number of deductions from these equations have already been recorded in § 14·6, and we need not repeat them here.

14·9 Projection from an apse. The special case of projection from an apse, where the angle α of Fig. 14·6a is a right angle, is perhaps worth a moment's consideration. (We may notice in passing that the discussion of § 14·4 is simplified if we project from an apse: in that case the constant A is zero.) We have seen that the orbit is an ellipse if $w_0^2 < 2\mu/c$, and that the major axis of this ellipse is given by equation (1) of § 14·6,

$$\tfrac{1}{2}w_0^2 - \frac{\mu}{c} = -\frac{\mu}{2a}.$$

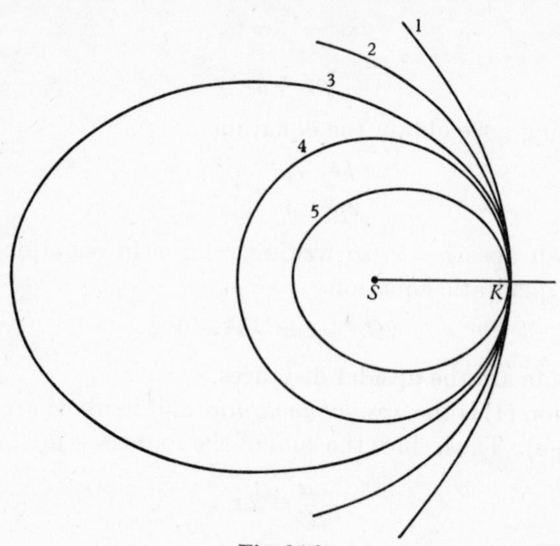

Fig. 14·9

If therefore $w_0^2 > \mu/c$, $c < a$, and in that case the apse K from which the particle is projected is the perihelion. Similarly, if $w_0^2 < \mu/c$, K is the aphelion. If $w_0^2 = \mu/c$ the orbit is a circle. We are thus able to construct the following table, giving the type of orbit for projection from an apse with different values of w_0:

Case 1. $w_0^2 > 2\mu/c$, hyperbola.
2. $w_0^2 = 2\mu/c$, parabola.
3. $2\mu/c > w_0^2 > \mu/c$, ellipse, perihelion at K.
4. $w_0^2 = \mu/c$, circle.
5. $\mu/c > w_0^2$, ellipse, aphelion at K.

The various cases in the table are illustrated in Fig. 14·9.

14·10 Accessibility, the enveloping ellipse. If we project the particle from a point K at distance c from S with velocity w_0, where $w_0^2 < 2\mu/c$, the orbit is an ellipse whose major axis is determined by equation (1) of § 14·6, and the energy level λ is the circle of radius $2a$ about S. But not all points inside this circle are accessible by projection from K with velocity w_0 (Fig. 14·10a). What region of this circle is accessible from K?

If we wish to project from K so that the orbit shall pass through a point Q, the second focus lies on a circle with K as centre, of radius $(2a-c)$; this circle touches the energy level λ, and lies inside λ. The second focus also lies on the circle with Q as centre which touches λ, and lies inside λ. If these two circles

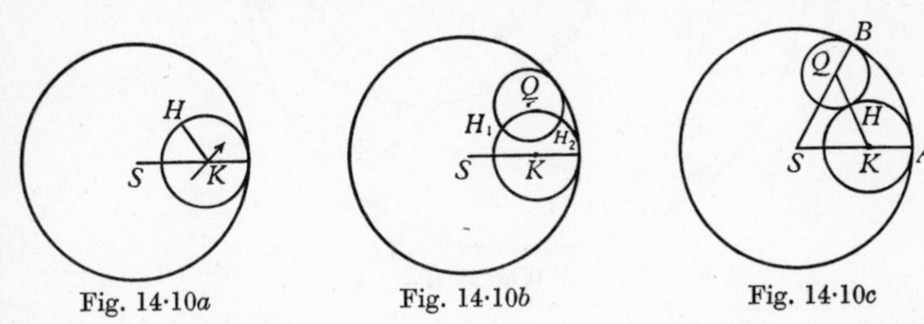

Fig. 14·10a Fig. 14·10b Fig. 14·10c

intersect in two points H_1, H_2 there are two possible orbits from K to Q, one having its second focus at H_1, the other at H_2 (Fig. 14·10b).

If the two circles do not intersect in real points, Q is not accessible from K.

If the two circles touch, Q is just accessible from K. There is a unique orbit from K to Q, the second focus being the point H where the two circles touch (Fig. 14·10c). It is easy to find the locus of such points Q which are just accessible from K. If SK, SQ meet λ in A and B we have

$$\begin{aligned} SQ+QK &= SQ+QH+HK \\ &= SQ+QB+KA \\ &= SB+KA \\ &= SA+KA, \end{aligned}$$

which is constant, and the locus of Q is an ellipse E with its foci at S and K, touching λ at A. Moreover, the orbit through

K and Q touches this ellipse E at Q, since the tangents to both ellipses at Q coincide in the external bisector to the angle SQK. E is therefore the envelope of the orbits. The region accessible from K consists of the ellipse E and its interior (Fig. 14·10d).

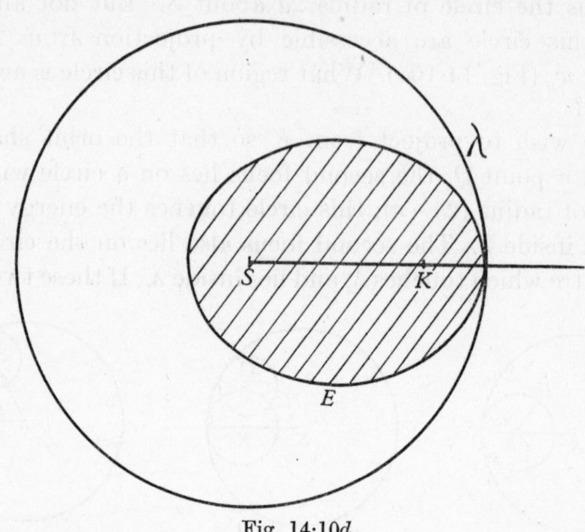

Fig. 14·10d

14·11 Relation between position and time in the elliptic orbit. We now find an explicit relation between the time and the position in the orbit. The basis of the calculation is the theorem of areas (§ 13·4). We shall find that there is a simple relation connecting the variable t, not with r or θ, but with the eccentric angle.

We shall regard the ellipse as the orthogonal projection of its auxiliary circle, and we use the fact that the ratio of areas is unaltered by orthogonal projection. Measuring t from an instant when the particle is at perihelion (the point A of the figure) we have

$$\frac{t}{\sigma} = \frac{\delta}{\Delta},$$

where δ is the area of the segment ASP of the ellipse, Δ the area of the ellipse. Let Q be the point on the auxiliary circle corresponding to P. Now

$$\frac{\delta}{\Delta} = \frac{\delta'}{\Delta'},$$

where δ' is the area of the segment ASQ of the auxiliary circle, Δ' the area of the circle. Now the segment ASQ is the segment ACQ minus the triangle CSQ, so

$$\delta' = \tfrac{1}{2}a^2\phi - \tfrac{1}{2}a^2 e \sin\phi,$$

where ϕ is the eccentric angle of P, i.e. the angle QCA. Thus we have

$$\frac{t}{\sigma} = \frac{\delta'}{\Delta'} = \frac{\phi - e\sin\phi}{2\pi},$$

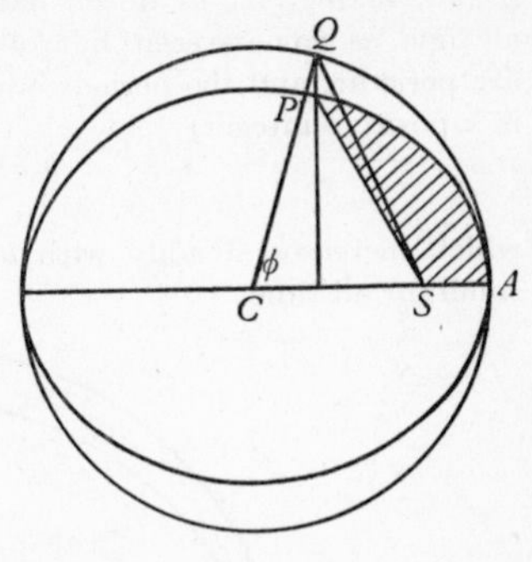

Fig. 14·11

or, writing $\sigma = 2\pi/n$, as in § 14·6, equation (5), we have

$$nt = \phi - e\sin\phi. \qquad (1)$$

This relation, known as Kepler's equation, is the fundamental relation connecting the position in the elliptic orbit with the time.

The polar coordinates (r, θ) are easily expressed in terms of ϕ,

$$r = a(1 - e\cos\phi),$$

$$\tan\frac{\theta}{2} = k\tan\frac{\phi}{2},$$

where $k^2 = \dfrac{1+e}{1-e}$. (SA is taken as initial line, so that θ is the angle ASP.) Thus r and θ, and hence also the Cartesian coordinates of P, are expressed directly in terms of ϕ, and indirectly in terms of t.

14·12 The disturbed circular orbit. We are now able to prove an important property (already mentioned in § 13·7) of the disturbed circular orbit.

A particle moves in a circle of radius R about S in the field μ/r^2. The angular velocity ω is $\sqrt{(\mu/R^3)}$, the period σ is $2\pi\sqrt{(R^3/\mu)}$, and the angular momentum $h = R^2\omega = \sqrt{(\mu R)}$.

Suppose now that at $t = 0$, when the particle is at the point P_0 of the circular orbit, a small radial impulse is applied (cf. § 13·7). The disturbed orbit is a nearly circular ellipse, and, since the angular momentum is unchanged, $R = l$. The line SP_0 is parallel to the minor axis of the ellipse. The period σ' of the elliptic motion is $2\pi\sqrt{(a^3/\mu)}$, so $\sigma' > \sigma$.

Now consider a point P moving in the original (circular) orbit, and a point P' moving in the disturbed (elliptic) orbit, both starting from P_0 at $t = 0$. We denote the polar coordinates of P at time t by (r, θ) and those of P' by (r', θ'), taking SP_0 as initial line. Then $|r-r'| = |a-r'|$ remains small for all time, as we have seen; but $|\theta-\theta'|$ cannot remain small, since both motions are periodic, and the periods are unequal. In fact, after a time $k\sigma'$ (where k is a positive integer)

$$\theta-\theta' = 2k\pi\left(\frac{\sigma'}{\sigma}-1\right),$$

which increases steadily with k. Thus the distance PP' does not remain small for all time.

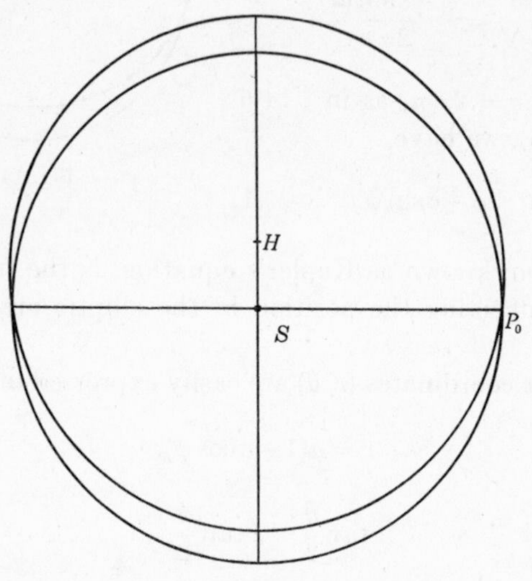

Fig. 14·12

The circular orbit is stable in the broad sense that the disturbed orbit lies, as a whole, near the original orbit. But it is not stable in the more exacting sense that $|PP'|$ remains small for all time.

It may be of interest to observe that for the isotropic oscillator (where the period of the disturbed orbit is equal to that of the original orbit) the circular orbit is stable in the more exacting sense. To prove this, let us take rectangular axes through S, and suppose that the particle is at $(a,0)$ at $t = 0$. Its velocity at this instant in the disturbed orbit has components $(\lambda na, na)$, and the coordinates of P' at time t are

$$\begin{cases} x = a\cos nt + \lambda a \sin nt, \\ y = a \sin nt. \end{cases}$$

The coordinates of P are $(a\cos nt, a\sin nt)$. Thus

$$|PP'| = |\lambda a \sin nt| \leqslant \lambda a,$$

and the circular orbit is therefore stable in the more exacting sense.

14·13 Parabolic orbit. Hitherto we have been concerned mainly with the case $\mu > 0, C < 0$. Now consider the case $\mu > 0, C = 0$. This case occurs (§§ 14·4 and 14·5) when the field is attractive and the velocity of projection is $\sqrt{(2\mu/c)}$. The orbit is a parabola with its focus S at the centre of force. The apsidal quadratic (equation (4) of § 14·8) reduces to the linear equation (for r)

$$2\mu r = h^2,$$

and this is not unexpected, since there is now only one apse. If $4a$ is the length of the *latus rectum* of the parabola

$$a = h^2/2\mu.$$

The directrix of the parabola can be found by a simple geometrical construction. Let the foot of the perpendicular

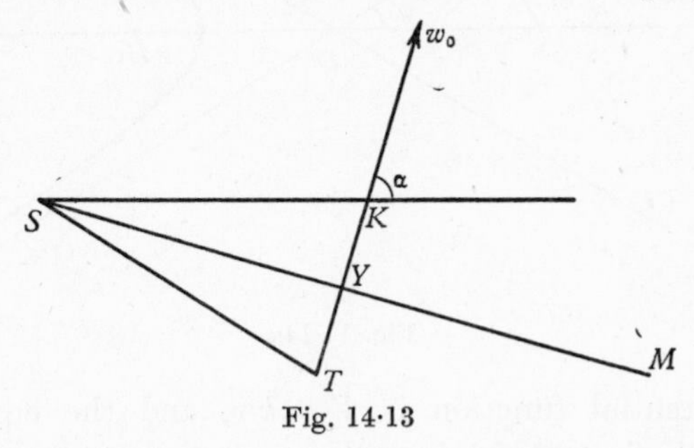

Fig. 14·13

from S on the tangent to the path at K be Y. If we produce SY to M, where $SY = YM$, the point M lies on the directrix. If we produce KY to T, where $KY = YT$, then ST is the axis of the parabola. The directrix is the line through M at right angles to ST.

14·14 Hyperbolic orbit. If the field is an attraction to S proportional to the inverse square of the distance, and if $C > 0$, the orbit is a branch of a hyperbola concave to S (§ 14·5). If the field is a repulsion from S proportional to the inverse square of the distance, in which case C is necessarily positive, the orbit is a branch of a hyperbola convex to S.

Let us consider for definiteness the case of a repulsion k/r^2 per unit mass (Fig. 14·14a). This problem is important in physics in connexion with the bombardment of atoms by α-particles.

If we wish to determine the size and shape of the hyperbolic orbit in terms of C and h the method of the apsidal quadratic, which we used for the elliptic orbit, is no longer applicable; in the hyperbolic orbit there is only one apsidal distance, and only one root of the apsidal quadratic is known in terms of a and e. However, the knowledge of this root will give us one relation, and we can easily find a second relation as follows.

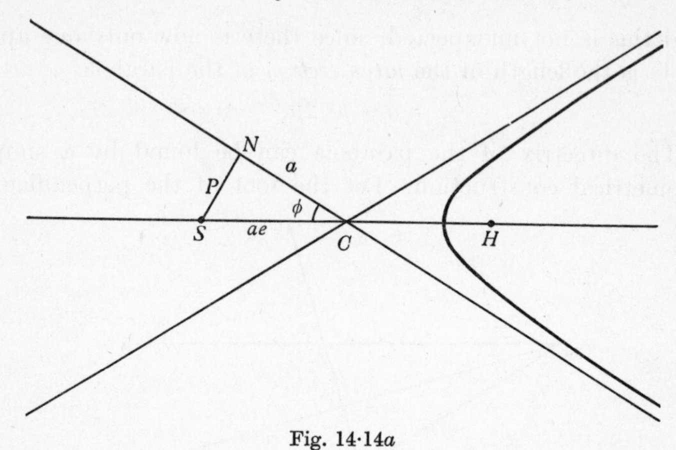

Fig. 14·14a

The potential function is $V = k/r$, and the equations of energy and of angular momentum are

$$\tfrac{1}{2}v^2 + \frac{k}{r} = C, \tag{1}$$

$$pv = h. \tag{2}$$

Hence, eliminating v, we find (cf. § 14·8)

$$\frac{h^2}{2p^2} + \frac{k}{r} = C. \tag{3}$$

If we write r in place of p in this equation we obtain a quadratic equation, the positive root of which is the apsidal distance $a(e+1)$. This gives us one equation connecting a and e

$$\frac{h^2}{2a^2(e+1)^2} + \frac{k}{a(e+1)} = C. \tag{4}$$

The second relation is found by letting $r \to \infty$ in equation (3); then $p^2 \to a^2(e^2-1)$, and we have

$$\frac{h^2}{2a^2(e^2-1)} = C. \tag{5}$$

From (4) and (5) we find

$$C = \frac{k}{2a}, \tag{6}$$

$$h^2 = ka(e^2-1) = kl, \tag{7}$$

and these are the equations determining a and e in terms of the initial conditions (cf. § 14·5).

The determination of the orbit when the particle is projected from a point K with velocity w_0 is now easy. If $SK = c$ we have

$$\tfrac{1}{2}w_0^2 + \frac{k}{c} = C = \frac{k}{2a},$$

which determines a, and then the distance KH is given by

$$SK - KH = 2a.$$

Remember that this time the direction of projection bisects the angle SKH internally (Fig. 14·14b). We thus find the second focus H, and the orbit is defined by the rule

$$SP - PH = 2a.$$

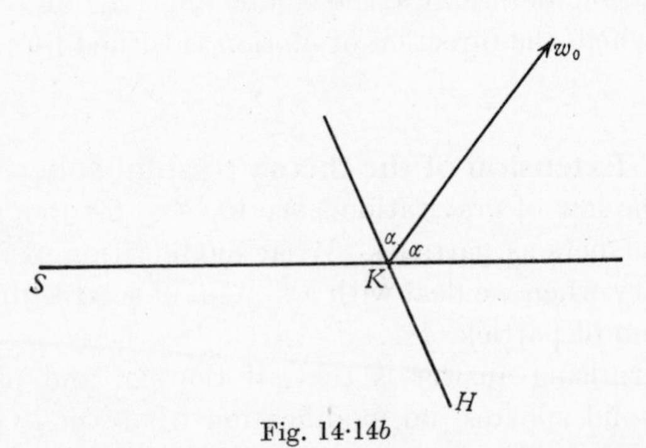

Fig. 14·14b

As we have seen already (§ 14·5) for an *attraction* μ/r^2 the orbit, if $C > 0$, is the branch of a hyperbola nearer to S, and the equations corresponding to (6) and (7) above can be found, by the method of the present paragraph, to be

$$C = \mu/(2a), \quad h^2 = \mu l.$$

We notice that, when c and w_0 are prescribed, we get the same values of a and e both for an attraction μ/r^2 and for a repulsion μ/r^2. This fact is easily established otherwise, for example by the method of § 14·5. For an attraction μ/r^2 the roots of the apsidal quadratic (equation (4) of §14·8) are $a(e-1)$ and $-a(e+1)$.

We return now to the problem of a repulsion k/r^2 per unit mass. In § 14·6, equation (10), we found the result

$$\tan\phi = PW^2/k$$

giving the angle between the asymptotes of the hyperbolic orbit; we conclude this paragraph with an alternative proof of this result. If N is the foot of the perpendicular from S on an asymptote, the length CN is equal to a. Therefore, if 2ϕ is the angle between the asymptotes, as indicated in Fig. 14·14a, we have

$$\tan\phi = P/a. \tag{8}$$

Now, using equations (1) and (6), we have

$$\tfrac{1}{2}W^2 = C = k/(2a), \tag{9}$$

and from equations (8) and (9) we have

$$\tan\phi = PW^2/k. \tag{10}$$

This equation determines the acute angle ϕ, and the angle through which the direction of motion is turned by the impact is $\pi - 2\phi$.

14·15 Extension of the theory to solid spheres.

14·15 Extension of the theory to solid spheres. We return to the law of gravitation. We have so far thought of the sun and planets as particles. What modification of the theory is necessary when we deal with a system of solid bodies instead of a system of particles?

The surprising answer is that, if the sun and planets are uniform solid spheres, no modification whatever is necessary. The motion of the centres of the spheres is given by the same laws as for particles, just as though the mass of each sphere were concentrated at a single point. (This is not true for solids of other shapes.)

Consider the attraction of a uniform solid sphere of mass M on a particle P of mass m. The attraction on P is a force

$\gamma Mm/r^2$ in the line PA, where A is the centre of the sphere, and r the distance AP. This is the same as if all the mass of the sphere were concentrated at A. This theorem is not difficult to establish, either by direct integration, or, even more simply, from Gauss' theorem; but we omit the proof, which is a little outside the scope of this book.

The attraction of the particle on the constituent parts of the solid sphere is equivalent to a single force $\gamma Mm/r^2$ at A, in the direction AP.

Suppose now we have two uniform solid spheres, masses M and M', with their centres at A and A'. The attraction of the first sphere on any particle of the second is the same as if all the mass of the first sphere were concentrated at A. Therefore the total action on the second sphere is a force $\gamma MM'/R^2$ at A', acting in the direction $A'A$, R now denoting the distance AA'. Thus if the sun and a planet are uniform solid spheres the gravitational action on each is a force at its centre, and the magnitude of the force is what would be given by the law of gravitation if the mass of each sphere were concentrated at its centre.

Now we shall prove later (Chapter XXI) that when a system of forces acts on a rigid body, and we reduce the system of forces to a force $\mathbf{P}$ at the centre of gravity G of the body and a couple, then the acceleration $\mathbf{f}$ of G is given by Newton's law

$$\mathbf{P} = M\mathbf{f}.$$

This is the same as we should get if a force $\mathbf{P}$ acted on a particle of mass M at G. Now the centre of gravity of a uniform solid sphere is at its centre, so the theory we have worked out for a solar system of particles is still valid for the motion of the centres when we replace the particles by uniform solid spheres.

Actually we can go a little further. The theory is valid if the spheres are not uniform, provided that in each sphere there is spherical symmetry, i.e. the density is a function only of the distance from the centre.

In fact the earth is not exactly spherical and is not solid throughout, and the next step would be to discuss the effect of these deviations from the uniform solid sphere. But that is another story.

EXAMPLES XIV

1. Prove that the earth's velocity of approach to the sun, when the earth in its orbit is at one extremity of the *latus rectum* through the sun, is approximately $18\frac{1}{2}$ miles per minute, taking the eccentricity of the earth's orbit as $1/60$ and 93,000,000 miles as the semi-axis major of the earth's orbit.

2. A particle P moves in a plane under an attraction μ/SP^2 per unit mass to a fixed point S of the plane. The particle is projected from a point at distance c from S at right angles to the radius vector, the velocity of projection being u. Writing α for cu^2/μ, show that if $\alpha < 2$ the orbit is an ellipse, of which the major semi-axis is $c/(2-\alpha)$ and the eccentricity is $|\alpha - 1|$. Show further that the point of projection is the perihelion or the aphelion according as $\alpha > 1$ or $\alpha < 1$.

3. A comet is describing a parabolic path about the sun. Show that at distance r the velocity v is given by $v^2 = 2\mu/r$, where μ/r^2 is the radial attraction per unit mass of the comet.

If the comet is 6×10^7 miles from the sun at its point of closest approach, show that the time that elapses between the instants at which it is at the ends of the *latus rectum* is about 114 days. It may be assumed that the earth's orbit is circular, of radius $9{\cdot}3 \times 10^7$ miles, and is described in $365\frac{1}{4}$ days.

4. A planet is describing an elliptic orbit with the sun in a focus, and at a certain point it receives a small tangential impulse so that its velocity changes from v to $v + \delta v$. Find the consequent small changes in a, in e, and in the orientation of the orbit.

5. A particle of mass m describes a parabola under an attraction to the focus; prove that the attraction at distance r is $m\mu/r^2$, and that the velocity is $(2\mu/r)^{\frac{1}{2}}$.

Two particles describe in equal times the arc of a parabola bounded by the *latus rectum*, one under an attraction to the focus, and the other with constant acceleration g parallel to the axis. Show that the acceleration of the first particle at the vertex of the parabola is $\frac{16}{9}g$.

6. Prove that the mean value (with respect to time) of the kinetic energy T in the Newtonian orbit is

$$\overline{T} = \tfrac{1}{2}mn^2a^2 = \mu m/(2a).$$

*7. A planet moves in an elliptic orbit about the sun in one focus. Express the average value (with respect to time) of each of the following quantities in terms of its maximum and minimum values:

(i) $1/r$, the inverse of the radius vector,

(ii) $\dot{\theta}$, the angular velocity,

(iii) r, the radius vector,

the sun being taken as origin. From the first of these results find the mean value of the potential energy, and deduce the mean value of the kinetic energy.

CHAPTER XV

THE GENERAL THEORY OF CENTRAL ORBITS

15·1 The fundamental equations, the apsidal distances and the differential equation of the orbit. We now consider the general theory of motion in a central field. We take the centre of attraction O as origin, and denote the potential of the field by $mV(r)$, so that $V(r)$ is the potential per unit mass. Using polar coordinates (r, θ) the equations of energy and of angular momentum are

$$\tfrac{1}{2}(\dot{r}^2 + r^2\dot{\theta}^2) + V(r) = C, \tag{1}$$

$$r^2\dot{\theta} = h. \tag{2}$$

Eliminating $\dot{\theta}$ we obtain the differential equation connecting time and distance from O,

$$\tfrac{1}{2}\dot{r}^2 = C - V(r) - \frac{1}{2}\frac{h^2}{r^2}. \tag{3}$$

This is a differential equation of a type we have studied in detail in Chapter VI. If, as frequently happens, r lies initially between consecutive simple real zeros r_1 and r_2 of the second member, where $0 < r_1 < r_2$, there is a libration motion in r between r_1 and r_2. The distances r_1 and r_2 are the apsidal distances, and the orbit lies in the annulus between the circles of radii r_1 and r_2, touching each in turn.

We next find a differential equation connecting r and θ, the differential equation of the orbit. During the motion

$$\left(\frac{dr}{d\theta}\right)^2 = \frac{\dot{r}^2}{\dot{\theta}^2}, \tag{4}$$

and substituting for $\dot{r}^2$ from equation (3), and for $\dot{\theta}^2$ from equation (2), we obtain

$$\left(\frac{dr}{d\theta}\right)^2 = \frac{2r^4}{h^2}\left\{C - V(r) - \frac{1}{2}\frac{h^2}{r^2}\right\}. \tag{5}$$

This is the differential equation of the orbit. It is again of the type studied in Chapter VI.

If the field is attractive, $V(r)$ increases always with r; as $r \to \infty$, $V(r)$ tends to a limit or to ∞, and as $r \to 0$, $V(r)$ tends to a limit or to $-\infty$. This fact gives us a simple classification of attractive fields into four types, according to the behaviour of $V(r)$ at 0 and at ∞. If $V(r)$ tends to a limit at 0 or at ∞ (but not at both) we can conveniently take this limit to be zero; to achieve this we have only to make a suitable choice of the arbitrary constant in V.

The constants C and h are determined by the initial conditions. If the particle is projected from a point K at distance c from O, with velocity w_0 in a direction making an angle α with OK produced (see Fig. 15·1), we have

$$C = \tfrac{1}{2}w_0^2 + V(c), \qquad (6)$$

$$h = cw_0 \sin\alpha. \qquad (7)$$

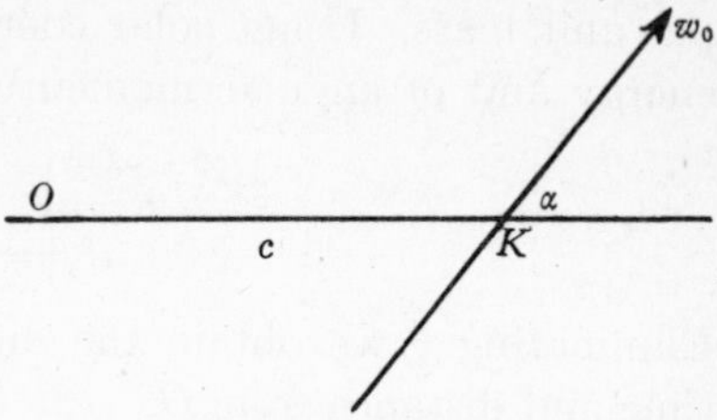

Fig. 15·1

We shall suppose that α is not 0 or π, for in that case the result would be the rectilinear motion already discussed in Chapter VI; we shall suppose $0 < \alpha < \pi$, so that $h > 0$.

If $V(r)$ increases steadily to zero as $r \to \infty$, the velocity from infinity w_r is given by

$$\tfrac{1}{2}w_r^2 + V(r) = 0,$$

so in this case we can write equation (6) in the form

$$C = \tfrac{1}{2}(w_0^2 - w_c^2).$$

15·2 The Newtonian orbit.

As a first application of the general theory we recover the results already found by more elementary methods in Chapter XIV. In this case

$$V = -\mu/r, \qquad (1)$$

and the equations (3) and (5) of § 15·1 become

$$\tfrac{1}{2}\dot{r}^2 = (2Cr^2 + 2\mu r - h^2)/(2r^2)$$
$$= Q(r)/(2r^2), \text{ say,} \qquad (2)$$

and
$$\left(\frac{1}{r^2}\frac{dr}{d\theta}\right)^2 = \frac{2C}{h^2}+\frac{2\mu}{h^2 r}-\frac{1}{r^2} = \frac{Q(r)}{h^2 r^2}. \tag{3}$$

(The equation $Q(r) = 0$ is of course the apsidal quadratic, equation (4) of § 14·8.)

The graph of $Q(r)$ is a parabola, unless $C = 0$, when it is a straight line. Let us be content to discuss in detail the case $C < 0$. If $\alpha \neq \frac{1}{2}\pi$, $\dot{r}^2 > 0$ at $r = c$, so that $Q(c) > 0$, and the graph of $Q(r)$ is as shown in Fig. 15·2*a*.

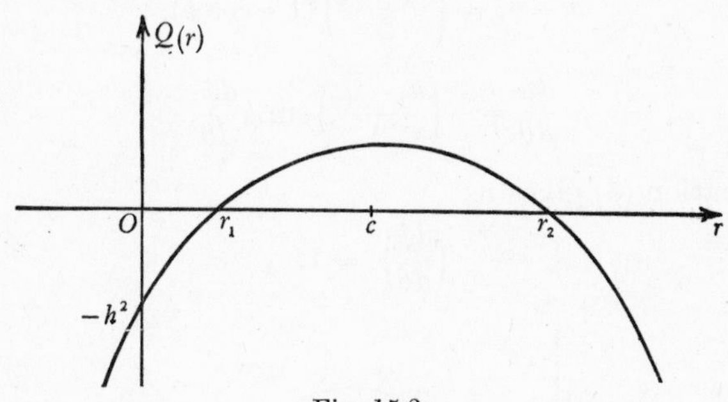

Fig. 15·2*a*

If $\alpha = \frac{1}{2}\pi$, so that the point of projection is an apse, the graph is similar, except that c coincides with r_1 or r_2; unless $w_0^2 = \mu/c$, when c is a double zero of $Q(r)$, and the orbit is a circle. (We may mention in passing that if $w_0^2 = \mu/c$ and $\alpha \neq \frac{1}{2}\pi$, the point of projection is at one end of the minor axis of the elliptic orbit; the reader will have no difficulty in proving this theorem.) We will exclude the exceptional case ($\alpha = \frac{1}{2}\pi$, $w_0^2 = \mu/c$) and assume in what follows that $Q(r)$ has distinct real positive zeros r_1 and r_2, where $r_1 < r_2$.

Consider first the differential equation of the orbit, equation (3). We can write this in the form

$$\left(\frac{1}{r^2}\frac{dr}{d\theta}\right)^2 = \left(\frac{1}{r_1}-\frac{1}{r}\right)\left(\frac{1}{r}-\frac{1}{r_2}\right), \tag{4}$$

or, say,
$$\left(\frac{du}{d\theta}\right)^2 = (u_1-u)(u-u_2), \tag{5}$$

where we have written u for $1/r$, u_1 and u_2 for $1/r_1$ and $1/r_2$.

The equation (5) is handled by the method of § 6·5. We introduce in place of u a variable ψ (Fig. 15·2b) defined by

$$u = \frac{u_1+u_2}{2} + \frac{u_1-u_2}{2}\cos\psi, \tag{6}$$

so that $\psi = 0$ at perihelion. Then

$$u_1 - u = \left(\frac{u_1-u_2}{2}\right)(1-\cos\psi),$$

$$u - u_2 = \left(\frac{u_1-u_2}{2}\right)(1+\cos\psi),$$

$$\frac{du}{d\theta} = -\left(\frac{u_1-u_2}{2}\right)\sin\psi\,\frac{d\psi}{d\theta},$$

and equation (5) gives us

$$\left(\frac{d\psi}{d\theta}\right)^2 = 1.$$

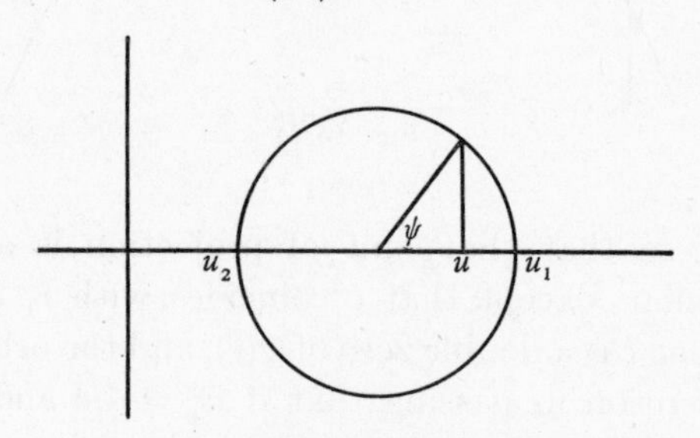

Fig. 15·2b

Hence $$\psi = \pm(\theta-\theta_0),$$

and equation (6), replacing now the symbol r in preference to u, becomes

$$\frac{1}{r} = \frac{1}{2}\left(\frac{1}{r_1}+\frac{1}{r_2}\right) + \frac{1}{2}\left(\frac{1}{r_1}-\frac{1}{r_2}\right)\cos(\theta-\theta_0). \tag{7}$$

This is the equation of the orbit, and represents an ellipse with one focus at O. If we write $r_1 = a(1-e)$ and $r_2 = a(1+e)$, equation (7) takes the form

$$\frac{l}{r} = 1 + e\cos(\theta-\theta_0),$$

where $l = a(1-e^2)$.

There is no need to repeat the argument which establishes the relations determining a and e in terms of C and h. These relations arise immediately, as in § 14·8, from the fact that the apsidal distances are the zeros of $Q(r)$. We may assume therefore the relations

$$C = -\mu/(2a), \tag{8}$$

$$h^2 = \mu l. \tag{9}$$

We next turn to the equation (2), the relation between position and time. Using equation (8), equation (2) becomes

$$\dot{r}^2 = \mu(r_2 - r)(r - r_1)/(ar^2). \tag{10}$$

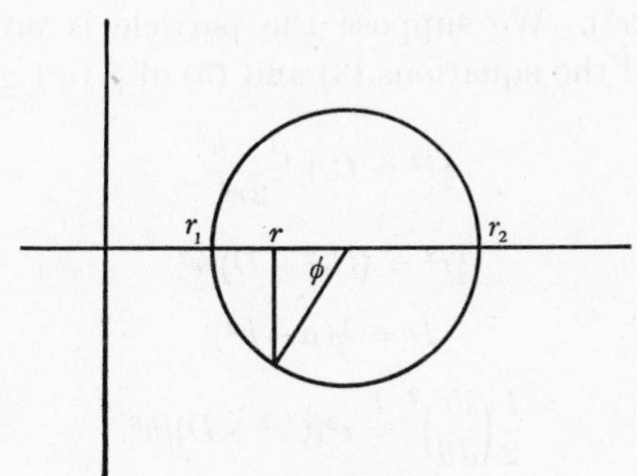

Fig. 15·2c

Now introduce a variable ϕ (Fig. 15·2c) defined by

$$r = \left(\frac{r_2 + r_1}{2}\right) - \left(\frac{r_2 - r_1}{2}\right)\cos\phi, \tag{11}$$

so that $\phi = 0$ at $r = r_1$. Then, by reasoning similar to that used above, we have

$$\frac{dr}{\sqrt{\{(r_2 - r)(r - r_1)\}}} = d\phi,$$

and equation (10) becomes

$$\sqrt{\left(\frac{\mu}{a}\right)} dt = r d\phi$$

$$= \left\{\left(\frac{r_2 + r_1}{2}\right) - \left(\frac{r_2 - r_1}{2}\right)\cos\phi\right\} d\phi,$$

whence
$$\sqrt{\left(\frac{\mu}{a}\right)} t = \left(\frac{r_2 + r_1}{2}\right)\phi - \left(\frac{r_2 - r_1}{2}\right)\sin\phi,$$

i.e.
$$nt = \phi - e\sin\phi,$$

where $n^2 = \mu/a^3$. This is the equation (1) of § 14·11. Of course, equation (11) is equivalent to

$$r = a(1-e\cos\phi),$$

which identifies ϕ as the eccentric angle. We have thus recovered the properties of the Newtonian elliptic orbit already established in Chapter XIV.

15·3 The inverse cube law. We now consider the motion of a particle attracted to a centre of force by a force varying as the inverse cube of the distance. Here $f(r) = \mu/r^3$, $V(r) = -\mu/(2r^2)$. We suppose the particle is projected as in Fig. 15·1, and the equations (3) and (5) of § 15·1 give us

$$\tfrac{1}{2}\dot{r}^2 = C + \frac{\mu-h^2}{2r^2}, \tag{1}$$

or, say,
$$\tfrac{1}{2}\dot{r}^2 = (Cr^2+D)/r^2, \tag{2}$$

where
$$D = \tfrac{1}{2}(\mu-h^2); \tag{3}$$

and
$$\frac{1}{2}\left(\frac{dr}{d\theta}\right)^2 = r^2(Cr^2+D)/h^2. \tag{4}$$

The values of C and h are

$$C = (c^2w_0^2-\mu)/(2c^2), \quad h = cw_0\sin\alpha.$$

It is evident that the problem is much simpler than that of the inverse square law—compare, for example, equation (2) with the corresponding equation (2) of § 15·2.

It is clear, from equations (2) and (4), that the essence of the matter lies in the signs of C and D. We notice first that if $D<0$, C must be positive, since the second member of equation (2) must not be negative; and this is evident otherwise, for $D<0$ implies $h^2>\mu$, i.e. $c^2w_0^2\sin^2\alpha>\mu$; *a fortiori* $c^2w_0^2>\mu$, and $C>0$. Similarly, if $D=0$, $C\geqslant 0$. We have therefore, in this problem, six types of orbits, arising from the six cases following:

(i) $D<0$, $C>0$; (ii) $D=0$, $C>0$; (iii) $D=0$, $C=0$; (iv) $D>0$, $C>0$; (v) $D>0$, $C=0$; (vi) $D>0$, $C<0$.

The classification of the orbits is now simple; equations (2) and (4) are of a type we have frequently encountered, and we can deduce the general nature of the orbit from a glance at the quadratic form Cr^2+D. When once the general nature of the orbit is known, the equation of the orbit and the (t, r) relation are found by quadratures. Case (iii) is the simplest; the orbit is the circle $r = c$. In (i) Cr^2+D is of the form $C(r^2-r_1^2)$; there is one apsidal distance, and it is a perihelion; the orbit lies outside the circle $r = r_1$. In (vi) Cr^2+D is of the form $|C|(r_2^2-r^2)$; there is one apsidal distance, and it is an aphelion; the orbit lies inside the circle $r = r_2$. In (ii), (iv) and (v), $\dot{r}$ retains its initial sign throughout, and there are no apses.

We can simplify the formulae a little by introducing a parameter $k = \mu/(c^2 w_0^2)$; we notice that k can take all positive values. With this notation the values of C and D, and the equations (2) and (4), become

$$C = \tfrac{1}{2}(1-k)w_0^2, \quad D = \tfrac{1}{2}(k-\sin^2\alpha)c^2w_0^2,$$

$$r^2\dot{r}^2 = \{(1-k)r^2+(k-\sin^2\alpha)c^2\}w_0^2, \tag{5}$$

$$\left(\frac{du}{d\theta}\right)^2 = \frac{1}{c^2\sin^2\alpha}\{(1-k)+(k-\sin^2\alpha)c^2u^2\}, \tag{6}$$

where $u = 1/r$. In the six cases enumerated above the parameters k and α satisfy the relations:

(i) $k<\sin^2\alpha$; (ii) $\sin^2\alpha = k<1$; (iii) $\sin^2\alpha = k = 1$;

(iv) $\sin^2\alpha<k<1$; (v) $\sin^2\alpha<k = 1$; (vi) $\sin^2\alpha<1<k$.

We now turn to the consideration of the various cases. The orbits obtained in this problem are called *Cotes' spirals*.

Case i. The equation (6) can be written

$$\left(\frac{du}{d\theta}\right)^2 = n^2(u_1^2-u^2), \tag{7}$$

where
$$n^2 = \frac{\sin^2\alpha-k}{\sin^2\alpha},$$

and $u_1 = 1/r_1$, where
$$\frac{r_1^2}{c^2} = \frac{\sin^2\alpha-k}{1-k}.$$

For the integration of equation (7), we introduce an auxiliary variable ϕ in place of u given by

$$u = u_1\cos\phi;$$

then equation (7) becomes

$$\left(\frac{d\phi}{d\theta}\right)^2 = n^2,$$

whence $$\phi = \pm n(\theta - \theta_0).$$

The equation of the orbit is

$$\frac{1}{r} = \frac{1}{r_1}\cos n(\theta - \theta_0),$$

or, if we measure θ from the perihelion,

$$r = r_1 \sec n\theta.$$

We see that $r \to \infty$ as $\theta \to \beta$, where $\beta = \pi/(2n)$. Since $n < 1$, $\beta > \frac{1}{2}\pi$. If $n > \frac{1}{2}$ the curve resembles a branch of a hyperbola; Fig. 15·3*a* illustrates the case $n = \frac{2}{3}$, $\beta = 3\pi/4$. In Fig. 15·3*b*, $n = \frac{2}{15}$ and $\beta = 15\pi/4$. In drawing the figures we notice that, if $\theta = \beta - \epsilon$, the perpendicular distance of O from an asymptote is

$$\lim_{\epsilon \to 0} r \sin \epsilon = \lim r_1 \frac{\sin \epsilon}{\cos n(\beta - \epsilon)} = \lim r_1 \frac{\sin \epsilon}{\sin n\epsilon} = \frac{r_1}{n}.$$

The relation between r and t in case (i) is found from equation (2) or equation (5), and is

$$r^2 \dot{r}^2 = 2C(r^2 - r_1^2),$$

giving $$r^2 - r_1^2 = 2C(t - t_0)^2,$$

if $t = t_0$ at perihelion.

The reader will be able to carry through the corresponding calculations in the other cases. We shall be content, for the sake of brevity, to indicate the nature of the orbits, leaving the verification and the detailed study of the motion to the reader.

Case ii. The orbit is the spiral

$$\theta = \left(1 - \frac{c}{r}\right)\tan\alpha;$$

the initial line has not been changed. The curve is essentially the spiral

$$r\theta = \text{constant}.$$

(See Example 3 of § 13·11.)

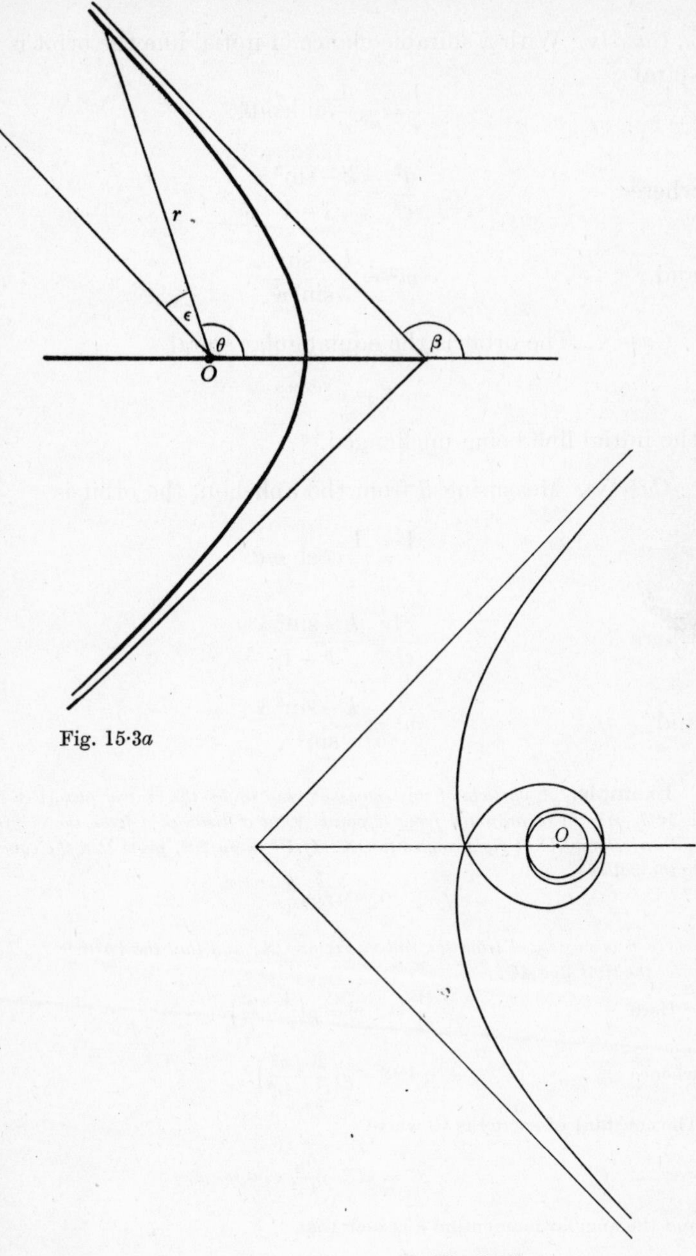

Fig. 15·3a

Fig. 15·3b

Case iii. The orbit is the circle $r = c$.

Case iv. With a suitable choice of initial line the orbit is the spiral

$$\frac{1}{r} = \pm\frac{1}{a}\sinh m\theta,$$

where
$$\frac{a^2}{c^2} = \frac{k-\sin^2\alpha}{1-k},$$

and
$$m^2 = \frac{k-\sin^2\alpha}{\sin^2\alpha}.$$

Case v. The orbit is the equiangular spiral

$$r = ce^{\theta\cot\alpha},$$

the initial line being unchanged.

Case vi. Measuring θ from the aphelion, the orbit is

$$\frac{1}{r} = \frac{1}{r_2}\cosh m\theta,$$

where
$$\frac{r_2^2}{c^2} = \frac{k-\sin^2\alpha}{k-1},$$

and
$$m^2 = \frac{k-\sin^2\alpha}{\sin^2\alpha}.$$

Example. *A particle of unit mass moving under the central attractive force* $\mu(4r^{-3}+a^2r^{-5})$ *is projected from a point A, at a distance a from the origin O, with a velocity U at right angles to OA. If* $U^2 = 9\mu/2a^2$, *prove that the equation to the path is*

$$r = a\cos\frac{\theta}{3},$$

where θ *is measured from the radius vector OA; and that the particle reaches O after the time* $3\pi a/4U$.

Here
$$\frac{dV}{dr} = f(r) = \mu\left(\frac{4}{r^3}+\frac{a^2}{r^5}\right),$$

whence
$$V = -\mu\left(\frac{2}{r^2}+\frac{a^2}{4r^4}\right).$$

The constant of energy is C, where

$$C = \tfrac{1}{2}U^2 - \frac{9\mu}{4a^2} = 0,$$

and the angular momentum h is such that

$$h^2 = a^2U^2 = 9\mu/2.$$

To find the equation of the orbit we can quote equation (5) of the general theory of § 15·1. But it is almost as quick to establish the (r, θ) relation for the particular case: thus

$$T = \tfrac{1}{2}(\dot{r}^2 + r^2\dot{\theta}^2) = \frac{1}{2}\left\{\left(\frac{dr}{d\theta}\right)^2 + r^2\right\}\dot{\theta}^2 = \frac{1}{2}\left\{\left(\frac{dr}{d\theta}\right)^2 + r^2\right\}\frac{h^2}{r^4},$$

and the equation of energy, $T+V=0$, becomes

$$\frac{9\mu}{4r^4}\left\{\left(\frac{dr}{d\theta}\right)^2 + r^2\right\} - \mu\left(\frac{2}{r^2} + \frac{a^2}{4r^4}\right) = 0,$$

whence
$$\left(\frac{dr}{d\theta}\right)^2 = \frac{1}{9}(a^2 - r^2).$$

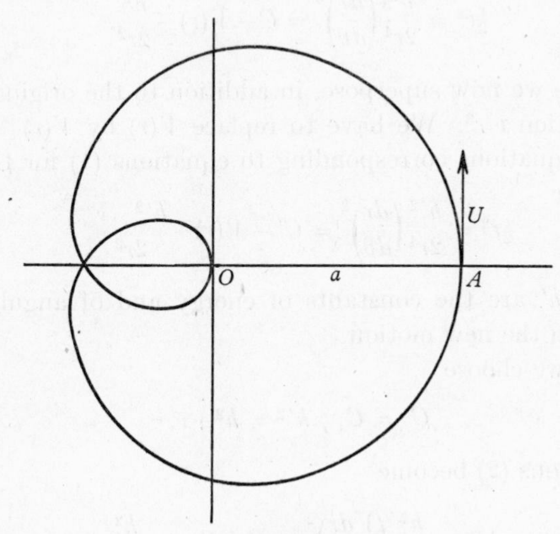

Fig. 15·3c

To integrate this equation, substitute $r = a\cos\phi$, and we have

$$\left(\frac{d\phi}{d\theta}\right)^2 = \frac{1}{9}, \quad \phi = \tfrac{1}{3}\theta,$$

remembering that $r = a$ when $\theta = 0$: the orbit is the curve $r = a\cos\tfrac{1}{3}\theta$ (Fig. 15·3c).

The particle reaches O when $\theta = 3\pi/2$; now

$$\dot{\theta} = \frac{h}{r^2} = \frac{U}{a\cos^2\tfrac{1}{3}\theta},$$

$$dt = \frac{a}{U}\cos^2\frac{\theta}{3}\,d\theta.$$

The time from A to O is therefore

$$\frac{a}{U}\int_0^{\frac{3}{2}\pi}\cos^2\frac{\theta}{3}\,d\theta = \frac{3a}{U}\int_0^{\frac{1}{2}\pi}\cos^2\xi\,d\xi = \frac{3\pi a}{4U}.$$

In this problem we have an example of a phenomenon mentioned in § 13·2 (ii)—the particle actually falls into the centre of force! Since the velocity of the particle is infinite when it reaches O the physical conditions become obscure at this point, and our calculation of the orbit is not valid for $\theta > 3\pi/2$ unless we introduce further conditions.

15·4 Additional attraction ν/r^3, Newton's theorem on revolving orbits. Suppose a particle to move in a central field of force. We have found the equations determining the motion (equations (2), (3) and (4) of § 15·1) to be

$$\tfrac{1}{2}\dot{r}^2 = \frac{h^2}{2r^4}\left(\frac{dr}{d\theta}\right)^2 = C - V(r) - \frac{h^2}{2r^2}. \tag{1}$$

Suppose we now superpose, in addition to the original field, an attraction ν/r^3. We have to replace $V(r)$ by $V(r) - \nu/(2r^2)$, and the equations corresponding to equations (1) for the new motion are

$$\tfrac{1}{2}\dot{r}^2 = \frac{h'^2}{2r^4}\left(\frac{dr}{d\theta}\right)^2 = C' - V(r) - \frac{h'^2 - \nu}{2r^2}, \tag{2}$$

where C', h' are the constants of energy and of angular momentum in the new motion.

If now we choose

$$C' = C, \quad h'^2 = h^2 + \nu, \tag{3}$$

the equations (2) become

$$\tfrac{1}{2}\dot{r}^2 = \frac{h^2}{2r^4}\left(\frac{1}{k}\frac{dr}{d\theta}\right)^2 = C - V(r) - \frac{h^2}{2r^2}, \tag{4}$$

where $k = h/h'$ (<1). Comparing equations (1) with equations (4) we see that the differential equation connecting r and t is the same in the new motion as in the old; the differential equation connecting r and $k\theta$ in the new motion is the same as the differential equation connecting r and θ in the original motion. If in the original motion

$$r = \phi(t), \quad r = \psi(\theta),$$

then in the new motion

$$r = \phi(t - t_0), \quad r = \psi\{k(\theta - \theta_0)\}.$$

This theorem was known to Newton.

We can easily arrange to satisfy the conditions (3). Suppose for definiteness that the original orbit has an apse at K, where $OK = c$, and the velocity at K is w_0. If in the modified field we project from K with velocity $\surd\{w_0^2 + (\nu/c^2)\}$ at right angles to OK, the conditions (3) are fulfilled. If the original orbit, with OK as initial line, is

$$r = \psi(\theta),$$

the new orbit is
$$r = \psi(k\theta).$$

To reach the value of r corresponding to $\theta = \theta_0$ in the original orbit, we must take $\theta = \theta_0/k$; this is greater than θ_0, since $k < 1$, and the general effect of the additional term in the attraction is a continuous rotation of the orbit. If the apsidal angle in the original motion is α, the new apsidal angle is α/k.

In particular the Newtonian orbit (equation (7) of § 15·2)

$$\frac{l}{r} = 1 + e\cos\theta$$

becomes, if we superpose an attraction ν/r^3,

$$\frac{l}{r} = 1 + e\cos k\theta,$$

the apsidal angle in the new orbit being π/k.

15·5 Another differential equation of the orbit. In § 15·1 we deduced the differential equation of the orbit from the integrals of energy and momentum. We can write

$$T = \tfrac{1}{2}(\dot{r}^2 + r^2\dot{\theta}^2) = \frac{1}{2}\left\{\left(\frac{dr}{d\theta}\right)^2 + r^2\right\}\dot{\theta}^2,$$

and, since $\dot{\theta} = h/r^2$, the equation $T + V = C$ becomes

$$\frac{1}{2}\left\{\left(\frac{1}{r^2}\frac{dr}{d\theta}\right)^2 + \frac{1}{r^2}\right\} = \frac{1}{h^2}\{C - V(r)\}, \tag{1}$$

which is equivalent to equation (5) of § 15·1. If we write u for $1/r$, and if $V(r) = W(u)$, we obtain

$$\frac{1}{2}\left\{\left(\frac{du}{d\theta}\right)^2 + u^2\right\} = \frac{1}{h^2}\{C - W(u)\}. \tag{2}$$

If we now differentiate with respect to u we find

$$\frac{d^2u}{d\theta^2}+u=-\frac{1}{h^2}\frac{dW}{du}. \tag{3}$$

Now
$$\frac{dW}{du}=\frac{dV}{dr}\frac{dr}{du}=-\frac{f(r)}{u^2},$$

and, if we write $F(u)$ for the attraction at distance $1/u$ from O, equation (3) becomes

$$\frac{d^2u}{d\theta^2}+u=\frac{F(u)}{h^2u^2}. \tag{4}$$

This is the new form of the differential equation of the orbit.

It is clear that in general the new form (4) is less useful than the old form (1), since the new equation is of the second order, whereas the old was of the first order; indeed, our first instinct on meeting the equation (4) might well be to integrate and recover the equation (2), which is equivalent to the old form. Nevertheless, equation (4) is of value in one particular case, namely, when the attraction has the form $\mu u^2+\nu u^3$; with this law of attraction the equation (4) becomes a linear differential equation with constant coefficients, and the solution is simple. The inverse square and inverse cube laws are included as particular cases.

For example, in the Newtonian orbit

$$F(u)=\mu u^2,$$

and the differential equation (4) becomes

$$\frac{d^2u}{d\theta^2}+u=\frac{\mu}{h^2}.$$

The solution of this is

$$u-\frac{\mu}{h^2}=A\cos(\theta-\theta_0),$$

or, say,
$$\frac{l}{r}=1+e\cos(\theta-\theta_0),$$

where
$$h^2=\mu l.$$

If we superpose an attraction νu^3 we have

$$F(u) = \mu u^2 + \nu u^3,$$

and the differential equation (4) becomes (writing h' for the new angular momentum),

$$\frac{d^2u}{d\theta^2} + u = \frac{\mu}{h'^2} + \frac{\nu}{h'^2} u,$$

i.e.

$$\frac{d^2u}{d\theta^2} + \left(1 - \frac{\nu}{h'^2}\right) u = \frac{\mu}{h'^2}.$$

The solution is $$u = \frac{\mu}{h'^2 - \nu} + B \cos n(\theta - \theta_0),$$

where $$n^2 = \frac{h'^2 - \nu}{h'^2}.$$

If $$h'^2 = h^2 + \nu,$$

as in equation (3) of § 15·4, we have

$$\frac{l}{r} = 1 + e' \cos k(\theta - \theta_0),$$

where $k = h/h'$, as before. We thus recover the effect of the added attraction as a rotation of the Newtonian orbit, which is a result already established as an application of the theorem of § 15·4.

15·6 Another proof. Another and very simple way of establishing the differential equation (4) of § 15·5 is to eliminate t between the equations

$$\ddot{r} - r\dot{\theta}^2 = -f(r), \tag{1}$$

$$r^2 \dot{\theta} = h.$$

If we think of r expressed as a function of t through its expression as a function of θ we have

$$\dot{r} = \frac{dr}{d\theta} \dot{\theta} = \frac{h}{r^2} \frac{dr}{d\theta} = -h \frac{du}{d\theta},$$

and $$\ddot{r} = -h \frac{d^2u}{d\theta^2} \dot{\theta} = -h^2 u^2 \frac{d^2u}{d\theta^2}.$$

This is a transformation of the first term in the first member of equation (1), and for the second we have

$$r\dot{\theta}^2 = rh^2u^4 = h^2u^3.$$

Thus equation (1) becomes

$$-h^2u^2\frac{d^2u}{d\theta^2} - h^2u^3 = -f(r),$$

and, if we write $F(u)$ for $f(r)$ as before, this becomes

$$\frac{d^2u}{d\theta^2} + u = \frac{F(u)}{h^2u^2}.$$

This is the required result.

15·7 The inverse problem—given the orbit, to find the field. A particle moves in a conservative central field. If we observe the orbit, what can we deduce about the field?

We have considered the classical problem of this type already, in § 14·2; we now consider briefly the more general problem. It is clear that we cannot expect a complete answer for all values of r. For example, if the orbit is a circle of radius a with O as centre, and is traversed with speed w_0, we can only deduce the value of $f(r)$ for $r = a$; and, more generally, if the apsidal distances are r_1 and r_2 $(0 < r_1 < r_2)$ we can deduce nothing about the field for $r < r_1$ or for $r > r_2$. However, in general we can find the field in the range (r_1, r_2).

If we write the equations of energy and of angular momentum in the forms

$$\tfrac{1}{2}v^2 + V(r) = C,$$

$$pv = h,$$

we obtain, on eliminating v,

$$\frac{1}{2}\frac{h^2}{p^2} + V(r) = C. \tag{1}$$

If therefore we can find the relation between p and r in the orbit—this relation is called the (p,r) equation of the curve—we can find $V(r)$ for the range of values of r covered during the motion.

Suppose as a concrete example that the orbit is a circle of radius a with its centre D at distance c ($<a$) from O (Fig. 15·7a). With the notation of the figure

$$a^2 - c^2 + r^2 = 2ar\cos\phi$$
$$= 2ap, \tag{2}$$

and equation (1) gives us

$$V(r) = C - \frac{2h^2a^2}{(a^2-c^2+r^2)^2}.$$

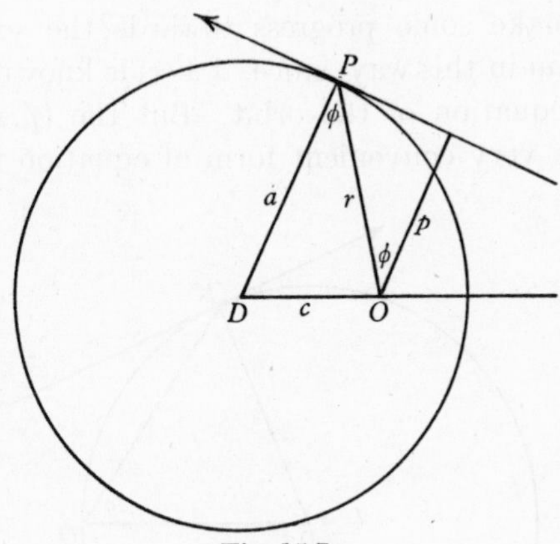

Fig. 15·7a

The law of attraction is

$$f(r) = \frac{dV}{dr} = \frac{8h^2a^2r}{(a^2-c^2+r^2)^3}. \tag{3}$$

Alternatively, considering the component of acceleration along the normal to the path, we have

$$\frac{v^2}{a} = f(r)\cos\phi = \frac{p}{r}f(r),$$

and since $v = h/p$, this gives us

$$f(r) = \frac{h^2r}{ap^3}. \tag{4}$$

On substituting for p from equation (2) we recover the same formula for $f(r)$ as in equation (3).

Perhaps the most interesting case is the limiting case in which $c = a$, and we have

$$f(r) = \mu/r^5,$$

where $\mu = 8h^2a^2$. In this case the (p, r) equation of the curve, equation (2), takes the simple form

$$r^2 = 2ap,$$

as is evident from the geometry.

We can make some progress towards the solution of the direct problem in this way, since, if $V(r)$ is known, equation (1) is the (p, r) equation of the orbit. But the (p, r) equation is clearly not a very convenient form of equation to work with.

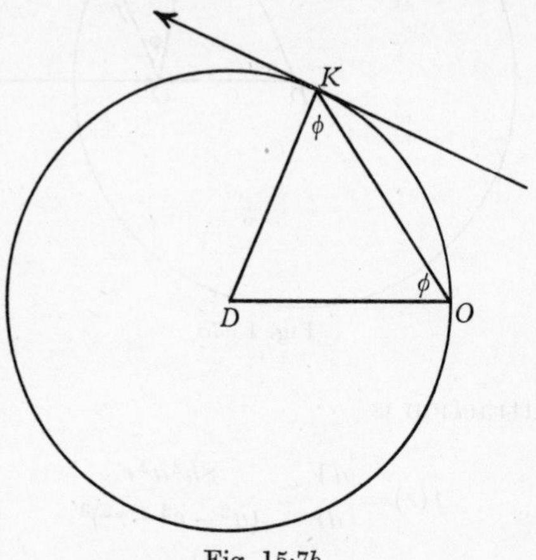

Fig. 15·7*b*

However, it is useful in a few very simple cases. Suppose, for example, the field is an attraction μ/r^5, and the particle is projected from a point K at distance c from O with velocity w_0, where

$$w_0^2 = \frac{\mu}{2c^4}.$$

This makes $C = 0$, and the (p, r) equation of the orbit, equation (1), is

$$\frac{h^2}{2p^2} = \frac{\mu}{4r^4}.$$

Thus (p, r) equation of the orbit is

$$r^2 = 2ap,$$

where $\mu = 8h^2a^2$. This represents a circle through the centre of force. It is easy to construct the circle, since KD is normal to the direction of projection, and DKO is an isosceles triangle (Fig. 15·7*b*).

15·8 Elliptic orbit when the centre of force is not a focus. We can deduce from § 15·7 a law of force under which an elliptic orbit can be described under an attraction towards any given point O inside the ellipse. We use an orthogonal projection in which the ellipse is the projection of a circle. We suppose a particle to move in the circle under an attraction to the point O_1 which corresponds to (i.e. which projects into) O, and we suppose also that corresponding points are reached in the two orbits at the same instant. (The last restriction implies that we shall not necessarily find the most general law of attraction to O under which the elliptic orbit can be described.)

We denote the coordinates of a point in the ellipse figure by (x, y), and the coordinates of the corresponding point in the circle figure by (x_1, y_1); O and O_1 are origins of coordinates in the two figures. The suffix 1 denotes in general the element in the circle figure corresponding to the element in the ellipse figure denoted by the same symbol without suffix. Now in the original motion in the ellipse

$$\frac{\ddot{x}}{x} = \frac{\ddot{y}}{y} = -\frac{f(\mathbf{r})}{r}, \tag{1}$$

and in the new motion in the circle

$$\frac{\ddot{x}_1}{x_1} = \frac{\ddot{y}_1}{y_1} = -\frac{f_1(\mathbf{r}_1)}{r_1}, \tag{2}$$

where $f_1(\mathbf{r}_1)$ is the attraction at the point $\mathbf{r}_1$ in the circular orbit. The function $f_1(\mathbf{r}_1)$ is known from § 15·7, and we wish to determine $f(\mathbf{r})$.

Now x_1 and y_1 are homogeneous linear functions of (x, y), so that

$$\frac{\ddot{x}_1}{x_1} = \frac{\ddot{x}}{x}, \quad \frac{\ddot{y}_1}{y_1} = \frac{\ddot{y}}{y}, \tag{3}$$

so, from equations (1), (2) and (3), we have

$$\frac{f(\mathbf{r})}{r} = \frac{f_1(\mathbf{r}_1)}{r_1}. \tag{4}$$

But by equation (4) of § 15·7

$$\frac{f_1(\mathbf{r}_1)}{r_1} = \frac{h_1^2}{a}\frac{1}{p_1^3}, \tag{5}$$

where p_1 is the perpendicular on the tangent to the circle. We have now to express this in terms of the original figure, and this we can easily do from the geometry (Fig. 15·8*b*). Let P_1M_1, C_1N_1 be the perpendiculars from P_1 and C_1 on the polar of O_1 with respect to the circle. Then

$$\frac{a}{p_1} = \frac{C_1N_1}{P_1M_1} = \frac{CN}{PM}, \tag{6}$$

since the ratio of the lengths of parallel segments is unaltered by orthogonal projection. Now

$$\frac{CN}{PM} = \frac{q_0}{q} \tag{7}$$

where (in the original figure with the ellipse) q is the perpendicular distance of P from the polar of O, and q_0 is the perpendicular distance of the centre C of the ellipse from the same line. Thus from equations (4), (5), (6) and (7), the required attraction is $f(\mathbf{r})$, where

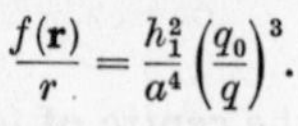

$$\frac{f(\mathbf{r})}{r} = \frac{h_1^2}{a^4}\left(\frac{q_0}{q}\right)^3.$$

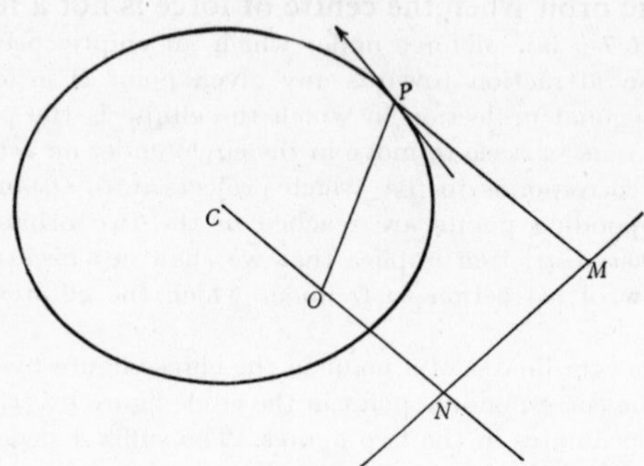

Fig. 15·8*a*

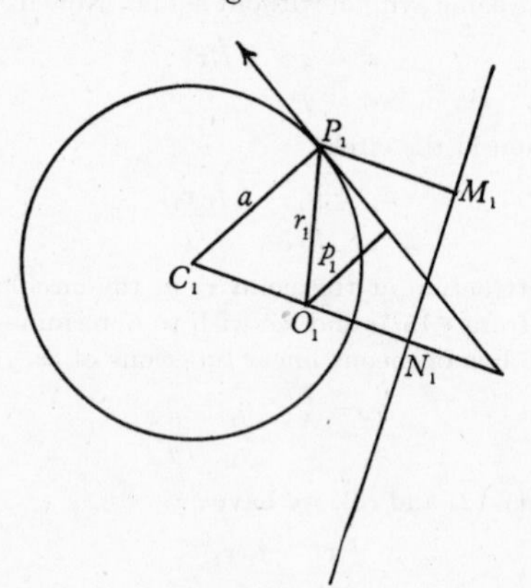

Fig. 15·8*b*

The necessary attraction to the centre of force O is

$$\mu r/q^3,$$

where q is the perpendicular distance of P from a certain straight line. We notice that this is equivalent to the Newtonian case if O is a focus of the ellipse.

The converse is true. The orbit under a central attraction $\mu r/q^3$, where q is the perpendicular distance from a line l, is a conic for which l is the polar of O (§ 16·3).

Remember that the attraction in this problem is a function of $\mathbf{r}$, not of r only; the field is not conservative, and there is no equation of energy.

EXAMPLES XV

1. A particle of unit mass moves on a smooth horizontal table and is attached to a fixed point of the table by an elastic string of natural length l and modulus of elasticity λ. The particle is held with the string just taut, and projected at right angles to the string with velocity $2\sqrt{(l\lambda/3)}$. Show that the greatest length of the string in the resulting motion is $2l$.

2. A particle of mass m is tied by a light elastic string, of natural length a, to a fixed point on a smooth horizontal table. Initially the particle is at rest on the table with the string just taut, and it is projected horizontally with velocity u at right angles to the string. Prove that during the subsequent motion the length r of the string satisfies the differential equation

$$\left(\frac{dr}{dt}\right)^2 = \frac{\lambda}{ma}\left(\frac{r-a}{r^2}\right)\left\{\frac{mu^2}{\lambda}a(r+a)-r^2(r-a)\right\},$$

where λ is the tension needed to double the length of the string.

Prove that, if λ is large, the maximum extension of the string during the motion is approximately $2mu^2/\lambda$.

3. A particle is attracted towards a centre of force S with a force μ/r^2 per unit mass. It is projected from a point P at a distance c from S with a velocity V at an angle $\frac{1}{2}\pi-\alpha$ with PS, where $0 \leqslant \alpha < \frac{1}{2}\pi$. If θ is the angle between SP and the radius vector from S to the particle establish the equation

$$\frac{d^2u}{d\theta^2}+u = \frac{\mu}{V^2c^2}\sec^2\alpha,$$

where $u = 1/r$.

Examine in particular the case $V = \sqrt{(\mu/c)}$. Prove that if $\alpha = 0$ the orbit is a circle, and that if $\alpha \neq 0$ it is an ellipse having its major axis parallel to the direction of projection. Find the polar equation of this ellipse, and prove that when $\theta = \pi$,

$$r = \frac{c}{1+2\tan^2\alpha}.$$

4. Obtain the differential equation to a central orbit in the form

$$\frac{d^2u}{d\theta^2}+u = \frac{P}{h^2u^2},$$

where u is the reciprocal of the radius vector, P is the acceleration to the centre of force and h is a constant.

A particle is projected from infinity with velocity V so as to pass a fixed point at a distance c if undisturbed. If it is attracted to the fixed point with an acceleration μu^2, show that the equation of the orbit is

$$u = \frac{\mu}{V^2c^2}+\frac{\cos\theta}{c}\sqrt{\left(1+\frac{\mu^2}{V^4c^2}\right)},$$

θ being measured from the apse.

5. A particle of mass m is attracted to a fixed point O by a force of magnitude $m\mu^2/r^3$ and is projected from a point A with velocity μ/OA in a direction making an angle α with OA. Prove that the tangent to the path at every point P of the path makes an angle α with OP.

6. A particle P, of mass m, moves under an attractive force λ/r^3 towards a fixed centre of force S, where λ is constant, and r is the distance SP. If P is projected with velocity $5\lambda^{\frac{1}{2}}/4l$ at right angles to SP when SP equals l, show that when P has receded to infinity the radius vector will have turned through an angle $\frac{5}{6}\pi$.

7. A particle is attracted to a fixed origin O by a force μ/r^3 per unit mass, where r denotes distance from O. It is projected from a point A, at distance a from O, with velocity v_0 at right angles to OA. Using polar coordinates, with O as origin, and OA as initial line, prove that during the motion

$$\dot{r}^2 = (a^2 v_0^2 - \mu)\left(\frac{1}{a^2} - \frac{1}{r^2}\right).$$

Prove also that, if $a^2 v_0^2 > \mu$, the orbit is the curve

$$r = a \sec(\theta/n),$$

where

$$\frac{1}{n^2} = 1 - \frac{\mu}{a^2 v_0^2}.$$

8. A particle is projected with velocity $\sqrt{(2\mu/3a^3)}$ at right angles to the radius vector at a distance a from a centre of attracting force μ/r^4 per unit mass. Find the path of the particle and show that the time it takes to reach the centre of force is $\frac{3}{8}\pi\sqrt{\left(\frac{3a^5}{2\mu}\right)}$.

9. A particle moves in a plane under an attraction μ/r^5 per unit mass towards a centre of attraction O in the plane. It is projected from a point A at distance a from O with velocity u at right angles to OA. Prove that the differential equation of the orbit is

$$\left(\frac{dr}{d\theta}\right)^2 = \frac{1}{a^2 u^2}(r^2 - a^2)\left\{\left(u^2 - \frac{\mu}{2a^4}\right) r^2 - \frac{\mu}{2a^2}\right\}.$$

Prove that in the special case when

$$u^2 = \frac{\mu}{2a^4}$$

the orbit is the circle on OA as diameter, and that the time from A to O is $\pi a^3/\sqrt{(8\mu)}$.

10. A particle moves in a circular orbit of radius a under the influence of an attractive force proportional to $r^{-\frac{5}{2}}$ directed towards the centre of the circle. When $\theta = \alpha$ its velocity is suddenly increased by a factor $2/\sqrt{3}$, without change of direction. Show that the polar equation of the orbit becomes

$$r = a \sec^4\left(\frac{\theta - \alpha}{4}\right).$$

11. A particle moves in a field of attraction $2\mu(r^{-3} - a^2 r^{-5})$ per unit mass. It is projected at $t = 0$ from an apse at distance a from the centre of attraction with velocity $\sqrt{(\mu/a^2)}$. Prove that during the subsequent motion

$$\dot{r}^2 = \mu(r^2 - a^2)/r^4, \quad \dot{\theta}^2 = \mu/r^4.$$

Find the polar equation of the orbit, and the relation between t and θ.

*12. A particle is attracted towards a centre of force S with a force $\lambda(r^{-2}+2cr^{-3})$ per unit mass, where r denotes distance from S. It is projected from a point P at distance c from S with a velocity $\frac{4}{3}\sqrt{\left(\frac{2\lambda}{c}\right)}$ at an angle $\frac{1}{3}\pi$ with PS. Prove that

$$\frac{d^2u}{d\theta^2}+\frac{1}{4}u = \frac{3}{8c},$$

where θ is the angle between SP and the radius vector to the particle, and $u = 1/r$.

Find the polar equation of the orbit and sketch the curve.

13. A particle traverses the orbit

$$r = a+b\cos\theta \quad (a>b>0)$$

under a central force to the origin. If the velocity of projection is the velocity from infinity (so that $C = 0$) prove that the potential per unit mass is

$$V = \mu\left(\frac{a^2-b^2}{2r^4}-\frac{a}{r^3}\right).$$

Conversely, starting from this law of force, and supposing the particle projected from the apse $r = a+b$, $\theta = 0$, with the velocity from infinity, prove that the orbit is the given curve. Prove that the force is attractive at all points of the orbit, and that consequently the orbit is everywhere concave to O, if $a<2b$.

CHAPTER XVI

OTHER PROBLEMS ON THE MOTION OF A PARTICLE IN A PLANE

16·1 Survey of the problems to be discussed. So far in our study of motion in a plane we have dealt almost entirely with motion, either free or under smooth constraints, in a conservative field of force. We now turn to problems of other types, for example, problems of motion in fields which are not conservative, or problems where the force is not simply a function of position, but depends also on the speed. The problems to be discussed in this chapter may be conveniently classified as follows: (*a*) constrained motion with friction, (*b*) free motion in a field which is not conservative, (*c*) gyroscopic forces, (*d*) motion in a resisting medium, where the medium offers a resistance opposed to the velocity, the magnitude of the resisting force being a given function of the speed.

16·2 Particle moving on a rough surface. We begin by considering a problem of constrained motion with friction—for example, a particle slides, under the action of gravity, on a rough surface, or a bead slides on a rough wire. Suppose, for definiteness, that we have a bead sliding on a wire in a vertical plane; in this case we do not encounter the embarrassment of a unilateral constraint. We measure the inclination ψ of the tangent to the curve from the downward vertical, and we suppose, for definiteness, that s increases with ψ. The friction opposes the motion, and is assumed to be μ times the normal component of reaction; the coefficient μ is a given function of position

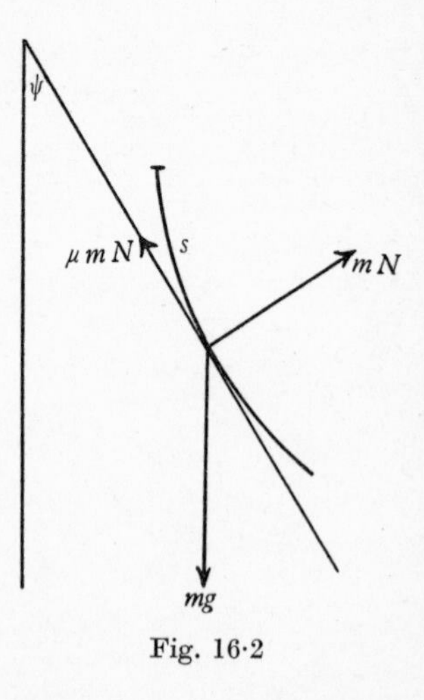

Fig. 16·2

on the wire. We shall be content to discuss the simplest and commonest case, that in which μ has the same value at all points of the wire. The force on the particle is the vector sum of three terms: (i) its weight, (ii) a normal component of reaction, say mN, (iii) a tangential frictional force $\mu m \mid N \mid$, which opposes the motion so long as motion continues. The last needs particular care in the matter of sign, because of the two ambiguities, the direction of the normal reaction and the direction of the velocity. We must be alive to the possibility that the normal component of reaction may change sign during the motion.

Consider, for definiteness, the particular case illustrated in Fig. 16·2, where the direction of the normal reaction on the bead is along the inward normal to the curve, and $v = \dot{s} > 0$. The equations of motion, tangential and normal, are

$$\frac{d}{ds}(\tfrac{1}{2}v^2) = g\cos\psi - \mu N, \tag{1}$$

$$\frac{v^2}{\rho} = N - g\sin\psi, \tag{2}$$

whence, eliminating N, we have

$$\frac{d}{ds}(\tfrac{1}{2}v^2) + \mu\frac{v^2}{\rho} = g(\cos\psi - \mu\sin\psi). \tag{3}$$

We now write $\frac{1}{2}v^2 = \eta$, and then, since $\dfrac{d\eta}{ds} = \dfrac{1}{\rho}\dfrac{d\eta}{d\psi}$, equation (3) becomes

$$\frac{d\eta}{d\psi} + 2\mu\eta = g\rho(\cos\psi - \mu\sin\psi).$$

Multiplying through by the integrating factor $e^{2\mu\psi}$ we find

$$\eta e^{2\mu\psi}\Big|_{\psi_1}^{\psi_2} = g\int_{\psi_1}^{\psi_2} \rho e^{2\mu\psi}(\cos\psi - \mu\sin\psi)\,d\psi,$$

where, in the integral, ρ is to be expressed as a function of ψ. The equation is valid only so long as the assumed conditions hold. Of course, if the motion starts from rest, we must have $\cos\psi > \mu\sin\psi$ initially, for otherwise no motion will take place.

We carry the calculation a stage further for the case of the circle $\rho = a$, supposing the bead to start from rest at the level of the centre. It is evident that the conditions assumed above are satisfied in this case, and $\eta = 0$ when $\psi = 0$, so we have

$$\tfrac{1}{2}v^2 = gae^{-2\mu\psi}\int_0^\psi e^{2\mu\theta}(\cos\theta - \mu\sin\theta)\,d\theta$$

$$= \frac{ga}{1+4\mu^2}\{3\mu\cos\psi + (1-2\mu^2)\sin\psi - 3\mu e^{-2\mu\psi}\}.$$

16·3 Non-conservative field. Here the force on the particle is a function of position in the plane, but there is no longer an integral of energy to help us towards a solution, and we must devise an attack suited to the particular problem.

Consider, as an example of this category, the motion of a particle in a central field $\mu r/p^3$, where r is the distance from the centre of force, and p is the distance from a fixed line ϖ in the plane (see § 15·8). Taking the line ϖ to be $x+c=0$ $(c>0)$, and supposing the particle to lie initially to the right of ϖ, so that $x+c>0$ initially, the equations of motion are

$$\ddot{x} = -\mu x/(x+c)^3, \tag{1}$$

$$\ddot{y} = -\mu y/(x+c)^3. \tag{2}$$

Let us determine the orbit.

Two first integrals of the equations are found immediately, one by integration of equation (1), the other from the conservation of angular momentum; they are

$$\tfrac{1}{2}\dot{x}^2 = \frac{\mu}{2c}\left\{a - \left(\frac{x}{x+c}\right)^2\right\}, \tag{3}$$

where a is a positive constant, and

$$x\dot{y} - y\dot{x} = h. \tag{4}$$

If we now introduce the projective transformation

$$\xi = (x+c)/x, \quad \eta = y/x,$$

the equations (3) and (4) become

$$x^4\dot{\xi}^2 = \mu c\left(a - \frac{1}{\xi^2}\right),$$

and
$$x^2\dot{\eta} = h,$$

whence
$$\left(\frac{d\xi}{d\eta}\right)^2 = \frac{\mu c}{h^2}\left(a - \frac{1}{\xi^2}\right). \tag{5}$$

The integration of equation (5) leads to

$$h^2(a\xi^2 - 1) = \mu c a^2(\eta - b)^2,$$

where b is a new constant of integration. This represents a conic in the (ξ, η) plane, and therefore, since the transformation is projective, it also represents a conic in the (x, y) plane. Explicitly the (x, y) equation is

$$ah^2(x + c)^2 = \mu c a^2(y - bx)^2 + h^2 x^2,$$

a conic for which ϖ is the polar line of O. The orbit does not intersect ϖ in real points. The constants a, h, b are determined by the initial conditions.

16·4 Gyroscopic forces. A gyroscopic force is a force perpendicular to the direction of motion, and proportional to the speed, say a force $mk\mathbf{v}'$.

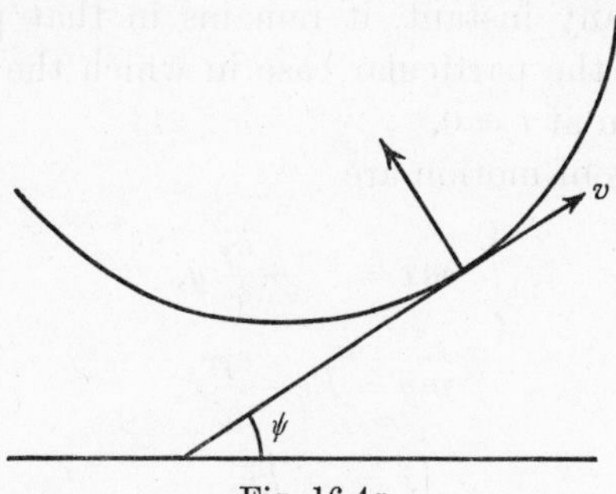

Fig. 16·4*a*

If the particle moves under the action solely of a gyroscopic force, its path is a circle traversed with constant speed; and the period of the motion is independent of the circumstances of projection. To prove these statements we notice that the tangential and radial equations of motion are

$$\dot{v} = 0, \quad v^2/\rho = kv.$$

Thus v is constant and ρ is constant, and the angular velocity $\dot{\psi} = v/\rho = k$. The period is $2\pi/k$, and is an absolute constant, independent of the circumstances of projection.

If the particle moves under the action of a conservative field and a gyroscopic force the presence of the gyroscopic force does not invalidate the equation of energy; the gyroscopic force does no work since it is always perpendicular to the direction of motion. In this problem, in contrast to the other problems discussed in this chapter, there is an integral of energy.

One of the most important problems of this class is that of the motion of a charged particle in 'crossed fields', i.e. in uniform electric and magnetic fields at right angles. Let the mass of the particle be m, its charge e, the electric field E in the direction Oy, and the magnetic field γ in the direction Oz. The electric field provides a force Ee in the direction Oy; the magnetic field gives rise to a force $\frac{e}{c}(\mathbf{v}\times\mathbf{H})$, where $\mathbf{v}$ is the velocity of the particle and $\mathbf{H}$ the magnetic field. In our case, with the magnetic field $(0, 0, \gamma)$, the force arising from the magnetic field is $\frac{e\gamma}{c}(\dot{y}, -\dot{x}, 0)$. If the particle is moving in the plane $z = 0$ at any instant, it remains in that plane always. Let us consider the particular case in which the particle is at rest at the origin at $t = 0$.

The equations of motion are

$$m\ddot{x} = \quad +\frac{e\gamma}{c}\dot{y},$$

$$m\ddot{y} = Ee-\frac{e\gamma}{c}\dot{x},$$

or, say,

$$\begin{cases}\ddot{x} = \quad k\dot{y},\\ \ddot{y} = g-k\dot{x},\end{cases}$$

where $g = Ee/m$, $k = e\gamma/mc$. If we differentiate the second equation with respect to t, and substitute for $\ddot{x}$ from the first, we shall obtain for y a linear differential equation with constant coefficients, and such an equation can be integrated by rule. However, a simpler method is available in this case, namely, to introduce the complex variable $z = x+iy$. This leads to a simple differential equation of the second order with constant coefficients, and the solution of this differential equation determines the motion.

If we multiply the second equation by i, and add to the first, we get

$$\ddot{z}+ik\dot{z} = ig.$$

This is a differential equation similar to equation (1) of § 7·3 (though here the dependent variable and the coefficients are complex) and the method of integration is the same as it was there. The solution, with $z = \dot{z} = 0$ at $t = 0$, is

$$z = \frac{g}{k^2}(kt + i - ie^{-ikt}).$$

If we write $a = g/k^2$, $\theta = kt$, this is

$$\begin{aligned} z &= a(\theta + i - ie^{-i\theta}) &(1)\\ &= a\{(\theta - \sin\theta) + i(1-\cos\theta)\}, \end{aligned}$$

giving

$$\left.\begin{aligned} x &= a(\theta - \sin\theta),\\ y &= a(1-\cos\theta). \end{aligned}\right\} \qquad (2)$$

The path is a cycloid with a line of cusps on Ox (cf. § 11·14); the figure is drawn for the case when e, E, γ are all positive.

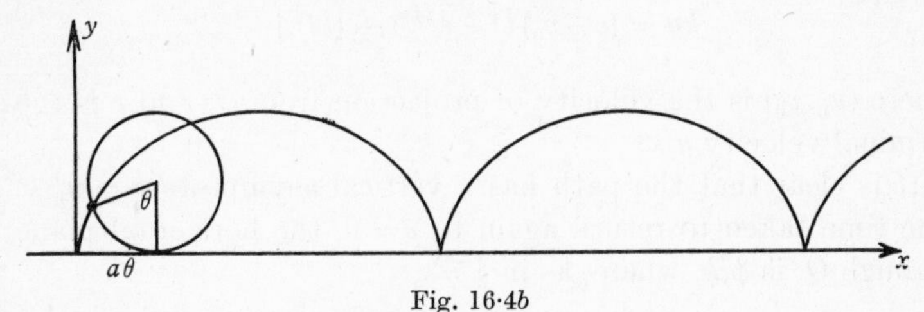

Fig. 16·4*b*

The interpretation of the path as the locus of a point on the rim of a rolling wheel can be seen from the equations (2), or directly from the formula (1) expressing the position of z in the plane of the complex variable in terms of the parameter θ. The important quantity in the physical application is the greatest distance of the particle from Ox, namely,

$$2a = 2mEc^2/(e\gamma^2).$$

16·5 Resisting medium, resistance proportional to speed. We now consider problems of motion under gravity in a resisting medium. The forces acting on the particle are its weight, and a frictional force in a direction directly opposed to the direction of motion. We begin with the simplest problem of this class, in which the resistance is proportional to the speed. The resisting force is $-mk\mathbf{v}$. The equation of motion in the uniform field $\mathbf{g}$ is

$$\ddot{\mathbf{r}} + k\dot{\mathbf{r}} = \mathbf{g}. \qquad (1)$$

This problem has been mentioned already in §§ 10·6 and 10·9, and the solution for the corresponding problem in one dimension has been found in §§ 7·2 and 7·3. The solution of the vector equation can be found in the same way, and we can work either with the vector formulae or with the Cartesian coordinates. The latter will be rather more convenient for our present purpose. The equations of motion are

$$\left.\begin{aligned} \ddot{x}+k\dot{x} &= 0, \\ \ddot{y}+k\dot{y} &= -g, \end{aligned}\right\} \tag{2}$$

where the axis Ox is horizontal, and the axis Oy is vertically upwards. The solutions, which we have found already, are

$$\left.\begin{aligned} kx &= u_0(1-e^{-kt}), \\ ky &= (c+v_0)(1-e^{-kt})-c(kt), \end{aligned}\right\} \tag{3}$$

where (u_0, v_0) is the velocity of projection from O, and c is the terminal velocity g/k.

It is clear that the path has a vertical asymptote $x = u_0/k$. The time taken to return again to $y = 0$, the horizontal plane through O, is ϕ/k, where, as in § 7·3,

$$\frac{1-e^{-\phi}}{\phi} = \frac{c}{c+v_0}. \tag{4}$$

The equations (3) express the coordinates (x, y) of a point on the path in terms of the parameter t, and the (x, y) equation of the path is easily seen to be

$$y = \left(\frac{c+v_0}{u_0}\right)x - \frac{c}{k}\log\left(\frac{u_0}{u_0-kx}\right) = \frac{v_0}{u_0}x - \frac{c}{k}\sum_{n=2}^{\infty}\frac{1}{n}\left(\frac{kx}{u_0}\right)^n.$$

We see from the equations (3) that, throughout the motion, $\ddot{y}/\ddot{x}$ retains the constant value $(c+v_0)/u_0$, so the direction of the acceleration is constant. It follows that the component of velocity perpendicular to this direction is constant; it is easy to verify that the value of this constant component of velocity is

$$\frac{cu_0}{\sqrt{\{(c+v_0)^2+u_0^2\}}}.$$

There is another way of seeing that the direction of the acceleration is always the same. If we differentiate equation (1) with respect to t, remembering that $\ddot{\mathbf{r}} = \mathbf{f}$, we find

$$\dot{\mathbf{f}} + k\mathbf{f} = \mathbf{0},$$

whence

$$\mathbf{f} = \mathbf{f}_0 e^{-kt}.$$

Thus $\mathbf{f}$ has the same direction throughout the motion, and its magnitude is proportional to e^{-kt}.

16·6 Envelope of the paths for a given speed of projection. Consider now the paths obtained by projection from O with given speed w_0. These paths have an envelope E which is the boundary of the region accessible from O with the given speed of projection. The most important result in connexion with this envelope is this: the tangent to each path at its point of contact with the envelope is perpendicular to the direction of projection from O.

The reader will recall that the same phenomenon has already been encountered in two other problems, in the theory of projectiles (§ 12·3) and in the theory of motion under a central attraction mn^2r (§ 12·5). It may be of interest at this point to give a proof of the general theorem which embraces all three problems. The form of the equation in the special cases will suggest the main idea of the general theorem; they are

Projectile
$$\begin{cases} x = u_0 t, \\ y = v_0 t - \tfrac{1}{2}gt^2; \end{cases}$$

Attraction mn^2r
$$\begin{cases} x = (u_0/n)\sin nt - a(1-\cos nt), \\ y = (v_0/n)\sin nt; \end{cases}$$

Projectile with resistance $k\mathbf{v}$
$$\begin{cases} kx = u_0(1-e^{-kt}), \\ ky = v_0(1-e^{-kt}) - c(e^{-kt} - 1 + kt). \end{cases}$$

THEOREM. *A curve is defined by the equations*

$$\left.\begin{aligned} x &= u_0 f(t) + g(t), \\ y &= v_0 f(t) + h(t), \end{aligned}\right\} \tag{1}$$

where

$$\left.\begin{aligned} f(t) = g(t) = h(t) = 0, \\ f'(t) = 1, \quad g'(t) = h'(t) = 0, \end{aligned}\right\} \tag{2}$$

at $t = 0$. The direction of the curve where it touches the envelope of the family given by

$$u_0^2 + v_0^2 = \textit{constant}$$

is at right angles to its direction at O.

If the point on the curve (1) where it touches the envelope is given by $t = \theta$, we have

$$\left.\begin{aligned} 0 &= du_0 f(\theta) + u_0 f'(\theta)\,d\theta + g'(\theta)\,d\theta, \\ 0 &= dv_0 f(\theta) + v_0 f'(\theta)\,d\theta + g'(\theta)\,d\theta, \end{aligned}\right\} \tag{3}$$

where we have thought of the point where the curve touches the envelope as the limit of the intersection of the curve with a neighbouring curve of the family. If we multiply the first of equations (3) by u_0, and the second by v_0, and add, we find

$$u_0\{u_0 f'(\theta) + g'(\theta)\} + v_0\{v_0 f'(\theta) + h'(\theta)\} = 0.$$

We can write this in the form

$$u_0 \dot{x}(\theta) + v_0 \dot{y}(\theta) = 0,$$

and this proves the result stated.

16·7 Maximum range on the horizontal plane through *O*.

If, when the velocity of projection w_0 is prescribed, the range OM is a maximum, the tangents to the path at O and at M are at right angles. If the time from O to M is ϕ/k, we have, at $t = \phi/k$, $y = 0$ and $u_0\dot{x} + v_0\dot{y} = 0$. These conditions are

$$v_0(1 - e^{-\phi}) = c(e^{-\phi} - 1 + \phi), \tag{1}$$

$$cv_0(1 - e^{-\phi}) = w_0^2 e^{-\phi}. \tag{2}$$

From these, eliminating v_0, we find at once the equation for ϕ,

$$1 - \phi = \lambda e^{-\phi}, \tag{3}$$

where $\lambda = 1 - (w_0/c)^2$. We suppose $w_0 < c$, so that $0 < \lambda < 1$. It is clear from the graphs of $1 - \phi$ and of $\lambda e^{-\phi}$ that this equation has just one positive root, $0 < \phi < 1$. When ϕ is known, v_0 is determined by equation (2), and the maximum range R is given by

$$k^2 R^2 = (w_0^2 - v_0^2)(1 - e^{-\phi})^2$$

This result can be simplified, for on substituting for v_0^2 from equation (2), we find

$$k^2R^2 = w_0^2(1-e^{-\phi})^2-(w_0^2e^{-\phi}/c)^2.$$

In this we substitute $(1-\phi)/\lambda$ for $e^{-\phi}$ (using equation (3)), which leads to

$$R = \frac{cw_0}{g}\sqrt{\left(\frac{c^2\phi^2-w_0^2}{c^2-w_0^2}\right)}.$$

16·8 Resistance proportional to the square of the speed. If the resistance is mkw^2 the horizontal equation of motion is

$$\frac{d}{dt}(w\cos\psi) = -kw^2\cos\psi, \tag{1}$$

where ψ is the inclination of the direction of motion to the horizontal line Ox (Fig. 16·8). The

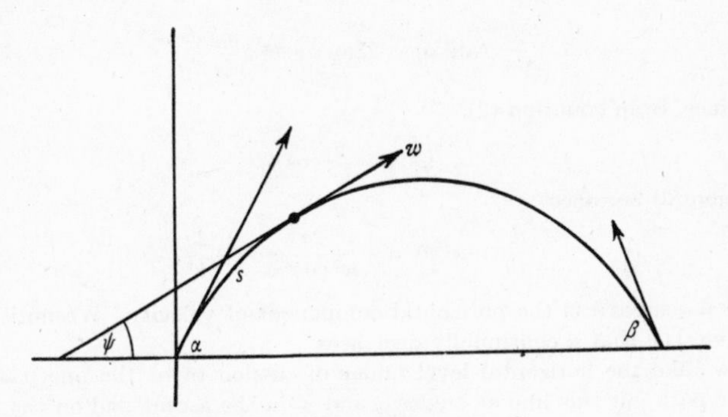

Fig. 16·8

equation can be written in the form

$$\frac{d}{ds}(w\cos\psi)+k(w\cos\psi) = 0, \tag{2}$$

whence
$$w\cos\psi = u_0e^{-ks}, \tag{3}$$

where u_0 is the horizontal velocity component when $s=0$.

For the equation of motion normal to the path we have

$$\frac{w^2}{\rho} = g\cos\psi. \tag{4}$$

We can now easily find the (s,ψ) equation of the path. Eliminating w from equations (3) and (4) we have (remembering $\rho = -ds/d\psi$)

$$\frac{g}{u_0^2}e^{2ks}ds = -\sec^3\psi\, d\psi,$$

and on integration we obtain as the equation of the path

$$\frac{g}{ku_0^2}(e^{2ks}-1) = \sec\alpha\tan\alpha - \sec\psi\tan\psi - \log\left(\frac{\sec\psi+\tan\psi}{\sec\alpha+\tan\alpha}\right),$$

where α is the value of ψ when $s = 0$.

16·9 Other laws of resistance. In the general case, when the resistance is $mf(w)$, we cannot make much progress towards a complete solution. If we resolve horizontally and normally, as in § 16·8, we obtain two equations

$$\begin{cases} \dfrac{d}{dt}(w\cos\psi) = -f(w)\cos\psi, & (1) \\ w^2 = g\rho\cos\psi, & (2) \end{cases}$$

the second of which does not involve the function f; it is the same as the equation (4) of § 16·8.

We content ourselves with one simple application of equation (2)—the proof that the path is steeper at any point on the descent than at the point on the same level on the ascent. This is, indeed, almost evident. To provide a formal proof we have

$$\frac{d}{dy}(\tan^2\psi) = 2\tan\psi\sec^2\psi\frac{d\psi}{ds}\frac{ds}{dy}, \tag{3}$$

and since, from equation (2),

$$\frac{d\psi}{ds} = -\frac{g}{w^2}\cos\psi, \tag{4}$$

equation (3) becomes

$$\frac{d}{dy}(\tan^2\psi) = -\frac{2g}{w^2\cos^2\psi} = -\frac{2g}{u^2}, \tag{5}$$

where $u = w\cos\psi$ is the horizontal component of velocity. We notice, from equation (1), that u continually decreases.

Now take the horizontal level under discussion to be the line $y = 0$, and let the path cut this line at angles α and β on the ascent and on the descent (Fig. 16·8). Let h be the greatest height of the path above Ox, so that $\psi = 0$ when $y = h$. We see, from equation (5), that

$$\tan^2\alpha = 2g\int_0^h \frac{1}{u^2}\,dy,$$

where u is the horizontal component of velocity at height y on the ascent. Similarly

$$\tan^2\beta = 2g\int_0^h \frac{1}{u^2}\,dy,$$

where this time u denotes the horizontal component of velocity at height y on the descent. Since u continually decreases the integrand is greater, for every value of y, in the second integral than in the first. Hence $\tan^2\beta > \tan^2\alpha$, and therefore, since both angles are acute, $\beta > \alpha$.

16·10 The three forms of the equation of energy. In § 7·7 we saw that if a particle moves on a straight line under the action of two forces, one of which is a function of x, we have a general form of the equation of energy

$$\frac{d}{dt}(T+\Omega) = X''v,$$

where X'' is the remaining force. We conclude this chapter with the extension of this result to the problem of a particle moving in a plane.

Suppose then that the particle moves in a plane under the action of (i) a conservative field whose potential function is V, and of (ii) another force $\mathbf{R}''$. The equations of motion, in Cartesian coordinates, are

$$m\ddot{x} = -\frac{\partial V}{\partial x} + X'',$$

$$m\ddot{y} = -\frac{\partial V}{\partial y} + Y'',$$

where X'', Y'' are components of $\mathbf{R}''$. If we multiply the equations by $\dot{x}, \dot{y}$ respectively, and add, we get

$$\begin{aligned}\frac{d}{dt}(T+\Omega) &= X''\dot{x} + Y''\dot{y} \\ &= \mathbf{R}''.\mathbf{v}, \qquad (1)\end{aligned}$$

where $\Omega(t) = V(x, y)$.

Ω is the value of the potential function at the point occupied by the particle at time t. The second member of equation (1) represents the rate of working by the forces not contained in V. Moreover, the presence of forces of constraint or of gyroscopic forces in addition to those mentioned will not invalidate equation (1).

Equation (1) contains the three forms of the equation of energy. The first, or primitive form, arises when $V = 0$; the equation is

$$\frac{dT}{dt} = \mathbf{R}''.\mathbf{v},$$

and expresses the fact that the rate of increase of kinetic energy is equal to the rate of working of the forces acting on the particle. The second, or classical, form is the familiar

$$T + \Omega = C,$$

already discussed in § 11·4. The third, or general, form, is given by equation (1). It is important to notice the aspect of the equation mentioned earlier in the analogous case of § 7·7. We can think of the particle-and-field together as a self-contained physical system; the energy of this system is not conserved, its rate of increase being measured by the rate of working of the extraneous forces.

EXAMPLES XVI

1. A bead threaded on a rough fixed circular wire whose plane is horizontal is projected with velocity V. Show that it will come to rest when the arc traversed is

$$\frac{a}{2\mu}\sinh^{-1}\left(\frac{V^2}{ga}\right),$$

where a is the radius of the wire, and μ is the coefficient of friction.

2. A bead slides on a circular wire which is fixed in a vertical plane, the coefficient of friction between the bead and the wire being μ. If the bead, starting from rest at the level of the centre of the circle, comes to rest at the lowest point, prove that

$$3\mu = e^{\mu\pi}(1-2\mu^2).$$

3. A particle slides down a rough cycloid, whose base is horizontal and vertex downwards, starting from rest at a cusp and coming to rest at the vertex. Prove that, if μ is the coefficient of friction, $\mu^2 e^{\mu\pi} = 1$.

*4. A particle of mass m moves on the outer surface of a fixed sphere of radius a and coefficient of friction $\frac{1}{2}$. It is projected horizontally from the highest point of the sphere with velocity $a\omega$. Prove that if $\omega < \omega_0$ the particle comes to rest on the sphere when the radius to the particle makes an angle α with the upward vertical, but that if $\omega_0 < \omega < n$ the particle leaves the sphere when the radius to the particle makes an angle β with the upward vertical, where

$$n^2 = \frac{g}{a}, \quad \omega_0^2 = \tfrac{1}{2}n^2(\sqrt{5}e^{-\lambda}-1), \quad \tan\lambda = \tfrac{1}{2},$$

α is the smaller root of the equation

$$\omega^2 = \tfrac{1}{2}n^2\{(3\sin\alpha+\cos\alpha)e^{-\alpha}-1\},$$

and

$$\omega^2 = \tfrac{1}{2}n^2\{3(\sin\beta+\cos\beta)e^{-\beta}-1\}.$$

Give a rough numerical estimate of ω_0^2/n^2.

*5. A bead of mass m slides on a rough circular wire of radius a, the coefficient of friction being μ. The force acting on the bead (other than the reaction of the wire) has tangential and outward radial components T and R respectively. If θ denotes the angle between the radius through the bead and a fixed radius, show that

$$ma\ddot{\theta} = T - \mu\dot{\theta}\left|\frac{R+ma\dot{\theta}^2}{\dot{\theta}}\right|.$$

Consider in particular the case in which the plane of the wire is vertical, the only forces on the bead are its weight and the reaction of the wire, and $\mu = 1/\sqrt{2}$. The bead is projected from the lowest point: show that in order that it may just reach the highest point the velocity of projection must be

$$\sqrt{\left\{\frac{2}{3}\sqrt{6ga\, e^{\sqrt{2[(\pi/2)+\alpha]}}}\right\}},$$

where α is the acute angle $\tan^{-1}\sqrt{2}$.

6. Prove that a particle of unit mass which moves under a central force μr describes an ellipse, and that the period is independent of the initial motion.

Assuming the motion to be in one plane, show that if, in addition to the central force, there is a force acting in the plane on the particle proportional to its velocity and in a direction at right angles to the direction of motion, the right angle being measured always in the same sense, show that the Cartesian equations of motion have the form

$$\ddot{x} - k\dot{y} + \mu x = 0,$$
$$\ddot{y} + k\dot{x} + \mu y = 0.$$

Show that uniform motion in a circle with the origin as centre is possible, the angular velocity having one or other of two values which are independent of the radius.

7. Discuss the problem of § 16·5 (motion under gravity in a medium offering a resistance proportional to the speed) using vector notation throughout. (The particle reaches the horizontal plane through O when $\mathbf{r}.\mathbf{g} = 0$.)

8. A particle, moving under gravity in a medium offering resistance per unit mass of κg times the velocity, has initial velocity w_0 in a direction inclined at an angle α above the horizontal. Determine the horizontal and vertical components of velocity at any subsequent time.

Show that the acceleration of the particle at any time during the motion is in a fixed direction determined by the initial conditions, and hence prove that there is a direction in which the component of velocity during the motion is equal to the fixed value $\frac{1}{\kappa}\cos\beta$, where β is the acute angle given by

$$\tan\beta = \tan\alpha + \frac{1}{\kappa w_0}\sec\alpha.$$

9. A particle moves under gravity in a resisting medium, the resistance being kw per unit mass when the speed is w. If it is projected at $t = 0$ from the origin O with velocity (u_0, v_0), prove the following statements:

(i) the hodograph of the path is the straight line whose equation is

$$u_0(y+c) = (c+v_0)x;$$

(ii) the highest point reached is at a height $u_0 v_0/(g+kv_0)$ above O;

(iii) the path has a vertical asymptote $x = u_0/k$;

(iv) the resultant force acting on the particle is constant in direction, and its magnitude is $(1-kx/u_0)$ times its initial value;

(v) the vertical distance of the particle from a line through the origin in the direction of the resultant force is gt/k.

10. A particle subject to gravity is projected at an angle α with the horizontal in a medium which produces a retardation equal to k times the velocity. It strikes the horizontal plane through the point of projection at an angle ω, and the time of flight is T. Prove that

$$\frac{\tan\omega}{\tan\alpha} = \frac{e^{kT}-1-kT}{e^{-kT}-1+kT},$$

and deduce that $\omega > a$.

11. A particle of mass m is projected horizontally with velocity U from the top of a tower which stands on a horizontal plane. The air resistance is mk times the velocity. What is the height of the tower if the particle hits the plane at a distance x from the base of the tower?

If x_0 is the distance from the base of the tower at which the particle would hit the plane if there were no resistance, and if k is small, prove that to a first approximation

$$x = x_0\left(1-\frac{kx}{3U}\right).$$

*12. A particle is projected *in vacuo* from a point A with velocity w_0 and elevation α. Prove that its range on the horizontal plane through A is

$$\frac{w_0^2}{g}\sin 2\alpha.$$

Prove that to obtain the same range in a slightly resisting medium the elevation must be increased by approximately

$$\frac{2kw_0}{3g}\sin\alpha\tan 2\alpha,$$

where k is the constant ratio of the resistance to the momentum of the particle. It may be assumed that α is not near to $\frac{1}{4}\pi$.

13. A particle is projected vertically upwards under gravity. The resistance of the air produces an acceleration opposite to the velocity and numerically equal to κv^2, where v is the velocity and κ a constant. If the initial velocity is V, and the square of $\kappa V^2/g$ can be neglected, show that the particle reaches its highest point in time $\frac{V}{g}-\frac{1}{3}\frac{\kappa V^3}{g^2}$, and that the greatest altitude reached is $\frac{V^2}{2g}-\frac{1}{4}\frac{\kappa V^4}{g^2}$.

If the initial velocity, in addition to the vertical component V, has a small horizontal component U, and the resistance follows the same law, show that when the particle returns to its original level its horizontal velocity is approximately $Ue^{-\kappa V^2/g}$.

14. A particle is projected horizontally with a velocity v_0 in a medium in which the resistance to its motion is kv^2 per unit mass, when its velocity is v. If ψ is the downward inclination of its path to the horizontal when it has traversed an arc s, show, by resolving along the horizontal, that

$$v\cos\psi = v_0 e^{-ks},$$

and, by resolving along the normal to the path, that

$$e^{2ks}\frac{ds}{d\psi} = \frac{v_0^2}{g}\sec^3\psi.$$

Hence find the (s, ψ) equation of the path.

15. Derive for a particle moving in a plane curve the expressions $v\dfrac{d\psi}{dt}$, $\dfrac{dv}{dt}$ for the normal and tangential components respectively of the acceleration.

A particle moving under gravity in a medium offering a constant resistance to its motion is projected with velocity u inclined at an angle $\frac{1}{2}\pi-\alpha$ to the upward vertical. Prove that when its velocity is v the angle which the direction of motion makes with the upward vertical is $\frac{1}{2}\pi-\psi$, where

$$v(1+\sin\alpha)^n\cos^{n+1}\psi = u(1+\sin\psi)^n\cos^{n+1}\alpha$$

and n is the ratio of the constant resistance to the weight of the particle.

16. A ball of coefficient of restitution e is dropped on to a level floor from a height h. Find how long it will continue to bounce.

If it is given initially a horizontal velocity v and is subjected to frictional impulses μ times the normal impulse at each bounce, how far will it travel before it ceases to bounce, assuming that v is large enough for a forward velocity to be maintained throughout?

*17 A particle moves under gravity in a resisting medium, the retardation being kw when the speed is w. It is projected from the origin O with given speed in a given vertical plane. Prove that the envelope of the paths for different directions of projection can be expressed in terms of the parameter v_0 in the form

$$kx = \frac{b}{b+v_0}\sqrt{(bc-v_0^2)},\quad ky = \frac{b(c+v_0)}{b+v_0} - c\log\left(\frac{b+v_0}{v_0}\right),$$

for $v_0^2 < bc$. The axis Oy is vertically upwards, c is the terminal velocity, and the given speed of projection is $\sqrt{(bc)}$.

CHAPTER XVII

MOTION OF TWO PARTICLES IN A PLANE, COLLISIONS

17·1 The two classical theorems. The theory of the motion of two particles P_1 and P_2 on a straight line, which we discussed in Chapter VIII, is easily extended to motion in a plane, and to motion in space. In this chapter we consider the motion of two particles in a plane. The two classical theorems of § 8·1 are easily generalized.

We denote the position vectors of the particles by $\mathbf{r}_1, \mathbf{r}_2$, their masses by m_1, m_2, the external forces by $\mathbf{Z}_1, \mathbf{Z}_2$, and the internal forces by $-\mathbf{R}, \mathbf{R}$; the line of action of $\mathbf{R}$ is P_1P_2. The equations of motion are

$$m_1\ddot{\mathbf{r}}_1 = \mathbf{Z}_1 - \mathbf{R}, \tag{1}$$

$$m_2\ddot{\mathbf{r}}_2 = \mathbf{Z}_2 + \mathbf{R}. \tag{2}$$

(i) We obtain immediately, by addition of the equations (1) and (2), the equation

$$m_1\ddot{\mathbf{r}}_1 + m_2\ddot{\mathbf{r}}_2 = \mathbf{Z}_1 + \mathbf{Z}_2, \tag{3}$$

i.e.

$$M\ddot{\mathbf{r}}_0 = \mathbf{Z}_1 + \mathbf{Z}_2, \tag{4}$$

where $M = m_1 + m_2$, and $\mathbf{r}_0$ is the position vector of the centre of gravity G. Equation (4) shows that the motion of G is the same as that of a particle of mass M acted on by a force $\mathbf{Z}_1 + \mathbf{Z}_2$. This is the first of the two classical theorems.

In particular, if $\mathbf{Z}_1 + \mathbf{Z}_2 = \mathbf{0}$, G moves uniformly in a straight line; an important special case is that in which $\mathbf{Z}_1 = \mathbf{Z}_2 = \mathbf{0}$.

(ii) If $\mathbf{s}$ is the position vector of P_2 relative to P_1 we have

$$\mathbf{s} = \mathbf{r}_2 - \mathbf{r}_1, \tag{5}$$

and therefore, from equations (1) and (2),

$$\ddot{\mathbf{s}} = \frac{1}{m_2}\mathbf{Z}_2 - \frac{1}{m_1}\mathbf{Z}_1 + \frac{1}{\mu}\mathbf{R}, \tag{6}$$

where $\dfrac{1}{\mu} = \dfrac{1}{m_1} + \dfrac{1}{m_2}$. This is the second of the two classical results.

We notice that we can easily express $\mathbf{r}_1$ and $\mathbf{r}_2$ in terms of $\mathbf{r}_0$ and $\mathbf{s}$ (cf. § 8·1). We have

$$\mathbf{r}_1 = \mathbf{r}_0 - \frac{m_2}{M}\mathbf{s},$$

$$\mathbf{r}_2 = \mathbf{r}_0 + \frac{m_1}{M}\mathbf{s}.$$

17·2 An important corollary. Just as in the similar problem of § 8·1, equation (6) of § 17·1 takes a particularly simple form in two special cases, namely, (i) if $\mathbf{Z}_1 = \mathbf{Z}_2 = \mathbf{0}$, or (ii) if $\mathbf{Z}_1/m_1 = \mathbf{Z}_2/m_2$. The first case is that in which there are no external forces, the second is that in which the external forces are parallel to one another and proportional to the masses. If one of these conditions is satisfied we find

$$\mu\ddot{\mathbf{s}} = \mathbf{R}. \tag{1}$$

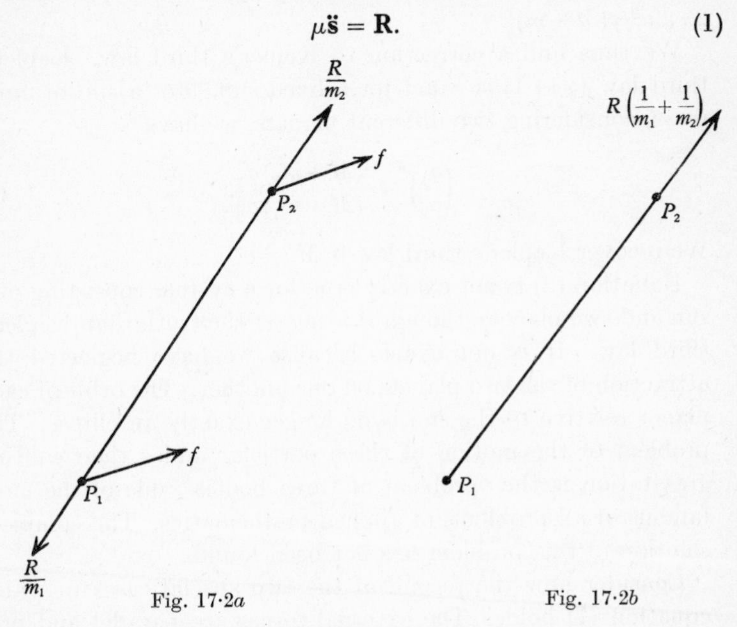

Fig. 17·2*a*

Fig. 17·2*b*

It is interesting to verify the result from a diagram. If the external forces are $m_1\mathbf{f}, m_2\mathbf{f}$, and R is the directed modulus of $\mathbf{R}$ in the sense P_1P_2, the actual accelerations of the particles are as marked in Fig. 17·2*a*. Therefore the acceleration of P_2 relative to P_1 is as marked in Fig. 17·2*b*, and we recover at once equation (1).

Since $\mathbf{R}$ is parallel to $\mathbf{s}$ the motion of P_2 relative to P_1, as determined by equation (1), is a problem of central orbits. Moreover, if R is a function of s ($= |\mathbf{s}|$) there exists an 'energy integral' for the relative motion.

As we have noticed, the first of the two special cases in which equation (1) is valid is that in which there are no external forces; the only forces on the particles are their mutual repulsion or attraction. In the Newtonian case the acceleration of P_2 relative to P_1 is $\gamma(m_1+m_2)/s^2$ in the line P_2P_1. The relative motion is that of a particle attracted to a fixed centre of force, the acceleration being k/s^2, where $k = \gamma(m_1+m_2)$. Thus if the mass M of the sun is finite, so that the sun can no longer be regarded as fixed, the period σ of a planet of mass m in its elliptic orbit about the sun is, by equation (5) of § 14·6, $2\pi\sqrt{\{a^3/\gamma(M+m)\}}$.

We thus find a correction to Kepler's third law. Kepler's third law (§ 14·1) is exact for a fixed sun. For a sun of finite mass, considering two different planets, we have

$$\left(\frac{\sigma_1}{\sigma_2}\right)^2 = \frac{(M+m_2)\,a_1^3}{(M+m_1)\,a_2^3}. \tag{2}$$

We recover Kepler's third law if $M \to \infty$.

Equation (2) is not exactly true for a system consisting of a sun and two planets, though it is nearer the truth than Kepler's third law. It is not exact, because we have neglected the attraction of the two planets on one another. The orbit of each planet relative to the sun is no longer exactly an ellipse. The problem of the motion of three particles under their mutual gravitation is the 'problem of three bodies', one of the most famous of all problems of applied mathematics. The complete solution of this problem has not been found.

Consider now the second of the two special cases in which equation (1) holds. The external forces are parallel and proportional to the masses. The most conspicuous example arises when we have a planet and satellite moving in the sun's field; the sun need not be thought of as fixed. If the distance between planet and satellite is small in comparison with their distances from the sun, the forces on the two bodies arising from the sun's gravitation are, at each instant, nearly parallel and proportional to the masses. As a first approximation the motion of the satellite relative to the planet is not affected by the sun's field.

17·3 Collisions. The term collision includes problems of two types.

(i) The first is the impact of smooth spheres. The classical example comes from the game of billiards; though the conditions assumed in our discussion are not quite those in the actual game, since we suppose the spheres to move without rotation. In this problem the second sphere is initially at rest. The special case in which the centres of the spheres move on the same straight line has already been considered in § 8·5.

(ii) The second problem is that of the interaction between two particles under a mutual attraction or repulsion. Initially, the particles are at a great distance apart, and the motion of each is nearly rectilinear and uniform. The attraction or repulsion is assumed to be a function of s, the distance between the particles, and the potential function V from which the attraction or repulsion is derived tends to zero as s tends to infinity. Finally, the particles are again at a great distance apart, and the motion of each is again nearly rectilinear and uniform. Since the relative motion is a central orbit motion with an energy integral (§ 17·2), the net effect of the collision is to turn the relative velocity through a certain angle without altering its magnitude. If the law of attraction or repulsion is an inverse-square law the relative orbit is a hyperbola (§ 14·4). The most important application is that of the collision of an α-particle with an atomic nucleus; in this problem, as in the classical billiard-ball problem, the second particle is usually assumed to be initially at rest.

17·4 Impact of smooth spheres. We denote the velocity before impact of the first sphere, of mass m_1, by $\mathbf{U}_1$, and the velocity before impact of the second sphere, of mass m_2, by $\mathbf{U}_2$. The corresponding velocities after the collision are denoted by $\mathbf{V}_1, \mathbf{V}_2$. Usually we shall be concerned with the problem in which the velocities before impact are coplanar with the line of centres, so that the problem is two-dimensional. In this case we denote the components of $\mathbf{U}_1$ by u_1 and w_1, u_1 being along the line of centres, and w_1 perpendicular to the line of centres; the corresponding components of $\mathbf{U}_2$ are denoted by u_2 and w_2.

Since the spheres are smooth the impulsive reaction on impact is in the line of centres, and the transverse components of velocity are unaltered. We denote the components of velocity along the line of centres after impact by v_1, v_2. Fig. 17·4*a* illustrates the motion immediately before impact, and Fig. 17·4*b* immediately after.

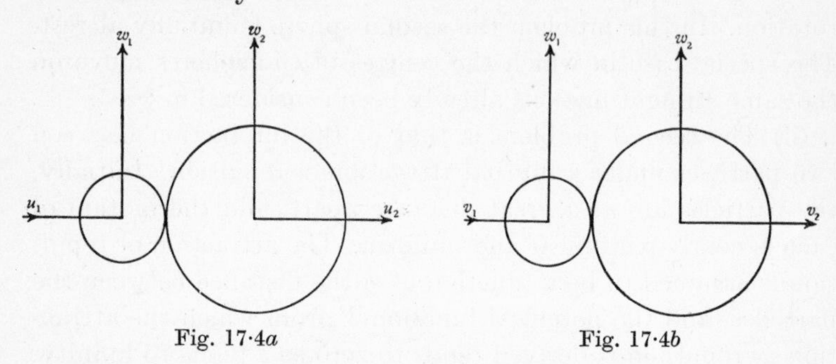

Fig. 17·4*a* Fig. 17·4*b*

The general form of Newton's law of impact is found experimentally to be that the component velocity of separation in the direction of the common normal at the point of impact is e times the same component of the velocity of approach. This law, and the law of conservation of momentum in the line of centres, imply that the equations (1) and (2) of § 8·5 for the linear problem are still valid for the component velocities along the line of centres. These equations are

$$m_1 v_1 + m_2 v_2 = m_1 u_1 + m_2 u_2,$$
$$v_2 - v_1 = e(u_1 - u_2),$$

and the values of v_1, v_2 have already been found from these equations in § 8·5.

Example 1. *The billiard-ball problem. If* $m_1 = m_2$, $e = 1$, *and the second sphere is initially at rest, then the velocities of the spheres after impact are at right angles.* To prove this we have only to observe that in this problem $w_2 = v_1 = 0$. (We have supposed $w_1 \neq 0$.

Example 2. *If* $m_1 > m_2$, $e = 1$, *and the second sphere is initially at rest, the greatest angle through which the direction of motion of the first sphere can be turned by the collision is the acute angle* $\sin^{-1}(m_2/m_1)$.

Let the direction of motion of the first sphere before the impact make an angle $\frac{1}{2}\pi-\alpha$ with the line of centres; and let the corresponding angle after the impact be $\frac{1}{2}\pi-\beta$. These angles are shown in the vector diagram, Fig. 17·4c.

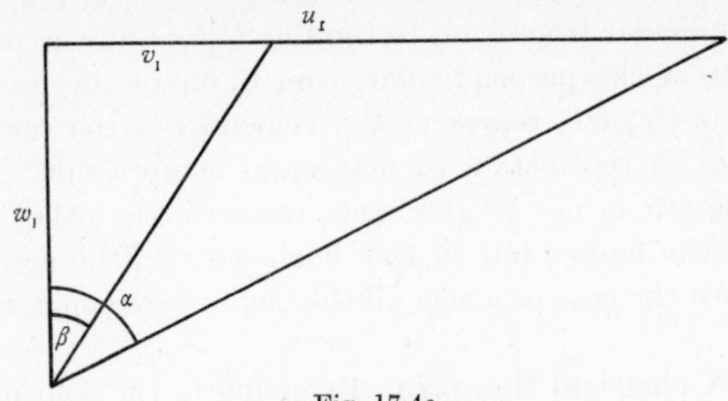

Fig. 17·4c

We have
$$\frac{\tan\alpha}{\tan\beta}=\frac{u_1}{v_1}=\frac{m_1+m_2}{m_1-m_2},$$
whence
$$\tan\beta=k\tan\alpha,$$
where
$$k=(m_1-m_2)/(m_1+m_2).$$
Thus
$$\tan(\alpha-\beta)=\frac{(1-k)\tan\alpha}{1+k\tan^2\alpha},$$
which has a maximum value when $k\tan^2\alpha=1$. The maximum value δ of $(\alpha-\beta)$ is given by
$$\tan\delta=(1-k)/2\sqrt{k},$$
$$\sin\delta=\frac{1-k}{1+k}=\frac{m_2}{m_1}.$$

17·5 The collision diagram. We construct a diagram in which the vector $\mathbf{U}_1$ is denoted by OU_1, the vector $\mathbf{U}_2$ by OU_2, and so on. The vectors $\mathbf{U}_1-\mathbf{V}_1$, $\mathbf{V}_2-\mathbf{U}_2$ have the same direction, the direction of the line of centres at the instant of impact; this direction is taken horizontal in the diagram (Fig. 17·5b).

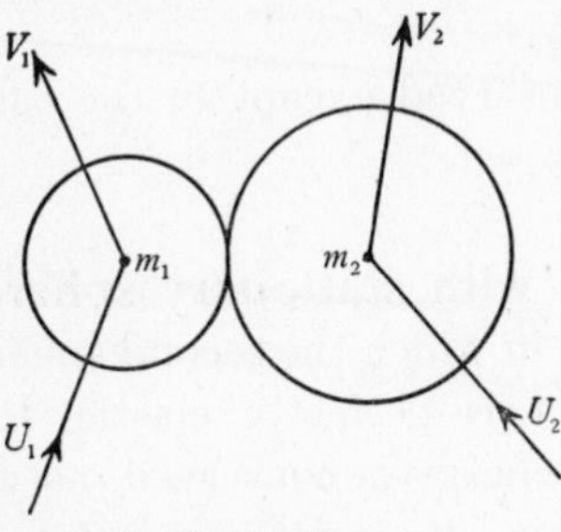

Fig. 17·5a

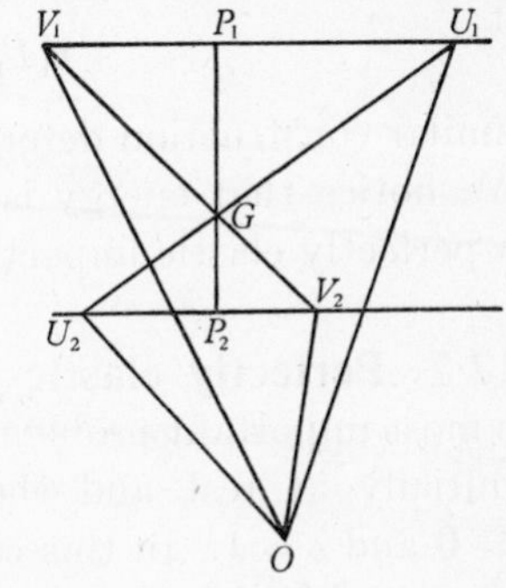

Fig. 17·5b

The point G divides U_1U_2 in the ratio $m_2 : m_1$, and the vector OG represents the velocity of the centre of gravity, a velocity which is *unaltered by the collision*. The point P_1 is the foot of the perpendicular from G on to the line U_1V_1, and the point P_2 is the foot of the perpendicular from G on to the line U_2V_2. The vectors OP_1, OP_2 represent the velocities of the centres of the spheres at the instant of maximum compression. In the general case O is not in the same plane as the other points named in the figure; but in this book we shall be concerned mainly with the case in which all the vectors are coplanar.

17·6 A classical theorem. Referring to the collision diagram, Fig. 17·5*b*, we see that

$$\frac{V_1P_1}{P_1U_1} = \frac{P_2V_2}{U_2P_2} = \frac{V_1P_1 + P_2V_2}{P_1U_1 + U_2P_2} = e,$$

where we have quoted Newton's law of impact. Hence we derive the theorem: *The increase of velocity of the centre of either sphere after maximum compression is e times the decrease of velocity before maximum compression.*

It follows that when U_1 and U_2 have been marked in the diagram, V_1 and V_2 can be determined by an elementary construction. This is what we should expect, since the diagram is merely an illustration of the equations determining the motion after the collision. Given U_1 and U_2 we first mark G, and then determine P_1 by dropping the perpendicular from G on to the line through U_1 in the direction of the line of centres. The point V_1 is then determined by producing U_1P_1 to V_1 such that

$$V_1P_1 = e.P_1U_1.$$

A similar construction determines V_2.

We notice that energy is not conserved except in the case of a perfectly elastic impact, when $e = 1$.

17·7 Perfectly elastic impact with stationary sphere. The most important problem is that in which the second sphere is initially at rest and the impact is perfectly elastic, i.e. $\mathbf{U}_2 = \mathbf{0}$ and $e = 1$. In this case the energy is conserved, as we have noticed already. The collision diagram now takes a simpler form. The point U_2 coincides with O, the figure is plane (the motion takes place in the plane determined by the vector $\mathbf{U}_1$ and the line of centres) and there is symmetry about P_1P_2. Fig. 17·7a represents the case $m_1 > m_2$, and Fig. 17·7b the case $m_1 < m_2$. The angles θ and ϕ are the inclinations of the final velocities of m_2 and of m_1 to the initial velocity of m_1.

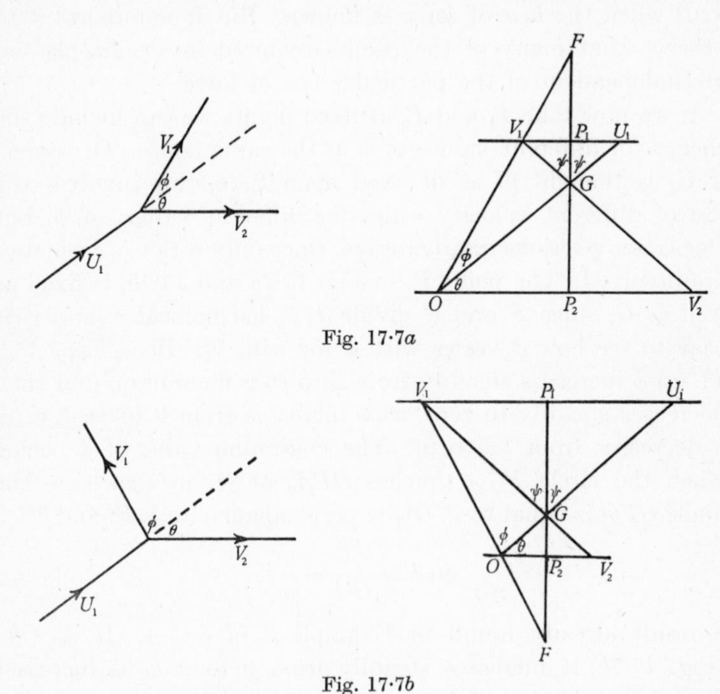

Fig. 17·7a

Fig. 17·7b

We see from the diagrams that the velocity of m_1 relative to G is unchanged in magnitude by the collision, and turned through an angle which we call 2ψ; moreover, the velocity of m_1 relative to m_2 is unchanged in magnitude and turned through the same angle 2ψ. The angles ψ and θ are complementary.

17·8 Collisions of the second kind. We now come to the crux of the whole theory. *The same figures also exhibit the result of a collision of the second kind*, where there are two particles under a mutual attraction or repulsion. The horizontal line in the figure, formerly the line of centres, is now the direction of the final velocity of m_2. The motion of m_1 relative to the centre of gravity is unchanged in magnitude by the collision, and turned through an angle which we call 2ψ. This angle 2ψ is also the angle through which the motion of m_1 relative to m_2 is turned by the interaction, and this angle can be calculated from the theory of central orbits (Chapter XIII) when the law of force is known. But it is important to observe that many of the results required in atomic physics are independent of the particular law of force.

If we now take P_1 and P_2 as fixed points we can include the theory for different values of ψ in the same figure. Of course, if $\mathbf{U}_1$ is thought of as of fixed magnitude, this involves the use of different velocity-scales for different values of ψ, but this is not a serious disadvantage, since only ratios of velocities are involved. The point F, in Fig. 17·7*a* and 17·7*b*, is fixed as well as G, since F and G divide P_1P_2 harmonically, and it is easy to see how ϕ varies with ψ (or with θ). If $m_1 > m_2$ (Fig. 17·7*a*) ϕ increases steadily from zero to a maximum, and then decreases steadily to zero, as ψ increases from 0 to $\frac{1}{2}\pi$ (i.e. as θ decreases from $\frac{1}{2}\pi$ to 0). The maximum value of ϕ occurs when the circle FGO touches OP_2V_2 at O; in that case the angle GFO is equal to θ, GV_1 is perpendicular to V_1F, and

$$\sin\phi = \frac{V_1G}{GO} = \frac{m_2}{m_1},$$

a result already found in Example 2 of § 17·4. If $m_1 < m_2$ (Fig. 17·7*b*) ϕ increases steadily from 0 to π as ψ increases from 0 to $\frac{1}{2}\pi$ (i.e. as θ decreases from $\frac{1}{2}\pi$ to 0).

17·9 Applications to atomic theory. We can read off from the collision diagram the various relations which are needed in the theory of the collisions of α-particles with atomic nuclei or with electrons. The fundamental relation is that giving the ratio of the masses in terms of the angles θ and ϕ; this relation is of great value in interpreting measurements of the branched tracks observed in an expansion chamber. We have immediately, on reference to Fig. 17·7*a* or Fig. 17·7*b*,

$$\frac{m_2}{m_1} = \frac{V_1G}{GV_2} = \frac{V_1G}{GO} = \frac{\sin\phi}{\sin(2\psi-\phi)} = \frac{\sin\phi}{\sin(2\theta+\phi)}. \tag{1}$$

Next we require the formulae for the final velocities of the two particles in terms of the initial velocity of m_1, and the observed angles θ and ϕ. We denote $|\mathbf{U}_1|$ by u_1, and so on, so that in the figures u_1 is the length OU_1. We have immediately

$$v_2 = OV_2 = 2OG\cos\theta = 2\frac{m_1}{m_1+m_2}u_1\cos\theta, \tag{2}$$

and
$$v_1 = OG\cos\phi \pm \sqrt{(GV_1^2 - OG^2\sin^2\phi)}$$
$$= \frac{m_1\cos\phi \pm \sqrt{(m_2^2 - m_1^2\sin^2\phi)}}{m_1+m_2}u_1. \tag{3}$$

Equation (2) allows us to calculate immediately the fraction of the kinetic energy which is transferred from the projected particle to the particle initially at rest. If the energy transferred is E_2 and the initial energy is E_0 we have, from equation (2),

$$\frac{E_2}{E_0} = \frac{m_2 v_2^2}{m_1 u_1^2} = \frac{4m_1 m_2}{(m_1+m_2)^2}\cos^2\theta,$$

giving E_2/E_0 in terms of the angle θ and the ratio of the masses. It will be seen that the fraction of the energy transferred is a maximum when the first particle is projected directly towards the second, $\theta = 0$, and the fraction transferred decreases steadily as θ increases from 0 to $\frac{1}{2}\pi$. The maximum value of the fraction transferred is

$$\frac{4m_1 m_2}{(m_1+m_2)^2} = 1 - \left(\frac{m_1 - m_2}{m_1 + m_2}\right)^2;$$

and this has a maximum value 1 when $m_1 = m_2$. In this case the impact is direct, and *all* the energy is transferred, as is well known (see equation (1) of § 8·6). In the extreme case $m_1 \gg m_2$ (as in the collision of an α-particle with an electron) the maximum transferred is approximately $4m_2/m_1$.

It is most important to observe that all the results in this paragraph are independent of the particular law of attraction or repulsion between the particles.

17·10 The special case of the inverse-square law.

If, finally, we introduce the assumption of a particular law of force between the particles, we can determine the angle ψ in terms of the initial conditions of the experiment. By the theorem of § 17·2 the motion of one particle relative to the other is the same as if the second were fixed, and the mass of the first reduced to μ. In particular, if the law of force is an inverse-square law, with an attraction or repulsion ZeE/s^2, the relative orbit is one branch of a hyperbola, and the angle between the asymptotes is 2θ, where (equation (9) of § 14·6)

$$\tan\theta = Pu_1^2/k,$$

where $k = ZeE/\mu$; here P denotes the 'impact parameter'—i.e. the initial distance of m_2 from the initial line of motion of m_1. Hence

$$\tan\psi = \cot\theta = \frac{ZeE}{Pu_1^2}\left(\frac{1}{m_1}+\frac{1}{m_2}\right).$$

EXAMPLES XVII

1. Two equal particles, each of mass m, connected by a fine string of length $2a$, are placed on a smooth horizontal plane, the string between them being straight. One of the particles is struck a blow B, in a direction at right angles to the string. Determine the magnitude and direction of the velocity of either particle at any instant; show that each particle describes a cycloid; and draw a figure showing the character of the motion of the system in general.

2. A particle A, of mass m_1, and a particle B, of mass m_2, move under no forces save a mutual reaction in the line AB. Show that the motion of B relative to A is the same as if A were fixed and the mass of B were reduced to $m_1m_2/(m_1+m_2)$.

Two particles A and B, each of mass m, attract each other with a force kr, where k is constant and r is the distance AB. At an instant when the particles are at rest at a distance a apart, B is projected with velocity na in a direction perpendicular to AB, where $n=\sqrt{(2k/m)}$. Prove that the subsequent motion of the particles is the same as that of the ends of a diameter of a circular disc which rolls, with uniform speed, on a straight line.

3. Two particles of masses m, m' are moving in the plane of xy, under an attraction R along the radius r joining the particles. Show that the centre of gravity moves with constant velocity in a straight line; and that if x, y are the rectangular coordinates of either particle with respect to the other, then

$$\frac{mm'}{m+m'}\frac{d^2x}{dt^2}=-R\frac{x}{r},\quad \frac{mm'}{m+m'}\frac{d^2y}{dt^2}=-R\frac{y}{r}.$$

If the relative orbit is a circle of radius r, described in a period T, prove that

$$R=\frac{mm'}{m+m'}\frac{4\pi^2r}{T^2}.$$

Assuming Newton's law of attraction, and that the moon describes a circle of radius r, relative to the earth, establish the equation

$$1+\frac{M}{E}=\frac{4r^3}{a^2lN^2},$$

where a is the earth's radius, l is the length of a pendulum whose period is 2 sec., M and E are the masses of the moon and earth respectively, and N is the number of seconds in the moon's period.

4. Three particles P, Q, R, each of mass m, attract each other with a force $\mu m^2 \times$ (distance). They move on a smooth horizontal plane and, when $t=0$, are at A, B, C, and are then moving with velocities equal in magnitude and direction to λBC, λCA, λAB. Show that their centre of gravity, G, remains at rest, and that each particle describes an ellipse about G in the periodic time $T=2\pi/\sqrt{(3\mu m)}$.

Show that the area of each ellipse is $\frac{2}{9}\lambda ST$, where S is the area of ABC.

5. If S, J and T are the masses of three celestial particles situated at the vertices of an equilateral triangle of side a, prove that the total force per unit mass on any one of them is proportional to its distance from the centre of mass of the system. Hence show that this triangular arrangement is a possible configuration of relative motion with steady angular speed of rotation, and that the period of the motion is $2\pi\sqrt{(a^3/\gamma M)}$, where $M=S+J+T$.

6. Two gravitating particles, of masses m and M, starting from rest at an infinite distance apart, are allowed to fall freely towards one another. Show that, when their distance apart is a, their relative velocity of approach is

$$\sqrt{\left(\frac{2\gamma(M+m)}{a}\right)},$$

γ being the constant of gravitation.

Impulses of magnitude I are applied to the particles in opposite directions perpendicular to their line of motion at the instant when their distance apart is a. Show how to determine p, their distance of closest approach during the subsequent motion, and prove that, if I is small, p is given approximately by

$$\frac{p}{a}=\frac{1}{2}\frac{I^2a}{\gamma}\frac{M+m}{M^2m^2}.$$

7. Two particles, each of mass m, moving in a plane, attract each other with a force of magnitude λr^{-2}, where λ is constant and r is the distance between the particles. If initially $r=r_0$, one particle is at rest, and the other is moving with velocity $2v_0$ at right angles to the join of the particles, where $v_0^2<\lambda/(mr_0)$, show that the stationary value of r other than the value r_0 is given by

$$mv_0^2r_0^2\left(\frac{1}{r}+\frac{1}{r_0}\right)=\lambda.$$

8. A small uniform solid sphere is let fall vertically under gravity, and after describing a distance h impinges at a point A on a smooth plane inclined at an angle α to the horizontal. Show that the particle ceases to rebound from the plane when it reaches a point B such that

$$AB=4he\sin\alpha/(1-e)^2.$$

9. Two equal smooth spheres moving along parallel lines in opposite directions with velocities u, v collide, the line of centres making an angle α with the direction of motion. If, after the impact, their lines of motion are perpendicular, show that

$$\left(\frac{u-v}{u+v}\right)^2 = \sin^2\alpha + e^2\cos^2\alpha,$$

where e is the coefficient of elasticity.

10. Show that if two spheres of equal mass moving head-on suffer a perfectly elastic collision they interchange velocities.

Two smooth spheres of equal mass, whose centres are moving with equal speeds in the same plane, collide in such a way that at the moment of collision the line of centres makes an angle $\frac{1}{2}\pi - \epsilon$ with the direction bisecting the angle α between the velocities before impact. Show that after impact the velocities are inclined at an angle $\tan^{-1}(\cos 2\epsilon \tan\alpha)$, the collision being perfectly elastic.

11. Show that the maximum deviation which can be produced in the direction of motion of a smooth sphere by any collision with an equal stationary sphere is

$$\tan^{-1}(1+e)(8-8e)^{-\frac{1}{2}},$$

where e is the coefficient of restitution.

12. A billiard table is in the shape of a rectangle $ABCD$ ($AD > AB$) with small pockets at the corners A, B, C, D and at the middle points X, Y of AD, BC respectively. The balls are smooth and equal in mass; their size is small compared with that of the table, so that their motion may be regarded as that of small equal spheres. At a certain instant a player Q finds his (white) ball at the middle point of the edge CD and the red ball at the centre of the table. Q plays his ball, which hits the red ball, sending it straight into the pocket at A. Prove that it is not possible, for any value of the coefficient of restitution, for Q's ball to go after the impact straight into the pocket at Y, and find the value of the coefficient of restitution if Q's ball goes after the impact straight into the pocket at B.

13. Three equal similar smooth spheres α, β, γ lie on a table, with their centres at the vertices A, B, C of a square $ABCD$. α is projected along the table so as to collide successively with β and γ. Assuming the impacts to be perfectly elastic, show that the direction of projection must lie within an angle $\tan^{-1}\left(\dfrac{2b^2}{l^2 - b\sqrt{(l^2-4b^2)}}\right)$, where b is the diameter of a sphere, and l a side of the square.

Show that after the collisions the direction of motion of α lies within one of two angles, one being $\cos^{-1}\dfrac{b}{l}$ and the other $\sin^{-1}\dfrac{b}{l} + \cos^{-1}\dfrac{2b}{l}$.

14. From the equations of conservation or otherwise show that after an elastic collision ($e = 1$) between two equal smooth spheres, of which one is initially at rest but free to move in any direction, the directions of motion of the two spheres are at right angles.

Show further that if the mass of the resting sphere is greater in the ratio $1+\epsilon : 1$ (ϵ small) then the angle between the directions of motion will exceed a right angle by

$$\tfrac{1}{2}\epsilon \tan\phi$$

approximately, where ϕ is the deflection of the moving sphere.

15. Two smooth and perfectly elastic spheres of masses 1 and 4 respectively are initially at rest under no forces. The more massive sphere is then projected in such a direction as to strike the other sphere and rebound. Prove that the direction of motion of the more massive sphere cannot be deflected by the collision through an angle greater than $14^\circ\ 29'$.

16. A moving sphere of mass M collides with a stationary sphere of mass m. The spheres are smooth and perfectly elastic. Prove that the acute angle θ between the initial velocity and the line of centres at impact and the angular deviation δ produced in the motion of the first sphere are related by the equation

$$M \sin\delta = m \sin(2\theta+\delta).$$

Find the greatest value of δ which can be produced in the two cases $M \geqslant m$ and $M < m$.

17. A small sphere is suspended from a fixed point by an inextensible string. Another small sphere of equal mass moving in a direction making 60° with the downward vertical impinges directly on the first sphere with velocity V. Prove that the velocity of the first sphere after impact is $\dfrac{2\sqrt{3}}{7}(1+e)V$, where e is the coefficient of restitution.

18. A wedge of mass M and angle α is sliding along a smooth horizontal plane with velocity V. A smooth uniform sphere of mass m is dropped vertically and strikes the wedge. Show that if the coefficient of restitution between the wedge and the table is zero, and between the sphere and the wedge is e, then the sphere must strike the wedge with velocity

$$\frac{2V(M-me\sin^2\alpha)}{m(1+e)\sin 2\alpha}$$

in order to stop the wedge.

What happens if the masses of the wedge and of the sphere satisfy the equation

$$M = me\sin^2\alpha\,?$$

19. A smooth wedge of mass M and angle α is free to slide on a horizontal plane. A small perfectly elastic ball of mass m bounces on the wedge, the motion being in a vertical plane of symmetry. Show that impacts occur after successive equal intervals of time.

Immediately before an impact the component velocity of the ball perpendicular to the face of the wedge is u_1, and the velocity of the wedge is v_1. Show that the corresponding velocities immediately before the next impact are

$$u_2 = \{(M+3m\sin^2\alpha)u_1 - 2mv_1\sin^3\alpha\}/(M+m\sin^2\alpha),$$

$$v_2 = \{2mu_1\sin\alpha + (M-m\sin^2\alpha)v_1\}/(M+m\sin^2\alpha).$$

20. A sphere collides obliquely with another sphere of equal mass which is initially at rest, both spheres being smooth and perfectly elastic. Show that their paths after collision are at right angles.

The centres of two such spheres B, C, each of 3 cm. radius, are at E, F, where $EF = 16$ cm. An equal sphere A is projected with velocity u at right angles to EF, and strikes first B and then C. Its final path is perpendicular to EF. Find the point of contact between A and B. Show that

$$v_A = 9u/25, \quad v_B = 20u/25, \quad v_C = 12u/25,$$

where v_A, v_B, v_C are the final velocities.

21. A smooth perfectly elastic sphere S_1 of radius a is at rest on a horizontal table. A second equal sphere S_2 is projected with velocity V in a given direction so that its centre would pass at distance b $(= 2a \sin\phi)$ from the centre of S_1, the line of centres at contact making an angle ϕ with the given direction $(-\frac{1}{2}\pi < \phi < \frac{1}{2}\pi)$. The component velocities of the spheres in the given direction are V_1 and V_2 after the collision. Prove that if a large number of such collisions take place for all values of ϕ between $\pm\frac{1}{2}\pi$, and all values of ϕ are regarded as equally probable, then the average value of V_1 or of V_2 will be $\frac{1}{2}V$; but if all values of b between $\pm 2a$ are regarded as equally probable, then the average value of V_1 will be $\frac{2}{3}V$, and of V_2 will be $\frac{1}{3}V$.

22. Two masses m_1, m_2 move under the force of their mutual attraction which is equal to μ/r^2, when their distance apart is r. When at a great distance from m_1 the mass m_2 is projected with velocity V along a straight line whose point of nearest approach to m_1 is at a distance c from m_1. Prove that the path of m_2 relative to m_1 is a part of a hyperbola with m_1 at a focus, and show that if c is small enough, the distance of closest approach of the two bodies is approximately $\dfrac{c^2 V^2 m_1 m_2}{2\mu(m_1+m_2)}$.

23. Two particles of masses m_1 and m_2 are initially at a great distance apart. Initially m_2 is at rest and m_1 is projected with velocity U_1 in a direction which passes m_2 at a perpendicular distance p. After the encounter the final velocities of m_2 and m_1 make angles θ and ϕ with the initial velocity of m_1. On the assumption that the mutual potential energy of the particles is a function of r, their distance apart, and tends to zero as r tends to infinity, prove that

$$\frac{m_2}{m_1} = \frac{\sin\phi}{\sin(2\theta+\phi)}.$$

Show further that the ratio of the final energy of m_2 to the initial energy of m_1 is

$$\frac{4m_1 m_2}{(m_1+m_2)^2}\cos^2\theta.$$

24. Two gravitating particles of masses M and m are initially a great distance apart; M is at rest and m has a velocity v in a direction which passes M at a perpendicular distance P. Prove that after the encounter when m has again receded to a great distance, M will have acquired a velocity

$$\frac{2vGm}{\sqrt{\{G^2(M+m)^2+P^2v^4\}}},$$

where G is the constant of gravitation.

25. Two particles of masses M, m and electric charges E, e which either attract or repel, encounter one another after approach from a great distance. If at a great distance the velocity of one relative to the other is V along a line distant P from the other particle, and 2ϕ is the angle through which the relative velocity is turned by the complete encounter, prove that

$$\tan\phi = \frac{Ee}{\mu V^2 P},$$

where

$$\frac{1}{\mu} = \frac{1}{m} + \frac{1}{M}.$$

Show further that if the two particles have equal masses and one is initially at rest the final velocities are $V\cos\phi$ and $V\sin\phi$.

CHAPTER XVIII

MOTION OF A LAMINA IN ITS PLANE

18·1 Coordinates defining the configuration. We now turn to the theory of the motion of a rigid body. In this book we shall be concerned mainly with two-dimensional problems—i.e. each particle of the body moves in one of a family of parallel planes. As we shall see, there is no loss of generality if we suppose all the particles to lie in, and to move in, the

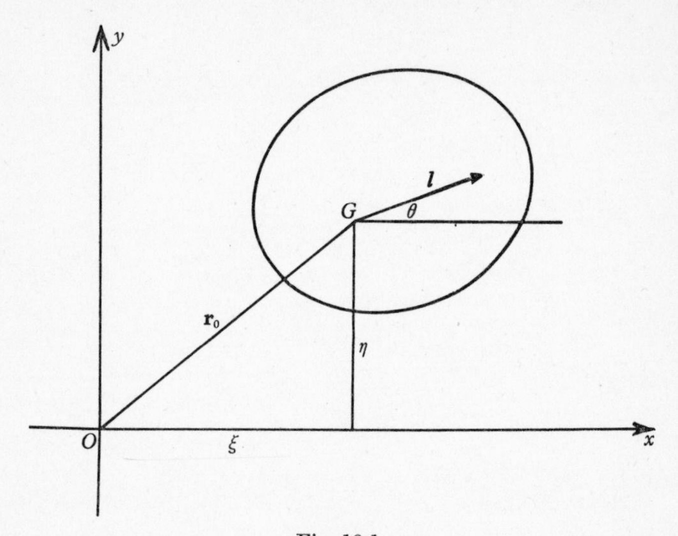

Fig. 18·1

same plane. We may thus suppose to begin with that the rigid body whose motion we study is a lamina. We consider, then, a lamina moving in its plane, and we begin with a study of the motion itself in isolation, without reference to the forces that maintain the motion.

To describe the configuration (i.e. position and orientation) of the lamina in the plane at a given instant we need to specify the position in the plane of a marked point G of the lamina, and the direction of a marked directed line l of the lamina, which we may suppose to pass through G. The point G and the line l are fixed in the lamina. We take a base of reference fixed in the plane, and we frequently need to use rectangular axes Ox, Oy fixed in the base (as in § 10·1). The configuration of the lamina is defined by the position vector $\mathbf{r}_0$ (or the coordinates ξ, η) of G, and the inclination θ of l to Ox.

At first sight it may appear that an ambiguity is involved in this description, since the lamina could be turned through 180° about l without altering the values of ξ, η, θ; but it is to be supposed that the lamina does not move out of the plane, the same face of the lamina being turned always towards the same side of the plane. The three coordinates ξ, η, θ therefore suffice to fix the configuration. The word *coordinates* is used in this context in a broader sense than in Cartesian geometry; the coordinates are parameters defining the configuration of the system. Three such coordinates are required in this case, and the dynamical system under discussion is said to have three degrees of freedom; a particle moving in a plane, where the configuration of the system is defined by the two coordinates (x, y), is said to have two degrees of freedom.

It will be convenient to speak of a point fixed in the lamina as a *molecule* of the lamina. In particular, the marked point G, used in specifying the configuration, is a molecule of the lamina; for our present purpose it need not be the centre of gravity, but when we come to consider the equations of motion we shall find that it is usually convenient to take the centre of gravity as the marked point.

When the lamina moves from one position in the plane to another the inclination θ of l to Ox changes, say from θ_1 to θ_2. The lamina is said to turn through an angle $\theta_2 - \theta_1$, or to suffer a rotation $\theta_2 - \theta_1$. The first thing to observe is that this angle $\theta_2 - \theta_1$ is independent both of the choice of the line fixed in the lamina, and of the axes Ox, Oy fixed in the plane. This is evident, because the choice of a different line or of different axes merely replaces θ by $\theta + \alpha$, where α is a constant. And in the same way, when we consider a continuous motion of the lamina, the angular velocity $\dot{\theta}$ or ω is independent of the axes chosen, and of the line l used to measure the orientation relative to the axes.

18·2 Any displacement is a rotation. We begin with an account of a famous theorem: any displacement of the lamina in the plane can be achieved by a pure rotation. In other words, if we consider any two configurations of the lamina in the plane, there is always one molecule of the lamina whose position in the plane is the same in both configurations. We could, as it were, pin this molecule in the position it occupies in the first configuration, and reach the second configuration by turning the lamina about the pin. Of course, if the lamina is of finite extent, the fixed point might lie outside the lamina; we should then have to think of it as a point rigidly connected with the lamina; but it is simpler in this work to think of the lamina as covering the whole plane. From another point of view, important in topology, we may express the theorem by the statement that the transformation defined by a rigid displacement of the plane has a fixed point.

The reader will have observed that an exception occurs if the rotation in the displacement is zero, since in that case each molecule of the lamina suffers the same displacement. We must suppose therefore in proving the theorem that the rotation has not the particular value zero. (We can avoid the occurrence of an exceptional case, if preferred, by means of the convention that, when there is no rotation, the fixed point is at infinity in the direction perpendicular to the displacement.) Many proofs of the theorem are known; we content ourselves with an account of two of these.

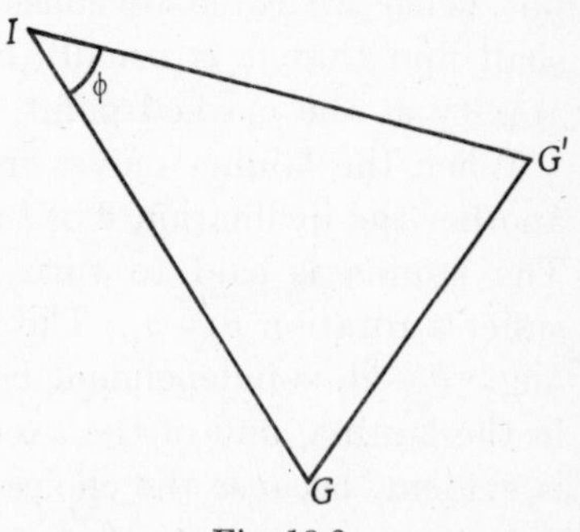

Fig. 18·2*a*

(i) Suppose that in the displacement a molecule G of the lamina moves to G', and the lamina turns in the positive sense through an angle ϕ. Construct the isosceles triangle IGG' whose vertical angle $G'IG$ is ϕ, and such that the rotation from IG to IG' is ϕ (Fig. 18·2*a*). Then I is a fixed point—i.e. the molecule of the lamina which is at I in the first configuration is also at I in the second.

The proof is simple. Since G moves to G', and the lamina turns through an angle ϕ, the line IG of the lamina moves to IG'; and since $IG = IG'$, I is occupied by the same molecule of the lamina in both configurations.

(ii) A molecule G of the lamina moves to G', and a molecule H moves to H'. Let GH meet $G'H'$ in M. The circles MGG', MHH' meet again in I (Fig. 18·2*b*). Then I is a fixed point.

For the triangles IGH, $IG'H'$ are obviously congruent, so the same molecule of the lamina is at I in both configurations.

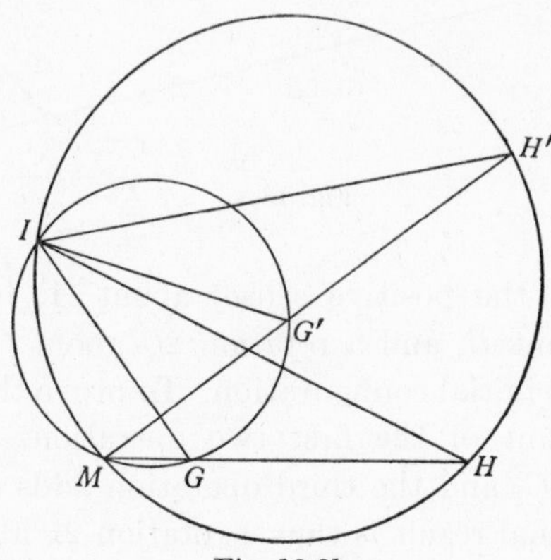

Fig. 18·2*b*

18·3 Resultant of two displacements.

We have seen that any displacement of the lamina can be thought of as a rotation. Let A and B be two points fixed in the plane, and let the first displacement be represented by a rotation through an angle α about A, and the second as a rotation through an angle β about B. The result of the two operations is a displacement of the lamina from its initial configuration, which must be representable by a rotation through an angle γ about some point C. The problem is to determine γ and C.

The answer to the first question is easy, for $\gamma = \alpha + \beta$. To find C we construct the triangle ABC as shown in Fig. 18·3, where the angle BAC is $\frac{1}{2}\alpha$, and the angle CBA is $\frac{1}{2}\beta$. The first operation brings the molecule of the lamina originally at C to C', where C' is the mirror image of C in AB, and the second operation brings it back to C. Thus C is the required fixed point in the resultant displacement.

The result just proved can be exhibited in a more picturesque form. Let A, B, C be three points of the plane, such that the circuit $ABCA$ is in the negative (clockwise) sense. Then a

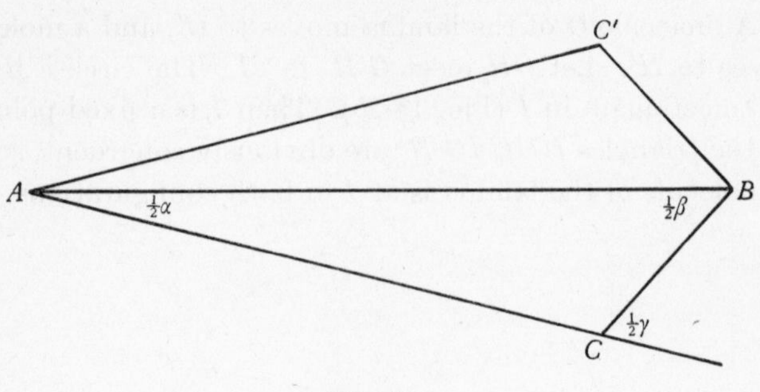

Fig. 18·3

rotation $2\hat{A}$ (in the positive sense) about A, followed by a rotation $2\hat{B}$ about B, and a rotation $2\hat{C}$ about C, will restore the lamina to its initial configuration. To prove this we observe that the resultant of the first two operations is a rotation $2(\hat{A}+\hat{B})$ about C, and the third operation adds a rotation $2\hat{C}$ about C. The final result is thus a rotation 2π about C, which proves the theorem.

18·4 Continuous motion, the velocity distribution in the lamina. We now consider, instead of a finite displacement of the lamina, a continuous motion. The coordinates ξ, η, θ vary with the time; they are always continuous functions of t, and for the present we shall suppose the motion such that ξ, η, θ possess continuous second derivatives. (Later on we shall consider problems involving impulses, in which, for exceptional values of t, the derivatives do not exist. But we set aside this case for the moment.)

The motion of the lamina is defined by the velocity $\dot{\mathbf{r}}_0$, or $\mathbf{V}$, of the marked point G, and the angular velocity $\dot{\theta}$, or ω. The value of ω, at any instant, is independent of the line l fixed in the lamina whose inclination to Ox determines the orientation. Any line fixed in the lamina has the same angular velocity ω.

The last remark is relevant to the problem of determining the velocity-distribution in the lamina, i.e. the determination of the velocity of any molecule P of the lamina in terms of $\mathbf{V}$ and ω. For the velocity of P is the vector sum of the velocity of G and the velocity of P relative to G (in the sense explained in § 10·5), and this relative velocity is a velocity $\rho\omega$ perpendicular to GP, where ρ is the length GP. The building up of the velocity of P as the vector sum of these two vectors is illustrated in Fig. 18·4a.

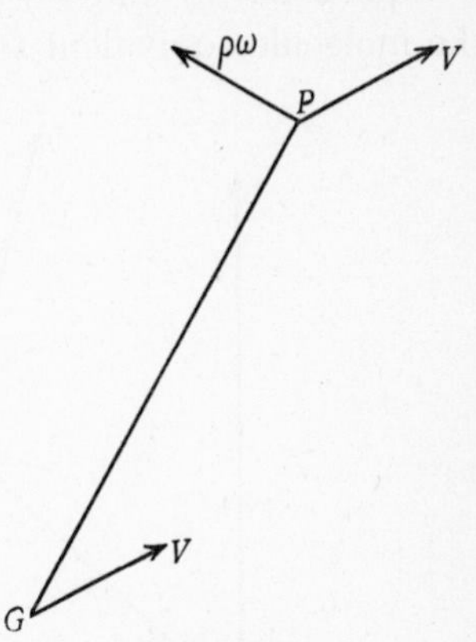

Fig. 18·4a

It is important to consider the same problem—the determination of the velocity-distribution—in another way. If $\mathbf{r}_1$ is the position-vector of a molecule P of the lamina (referred to the fixed axes in the plane) we have

$$\mathbf{r}_1 = \mathbf{r}_0 + \rho\mathbf{E}, \tag{1}$$

where $\mathbf{E}$ is a unit vector in the direction GP. Now GP is a line fixed in the lamina, so its angular velocity is ω, and (equation (1) of § 10·1),

$$\dot{\mathbf{E}} = \mathbf{E}'\omega. \tag{2}$$

Thus
$$\dot{\mathbf{r}}_1 = \dot{\mathbf{r}}_0 + \rho\omega\mathbf{E}' = \mathbf{V} + \rho\omega\mathbf{E}', \tag{3}$$

and this is precisely the result illustrated in Fig. 18·4a.

There are other important forms of equation (3). Let Gx', Gy' be rectangular axes fixed in the lamina, and let (x', y') be the coordinates of the molecule P relative to these axes (Fig. 18·4b). Let us denote by $\mathbf{A}, \mathbf{B}$ $(= \mathbf{A}')$ unit vectors in the directions Gx', Gy' so that $\dot{\mathbf{A}} = \omega\mathbf{B}, \dot{\mathbf{B}} = -\omega\mathbf{A}$. Then

$$\mathbf{r}_1 = \mathbf{r}_0 + x'\mathbf{A} + y'\mathbf{B}, \tag{4}$$

whence
$$\dot{\mathbf{r}}_1 = \dot{\mathbf{r}}_0 - \omega y'\mathbf{A} + \omega x'\mathbf{B}. \tag{5}$$

Finally, write
$$\dot{\mathbf{r}}_0 = \mathbf{A}u + \mathbf{B}v,$$

so that (u, v) are the components in the directions Gx', Gy' of the velocity of G in the plane. Equation (5) now becomes

$$\dot{\mathbf{r}}_1 = (u - \omega y')\mathbf{A} + (v + \omega x')\mathbf{B}. \tag{6}$$

The components of the velocity of P in the directions Gx', Gy' are

$$u - \omega y', \quad v + \omega x'. \tag{7}$$

Equations (5) and (6) give other forms for the velocity of the molecule equivalent to that given in (3).

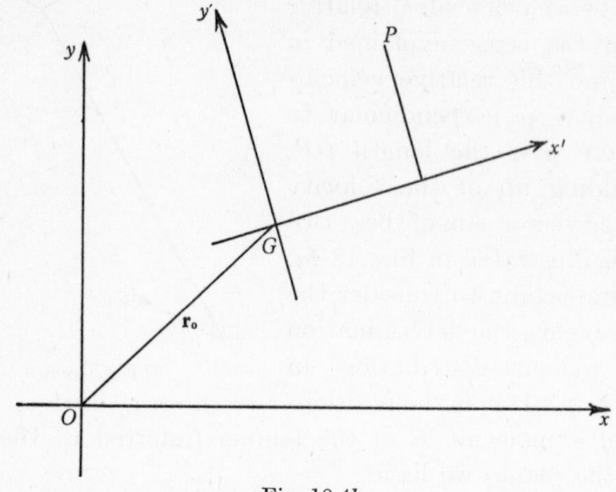

Fig. 18·4b

18·5 The first wandering point theorem. We are now in a position to prove an important and elegant theorem. We consider a point P, not fixed in the lamina, but moving in an arbitrary way. The theorem exhibits a simple but deep-seated relation between the velocity of the point P in the plane and its velocity relative to the lamina, i.e. its velocity as estimated by an observer who takes the lamina itself as his base of reference.

We have already met the notion of relative velocity. For example, in § 10·5, we spoke of 'the velocity of B relative to A', a phrase which involves the idea of motion relative to a new base of reference, the new base being in motion relative to the original base *but without rotation*. In the theorem we now consider the idea of relative motion is less simple, since this time the new base of reference is rotating relative to the old.

If $\mathbf{r}$ is the position vector of P measured in the axes Ox, Oy fixed in the plane we have, with the notation of the preceding paragraph,

$$\mathbf{r} = \mathbf{r}_0 + x'\mathbf{A} + y'\mathbf{B}, \tag{1}$$

which is reminiscent of equation (4) of § 18·4, but differs from it since x', y' are not now constants. Differentiating with respect to t we have

$$\dot{\mathbf{r}} = (\dot{\mathbf{r}}_0 - \omega y'\mathbf{A} + \omega x'\mathbf{B}) + (\dot{x}'\mathbf{A} + \dot{y}'\mathbf{B}). \tag{2}$$

Now the first bracket in the second member is $\dot{\mathbf{r}}_1$, the velocity of the molecule of the lamina at P (equation (5) of § 18·4), and the second bracket is $\mathbf{v}_r$, the velocity of P relative to the lamina. Thus

$$\dot{\mathbf{r}} = \dot{\mathbf{r}}_1 + \mathbf{v}_r,$$

and this is the result we wish to establish. *The velocity of the wandering point P is the vector sum of the velocity of the molecule of the lamina at P and the velocity of P relative to the lamina.*

If we write $\dot{\mathbf{r}}_0 = \mathbf{A}u + \mathbf{B}v$ as in the last paragraph, equation (2) becomes

$$\dot{\mathbf{r}} = \mathbf{A}(u + \dot{x}' - y'\omega) + \mathbf{B}(v + \dot{y}' + x'\omega).$$

This tells us that the components of the velocity of P in the directions Gx', Gy' are

$$u + \dot{x}' - y'\omega, \quad v + \dot{y}' + x'\omega. \tag{3}$$

This is a fundamental result in the theory of moving axes; but in this book we shall use moving axes only occasionally.

18·6 The instantaneous centre. We return now to the discussion of the velocity-distribution in the lamina. It is easy to see that, at each instant, there is precisely one molecule of the lamina which is instantaneously at rest; this is the analogue for continuous motion of the theorem of § 18·2. If we take GP perpendicular to $\mathbf{V}$, and $\rho = V/\omega$, the velocity of the molecule at P is zero (Fig. 18·6a). The point occupied by the molecule P which is instantaneously at rest is the instantaneous centre I. (There is again the exceptional case when $\omega = 0$; and again we may avoid the exception by saying that in this case the instantaneous centre is at infinity.)

We can prove the theorem in other ways, e.g. from the equation (3) or from the formulae (7) of § 18·4. From equation (3) we see that $\dot{\mathbf{r}}_1$ vanishes if, and only if, $\mathbf{E}$ is in the direction of $\mathbf{V}'$ and $\rho\omega = V$. From the formulae (7) we see that the co-ordinates of the instantaneous centre with respect to the axes Gx', Gy' are $(-v/\omega, u/\omega)$.

As an illustration, if a wheel rolls without skidding on a level road, the instantaneous centre is at the point of contact of the wheel with the road.

Corollary 1. The description of the velocity-distribution in the lamina at a given instant (§ 18·4) is greatly simplified by use of the instantaneous centre. Let P be any molecule of the lamina, and P_1 the molecule of the lamina at I. The velocity

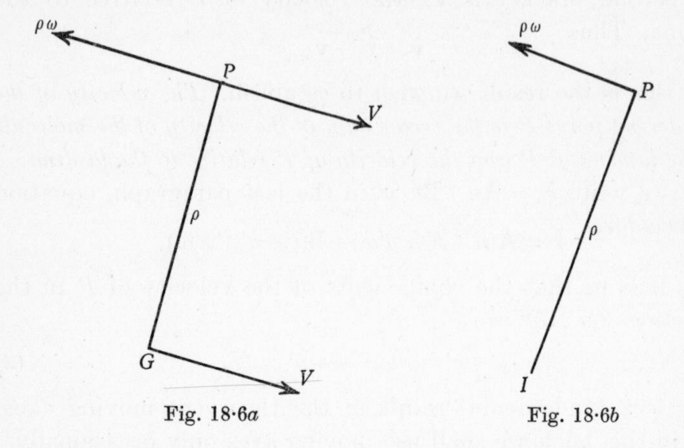

Fig. 18·6*a* Fig. 18·6*b*

of P is the vector sum of the velocity of P_1 and the velocity of P relative to P_1: the first of these is zero, and the second is a vector of magnitude $\rho\omega$ perpendicular to IP. Thus Fig. 18·4*a*, illustrating the velocity-distribution, may be replaced by the simpler Fig. 18·6*b*.

Corollary 2. The position of I is immediately determined if we know the directions of motion of two molecules A and B of the lamina. For the lines drawn through A and B perpendicular to the respective velocities intersect in I.

18·7 The centrodes. The point I has a locus in the plane, called the space centrode, and a locus in the lamina, called the body centrode. For example, in the simple case of the rolling wheel just mentioned, the space centrode is the straight line on which the wheel rolls, and the body centrode is the rim of the wheel.

We are now in a position to prove the fundamental theorem in the theory of the motion of a lamina. *The body centrode rolls, without slipping, on the space centrode, with angular velocity* ω. The most general motion of the lamina can be generated by the rolling of a curve fixed in the lamina on a curve fixed in the plane.

The proof follows immediately from the theorem of the wandering point (§ 18·5). If we apply the theorem to the point I we have

$$\dot{\mathbf{r}} = \mathbf{v}_r \tag{1}$$

since, by definition of I, $\dot{\mathbf{r}}_1 = 0$. Now $\dot{\mathbf{r}} = \dot{s}_0 \mathbf{T}_0$, where s_0 is the arc-length of the space centrode, and $\mathbf{T}_0$ is a unit vector in the direction of the tangent to the space centrode; and similarly $\mathbf{v}_r = \dot{s}_1 \mathbf{T}_1$, where s_1 is the arc-length of the body centrode, and $\mathbf{T}_1$ a unit vector along the tangent. Thus, from equation (1),

$$\dot{s}_0 = \dot{s}_1, \quad \mathbf{T}_0 = \mathbf{T}_1. \tag{2}$$

At each instant the body centrode touches the space centrode, and in any interval of time the corresponding arc-lengths of the two curves are equal. This means that the body centrode rolls, without slipping, on the space centrode, and this is the result we set out to establish.

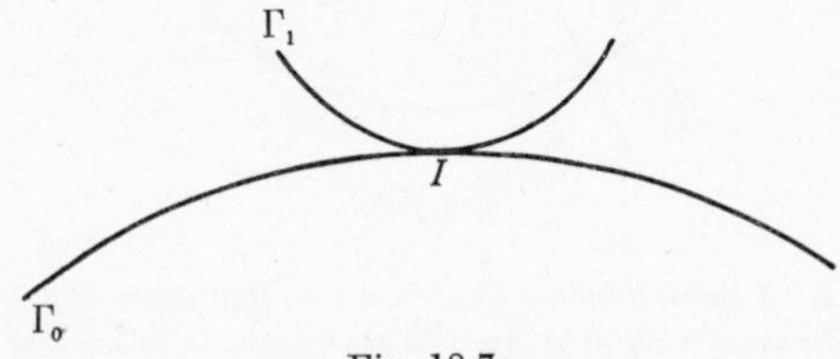

Fig. 18·7a

The standard figure is conventionally drawn as in Fig. 18·7a, with the common tangent at I horizontal, the space centrode Γ_0 being concave downwards, and the body centrode Γ_1 concave upwards; and ω is taken positive, so that I is moving to the left in the figure. But we must be prepared, in practice, for other arrangements of the curvatures, and for negative values of ω.

Example 1. *A lamina moves in such a way that points A and B fixed in the lamina move on lines a and b fixed in the plane; to determine the centrodes.*

From Corollary 2 of § 18·6 the point I is the intersection of the line through A perpendicular to a and the line through B perpendicular to b (Fig. 18·7b).

Denote the fixed length AB by k, the fixed angle (say the acute angle) between a and b by α, and the intersection of a and b by O. Then OI is a diameter of the circumcircle of the triangle OAB, and the radius R of this circle is

$$R = k/(2\sin\alpha),$$

and is constant. Thus the space centrode is the circle of radius $2R$ with O as centre; and the body centrode is a circle of radius R through A and B. The motion of the lamina is produced by the rolling of a circle of radius R fixed in the lamina on the inside of a circle of radius $2R$ fixed in the plane. We could almost have foreseen this result from the well-known theorem that when a circle of radius R rolls on the inside of a fixed circle of radius $2R$, the locus of any point on the rolling circle is a straight line.

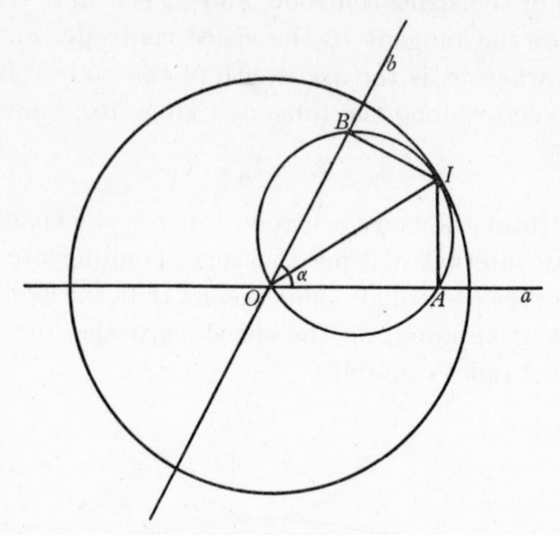

Fig. 18·7*b*

Example 2. *A lamina moves in such a way that a line a fixed in the lamina always passes through a point A fixed in the plane; to prove that at each instant the instantaneous centre I lies on the line through A perpendicular to a.*

To prove this we have only to observe that the velocity of the particle of the lamina at A is in the direction of a. Or use Corollary 2 of § 18·6, thinking of the motion of the 'fixed' plane relative to the lamina.

18·8 Velocity of *I*. We now establish a formula for the velocity of I on the space centrode, $\dot{s}_0$, in terms of the curvatures of the centrodes and the angular velocity of the lamina. As we have seen (equation (2) of § 18·7), $\dot{s}_1 = \dot{s}_0$, so the same formula also determines the velocity of I on the body centrode, as measured by an observer who takes the lamina as his base of reference. (The reader should take particular care at this point to distinguish between the velocity of I under discussion, and the velocity of the molecule of the lamina at I; the latter is of course zero.)

Fig. 18·8a shows the configuration at time t. At time $t+\tau$ the instantaneous centre will have moved to the point M on the space centrode Γ_0, and the point N on the body centrode Γ_1 will have moved into coincidence with M. The arc-lengths

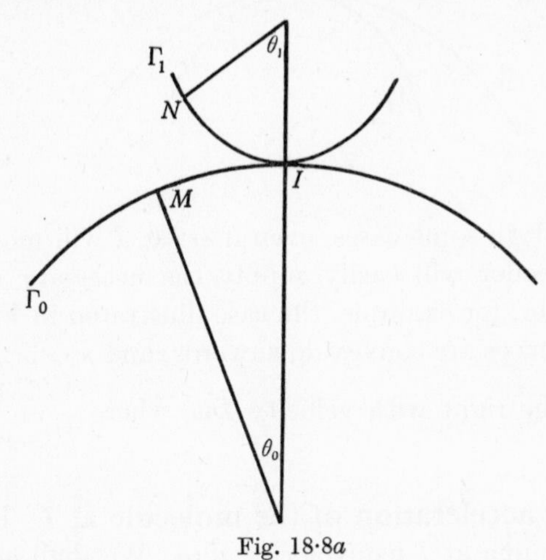

Fig. 18·8a

IM and IN are equal, say each is σ. The normals to the centrodes at M and N make angles θ_0 and θ_1 with the common normal at I. The velocity $\dot{s}_0$ that we wish to calculate is $\lim \sigma/\tau$.

Now
$$\frac{\sigma}{\tau}=\frac{\dfrac{\theta_0+\theta_1}{\tau}}{\dfrac{\theta_0+\theta_1}{\sigma}}. \tag{1}$$

Since the lamina turns through an angle $\theta_0+\theta_1$ in the interval from t to $t+\tau$, the numerator tends to ω as $\tau\to 0$, while the denominator tends to $1/\rho_0+1/\rho_1$, where ρ_0, ρ_1 are the radii of curvature of the centrodes at I. If we write

$$\frac{1}{D}=\frac{1}{\rho_0}+\frac{1}{\rho_1},$$

equation (1) leads, on letting $\tau\to 0$, to the equation

$$\dot{s}_0 = D\omega. \tag{2}$$

This is the result we set out to prove.

It is particularly to be observed that the formulae of this paragraph refer to the standard figure (Fig. 18·8*a*). In a problem in which the curvature of either centrode is in the sense opposite to that in the standard figure the formulae will need

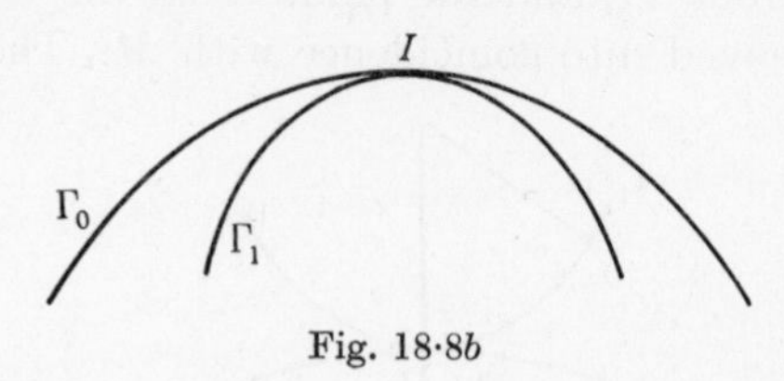

Fig. 18·8*b*

to be modified; in some cases, even if $\omega > 0$, I will move to the right. The reader will easily supply the necessary modifications. Consider, for example, the case illustrated in Fig. 18·8*b*, where both curves are convex downwards, and $\rho_0 > \rho_1$. If $\omega > 0$, I moves to the right with velocity $D\omega$, where $\dfrac{1}{D} = \dfrac{1}{\rho_1} - \dfrac{1}{\rho_0}$.

18·9 The acceleration of the molecule at *I*. The molecule of the lamina at I has velocity zero. We shall now prove that its acceleration is $D\omega^2$ normal to the centrodes (vertically upwards in the standard figure).

In general the molecule of the lamina at I is at a cusp of its path, since the molecule is at a point fixed on the body centrode, and this curve rolls on the space centrode. The tangent at the cusp is normal to the centrodes, and this defines the direction of the acceleration; for when a particle moves from rest at P the tangent to its path at P is in the direction of its acceleration.

The magnitude of the acceleration is $\lim v/\tau$, where v is the velocity of the molecule at time $t+\tau$; now, with the notation of the preceding paragraph, $v = h\omega'$, where h is the chord NI of the body centrode, and ω' the angular velocity of the lamina at time $t+\tau$. Thus the acceleration of the molecule at I at time t is

$$\lim_{\tau\to 0} \frac{h\omega'}{\tau} = \lim \frac{h}{\sigma}\frac{\sigma}{\tau}\omega' = D\omega^2,$$

since $h/\sigma \to 1$, $\sigma/\tau \to D\omega$, and $\omega' \to \omega$, as $\tau \to 0$ (§ 18·8).

18·10 The acceleration distribution in the lamina, the inflexion circle. In § 18·4 we discussed the velocity distribution in the lamina, i.e. we found the velocity of each molecule of the lamina in terms of the velocity of a marked point and the angular velocity of the lamina. As we have remarked already (§ 18·6, Cor. 1), the velocity distribution at a particular instant is most simply described by taking as the marked point the molecule of the lamina at I.

We now consider the acceleration distribution, i.e. we find the acceleration of any molecule of the lamina in terms of the

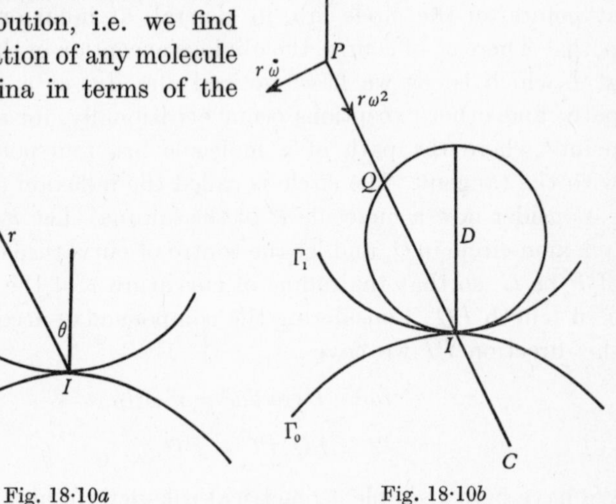

Fig. 18·10*a*

Fig. 18·10*b*

acceleration of a marked point, and the angular velocity and angular acceleration of the lamina. Any molecule of the lamina may be used as the marked point; we may as well use the molecule at I, and we then obtain, for the velocity and acceleration distributions, the results shown in Figs. 18·10*a* and 18·10*b*; (r, θ) are polar coordinates of the molecule P of the lamina, using I as origin, and the normal to the centrodes as initial line. The velocity of the molecule P is $r\omega$ perpendicular to IP. The acceleration of the molecule P is the vector sum of the acceleration of the molecule of the lamina at I (which is $D\omega^2$ in the direction normal to the centrodes) and the relative acceleration; the last has components $r\dot{\omega}$ perpendicular to IP, and $r\omega^2$ along PI. We thus obtain the acceleration distribution shown in Fig. 18·10*b*.

The component of acceleration along PI is zero if

$$r = D\cos\theta,$$

i.e. if P lies on a circle of diameter D touching the centrodes at I. The molecules of the lamina at points on this circle are at points of their paths where the acceleration is in the same direction as the velocity; the normal component of acceleration is zero. Now this normal component is $v^2\kappa$, where κ is the curvature of the path, so $\kappa = 0$. The molecules of the lamina at points on the circle are, in general, at inflexions on their paths. There is, of course, the obvious exception of the molecule at I which is, as we have noticed already, at a cusp of its path; and other exceptions occur occasionally, for example at points where the path of a molecule has four-point contact with the tangent. The circle is called the inflexion circle.

Consider now a molecule P of the lamina. Let PI meet the inflexion circle in Q, and let the centre of curvature of the path of P be C, so that the radius of curvature ρ of the path of P is of length PC. Considering the component of acceleration in the direction PI we have

$$r\omega^2 - D\cos\theta\omega^2 = r^2\omega^2/\rho,$$

i.e.
$$PQ.PC = PI^2. \tag{1}$$

We have here a simple geometrical rule determining the centre of curvature of the path of any molecule of the lamina; Fig. 18·10*b* is drawn for the case where P is outside the inflexion circle, but it is easy to verify that the rule is valid in all cases.

We notice that the component of acceleration of P along the normal to its path, i.e. in the direction PI, is towards I if P lies outside the inflexion circle and away from I if P lies inside the inflexion circle. It follows that the paths of molecules outside the inflexion circle are concave to I, and the paths of molecules inside the inflexion circle are convex to I.

18·11 A stability problem. We can now solve very simply the problem of the stability when a cylinder of any shape rests on a fixed rough cylinder of any shape (Fig. 18·11). In the position of equilibrium the centre of gravity G of the rolling cylinder is vertically above the point of contact I. Now if we consider a displacement of the rolling cylinder we have a motion in which the outlines of the cylinders are themselves the centrodes, and there is stability if the path of G is convex to I. This is true if G lies inside the inflexion circle, i.e. if the height of G above I is less than $D\cos\theta$. If G lies outside the inflexion circle, there is instability. The critical case, when G lies *on* the circle, is also unstable in general, since G then lies at an inflexion of its path.

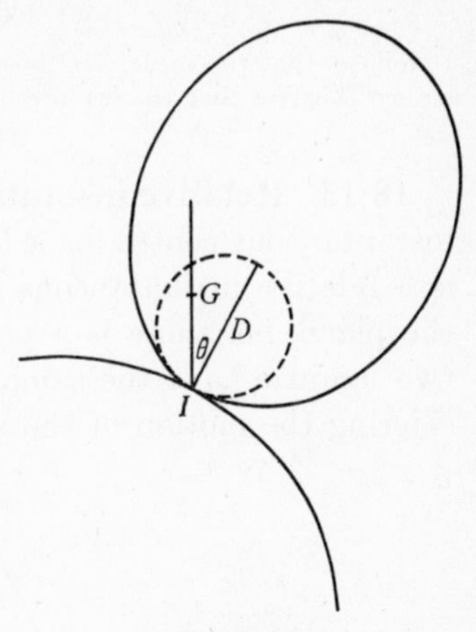

Fig. 18·11

18·12 The molecule of zero acceleration.

If the molecule of the lamina P lies on the inflexion circle, the component of its acceleration along PI is zero. There is just one molecule J for which the component of acceleration perpendicular to PI is also zero, i.e. *there is one molecule J of the lamina with zero acceleration.* For J the three vectors marked in Fig. 18·10b have a zero sum, and can be represented by the sides of a triangle, whence

$$\tan\theta = \dot{\omega}/\omega^2.$$

This equation gives a unique value of θ in $(-\frac{1}{2}\pi, \frac{1}{2}\pi)$.

Example. *A wheel of radius a rolls with uniform angular velocity on a straight line.*

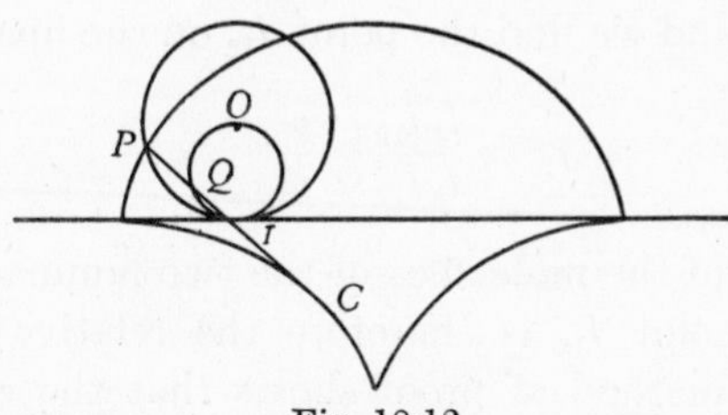

Fig. 18·12

The rim of the wheel is the body centrode, the straight line is the space centrode. Moreover, $D = a$, and the inflexion circle is the circle on OI as diameter, where O is the centre of the wheel (Fig. 18·12). Since $\dot{\omega} = 0$, $J \equiv O$, as is evident.

Consider now a molecule P on the rim of the wheel; its path is a cycloid. The radius of curvature PC of the path of P is given by equation (1) of § 18·10

$$PQ.PC = PI^2.$$

Since $PQ = \frac{1}{2}PI$, this gives us

$$PC = 2PI.$$

It follows that the locus of C is an equal cycloid; we thus recover the well-known theorem that the evolute of a cycloid is an equal cycloid.

18·13 Relative instantaneous centre. Just as there is an instantaneous centre for a lamina moving in a plane, so there is a relative instantaneous centre when two laminae move in the plane, i.e. there is a point at which the molecules of the two laminae have the same velocity. This is evident by considering the motion of the second lamina relative to the first.

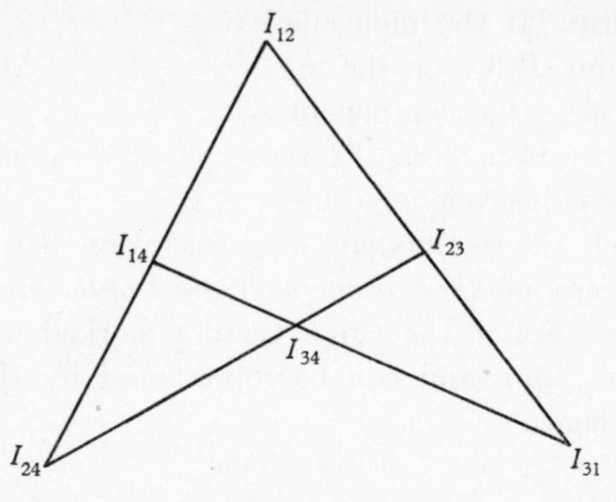

Fig. 18·13

Alternatively, if the instantaneous centres of the two laminae are I_1 and I_2, and we find the point I_{12} on the line $I_1 I_2$ such that

$$\frac{I_1 I_{12}}{I_2 I_{12}} = \frac{\omega_2}{\omega_1},$$

the velocities of the molecules of the two laminae at I_{12} are the same. This point I_{12} is therefore the relative instantaneous centre. This method of proof shows that the relative instantaneous centre is collinear with the instantaneous centres of the two laminae.

If three laminae move in the plane the three relative instantaneous centres for pairs of laminae are collinear. This we can see very simply by considering the motion of the second and third laminae relative to the first, and using the result just proved.

Finally, we observe that if four laminae move in a plane the six relative instantaneous centres for pairs of laminae lie at the vertices of a complete quadrilateral (Fig. 18·13).

18·14 The second wandering point theorem, the theorem of Coriolis. In § 18·5 we considered the velocity of a wandering point P; we now consider its acceleration. As in § 18·5, using the same notation, we have

$$\mathbf{r} = \mathbf{r}_0 + x'\mathbf{A} + y'\mathbf{B}, \tag{1}$$

$$\dot{\mathbf{r}} = \dot{\mathbf{r}}_0 - \omega y'\mathbf{A} + \omega x'\mathbf{B} + \dot{x}'\mathbf{A} + \dot{y}'\mathbf{B}. \tag{2}$$

We now differentiate again with respect to t, and we find

$$\ddot{\mathbf{r}} = (\ddot{\mathbf{r}}_0 - \omega^2 x'\mathbf{A} - \dot{\omega}y'\mathbf{A} - \omega^2 y'\mathbf{B} + \dot{\omega}x'\mathbf{B}) + 2\omega(-\dot{y}'\mathbf{A} + \dot{x}'\mathbf{B}) + (\ddot{x}'\mathbf{A} + \ddot{y}'\mathbf{B}). \tag{3}$$

Now the first bracket on the right is the acceleration of the molecule of the lamina at P, and the last bracket is the acceleration of P relative to the lamina, i.e. its acceleration as estimated by an observer who takes the lamina as his base of reference. There remains the extra term $2\omega\mathbf{v}_r'$, where $\mathbf{v}_r$ denotes the velocity of P relative to the lamina, and the accent indicates as usual rotation through a right angle. If we use the notation of the vector product (§ 13·5) we can also write the extra term in the form $2\boldsymbol{\omega}\times\mathbf{v}_r$, where $\boldsymbol{\omega}$ is a vector of magnitude ω in the direction Oz.

The first wandering point theorem was concerned with the velocity of P, and we find that its analogue for acceleration is not true. The acceleration of P is not merely the vector sum of the acceleration of the molecule and the acceleration relative to the lamina; to these we must add a third term $2\omega\mathbf{v}_r'$. This is the theorem of Coriolis for the case of motion in a plane.

Finally, we can find the formulae for acceleration analogous to the formulae for velocity previously found in § 18·5 (formulae (3)). The components of the velocity of P in the directions Gx', Gy' are

$$u + \dot{x}' - \omega y', \quad v + \dot{y}' + \omega x'.$$

If now we denote the components of the acceleration of G in the directions Gx', Gy' by (f, g) we find, from equation (3), the components of the acceleration of P in these directions,

$$f + \ddot{x}' - 2\omega\dot{y}' - \omega^2 x' - \dot{\omega}y', \quad g + \ddot{y}' + 2\omega\dot{x}' - \omega^2 y' + \dot{\omega}x'.$$

These formulae are fundamental in the theory of moving axes; but the use of moving axes is seldom necessary in two-dimensional work, and in any case is rather outside the scope of this book. (Some further comments on the subject of moving axes will be found in § 24·8.)

18·15 Bead on a rotating wire. A bead of mass m slides on a smooth plane rigid wire, which rotates in its plane, with uniform angular velocity ω, about a point O rigidly connected with the wire. We wish to discuss the motion of the bead relative to the wire.

We shall find that we can take account completely of the rotation of the wire by adding the centrifugal force $mr\omega^2$ outwards from O. In other words, the motion relative to the wire is the same as if the wire were at rest and the forces acting on the bead were augmented by the centrifugal force. In particular, if there are no forces (other than the reaction of the wire) we can find the motion relative to the wire from the equation of energy (with $V = -\frac{1}{2}m\omega^2 r^2$) for motion on the fixed wire.

The result is a simple corollary to the theorem of Coriolis. We can think of the wire as fixed in a lamina rotating about O. The acceleration of P is compounded of its acceleration relative to the lamina, the acceleration of the molecule of the wire at P ($r\omega^2$ inwards), and the additional term. But this additional term is normal to the wire, and does not influence the motion (though it does affect the reaction). The result follows.

It is easy and entertaining to establish the result again by more elementary methods. With the notation of the Fig. 81·15a, in which A is a fixed point of the wire, the radial and transverse components of acceleration are

$$\begin{cases} \ddot{r}-r(\dot{\theta}+\dot{\phi})^2, \\ r(\ddot{\theta}+\ddot{\phi})+2\dot{r}(\dot{\theta}+\dot{\phi}), \end{cases}$$

and, since $\dot{\phi} = \omega$, these can be written

$$\begin{cases} (\ddot{r}-r\dot{\theta}^2)-2\omega r\dot{\theta}-r\omega^2, \\ (r\ddot{\theta}+2\dot{r}\dot{\theta})+2\omega\dot{r}. \end{cases}$$

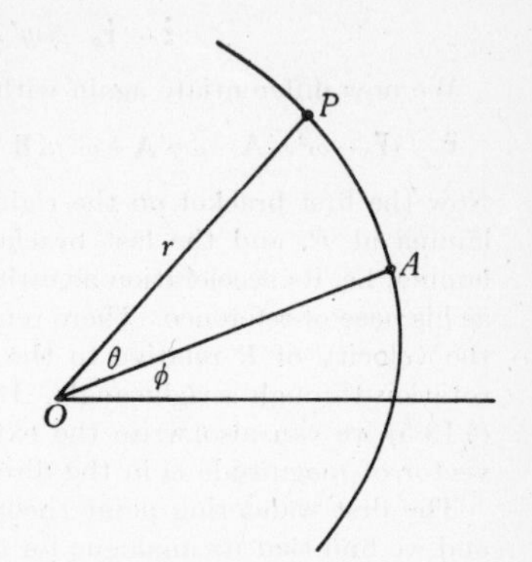

Fig. 18·15a

The equations of motion are therefore

$$\begin{cases} \ddot{r}-r\dot{\theta}^2 = R+2\omega r\dot{\theta}+r\omega^2, & (1) \\ r\ddot{\theta}+2\dot{r}\dot{\theta} = S-2\omega\dot{r}, & (2) \end{cases}$$

where (mR, mS) are the radial and transverse components of the reaction of the wire on the bead. Now this reaction is normal to the wire, and the terms $(2\omega r\dot{\theta}, -2\omega\dot{r})$ also represent a vector normal to the wire, so only the last term in the first equation affects the relative motion. This is the result already mentioned.

Now the radial force $r\omega^2$ is derivable from the potential function $-\frac{1}{2}r^2\omega^2$. Thus, for the relative motion we have an integral of energy

$$\tfrac{1}{2}v^2-\tfrac{1}{2}r^2\omega^2 = C, \tag{3}$$

where v denotes the velocity of the bead relative to the wire. Alternatively, we can establish the equation (3) from (1) and (2); if we multiply (1) by $\dot{r}$ and (2) by $r\dot{\theta}$ and add, remembering $R\dot{r}+Sr\dot{\theta} = 0$, we recover (3).

Example. *A bead slides on a smooth circular wire which can rotate about a point O of itself. The system is initially at rest with the bead at the other end of the diameter through O. At the instant $t = 0$ the wire starts to rotate about O with uniform angular velocity ω. Find the position of the bead on the wire at any subsequent time.*

With the notation of Fig. 18·15b we have

$$2a^2\dot{\theta}^2 - 2a^2\cos^2\theta\omega^2 = C,$$

and since $\dot{\theta} = \omega$ when $\theta = 0$ the value of the constant is zero, and

$$\dot{\theta} = \omega\cos\theta.$$

We have already encountered a similar relation in § 11·10 (iv), and, as there, the relation between θ and t is

$$\sin\theta = \tanh\omega t.$$

We see that $\theta \to \frac{1}{2}\pi$ as $t \to \infty$; the bead never reaches O.

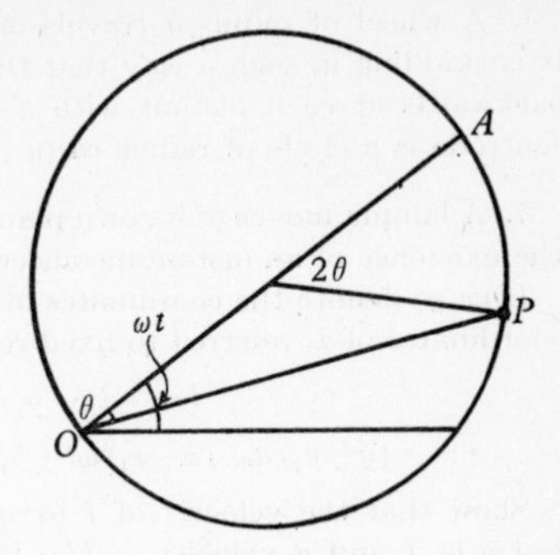

Fig. 18·15b

EXAMPLES XVIII

1. Prove that in general any finite displacement of a lamina in its plane can be obtained by a pure rotation of the lamina about a point I.

If a particle of the lamina moves from A to A', and the lamina rotates through an angle α, prove that I is the vertex of an isosceles triangle IAA' whose vertical angle is α.

A circular disc rolls without slipping on the outside of an equal circular disc which is fixed. If the centre of the rolling disc moves from A to A', prove that I is the circumcentre of the triangle OAA', where O is the centre of the fixed disc.

2. Two maps of the same district on the same scale are placed at random, face upwards one over the other, on a table. Prove that there is one point of the table at which the same place is represented on both maps.

Prove that the same is true if the maps are on different scales.

3. A lamina moves in a plane. Prove that the motion can be generated by the rolling of a curve fixed in the lamina (the body centrode) on a curve fixed in the plane (the space centrode).

A is a point fixed in the lamina, and l is a line fixed in the lamina and passing through A. The lamina moves so that A always lies on a line λ fixed in the plane, and l always passes through a point O fixed in the plane. Find the centrodes.

4. A circle and a tangent to it are given. A rod moves so that it touches the circle and one end lies always upon the tangent. Prove that both the centrodes are parabolas.

5. A lamina moves in a plane. A point A fixed in the lamina moves on a line Ox fixed in the plane, and a line AB fixed in the lamina always touches a circle, with centre O and radius a, which is fixed in the plane. Prove that the polar equation of the space centrode is

$$r\cos^2\theta = a.$$

Find the polar equation of the body centrode.

6. A wheel of radius a travels on a straight road with constant speed v. It is skidding in such a way that the lowest molecule of the wheel is moving backwards at each instant with a constant speed u. Prove that the body centrode is a circle of radius $va/(u+v)$, and find the space centrode.

7. A lamina moves in its own plane with angular velocity $\omega(\neq 0)$. Establish the existence of an instantaneous centre of rotation I.

If (x, y) denote the coordinates of a molecule of the lamina, and (x_1, y_1) the coordinates of I, referred to fixed rectangular axes, prove that

$$\dot{x} = -(y-y_1)\,\omega, \quad \dot{y} = (x-x_1)\,\omega,$$

$$\ddot{x} = -(y-y_1)\,\dot{\omega}-(x-x_1)\,\omega^2+\dot{y}_1\,\omega, \quad \ddot{y} = (x-x_1)\,\dot{\omega}-(y-y_1)\,\omega^2-\dot{x}_1\,\omega.$$

Show that the velocity of I in space is compounded of a velocity $JI\dot{\omega}/\omega$ towards J and a velocity $-JI\omega$ perpendicular to JI, where J denotes the point of zero acceleration.

8. A cylinder of radius a rolls with its axis horizontal on a perfectly rough plane inclined at an angle α to the horizontal. The cylinder is eccentrically loaded, so that its centre of gravity G is distant r from the axis. Show that, if $r > a \sin \alpha$, equilibrium is possible for two positions of G relative to the axis, and discuss the stability of each position (i) by considering the maxima and minima of the height of G, (ii) from the theory of the inflexion circle. If the cylinder instead of resting on a plane rests on the inside of a rough circular cylinder of radius b ($> a$) at a point where the inclination to the horizon is α, prove that *both* positions of equilibrium are stable if

$$b < a + \frac{a^2 \cos \alpha}{\sqrt{(r^2 - a^2 \sin^2 \alpha)}}.$$

9. Two points P and Q describe elliptic harmonic motion on similar and similarly situated concentric ellipses, with the same period but in opposite directions, crossing the common major axis simultaneously on the same side of the centre. The major and minor axes of the ellipses are of length a, b and $\lambda a, \lambda b$ respectively ($\lambda > 1$). Show that there is a value of λ for which both the distance PQ and the angular velocity of the line PQ remain constant throughout the motion.

Show further that for this value of λ the points of intersection of the line PQ with the axes are at fixed distances from P and Q.

Show that the motion of P is that of a molecule of a circular disc of radius r which rolls with uniform angular velocity, and without slipping, on the inside of a circle of radius $2r$. How is the motion of Q related to the moving lamina?

10. A circular disc of radius a rolls without slipping on a straight line. Find the inflexion circle, and find the position of J on the circle at an instant when the angular velocity of the disc is ω and its angular acceleration is ϕ.

Consider in particular the case in which the disc moves from rest with uniform acceleration. Prove that the locus of J in the plane, referred to Cartesian axes, can be expressed in the form

$$y(3a-2y)^2 = 4x^2(a-y),$$

and that its locus in the lamina, referred to polar coordinates, can be expressed parametrically in the form

$$r = a \cos \psi, \quad \theta = \psi + \tfrac{1}{2} \tan \psi.$$

CHAPTER XIX

FORCES ON A RIGID BODY

19·1 The idea of equivalence, reduction of the system. We begin by recalling the main ideas, familiar in statics, of the theory of a system of forces acting in one plane on a rigid body.

The first point to notice is that we are now dealing with a system of *localized* vector quantities. To describe a force of the system we need to specify, not only its magnitude and direction, but also the point at which it acts. We thus have to meet a new difficulty which did not emerge in the statics and dynamics of a single particle. The technique for dealing with this difficulty will be developed in this paragraph.

The basis of the theory is the notion of *equivalence*. Two different systems, built up from entirely different sets of forces, may yet have the same effect on the body as a whole. The example dealt with most conspicuously in statics is that of a system equivalent to the nul system. Such a system is said to be 'in equilibrium'. All systems in equilibrium are equivalent to one another, and in particular to the nul system. (It is hardly necessary to remind the reader that the notion of equivalence is only concerned with the effect on the body as a whole. If we consider the internal stresses in the body—e.g. the bending moment at a point of a rigid rod—we notice that two equivalent systems may set up entirely different states of internal stress. In particular a system in equilibrium does not in general leave the body with no internal stress.)

The primary object in the statics of a rigid body is the setting up of necessary and sufficient conditions for the equivalence of two systems; including as a special case necessary and sufficient conditions for equilibrium. We recall the theory in its broad outlines.

In general two (coplanar) forces acting on a rigid body are equivalent to a single force. But there is one conspicuous exception. If the two forces act in parallel lines, are equal in magnitude, and in opposite senses, they cannot be replaced by a single force; two such forces form an entity, a *couple*, which cannot be further simplified. We are led to an approach to the subject involving the idea of two basic elements, forces and couples, which leads to a radical simplification of the whole theory.

The three fundamental properties of couples, proofs of which are no doubt already familiar to the reader, are these:

(1) a couple has the same moment about all points of the plane; this invariant is called the moment of the couple;

(2) any two couples of equal moment are equivalent;

(3) two couples acting simultaneously on a rigid body (in the same plane) are equivalent to a single couple, whose moment is the (algebraic) sum of the moments of the individual couples; and more generally a number of couples acting simultaneously on a rigid body (all in the same plane) are equivalent to a single couple whose moment is the algebraic sum of the moments of the individual couples.

The upshot of the last of these properties is that couples acting in the same plane on a rigid body are scalar quantities. As we have already mentioned (§ 13·5), in the more general theory of forces in space couples are found to be vector quantities; in the particular case of a coplanar system couples can masquerade as scalar quantities, because their vectors are all parallel to the same line (§ 1·8).

The introduction of the notion of a couple leads easily to the solution of the problem of the *reduction* of the system. It is the essential tool needed to deal with the difficulty arising from the fact that we are now dealing with localized vectors. For the sake of completeness we give an outline of the procedure.

We choose a point O of the plane as origin. A force $\mathbf{F}$ of the system acting at a point A is equivalent to a force $\mathbf{F}$ *acting at O* together with a couple. The moment of this couple is the moment about O of the force $\mathbf{F}$ acting at A; its magnitude is the directed magnitude of $\mathbf{r}\times\mathbf{F}$ (where $\mathbf{r}$ is the displacement vector of A relative to O); or, more simply, it is $\pm p|\mathbf{F}|$, where p is the perpendicular distance OM of O from the line of action of the force; the sign is + when the twist through a right angle from OM to $\mathbf{F}$ is positive (Fig. 19·1). If the force has components (X, Y) and acts at (x, y), the axes being rectangular, its moment about O can also be expressed as $xY-yX$ (cf. § 13·2).

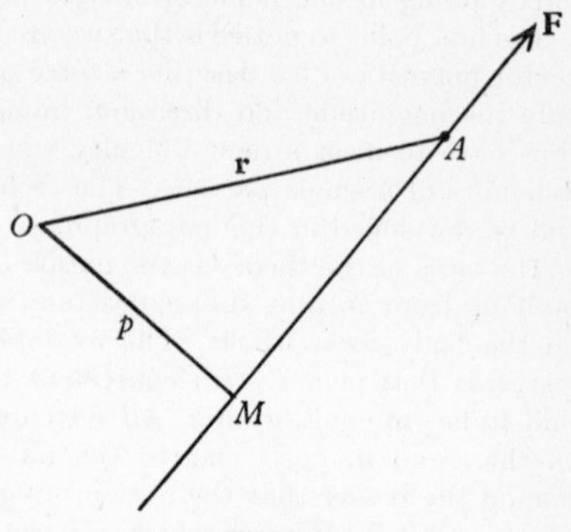

Fig. 19·1

The heart of the matter is that we have replaced the vector quantity *localized at A* by a vector quantity acting at the standard point of reference O, together with a couple.

Proceeding in this way with all the forces of the system, we obtain a number of forces *all acting at the same point O*, and a number of couples. The forces acting at O are equivalent to a single force acting at O, the couples are equivalent to a single couple. The system is thus reduced to a force at O and a couple. The force $\mathbf{R}$ is the vector sum of all the forces of the system, and the moment N of the couple is the moment of the system about O.

If we wish, we can carry the reduction a step further. If $\mathbf{R}\neq\mathbf{0}$ we can replace the force $\mathbf{R}$ at O and the couple by a single force $\mathbf{R}$ acting at such a point that its moment about O is N. The system is thus reduced to a single force. If $\mathbf{R}=\mathbf{0}$ the system is equivalent to a couple. Thus, a coplanar system of forces is equivalent either to a single force or to a couple.

For our present purpose, however, it is only the first step of the reduction that is needed, the replacing of the system by a force at O and a couple.

We can now answer the question proposed, as to necessary and sufficient conditions for the equivalence of two systems of forces. We reduce the first system to a force $\mathbf{R}_1$ at O and a couple of moment N_1, the second to a force $\mathbf{R}_2$ at O and a couple N_2. The two systems are equivalent if and only if

$$\mathbf{R}_1=\mathbf{R}_2, \quad N_1=N_2. \tag{1}$$

Expressed otherwise, the systems are equivalent if the vector sum of the forces is the same in each, and if both systems have the same moment about one point of the plane.

It is easy to express these conditions analytically. We take rectangular axes Ox, Oy through O; we denote by X_1, Y_1 the algebraic sum of the components of the forces of the first system in the directions Ox, Oy, and by N_1 the moment of the first system about O. (As we have seen, the first system is equivalent to a force (X_1, Y_1) at O, together with a couple N_1.) Using the corresponding symbols for the second system, the necessary and sufficient conditions for the equivalence of the two systems are

$$X_1 = X_2, \quad Y_1 = Y_2, \quad N_1 = N_2. \tag{2}$$

19·2 Other localized vector quantities.

A system of forces acting in one plane on a rigid body can be reduced to a force acting at a point O of the plane and a couple. The technique required for this reduction (which has been reviewed in the preceding paragraph) is applicable also to other localized vector quantities; we shall need it altogether four times in this book. The first we have just met. The second occurs when we deal with the *kineta* of the particles of a rigid body. These are localized vector quantities, just like the forces already discussed, and they can be reduced in the same way. We obtain a single kineton at O, which is the vector sum of the kineta of the separate particles, together with a scalar quantity, analogous to the couple obtained above, whose magnitude is the sum of the moments about O of the kineta of the separate particles. We need a nomenclature for the vector quantity and the scalar quantity so obtained, and, by analogy with the original case of a system of forces, we may speak of a $\mathscr{K}$-force at O (or kineton at O) and a $\mathscr{K}$-couple.

The third application will be to a system of impulses acting in one plane on a rigid body. The system can be reduced to an impulse at O and an impulse-couple. In the fourth application the vector quantities are the momenta of the particles. The system is equivalent to a single momentum at O, and a scalar quantity analogous to a couple. Borrowing again a nomenclature from the theory of the reduction of a system of forces, we may say that the system of momenta is equivalent to an $\mathscr{M}$-force (or linear momentum) at O, and an $\mathscr{M}$-couple (or angular momentum).

19·3 Change of origin. We have seen that a given coplanar system of forces is equivalent to a force $\mathbf{R}$ acting at a chosen origin O, together with a couple N. It is vitally important to be able to find the corresponding values $\mathbf{R}', N'$ when we choose a new origin O' (Fig. 19·3). To do this we notice first that $\mathbf{R}$ is invariant, for it is simply the vector sum of all the forces of the system; thus

$$\mathbf{R}' = \mathbf{R}.$$

Then to find N' we have only to observe that N' is the moment of the whole system about O', so

$$N' = N + \Gamma,$$

where Γ is the moment about O' of a force $\mathbf{R}$ acting at O.

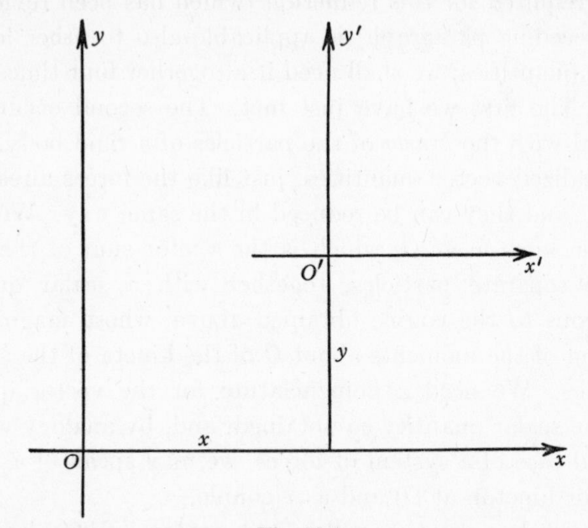

Fig. 19·3

Analytically, suppose the system is equivalent to a force X, Y at O, and a couple N. If we reduce the same system with a new origin O', whose coordinates relative to O are x, y, we have, using axes through O' parallel to the original axes,

$$X' = X, \quad Y' = Y, \quad N' = N - xY + yX. \tag{1}$$

The same law of transformation is of course also applicable to the other systems of localized vector quantities mentioned—to kineta, and impulses, and momenta.

19·4 Work done in a small displacement. The reader will recall that one of the fundamental properties of equivalent systems is that the work done by two equivalent systems in an infinitesimal displacement of the body is the same. Another form of expression of the same theorem is that the power (rate of working) of two equivalent systems in any motion of the body is the same.

Suppose then that a given system of forces acts on the body. We can reduce the system to a force $\mathbf{R}$ at a marked point O of the body, and a couple N. Suppose now that the body suffers an infinitesimal displacement, in which the molecule O suffers a displacement $\delta\mathbf{r}_0$, and the body turns through an angle $\delta\theta$. The work done by the original system is equal to the work done by the reduced system, and this is

$$\mathbf{R}.\delta\mathbf{r}_0 + N\delta\theta. \tag{1}$$

In Cartesian coordinates, denoting the coordinates of the marked point by ξ, η, and the components of $\mathbf{R}$ by X, Y, this formula becomes

$$X\delta\xi + Y\delta\eta + N\delta\theta. \tag{2}$$

An important special case is that in which the original system of forces is in equilibrium. In that case the work done in an arbitrary infinitesimal displacement of the body is zero. The converse also is true: if the work done in an arbitrary displacement is zero, the system of forces is in equilibrium. The theorem and converse constitute the principle of Virtual Work for a single rigid body.

We can express the same theorem in terms of velocities instead of displacements. A necessary and sufficient condition for equilibrium is that the rate of working in an arbitrary motion of the body is zero.

Corollary. A special case of the formulae (1) and (2) is worthy of notice, namely, that in which the marked point is the molecule I of the lamina whose position does not change in the displacement. In that case the formulae reduce to

$$N_1\delta\theta,$$

where N_1 is the moment of the system about I.

19·5 The dynamics of a rigid body. The fundamental principles of mechanics were developed by Newton for the problem of a single particle, and expounded in his famous treatise, the *Principia* (1687). The dynamics of a rigid body was first systematically studied some half-century later, and it was natural to think of the rigid body as an aggregate of particles. We think of the mass of a rigid body, not as distributed continuously through the body, but as concentrated in a large but finite number of points. The model of a rigid body that we employ in the initial stages in dynamics is that of a number of massive particles set in a rigid and imponderable frame.

The particles, though finite in number, are to be thought of as so numerous and so closely packed that they approximate to a continuous distribution of mass. For example, we shall speak of a *uniform* lamina, by which we mean one in which the mass in any small area (by which must be understood a physically small, but not a mathematically small, area) is very nearly proportional to the area. And similarly we shall speak of a uniform solid. For such bodies the calculation of the position of the centre of gravity and of a moment of inertia can be carried out to a high degree of accuracy by supposing the mass to be spread continuously and uniformly through the area or volume. The summations which appear in the definition of G and of a moment of inertia for a set of particles can be replaced by integrations through the area or volume.

19·6 Moments and products of inertia. In the dynamics of a rigid body we shall frequently meet with the notion of the moment of inertia of the body about a given line. It will be convenient at this point to make a brief digression from the main line of thought in order to remind the reader of the most important results.

Consider first a set of discrete particles; we need not assume to begin with that they lie in a plane. We denote the coordinates of a typical particle, of mass m, by (x, y, z). We can define the moment of inertia of the system with respect to a point, a line, or a plane. The moment of inertia is found by multiplying together the mass of each particle and the square of its distance from the point, or line, or plane, and summing over the particles. Thus the moments of inertia with respect to the point O, the line Oz, and the plane Oyz, are

$$K = Sm(x^2+y^2+z^2), \quad C = Sm(x^2+y^2), \quad P = Smx^2,$$

where the symbol S denotes here, and generally, summation over the particles. The important idea for our present purpose is that of the moment of inertia about a line; the others are only of subsidiary importance.

We denote the moments of inertia about the axes by A, B, C,

$$A = Sm(y^2+z^2), \quad B = Sm(z^2+x^2), \quad C = Sm(x^2+y^2),$$

and we write $\quad P = Smx^2, \quad Q = Smy^2, \quad R = Smz^2.$

We also define the products of inertia F, G, H with respect to the given trihedral of reference by the equations

$$F = Smyz, \quad G = Smzx, \quad H = Smxy.$$

A, B, C are of course positive, but F, G, H may have either sign. We have the obvious relations

$$K = \tfrac{1}{2}(A+B+C) = P+Q+R.$$

If the particles lie in the plane $z = 0$ we have

$$K = C = A+B.$$

In two-dimensional problems it is moments of inertia about axes perpendicular to the plane of motion that we need. When the particles all lie in the plane $z = 0$, for example, for a lamina moving in this plane, the moment of inertia about an axis perpendicular to this plane is the same thing as the moment of inertia about the point where the axis meets the plane.

If the moment of inertia about a given line is Mk^2, where $M(= Sm)$ is the mass of the body, we call the length k the *radius of gyration* about the given line.

The theorem of parallel axes. We denote the coordinates of the centre of gravity by (ξ, η, ζ), and we write

$$x = \xi+\alpha, \quad y = \eta+\beta, \quad z = \zeta+\gamma,$$

so that (α, β, γ) are coordinates relative to G. The moment of inertia about Oz is

$$\begin{aligned} C &= Sm(x^2+y^2) \\ &= Sm\{(\xi+\alpha)^2+(\eta+\beta)^2\} \\ &= M(\xi^2+\eta^2)+Sm(\alpha^2+\beta^2), \end{aligned}$$

where M is the mass of the system $(M = Sm)$, and we have used the equations $Sm\alpha = 0, Sm\beta = 0$. The last term in the second member is the moment of inertia about a line through G parallel to Oz, and we deduce the rule: the moment of inertia about any line is the moment of inertia about the parallel line through G plus Mr^2, where M is the mass of the system, and r the distance of G from the given line. For lines in a given direction the moment of inertia is least for the line through G. The theorem shows that it will suffice to know the moments of inertia about the lines through G, and then the moments of inertia about all lines can be written down. There is an analogous theorem for the products of inertia

$$\begin{aligned} F &= Smyz \\ &= Sm(\eta+\beta)(\zeta+\gamma) \\ &= M\eta\zeta+Sm\beta\gamma, \end{aligned}$$

but we shall not need this theorem in this book.

Uniform solids. When we consider a continuous distribution of mass—a rod, a lamina, or a solid—we replace the summations by the corresponding integrations, associating with the element of length dr, of area dS, or of volume $d\tau$, a mass λdr, σdS, or $\rho d\tau$. A case of particular importance is that in which the density is uniform (λ, σ, ρ are constants) and the shape has a particular form—if a lamina, rectangular or circular or elliptic; if a solid, rectangular or spherical or ellipsoidal. The body has three planes of symmetry, which we can take for the moment as the coordinate planes, and then F, G, H all vanish, and the values of A, B, C—the moments of inertia about the axes—are given by the simple mnemonic known as *Routh's rule*. This rule gives for the moments of inertia the formula Ms^2/n, where s^2 is the sum of the squares of the lengths of the semi-axes of the body at right angles to the given axis, and n is 3, 4, or 5 according as the body is rectangular, elliptical, or ellipsoidal.

The most important cases for our immediate purpose are:

(i) for a uniform rod of length $2a$ the moment of inertia about the centre (which is the same thing as the moment of inertia about a line through the centre at right-angles to the rod) is $Ma^2/3$;

(ii) for a uniform circular disc of radius a the moment of inertia about the centre (which is the same thing as the moment of inertia about a line through the centre perpendicular to the plane of the disc) is $Ma^2/2$; the same formula gives the moment of inertia of a uniform right circular cylinder of radius a about its axis.

Other special cases of the rule are:

(iii) for a uniform rectangular lamina, whose sides are of lengths $2a$, $2b$, the moment of inertia about the centre is $M(a^2+b^2)/3$;

(iv) for a uniform circular disc of radius a the moment of inertia about a *diameter* is $Ma^2/4$;

(v) for a uniform solid sphere of radius a the moment of inertia about a diameter is $2Ma^2/5$.

The proofs of the results just stated are simple, and we shall be content to work out a few particular cases. For the rod the moment of inertia about one end (which is the same thing as the moment of inertia about a line through one end at right angles to the rod) is

$$I = \int_0^{2a} (M/2a)r^2\,dr = 4Ma^2/3,$$

so (by the theorem of parallel axes) the moment about the centre is $Ma^2/3$. For the circular disc (which we may take in the plane $z = 0$ with its centre at O) we have

$$C = \int_{r=0}^{r=a} r^2\,d\mu,$$

where μ is the mass inside the concentric circle of radius r, $\mu = (r^2/a^2)M$, so that

$$C = (M/a^2)\int_0^{a^2} \xi\,d\xi = Ma^2/2.$$

In this case $C = 2A$, so $A = Ma^2/4$. Finally, for the uniform sphere, taking axes through the centre, we have

$$3A = A+B+C = 2K = 2\int_0^a r^2\,d\mu,$$

where $\mu = (r^3/a^3)M$, giving

$$A = 2Ma^2/5.$$

Lamina, the momental ellipse. Consider now a lamina in the plane $z = 0$, and consider the moment of inertia about a line through O *in the plane of the lamina*. If this line OP makes an angle θ with Ox we have for the moment of inertia about OP

$$\begin{aligned} I &= Sm(y\cos\theta - x\sin\theta)^2 \\ &= A\cos^2\theta - 2H\cos\theta\sin\theta + B\sin^2\theta. \end{aligned}$$

If we take the point P on OP at distance r from O, such that $Ir^2 = 1$, the locus of P is

$$Ax^2 - 2Hxy + By^2 = 1.$$

This represents an ellipse with O as centre, and this ellipse has the property that the moment of inertia about any radius OP is $1/OP^2$: it is called the *momental ellipse* for the point O.

The axes of the ellipse, with the axis Oz, are the *principal axes* of inertia at O. For these axes the products of inertia F, G, H are all zero, and the moments of inertia are called the principal moments of inertia at O.

We have proved the existence of the principal axes of inertia only for the special case when all the mass lies in one plane, and O lies in this plane. But there is a corresponding theorem in three dimensions; for any distribution of mass, and for any point O of space, there exists a trihedral of directions through O for which the products of inertia are all zero. The axes of this trihedral are the principal axes of inertia at O. They are of course not necessarily unique; for example, if O is the centre of a uniform sphere or cube any rectangular axes through O are principal axes.

Equimomental bodies. We can have different distributions of mass which have the same moments of inertia about all lines; such distributions are said to be equimomental. Perhaps the simplest example is that of the two distributions (i) a uniform rod of mass M, and (ii) three particles, one of mass $2M/3$ at the centre of the rod, and the others, of mass $M/6$, at the ends. It is clear that two systems are equimomental if they have the same mass, the same centre of gravity G, the same principal axes at G, and the same principal moments of inertia at G.

19·7 External and internal forces.

We return now to the study of the dynamics of a rigid body, which we have agreed to picture as a swarm of particles held in a rigid configuration by an imponderable frame. Perhaps the simplest picture of this frame is that of a rigid framework of light rods, the massive particles being attached at the joints where a number of the rods are freely hinged together. With this picture the total force on any particle at any instant is built up of two classes, the first class being the tensions and thrusts in the rods, the second class being the weights of the particles and any other forces on them arising from agents outside the body. The first class are called *internal* forces and the second class *external* forces. Usually we can ignore the gravitational attractions of the particles of the body on one another, but if we do include them they also belong to the class of internal forces.

Now *the internal forces form a system in equilibrium.* This is obvious if we use the model just mentioned, because a thrust or tension in any rod exerts equal and opposite forces on the particles at the two ends. If we use another model of the imponderable framework the result may not be so obvious, but a similar result is assumed to be true in all cases. The forces on the particles consist of two classes, internal and external; the internal forces are the forces exerted on the particles by the rigid frame, or in other words the forces maintaining the rigid configuration, and these forces as a whole constitute a system in equilibrium.

The first consequence of this fact that we have to notice is that the work done by the internal forces in any displacement of the body is zero (§ 19·4). Of course, the displacement contemplated here is a natural displacement in which the body remains rigid.

We have spoken of external forces (such as the weights) acting on the particles of the rigid body, but sometimes it is convenient to allow of the possibility of external forces acting on the imponderable frame, and not necessarily at a point of the frame where one of the particles is situated. But this difficulty turns out not to be serious. It is true that when we write the equation of motion we shall suppose in the first instance that the external forces do act on the particles; but we soon discover that what we are primarily interested in is an equivalent system to which the external forces can be reduced, and it is natural to assume that the equations of motion are the same for any two equivalent systems, even mass,*m*though in one of them the forces, or some of them, act the particles but at other points of the imponderable frame.

19·8 Lamina in a field of force, the work done in a small displacement. We return now to a subject already introduced in § 19·4, the work done by the external forces in a small displacement.

Suppose first that we have a lamina moving in a vertical plane, the external forces being the weights of the particles. We take rectangular axes in the plane, with Oy vertically upwards, and denote the coordinates of a typical particle, of mass m, by (x, y). The work done by the external forces (i.e. by the weights) in a small displacement is

$$-Smgdy,$$

where (as in § 19·6) the symbol S denotes summation over the particles. But

$$Smy = M\eta,$$

where $M(=Sm)$ is the mass of the lamina, and (ξ, η) its centre of gravity. Thus the work done can be expressed in the form $-Mgd\eta$.

It is perhaps worth while to notice that the result can be established in another way. By the theorem of § 19·4 the work done by two equivalent systems in a small displacement is the same. Now the system of forces in question, the weights of the particles, is equivalent to a force Mg downwards at G; thus the work done in the small displacement is $-Mgd\eta$ as before. The two lines of argument are of course closely related to one another.

In this problem, therefore, the work done by the external forces in a small displacement is the perfect differential of a function of the configuration, namely, of the function $-Mg\eta$. The same phenomenon occurs in any problem where the forces on the particles are derived from a conservative field of force; in any such problem the work done by these forces can be expressed in the form $-dV$, where V is a function of (ξ, η, θ) which can be calculated when ϕ, the potential of the field per unit mass, is known.

To prove this statement we notice that the external force on a particle is $-m \operatorname{grad} \phi$, and the work done by this force in a small displacement $d\mathbf{r}$ is

$$-m \operatorname{grad} \phi . d\mathbf{r} = -m d\phi,$$

by equation (2) of § 3·9. Thus, summing over the particles, the work done in an arbitrary small displacement is

$$-Smd\phi = -dV,$$

whence

$$V = Sm\phi. \tag{1}$$

To find V we multiply the mass of each particle by the value of ϕ at the point which it occupies, and sum over the particles. V is a function of the configuration (position and orientation) of the body, and, as we have noticed, it can be expressed in terms of (ξ, η, θ) when the function ϕ is given.

The term 'small displacement', which we have used in the preceding discussion, deserves some amplification; in this and similar contexts it must be interpreted in terms of differentials.

The coordinates (x, y, z) of the particle are certain simple functions of (ξ, η, θ), and by $d\mathbf{r}$ we mean the displacement (dx, dy, dz) where

$$dx = \frac{\partial x}{\partial \xi} d\xi + \frac{\partial x}{\partial \eta} d\eta + \frac{\partial x}{\partial \theta} d\theta, \tag{2}$$

with similar formulae for dy and dz. If we make arbitrary changes $(d\xi, d\eta, d\theta)$ in (ξ, η, θ) there is a consequent displacement $(\delta x, \delta y, \delta z)$ of the particle, and δx, for example, is not exactly equal to dx. What we can assert is that δx is nearly equal to dx when $\rho[=\sqrt{(d\xi^2 + d\eta^2 + d\theta^2)}]$ is small. More precisely δx differs from dx by an amount $\epsilon\rho$, where $\epsilon \to 0$ with ρ.

The differentials can be avoided altogether if desired by thinking in terms of velocity instead of displacement, and power instead of work. Thus if we think of the derivatives $(\dot{\xi}, \dot{\eta}, \dot{\theta})$ instead of the displacements $(d\xi, d\eta, d\theta)$ the formula for $\dot{x}$ corresponding to equation (2), namely,

$$\dot{x} = \frac{\partial x}{\partial \xi} \dot{\xi} + \frac{\partial x}{\partial \eta} \dot{\eta} + \frac{\partial x}{\partial \theta} \dot{\theta},$$

is exact. If we write

$$V(\xi, \eta, \theta) = Sm\phi,$$

as before, the rate of working of the forces is

$$-Sm\left(\frac{\partial \phi}{\partial x}\dot{x} + \frac{\partial \phi}{\partial y}\dot{y} + \frac{\partial \phi}{\partial z}\dot{z}\right) = -\left(\frac{\partial V}{\partial \xi}\dot{\xi} + \frac{\partial V}{\partial \eta}\dot{\eta} + \frac{\partial V}{\partial \theta}\dot{\theta}\right). \tag{3}$$

The same point arises in the discussion of the principle of Virtual Work in statics. The principle is usually enunciated in terms of the work done by certain forces in a small displacement of the system, the small displacement being interpreted, as here, in terms of differentials. But the differentials can be avoided by enunciating the principle in terms of velocity and power instead of displacement and work. In that form the principle is called the principle of Virtual Velocities (cf. § 19·4).

It may be said, however, as a matter of general policy in work of this kind, that the differentials are to be preferred on the whole to the velocities. The ultimate significance of the two theories is the same, but the differentials are usually rather easier to handle, as well as being perhaps more natural physically.

Example. *To find V when the field is an attraction n^2r per unit mass towards O.*

Here

$$\phi = \tfrac{1}{2}n^2r^2$$

and

$$V = Sm\phi = \tfrac{1}{2}n^2Smr^2 = \tfrac{1}{2}n^2M(r_0^2+k^2),$$

where r_0 is the distance OG, and Mk^2 is the moment of inertia of the lamina about G. The constant term $\frac{1}{2}n^2Mk^2$ is of no consequence, and we may write instead

$$V = \tfrac{1}{2}Mn^2r_0^2 = \tfrac{1}{2}Mn^2(\xi^2+\eta^2).$$

The work done by the forces of the field in a small displacement is

$$-dV = -Mn^2r_0dr_0.$$

Again, it is easy to verify this result directly from the theorem of § 19·4 By a well-known property of the centre of gravity the external forces are equivalent to a force Mn^2r_0 through G in the direction GO. The work done in a small displacement is therefore $-Mn^2r_0dr_0$ as before.

19·9 Lamina in a field of force, the rate of working.

If we substitute in V the values of (ξ, η, θ) at time t in the actual motion of the body we obtain a function Ω

$$\Omega(t) = V(\xi, \eta, \theta)$$

which measures the value of V in the configuration occupied by the lamina at time t. Now

$$\frac{d\Omega}{dt} = \frac{\partial V}{\partial \xi}\dot{\xi} + \frac{\partial V}{\partial \eta}\dot{\eta} + \frac{\partial V}{\partial \theta}\dot{\theta},$$

so, in virtue of equation (3) of § 19·8, the rate of working as the body moves is $-d\Omega/dt$. (Cf. §§ 5·3 and 11·4.)

19·10 d'Alembert's principle.

As we have noticed already in § 19·5, the fundamental principles of dynamics, as given by Newton, dealt only with the motion of a particle. To deal with the motion of a rigid body a new principle is needed; this new principle was given by d'Alembert. It is a surprisingly simple principle, and with its aid we are able easily to form the equations of motion of a rigid body.

We consider the kineta of the particles of the rigid body. These form a system of localized vector quantities of the kind discussed in §§ 19·1–19·3. d'Alembert's principle asserts that *the kineton system is equivalent at each instant to the external force system.* Here the word *equivalent* is used in the same sense in which it is used in the statics of a rigid body, as in § 19·1.

The proof of this principle is immediate. The kineton of each particle is equal to the vector sum of the forces, external and internal, acting on it. Thus, when we consider the body as a whole, the kineton system is equivalent to the system of all the forces, external and internal, acting on the particles. But the internal forces, when we consider the body as a whole, form a system in equilibrium, i.e. a system equivalent to the nul system. Therefore the kineton system is equivalent to the external force system, and this is what we wish to establish.

We shall be concerned mainly with problems in two dimensions, and we will suppose to begin with that the rigid body with which we are dealing is a lamina moving in a plane.

An immediate corollary of d'Alembert's principle is that the external force system and the *reversed* kineton system together form a system in equilibrium. Here the phrase *in equilibrium* is used in the same sense in which it is used in the statics of a rigid body.

Example. *The end A of a uniform rod AB is freely attached to a fixed support, and the rod moves as a conical pendulum with angular velocity ω. Prove that, if the plane in which B moves is at a depth h below A, $\omega^2 = 3g/2h$.*

Suppose the rod makes an angle θ with the downward vertical. The reversed kineton for a particle of the rod is $mr \sin \theta . \omega^2$ horizontally and outwards, where m is the mass of the particle and r its distance from A. In virtue of the last remark above we can regard the problem as a statical one, and taking moments about A we find

$$\begin{aligned} Mga \sin \theta &= Sr \cos \theta . mr \sin \theta . \omega^2 \\ &= \cos \theta \sin \theta . \omega^2 Smr^2 \\ &= \tfrac{4}{3} Ma^2 \cos \theta \sin \theta . \omega^2, \end{aligned}$$

where M is the mass and $2a$ the length of the rod. Hence, since $\sin \theta \neq 0$,

$$\omega^2 = \frac{3g}{4a \cos \theta} = \frac{3g}{2h}.$$

EXAMPLES XIX

1. Find the position of the mass-centre of a uniform plate in the form of a quadrant of a circle of radius a.

If the mass of the plate is M, prove that the moment of inertia about an axis through the mass-centre parallel to one of the straight edges of the plate is approximately $0{\cdot}07 Ma^2$.

2. Perpendicular axes Ox, Oy are taken in a plane lamina at a point O. Show that the moment of inertia of the lamina about a line passing through O and making an angle θ with Ox is $A\cos^2\theta - 2H\cos\theta\sin\theta + B\sin^2\theta$, where A, B are the moments of inertia about Ox, Oy and H is the product of inertia relative to these axes. Consider the following riders:

(i) Find the values of A, B and H for a uniform lamina of mass M in the form of an isosceles triangle OXY, right-angled at O, with axes Ox, Oy in the directions of OX, OY respectively, the sides OX, OY being each of length a.

(ii) Find A, B and H for a uniform lamina in the form of an ellipse of mass M and axes $2a, 2b$ when Ox, Oy are the bisectors of the angles between the major and minor axes of the ellipse.

(iii) A uniform rectangular plate, whose sides are of lengths $2a, 2b$, has a portion cut out in the form of a square whose centre is the centre of the rectangle and whose mass is half the mass of the plate. Show that the axes of greatest and least moment of inertia at a corner of the rectangle make angles θ, $\frac{1}{2}\pi + \theta$, with a side, where

$$\tan 2\theta = \frac{6ab}{5(a^2 - b^2)}.$$

3. If A, H, B are the usual constants of inertia of a lamina referred to perpendicular axes Ox, Oy in the plane of the lamina, and if (x, y) are the coordinates of a fixed point of the lamina, show that $Ax^2 - 2Hxy + By^2$ is unchanged if the axes are rotated about O.

Prove that the radius of gyration of a uniform equilateral triangular lamina is the same about any straight line through its centre and in its own plane.

4. Prove that a uniform triangular lamina of mass M, and a system of three particles each of mass $\frac{1}{3}M$ situated at the mid-points of the sides, have the same moment of inertia about any axis in the plane.

5. Find the moment of inertia of a uniform rod of mass m and length $2c$ about any axis for which the line of shortest distance to the given rod cuts the rod at its middle point. (Take θ for the angle between the axis and the rod.)

Show by direct integration of the previous result that the moment of inertia of a plane uniform elliptic lamina of mass M and semi-axes a and b about an axis through its centre in the plane normal to the lamina containing the axis a and making an angle ϕ with that axis is

$$\tfrac{1}{4}M(b^2 + a^2\sin^2\phi).$$

6. Find the radius of gyration

(i) of a uniform square of side $2a$ about an axis through the centre parallel to a side;

(ii) of the same square about an axis through the centre perpendicular to its plane;

(iii) of a uniform pyramid with a square base of side $2a$ about the line joining the apex to the centre of the base, this line being perpendicular to the base;

(iv) of the same pyramid about a line through a corner of the base and perpendicular to the base.

7. A smooth narrow tube is in the form of a circular hoop of radius a. It contains a length of chain which subtends an angle 2α at its centre ($\alpha < \pi$). The hoop is rotated with constant angular velocity ω about a fixed vertical diameter. Show that if $g < \omega^2 a |\cos\alpha|$ there are more than two positions of equilibrium for the chain, and only two otherwise.

8. A uniform straight rod AB of length $3a$ and mass $3am$ is rigidly attached at B to a uniform straight rod BC of length $4a$ and mass $4am$. The rods are at right angles. Small rings at A and C attach the rods smoothly to a straight rod about which the system rotates with constant angular velocity ω. Calculate the reaction at C and the bending moment and stress components along and perpendicular to the rod BC at the point of its length at a distance x from C, neglecting the effect of gravity.

9. A point P describes a horizontal circle of radius a about a fixed point O with constant angular velocity ω. A heavy uniform rod of length l has one end pivoted at P in such a way that it can move only in the vertical plane through OP. If the rod is inclined at an angle θ to the downward vertical, show that for relative equilibrium

$$\sin\theta + A = B\tan\theta,$$

where

$$A = \frac{3a}{2l}, \quad B = \frac{3g}{2\omega^2 l}.$$

Show that if $A > 1$, there are only two positions of equilibrium, but that if $A < 1$ there are four or two positions of equilibrium according as ω is greater or less than a critical value ω_0. If $A = \cos^3\alpha$, prove that $\omega_0^2 = g\cot^3\alpha/a$.

CHAPTER XX

MOTION ABOUT A FIXED AXIS

20·1 Motion about a fixed axis, the equation of motion. The simplest problem of rigid dynamics is that of the motion of a body which is free to rotate about a fixed axis, such as a flywheel, or a rigid pendulum. We will suppose that the body is a lamina, one particle of which is pinned in position at a point O, the lamina being free to rotate in its plane about O.

The external forces on the lamina consist of (i) the given forces, which can be reduced to a force $\mathbf{R}$ at O, and a couple N, (ii) the action of the pivot on the lamina, which we assume to be a single force $\mathbf{S}$ at O (Fig. 20·1*a*). The kineton system is shown in Fig. 20·1*b*; r denotes the distance of a typical particle P from O, $\dot{\theta}$ the angular velocity of the lamina.

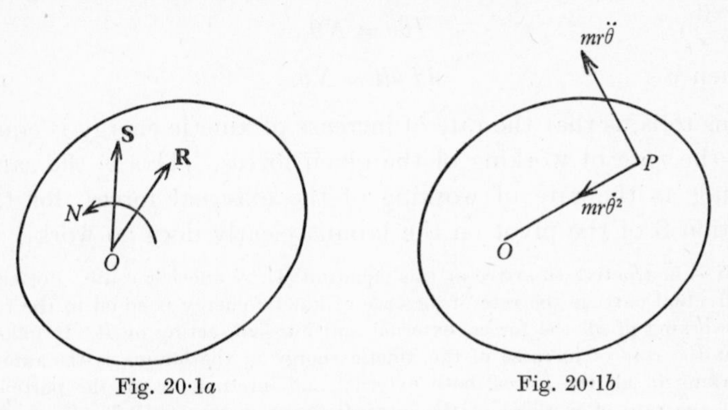

Fig. 20·1*a* Fig. 20·1*b*

Now the external force system is equivalent to the kineton system. Both systems have the same moment about O, and this alone is sufficient to establish the equation of motion, for it gives us

$$N = Sr \,.\, mr\ddot{\theta} = (Smr^2)\,\ddot{\theta}. \tag{1}$$

(The symbol S denotes summation over the particles, as in § 19·6.) Now Smr^2 is a constant of the lamina, its moment of inertia about O (or, more strictly, its moment of inertia about an axis through O perpendicular to the plane). We denote this moment of inertia by I. We have at once the required equation of motion

$$I\ddot{\theta} = N. \tag{2}$$

The system has one degree of freedom, since the configuration is described completely by the one coordinate θ, and the reader will notice the analogy between equation (2) and the fundamental equation for the motion of a particle on a straight line. The symbol N in the second member of equation (2), the moment of the given forces about O, is supposed given; in the most general case it may be a function of $t, \theta, \dot{\theta}$.

20·2 Motion about a fixed axis, the equation of energy. The kinetic energy T of the lamina is defined as the sum of the kinetic energies of the separate particles. Since the velocity of the typical particle is $r\dot{\theta}$ perpendicular to OP we have

$$T = S\tfrac{1}{2}mr^2\dot{\theta}^2 = \tfrac{1}{2}I\dot{\theta}^2.$$

From the equation of motion (equation (2) of § 20·1) we have

$$I\dot{\theta}\ddot{\theta} = N\dot{\theta},$$

whence

$$dT/dt = N\dot{\theta}. \tag{1}$$

This tells us that the rate of increase of kinetic energy is equal to the rate of working of the given forces. (This is the same thing as the rate of working of the external forces, for the action **S** of the pivot on the lamina clearly does no work.)

It is instructive to arrive at this equation (1) by another route. For each individual particle the rate of increase of kinetic energy is equal to the rate of working of all the forces, external and internal, acting on it. It follows that the rate of increase of the kinetic energy of the lamina is the rate of working of all the forces, both external and internal, on all the particles. But the rate of working of the internal forces is zero (§ 19·7). Hence the rate of increase of kinetic energy is equal to the rate of working of the external forces, and this is the same thing as the rate of working of the given forces. We thus arrive again at equation (1).

If the given forces are derived from a conservative field, the equation takes a simpler form. We introduce the function V (§ 19·8), and observe that $N = -dV/d\theta$. The function Ω (§ 19·9), defined by the equation

$$\Omega(t) = V(\theta),$$

represents the value of V in the position occupied by the lamina at time t. Then

$$\frac{d\Omega}{dt} = \frac{dV}{d\theta}\dot{\theta} = -N\dot{\theta},$$

so that (as in § 19·9) the expression $-d\Omega/dt$ measures the rate of working of the external forces. Equation (1) now takes the form

$$\frac{d}{dt}(T+\Omega) = 0.$$

Hence
$$T+\Omega = C, \tag{2}$$

and this is the desired form. We can also write it in the form

$$T+V = C,$$

where V now represents the value for the position of the lamina at time t.

20·3 The rigid or compound pendulum. The most famous problem, and one of the simplest problems, of this class is that of the rigid pendulum. The lamina moves in a vertical plane about a pivot O, and the given forces (i.e. the external forces other than the action of the pivot on the lamina) are the weights of the particles. In the position of stable equilibrium the centre of gravity G lies vertically below O. Suppose that at time t during the motion the line OG makes an angle θ with the downward vertical (Fig. 20·3). The weights of the particles are equivalent to a single force Mg downwards at G, and the equation of motion is

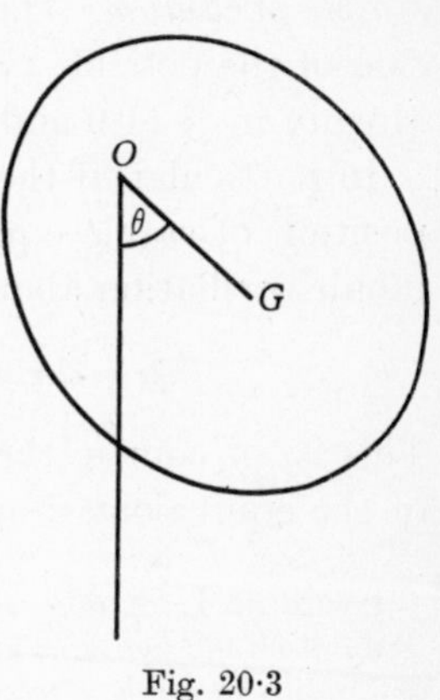

Fig. 20·3

$$I\ddot{\theta} = -Mgh\sin\theta, \tag{1}$$

where h denotes the length OG. The equation can be written in the form

$$\ddot{\theta}+n^2\sin\theta = 0, \tag{2}$$

where
$$n^2 = Mgh/I.$$

By integration of equation (2) we obtain

$$\tfrac{1}{2}\dot{\theta}^2 = n^2(\cos\theta + p), \tag{3}$$

where p is a constant.

Instead of finding equation (3) by integration of equation (2), we could have written it down at once from the equation of energy, equation (2) of § 20·2. For here

$$V(\theta) = -Mgh\cos\theta,$$

and the equation of energy is

$$\tfrac{1}{2}I\dot{\theta}^2 - Mgh\cos\theta = C,$$

which is equivalent to (3).

Now the equations (2) and (3) are precisely the same equations as those for the motion of a simple pendulum of length l, where

$$l = I/Mh, \tag{4}$$

(see equation (1) of § 11·9 and equation (1) of § 11·10). Thus we see, from either, that the motion of the rigid pendulum is similar to that of a simple pendulum, the so-called *equivalent simple pendulum*. There is no need to repeat here the enumeration of the possible types of motion which have been discussed already in §§ 11·9 and 11·10.

In particular, if the pendulum is slightly disturbed from the position of stable equilibrium, it executes a small nearly harmonic oscillation about $\theta = 0$ of period

$$\sigma = 2\pi/n = 2\pi\surd(l/g) = 2\pi\surd\{I/(Mgh)\}.$$

This is, of course, the same as the period of small oscillations of the equivalent simple pendulum.

Example 1. *A thin uniform rod AB of mass m and length* $2a$ *can turn freely about the end A which is fixed, and a uniform circular disc of mass* $12m$ *and radius* $\tfrac{1}{3}a$ *can be clamped to the rod so that its centre C is on the rod. Show that, for oscillations in which the plane of the disc remains vertical, the length of the simple equivalent pendulum lies between* $2a$ *and* $\tfrac{2}{3}a$.

If we denote the length AC by x the length of the simple equivalent pendulum is

$$l = I/Mh,$$

where

$$I = \tfrac{4}{3}ma^2 + 12m\left(x^2 + \frac{a^2}{18}\right)$$
$$= m(12x^2 + 2a^2)$$

and $$Mh = ma + 12mx$$
$$= m(12x+a).$$

Thus $$l = \frac{12x^2+2a^2}{12x+a},$$

and we wish to study how this varies in the range $0 \leqslant x \leqslant 2a$. To do this we write the formula for l in the form

$$\frac{1}{12}\left\{(12x+a) - 2a + \frac{25a^2}{12x+a}\right\} = \frac{5a}{12}\left(y + \frac{1}{y} - \frac{2}{5}\right),$$

where we have written y for $\dfrac{12x+a}{5a}$. Now for positive values of y the function $y + (1/y)$ has a minimum value 2 for $y = 1$, so l has a minimum value $\frac{2}{3}a$ for $x = \frac{1}{3}a$. The greatest value of l occurs at one end of the range, at $x = 0$ or at $x = 2a$, and in fact the value of l is $2a$ at both ends.

Example 2. *The period of a complete oscillation of a compound pendulum is* 2 *sec.* (*Such a pendulum is called a 'seconds pendulum'.*) *The mass of the pendulum is* M *and the centre of mass is* 3 *ft. below the oscillation axis. A particle of mass* m *is attached to the bottom part of the pendulum at a distance* 4 *ft. below the axis. It is then found that the pendulum loses at the rate of* 20 *seconds per day. Find the ratio of* M *to* m.

In this kind of work it is best not to put in the numbers too soon, and not to approximate too soon. If O is the point of suspension, denote the length OG by h, and the distance OP to the particle P by l; it is assumed that P is in line with O and G. Then

$$2\pi\sqrt{\left(\frac{I}{Mgh}\right)} = 2, \quad 2\pi\sqrt{\left(\frac{I'}{M'gh'}\right)} = \frac{n}{n-20}\,2,$$

where I', M', h' refer to the solid with the particle attached, and $n(= 24 \times 60 \times 60)$ is the number of seconds in a day. We have

$$I' = I + ml^2, \quad M'h' = Mh + ml,$$

so
$$\left(1 - \frac{20}{n}\right)^{-2} = \frac{\pi^2 I'}{M'gh'} = \frac{\pi^2 I + \pi^2 ml^2}{(Mh+ml)\,g}$$
$$= \frac{Mgh + \pi^2 ml^2}{(Mh+ml)g}$$
$$= \frac{(Mh/ml) + (\pi^2 l/g)}{(Mh/ml)+1}.$$

The first member is a little greater than 1, and we denote it by $1+\theta$, and then

$$\frac{M}{m} = \frac{l}{h}\,\frac{(\pi^2 l/g) - 1}{\theta} - \frac{l}{h};$$

this is exact so far.

Now we put in the numbers, $h = 3$, $l = 4$, $g = 32{\cdot}2$, and, to a sufficient approximation, $\theta = \dfrac{40}{n} = \dfrac{1}{2160}$. These give 650 as the approximate value of M/m.

20·4 Kater's pendulum. We return now to the study of the period of a rigid pendulum; if the moment of inertia of the lamina about G is Mk^2, we have

$$I = M(h^2+k^2),$$

whence (equation (4) of § 20·3)

$$l = h+(k^2/h). \tag{1}$$

The minimum value for l is $2k$, got by taking $h = k$. This is always possible; there must be a point O in the lamina at which $h = k$, because if h were less than k for all points of the lamina, the moment of inertia about G would be less than Mk^2, and if h were greater than k for all points of the lamina, the moment of inertia about G would be greater than Mk^2.

The equation (1) can be written

$$h^2-lh+k^2 = 0,$$

and if we prescribe any value of l greater than the minimum value $2k$ this equation has two real positive roots, say h_1 and h_2. The sum of these roots is l, and their product is k^2. If we

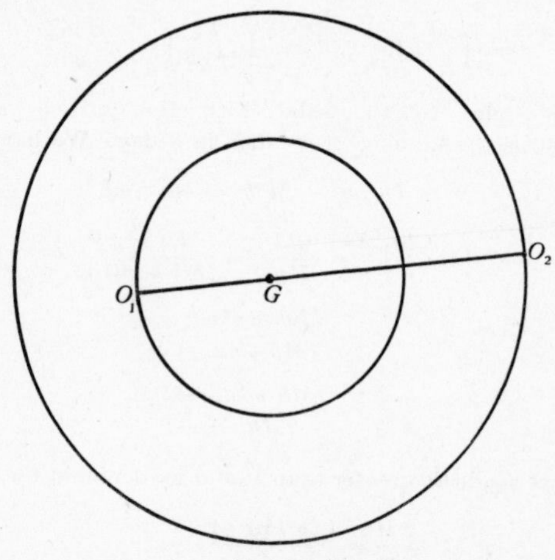

Fig. 20·4

take the axis of rotation O at any point on either of the circles having G as centre and radii h_1 and h_2, the length of the equivalent simple pendulum is l.

Consider then two points O_1 and O_2 of the lamina, collinear with, and on opposite sides of, G, and such that the period σ of a small oscillation is the same whether O_1 or O_2 be used as the point of support. Then O_1O_2 is the length of the simple pendulum for which the period of a small oscillation is σ (Fig. 20·4).

This result is the basis of the theory of Kater's pendulum, used in surveys of the variation of g at different places. The pendulum consists essentially of a steel bar, which can be swung about either of two axes. These axes are normal to the face of the bar, and lie in a plane of symmetry. They are at different distances on opposite sides of the centre of gravity, and the position of one of the axes is capable of adjustment. The position is adjusted so that the time of swing σ about the two axes is the same. The distance between the axes is then the length l of the equivalent simple pendulum. The value of g at the place is then given by

$$g = l(2\pi/\sigma)^2.$$

The advantage of the method is of course that the value of the moment of inertia of the bar, and the exact position of its centre of gravity, are not required.

20·5 Motion relative to a seized point.

In the preceding paragraphs we have discussed the motion of a lamina

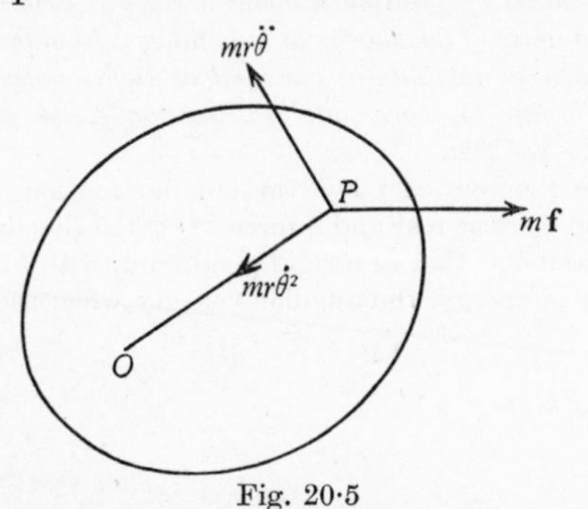

Fig. 20·5

about a fixed pivot. We now investigate the necessary modification required to determine the motion of the lamina relative to the pivot O, when the pivot, instead of being at rest, is moved about in an arbitrary way in the plane.

The requisite modification is very simple. The acceleration of a particle P at time t is the vector sum of the acceleration $\mathbf{f}$ of O and the acceleration of P relative to O. The kineton system is therefore as shown in Fig. 20·5, replacing the simpler figure (Fig. 20·1*b*) for the case where the pivot is fixed. The kineton system is equivalent to the external force system.

We can now easily see how the equation for the rotational motion is to be found. If we superpose a term $-m\mathbf{f}$ on each particle, both in the kineton system and in the external force system, the resulting systems are still equivalent. The effect on the kineton system is to reduce it to that shown in Fig. 20·1*b*, just as though O were at rest; and the effect on the external force system is to add a force $-M\mathbf{f}$ through G.

Thus the equation for the motion relative to O is found by adding a force $-M\mathbf{f}$ through G to the external forces, and then writing the equation of motion as though O were at rest.

Example 1. *A pendulum hangs in a lift which is rising with uniform acceleration f; to find the effect of the motion of the lift on the pendulum.*

In virtue of the theorem just proved, we have only to replace Mg acting downwards at G by $M(f+g)$, and then use the theory of § 20·3. For example, the period of a small oscillation becomes

$$2\pi\sqrt{[l/(f+g)]}.$$

Example 2. *The lock of a railway-carriage door will only engage if the angular velocity of the closing door exceeds ω. The door swings about vertical hinges and has a radius of gyration k about a vertical axis through the hinges, while the centre of gravity of the door is at a distance a from the line of the hinges. Show that if the door be initially at rest and at right angles to the side of the train which then begins to move with acceleration f, the door will not close unassisted unless $f > \frac{1}{2}\omega^2 k^2/a$.*

In virtue of the theorem of § 20·5 the angular motion of the door is the same as if the train were at rest and a force Mf acted (backwards) through G; the motion is the same as that of a rigid pendulum, with f replacing g. Thus, from the equation of energy, the angular velocity when the door slams is Ω, where

$$\tfrac{1}{2}Mk^2\,\Omega^2 = Mfa,$$

and $\Omega > \omega$ if $f > \frac{1}{2}\omega^2 k^2/a$.

EXAMPLES XX

1. A rotating flywheel is acted on by a driving couple $A\sin^2 pt$, and there is a constant frictional couple $\frac{1}{2}A$ opposing the motion. Find the least moment of inertia required to make the difference between the greatest and least angular velocities less than $p/100$.

2. A uniform solid circular cylinder of radius a and mass M is free to turn about its axis. One end of a light elastic string, of unstretched length $a \tan \beta$, is attached to a point on the cylinder, and the other end of the string is fixed to a point distant $a \sec \beta$ from the axis. The string lies in a plane which is at right angles to the axis, and frictional effects may be neglected. Show that, if the system is released from rest in a position in which the portion of the string in contact with the cylinder is of length $a\gamma$, the periodic time of the subsequent motion is

$$\left(2\pi + 4\frac{\beta}{\gamma}\right)\sqrt{\frac{M}{2k}},$$

where k is the increase in tension produced by unit increase in length.

3. A uniform circular cylinder of mass M and radius a is free to turn about its axis which is horizontal. A thin uniform cylindrical shell of mass $\frac{1}{2}M$ and radius a is fitted over the cylinder. At time $t = 0$ the angular velocity of the cylinder is Ω, whilst the angular velocity of the shell is zero. If the shell exerts on the cylinder a frictional couple of magnitude $k[\omega(t) - \varpi(t)]$, where $\omega(t), \varpi(t)$ are the angular velocities about the axis at time t of the cylinder and the shell respectively, prove that

$$2\omega(t) = \Omega[1 + e^{-4kt/Ma^2}].$$

Find the corresponding expression for $\varpi(t)$.

4. A uniform horizontal circular disc of mass m and radius a can rotate about a vertical axis through its centre, and the rotation is controlled by a spring which exerts a restoring couple of moment $kmga\theta$ when the disc is turned through an angle θ from its equilibrium position. A particle of mass m rests on the disc and is connected to the axis of rotation by a light horizontal straight rod of length $\frac{1}{2}a$ which can turn freely round the axis. The coefficient of friction between the particle and the disc is μ. The disc is turned through an angle β from its equilibrium position, and is released from rest. If $\beta > 3\mu/2k$, show that initially the particle slips on the disc, and that if slipping first ceases after time t, and if $p^2 = 2kg/a$, then pt is the root between 0 and π of the equation

$$\frac{\sin pt}{pt} = \frac{\mu}{k\beta - \frac{1}{2}\mu}.$$

5. A uniform rectangular lamina having vertices O, A, B, C (taken in order) lies in a given vertical plane and is free to swing about the fixed point O. A particle of mass m is fixed to the lamina at A and the system is released from rest in the position in which OA is horizontal. The system is next instantaneously at rest when OA is vertical. Express m in terms of M, a, b, where M is the mass of the lamina, $OA = 2a$, $AB = 2b$, and b is greater than a.

Find the length of the simple equivalent pendulum.

6. AB and CD are two rods of lengths $2a$ and b respectively, the mass of each rod being m per unit length. The rods are rigidly joined together at right angles at C, the middle point of AB, and the system is free to rotate in a vertical plane about D. If the system is held with AB vertical and then let go, calculate the angular velocity when AB is horizontal.

7. A rigid lamina, whose centre of gravity is G, can rotate in a vertical plane about a horizontal axis through a point O of the lamina. It is held with OG inclined at 30° above the horizontal and released. Owing to a frictional couple of constant magnitude at the bearing the first swing carries OG to a position only 45° beyond the vertical. Prove that in the next swing the lamina will come to rest before OG reaches the vertical.

8. A flywheel is balanced upon a horizontal knife-edge parallel to the axis of the wheel and inside the rim at a distance of 30 in. from the axis of the wheel. The wheel oscillates with a period of 2·6 sec.; find the radius of gyration of the wheel about its axis.

9. State the parallel axis theorem concerning moments of inertia.

A flat uniform circular disc of radius a has a hole in it of radius b whose centre is at distance c from the centre of the disc $(c<a-b)$. The disc is free to oscillate in a vertical plane about a smooth horizontal circular rod of radius b passing through the hole. Show that the length of the equivalent simple pendulum is $c+\frac{1}{2}\frac{(a^4-b^4)}{a^2c}$.

10. A uniform wire of length $3a$ is bent into the form of a sector of a circle with the two radii OA and OB and the arc AB each of length a. The point O is fixed, and the wire swings in a vertical plane. Show that the period of small oscillations is

$$2\pi\sqrt{\left(\frac{5a}{3g(\cos\frac{1}{2}+2\sin\frac{1}{2})}\right)}.$$

11. The mid-points of two adjacent edges of a uniform square lamina of side $2a$ are smoothly jointed to two fixed points in the same horizontal line. Find the length of the simple equivalent pendulum for an oscillation about the position of stable equilibrium.

12. A rigid body of mass M can turn freely round a horizontal axis about which its moment of inertia is I. Prove that small oscillations under gravity are of period

$$2\pi\left(\frac{I}{Mgh}\right)^{\frac{1}{2}},$$

where h is the length of the perpendicular GA drawn from G, the centre of gravity, to the axis of rotation.

The period of small oscillation is to be halved by attaching a heavy particle to the body at some point P in the same straight line AG; show that the mass of the particle required is a minimum if $AP = I/8Mh$.

13. A lamina swings in a vertical plane about a fixed point O of itself. Prove that the length of the simple equivalent pendulum is

$$h+(k^2/h),$$

where h is the distance of the centre of gravity G from O, and k is the radius of gyration of the lamina about G.

A and B are two points of the lamina collinear with G, on opposite sides of G, and the distance AB is c. When the lamina swings about A as axis the length of the simple equivalent pendulum is $c-a$, and when it swings about B the length of the simple equivalent pendulum is $c-b$. Prove that the distance AG is $bc/(a+b)$, and find k^2.

14. Three thin uniform rods, each of length l, are rigidly fastened together to form an equilateral triangle, which hangs in a vertical plane from a point of one of the rods. Show that if the triangle performs small oscillations in its own plane about the equilibrium position, the period is least when the point of suspension is at a distance $l/(2\sqrt{3})$ from the centre of one of the rods.

15. An arc of a circle is formed of thin wire (whose density may or may not be uniform) and hangs from a point P of the arc. Show that, if in the position of equilibrium Q is the point of the circle vertically below P, then PQ is the length of the simple equivalent pendulum when the wire oscillates about P in its own plane.

16. A uniform rod AB, of length $2a$ and mass M, is freely attached to a fixed support at A, and swings as a pendulum in a vertical plane through A. Determine the frequency of the small oscillations.

Prove that if a particle of mass kM is attached to the rod at B, the frequency is reduced in the ratio

$$\sqrt{\left(\frac{1+2k}{1+3k}\right)}.$$

If the rod is the pendulum of a clock which originally keeps good time, but loses one minute in a day after the particle is attached, prove that

$$k=\frac{2n-1}{n^2-6n+3},$$

where n is the number of minutes in a day.

17. A pendulum consists of a thin uniform rod of mass M pivoted at its mid-point and of a regulating nut of mass m, which can be screwed to any desired position on the rod. The nut may be treated as a particle. Prove that if $M>3m$ the period is always lengthened when the nut is raised slightly, but if $M<3m$ the period is lengthened or shortened according to the position of the nut.

18. A uniform rod AB of length $2a$ slides with its ends on a fixed smooth vertical circular wire whose centre is O. If b denotes the distance of the centre C of the rod from O, and θ the angle which OC makes with the vertical, prove that

$$\dot{\theta}^2=\frac{6bg}{3b^2+a^2}(\cos\theta-\cos\phi),$$

ϕ being the maximum value of θ.

Hence show that if ϕ is small, the period of a complete oscillation of the rod is

$$2\pi\sqrt{\left(\frac{3b^2+a^2}{3bg}\right)}.$$

19. Two wheels A, B of radii a, b and of moments of inertia I_1, I_2 respectively, are mounted on parallel axles; a light non-slipping belt passes round both wheels. A couple G is applied to wheel A about its axis. Find the angular acceleration of each wheel and the difference in the tensions in the free portions of the belt.

20. A uniform straight rod AB of mass M and length $2l$ is freely pivoted at its lower end A to a point in the side of a truck, and makes an angle α with the vertical. A light inextensible string of length $2l$ joins B to a point C of the truck vertically above A. Find the change in the tension of the string

if the truck begins to move in the plane ABC with a horizontal acceleration f towards AC from B.

21. A uniform straight rod AB of length $2l$ is initially horizontal and at rest, and falls in a vertical plane under the action of gravity. Find the angular velocity of the rod when it has turned through an angle θ (i) if A is held fixed, (ii) if A is forced to move along the initial direction of AB with constant acceleration f.

22. A uniform rod AB, of length l, lies on a smooth horizontal table, and is initially rotating with angular velocity Ω about one end A, which is at rest. If A is then constrained to move with constant acceleration f in a horizontal direction perpendicular to the original direction of AB, find the condition that the rod will make complete revolutions.

23. A compound pendulum is swinging freely about a horizontal axis with angular amplitude α (not necessarily small). The length of the equivalent simple pendulum is l. When the angular velocity of the pendulum is instantaneously zero, the axis is given a horizontal acceleration f in a direction perpendicular to itself. Prove that the least value of f for which the body can in the subsequent motion attain the position in which the centre of mass is vertically above the axis is $g \cot \frac{1}{2}\alpha$.

24. The point of suspension of a pendulum is A, and A is caused to move along a horizontal straight line OX. The centre of gravity of the pendulum is G, and $AG = l$. The radius of gyration about any axis through G perpendicular to AG is k. The pendulum can move in the vertical plane containing OX. At time t, $OA = x$, and the angle between AG and the vertical is θ, supposed positive when GAX is obtuse. Prove that

$$(l^2+k^2)\frac{d^2\theta}{dt^2}+lg\sin\theta = l\cos\theta\frac{d^2x}{dt^2}.$$

If d^2x/dt^2 has the constant value f, show that the pendulum can maintain a constant inclination α to the vertical, where $\tan\alpha = f/g$, and that the periodic time of small oscillations about this position is

$$2\pi\left\{\frac{l^2+k^2}{lg}\cos a\right\}^{\frac{1}{2}}.$$

25. A uniform trap-door swinging about a horizontal hinge is closed by a spring coiled about the hinge. The spring is coiled so that it is just able to hold the door shut in the horizontal position. The horizontal opening which the door closes is in a body which is mounting with uniform acceleration f. Show that, if $f = \left(0{\cdot}57+\dfrac{1{\cdot}23}{\alpha}\right)g$, the door starting from the vertical position will just reach the horizontal position, α being the angle through which the spring is coiled when the door is in the horizontal position.

26. A lift has an acceleration f vertically upwards. A frictionless pulley, of mass M, radius a, and radius of gyration k, is suspended from the roof of the lift, and over it passes a light cord, which carries at its ends masses m_1 and m_2. If the masses are released from rest relative to the lift, find the tensions in the two portions of the string during the subsequent motion. It is assumed that there is no slipping between the string and the pulley.

What force will the pulley exert on the lift?

CHAPTER XXI

MOTION OF A RIGID BODY IN A PLANE

21·1 Motion in a plane, reduction of the kineton system. We now consider the general problem of motion in a plane. We take the centre of gravity G as the marked point in the lamina. If the acceleration of G at time t is $\mathbf{f}$, the kineton system is as shown in Fig. 21·1a; the terms mf are all in the direction of $\mathbf{f}$, and $f = |\mathbf{f}|$. Our first task is the reduction

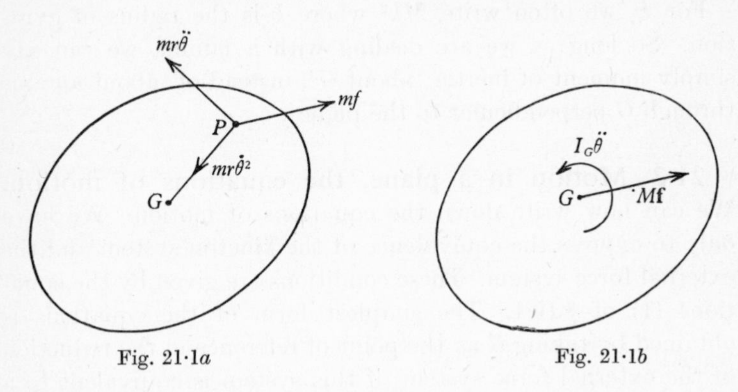

Fig. 21·1a Fig. 21·1b

of this kineton system to a $\mathscr{K}$-force at G and a $\mathscr{K}$-couple. We shall need the well-known lemma

$$Sm\mathbf{r} = 0,$$

where $\mathbf{r}$ is the position vector of a particle of mass m relative to G.

Consider then the reduction of the kineton system for the rigid body. The kineton of a typical particle of the body, as shown in Fig. 21·1a, is obtained from the theorem that the acceleration of the particle is the acceleration of G plus the relative acceleration. It will be convenient to consider separately the components of the three types:

(i) The terms $mr\dot{\theta}^2$ in PG form a system in equilibrium, for we can think of them as all transferred to G, and their vector sum is zero by the lemma just mentioned.

(ii) Next consider the terms $mr\ddot{\theta}$ perpendicular to GP. Their vector sum is zero, and they are equivalent to a $\mathscr{K}$-couple of moment

$$Sr\,.\,mr\ddot{\theta} = I_G\ddot{\theta},$$

where I_G is the moment of inertia of the lamina about an axis through G perpendicular to the plane.

(iii) The terms mf, all in the direction of $\mathbf{f}$, are equivalent to a $\mathscr{K}$-force $M\mathbf{f}$ at G, where M is the mass of the rigid body, $M = Sm$.

Thus the kineton system is equivalent to a $\mathscr{K}$*-force* $M\mathbf{f}$ at G, *and a* $\mathscr{K}$*-couple* $I_G\ddot{\theta}$ (Fig. 21·1*b*). This is the fundamental theorem in the dynamics of a rigid body moving in a plane.

For I_G we often write Mk^2 where k is the radius of gyration. So long as we are dealing with a lamina we can say simply moment of inertia 'about G', instead of 'about an axis through G perpendicular to the plane'.

21·2 Motion in a plane, the equations of motion.

We can now write down the equations of motion. We have only to express the equivalence of the kineton system and the external force system. These conditions are given by the equations (1) of § 19·1. The simplest form of the equations is obtained by taking G as the point of reference in the reduction of the external force system; if this system is equivalent to a force $\mathbf{R}$ at G and a couple N_G we have

$$M\mathbf{f} = \mathbf{R}, \quad I_G\ddot{\theta} = N_G.$$

These equations merit a moment's consideration.

(i) If the vector sum of the external forces is $\mathbf{R}$, the acceleration $\mathbf{f}$ of G is given by the equation

$$M\mathbf{f} = \mathbf{R}. \tag{1}$$

This means that the motion of G is the same as that of a particle of mass M acted on by a force $\mathbf{R}$; the motion of G is the same as if all the external forces acted at G, and all the mass were concentrated at G.

(ii) If the moment of the external forces about G is N_G, the angular motion of the lamina is given by the equation

$$I_G\ddot{\theta} = N_G, \quad \text{or} \quad Mk^2\ddot{\theta} = N_G. \tag{2}$$

This shows that the motion relative to G can be found by the theory of motion about a fixed axis, just as though G were fixed in space.

Finally, we write the equations for fixed rectangular axes Ox, Oy in the plane. The coordinates of G are (ξ, η), and the external force system is equivalent to a force (X_0, Y_0) at O and a couple N_0. Then

$$M\ddot{\xi} = X_0, \tag{3}$$

$$M\ddot{\eta} = Y_0, \tag{4}$$

$$M(\xi\ddot{\eta} - \eta\ddot{\xi}) + I_G\ddot{\theta} = N_0. \tag{5}$$

Here X_0 is the sum of the components of the external forces parallel to Ox, Y_0 the sum of the components parallel to Oy, and N_0 the moment of the system about O; X_0 and Y_0 are the components of $\mathbf{R}$.

Example 1. *A disc, of mass M and radius a, rolls without slipping down a line of greatest slope of a plane inclined at an angle α to the horizontal. The centre of gravity of the disc is at its centre. To find the motion.*

The centre G of the disc moves parallel to the plane; let its acceleration at time t be f down the plane. The kineton system is shown in Fig. 21·2*a*,

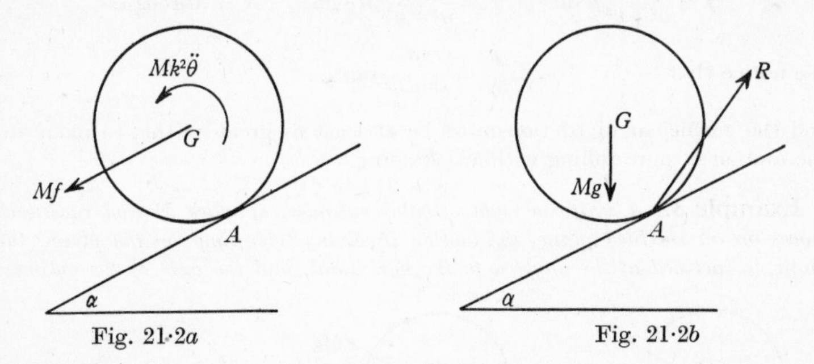

Fig. 21·2*a* Fig. 21·2*b*

and since there is no slipping $\ddot{\theta} = f/a$. The external force system is shown in Fig. 21·2*b*; it consists merely of the weight of the disc at G, and the reaction $\mathbf{R}$ of the plane on the disc at the point of contact A. The systems are equivalent, and taking moments about A we have

$$Mfa + Mk^2 f/a = Mga \sin\alpha,$$

whence

$$f = \frac{g \sin\alpha}{1 + (k^2/a^2)}.$$

Thus it turns out that f is constant. This constant value is smaller than that for a particle sliding down a smooth plane of the same slope. For a uniform disc, $k^2 = \frac{1}{2}a^2$, and $f = \frac{2}{3}g \sin\alpha$; for a uniform hoop, $k^2 = a^2$, and $f = \frac{1}{2}g \sin\alpha$.

In solving problems on the motion of a rigid body in a plane, the reader is recommended to draw separate figures showing the kineton system and the external force system, as in this example. The marking of both systems in the same figure is liable to lead to confusion, and is to be deprecated.

Example 2. *To find the reaction of the plane on the disc in the same problem.* We denote the components of reaction along and normal to the plane by F and N, replacing Fig. 21·2*b* by Fig. 21·2*c*. Resolving parallel and perpendicular to the plane we have

$$Mg\sin\alpha - F = Mf, \qquad (6)$$

$$Mg\cos\alpha - N = 0.$$

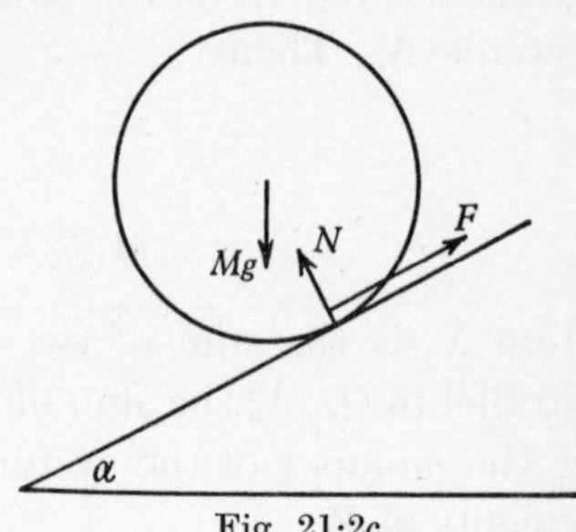

Fig. 21·2*c*

From these we determine F and N, since f is known. However, it is perhaps worth while to give a self-contained solution. The third equation is obtained by taking moments about any point. For example, if we take moments about A we obtain the equation used in the previous solution. If instead we take moments about G we obtain the simpler equation

$$Fa = Mk^2f/a. \qquad (7)$$

The equations (6) and (7) determine F and f, and we have finally

$$f = \frac{a^2}{a^2+k^2}g\sin\alpha, \quad F = \frac{k^2}{a^2+k^2}Mg\sin\alpha, \quad N = Mg\cos\alpha.$$

We notice that

$$\frac{F}{N} = \frac{k^2}{a^2+k^2}\tan\alpha,$$

and the coefficient of friction must be at least as great as this to maintain the motion of pure rolling without skidding.

Example 3. *A uniform right circular cylinder, of mass M and radius a, moves on an inclined plane, the motion involving 'slipping' on the plane; the plane is inclined at an angle α to the horizontal, and the axis of the cylinder*

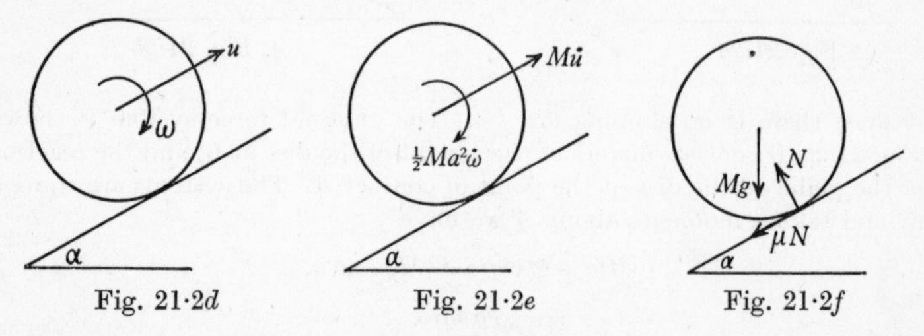

Fig. 21·2*d* Fig. 21·2*e* Fig. 21·2*f*

remains horizontal throughout. The coefficient of friction between plane and cylinder is μ. The cylinder is projected up the plane, the initial values of u and ω (u being the velocity of a point on the axis measured up the plane, and ω the angular velocity taken clockwise as in Fig. 21·2d) being u_0 and ω_0. Given that $u_0 > a\omega_0$, $u_0 > 0$, to discuss the subsequent motion.

We can treat the cylinder as a uniform circular disc in a vertical plane through a line of greatest slope. The kineton system is equivalent to a $\mathscr{K}$-force $M\dot{u}$ at G, and a clockwise $\mathscr{K}$-couple $\frac{1}{2}Ma^2\dot{\omega}$.

The external force system needs a little care. So long as the disc skids the friction has its limiting value and acts in a direction opposite to the direction of motion of the point of the disc in contact with the plane. In our problem the skid velocity $v = u - a\omega$ is positive at $t = 0$, and by continuity $v > 0$ for some interval starting at $t = 0$. Therefore, in the initial stage of the motion the friction acts down the plane. The kineton system and the external force system are shown in Figs. 21·2e and 21·2f. From the equivalence of these two systems we obtain at once the equations $N = Mg\cos\alpha$ and

$$\begin{cases} M\dot{u} = -Mg\sin\alpha - \mu Mg\cos\alpha, \\ \frac{1}{2}Ma^2\dot{\omega} = \mu Mga\cos\alpha. \end{cases}$$

Thus during the initial stage of the motion, u steadily decreases, and ω steadily increases, and this continues for an interval of time t_1 until $u = a\omega$. Explicitly

$$u = u_0 - gt(\sin\alpha + \mu\cos\alpha),$$

$$a\omega = a\omega_0 + 2\mu gt\cos\alpha,$$

and hence
$$v = u - a\omega = (u_0 - a\omega_0) - gt(\sin\alpha + 3\mu\cos\alpha).$$

Thus
$$gt_1 = \frac{u_0 - a\omega_0}{\sin\alpha + 3\mu\cos\alpha},$$

and the value of u at $t = t_1$ is

$$u_1 = \frac{2\mu u_0\cos\alpha + a\omega_0(\sin\alpha + 3\mu\cos\alpha)}{\sin\alpha + 3\mu\cos\alpha}.$$

In the interval $(0, t_1)$ the skid velocity v decreases steadily to zero. At $t = t_1$ both u and ω are continuous, and we must now find what happens for $t > t_1$. There are three possibilities: (i) that v, after falling to the value zero, increases again, so that $v > 0$ for some interval $t > t_1$, (ii) that v remains zero, and the disc rolls without skidding, and (iii) that v continues to decrease, $v < 0$ for some interval $t > t_1$.

But the first of these is immediately seen to be impossible. If v increases from zero the friction still acts down the plane, the equations of motion already found are valid, and $\dot{v} < 0$ as before; this contradicts the assumption that v increases.

If (ii) the disc rolls without skidding, taking the friction F (no longer limiting) up the plane, we have

$$\begin{cases} M\dot{u} = F - Mg\sin\alpha, \\ \frac{1}{2}Ma^2(\dot{u}/a) = -Fa, \end{cases}$$

giving $\dot{u} = -\frac{2}{3}g\sin\alpha$, $F = \frac{1}{3}Mg\sin\alpha$. This is only possible if $\mu \geqslant \frac{1}{3}\tan\alpha$.

If (iii) v continues to decrease, the friction is now $\mu Mg\cos\alpha$ *up* the plane, and we have

$$\begin{cases} M\dot{u} = Mg(\mu\cos\alpha - \sin\alpha), \\ \frac{1}{2}Ma^2\dot{\omega} = -\mu Mga\cos\alpha, \end{cases}$$

and $\dot{v} = \dot{u} - a\dot{\omega} = (3\mu\cos\alpha - \sin\alpha)g$. Our assumption that v decreases from zero is not valid unless

$$\mu < \tfrac{1}{3}\tan\alpha.$$

Thus all the possibilities are accounted for. If $\mu \geqslant \frac{1}{3}\tan\alpha$ the disc rolls without skidding for $t > t_1$. If $\mu < \frac{1}{3}\tan\alpha$ it continues to skid always; u continues to decrease, ω continues to decrease, and v continues to decrease, for all $t > t_1$.

There is a somewhat less formal attack on the problem for $t > t_1$. This is to assume provisionally that there is pure rolling, then to find the frictional force required for this motion, and then to assume, if this does not exceed the limiting value, that the cylinder rolls. This can be summed up in the maxim 'it will roll if it can', which is useful in practice; but a complete justification requires us to prove, as above, that the alternative assumptions are untenable.

Example 4. *A uniform solid circular cylinder, of mass M and radius a, rolls without slipping on the inside of a fixed hollow circular cylinder of radius b. The axes of the cylinders are horizontal and parallel. To investigate the motion; and to find the least value required for the coefficient of friction if the motion starts from rest with the point of contact at an angular distance* $\alpha (< \frac{1}{2}\pi)$ *from the lowest point of the fixed cylinder.*

We denote the inclination to the vertical of the plane containing the axes of the cylinders by θ, measured so that $\theta = 0$ when the point of contact is the lowest point of the fixed cylinder. We form the equations of motion as for a uniform circular disc rolling on a fixed circle. In the figure (Fig. 21·2*g*) A is the lowest point of the fixed circle, and C the point on the rim of the disc which is in contact with A when the disc is in its lowest position. The line GC, fixed in the disc, makes an angle ϕ with the downward vertical, so that the angular velocity of the disc (clockwise) is $\dot{\phi}$.

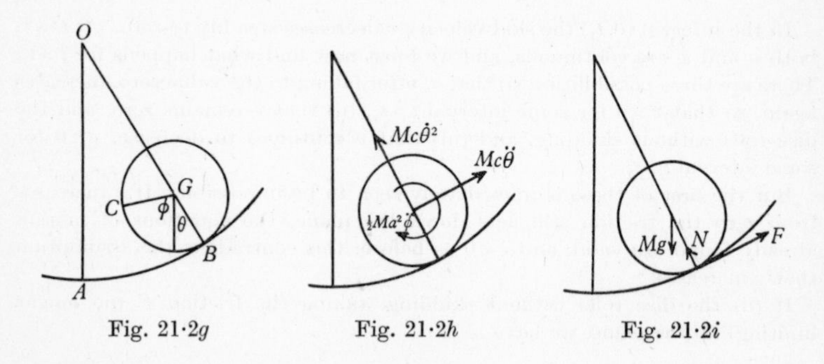

Fig. 21·2*g* Fig. 21·2*h* Fig. 21·2*i*

Since the motion is one of pure rolling, the arc AB of the circle is equal to the arc CB of the disc, which gives us

$$b\theta = a(\theta + \phi),$$

or, writing c for $b - a$,

$$a\phi = c\theta. \tag{8}$$

This is the so-called *rolling condition*, the relation expressing the fact that there is no skidding. Its effect is to reduce the number of variables required to describe the configuration to one. We are here dealing with a dynamical system with one degree of freedom.

The rolling condition can be found in various ways. For example, the velocity v of G is $c\dot{\theta}$ perpendicular to OG; but it is also $a\dot{\phi}$, since B is the instantaneous centre of rotation of the disc. Thus

$$v = a\dot{\phi} = c\dot{\theta},$$

and this is another form of the rolling condition. It shows at once that $a\phi - c\theta$ remains constant, and this constant value must be zero, since θ and ϕ vanish together.

The kineton system is shown in Fig. 21·2h and the external force system in Fig. 21·2i. Taking moments about B we have

$$Mga\sin\theta = -Mac\ddot{\theta} - \tfrac{1}{2}Ma^2\ddot{\phi},$$

and since $a\ddot{\phi} = c\ddot{\theta}$ this becomes

$$\ddot{\theta} + \frac{2g}{3c}\sin\theta = 0. \tag{9}$$

This equation is the same as that for a simple pendulum of length $\frac{3}{2}c$, and this holds so long as the motion is one of pure rolling. For this assumption to be verified we must assure ourselves that $N \geqslant 0$ throughout the motion; and further, if the surfaces are imperfectly rough, that $|F| \leqslant \mu N$ throughout.

From equation (9), remembering that

$$\ddot{\theta} = \frac{d}{d\theta}(\tfrac{1}{2}\dot{\theta}^2),$$

we find
$$\dot{\theta}^2 = \frac{4g}{3c}(\cos\theta - \cos\alpha), \tag{10}$$

since the motion is an oscillation of amplitude α.

Next, we express N and F in terms of θ. Resolving in the direction GO we have

$$N - Mg\cos\theta = Mc\dot{\theta}^2,$$

and substituting the value of $c\dot{\theta}^2$ in terms of θ from equation (10) we have

$$N = \tfrac{1}{3}Mg(7\cos\theta - 4\cos\alpha).$$

Again, taking moments about G, we have

$$\begin{aligned} Fa &= -\tfrac{1}{2}Ma^2\ddot{\phi} \\ &= -\tfrac{1}{2}Mac\ddot{\theta}, \\ F &= \tfrac{1}{3}Mg\sin\theta. \end{aligned}$$

The condition $N > 0$ is satisfied in the range $0 \leqslant \theta \leqslant \alpha$ (it is clearly sufficient to consider positive values of θ) as we should expect with $\alpha < \frac{1}{2}\pi$. The condition $F \leqslant \mu N$ is satisfied if the greatest value of F/N, i.e. of

$$\frac{\sin\theta}{7\cos\theta - 4\cos\alpha},$$

in the range $0 \leqslant \theta \leqslant \alpha$ is not greater than μ. Now as θ increases from 0 to α, the numerator of this fraction continually increases, and the denominator continually decreases, so the fraction continually increases, and reaches its greatest value when $\theta = \alpha$. This greatest value is $\frac{1}{3}\tan\alpha$, and the condition for pure rolling is satisfied if $\mu \geqslant \frac{1}{3}\tan\alpha$.

Example 5. *The loaded cylinder. A particle of mass m is attached to a point on the surface of a uniform solid cylinder, of mass M and radius a, at a point mid-way between the ends; the loaded cylinder rolls on a perfectly rough horizontal plane. To find the equation of motion.*

We suppose that at time t the axial plane through the particle makes an angle θ with the downward vertical, as shown in Fig. 21·2*j*. To form the equation of motion we have to make use of the equivalence of kineton system and external force system.

To reduce the kineton system we might think of the loaded cylinder as a rigid body with its centre of gravity G not on the axis of the cylinder. But it is much simpler to superpose on the kineton for the uniform cylinder the kineton of the added particle. We thus obtain, for the kineton and the external force systems, the results shown in Figs. 21·2*j* and 21·2*k*.

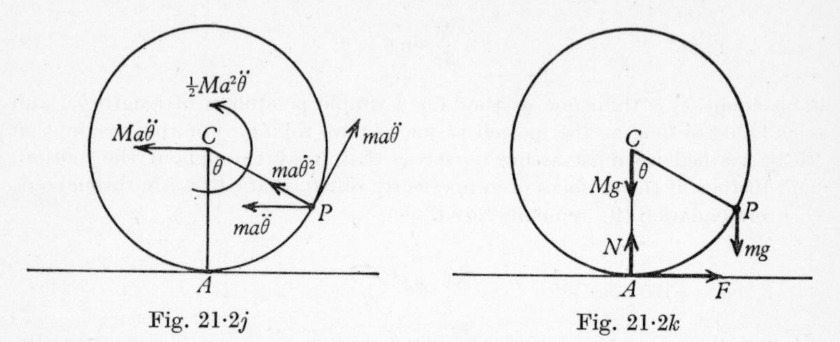

Fig. 21·2*j* Fig. 21·2*k*

Taking moments about the point of contact with the plane we have

$$-mga\sin\theta = \tfrac{1}{2}Ma^2\ddot{\theta} + Ma^2\ddot{\theta} + a\sin\theta(ma\dot{\theta}^2) + 2a(1-\cos\theta)(ma\ddot{\theta});$$

on rearranging we obtain the equation of motion

$$(3M/2m)\ddot{\theta} + 2(1-\cos\theta)\ddot{\theta} + \sin\theta.\dot{\theta}^2 + n^2\sin\theta = 0,$$

where $n^2 = g/a$.

Example 6. *A uniform circular disc, of mass M and radius a, is suspended in a vertical plane by two vertical strings attached at points A, B on the rim of the disc the radius vector to which makes an angle α with the upward vertical. One of the strings is cut; find the tension in the other immediately afterwards.*

This is a problem of *initial motion*. The characteristic of such a problem is that the velocity components are all zero at the instant considered, but the acceleration components are not.

Now the initial acceleration of A is horizontal; for A moves on a circle, and the angular velocity in the circle is zero. We denote the acceleration of A by f to the right (Fig. 21·2*l*), and the angular acceleration of the disc (clockwise) by ϕ; all of this refers to an instant just after the string at B has been cut. The acceleration of G relative to A is $a\phi$ perpendicular to AG, and the kineton system is as shown in Fig. 21·2*l*. The external force system is shown in Fig. 21·2*m*. From the equivalence of these systems we find

$$\begin{cases} f = a\phi\cos\alpha, \\ Mg - T = Ma\phi\sin\alpha, \\ Ta\sin\alpha = \tfrac{1}{2}Ma^2\phi. \end{cases}$$

From these equations we readily determine f, ϕ and T,

$$T = \frac{Mg}{1+2\sin^2\alpha} = \frac{Mf}{\sin 2\alpha} = \frac{Ma\phi}{2\sin\alpha}.$$

Before the other string is cut the tension is $\frac{1}{2}Mg$, so there is no change in the tension if α is $\frac{1}{4}\pi$.

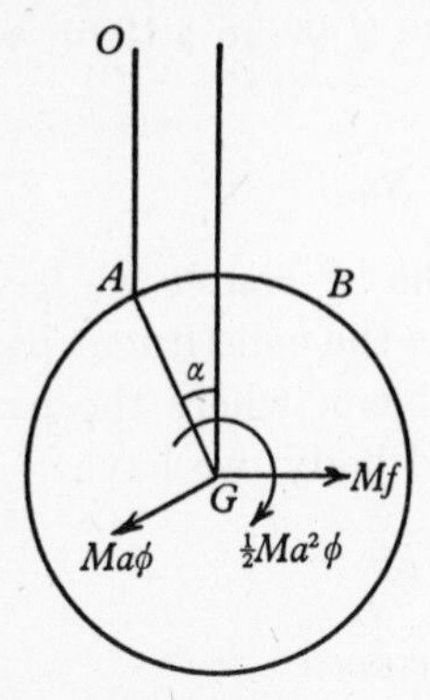

Fig. 21·2l

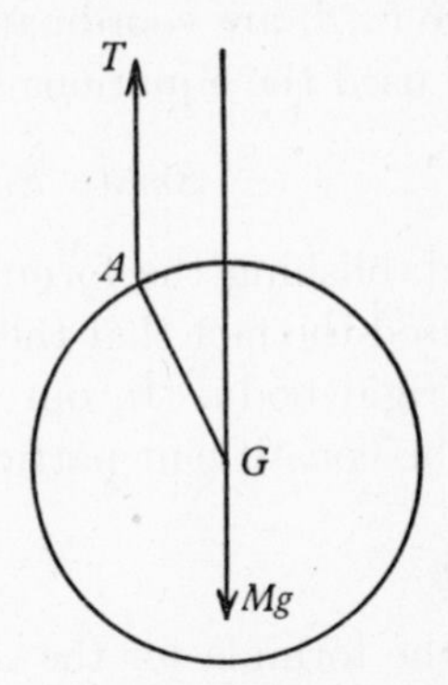

Fig. 21·2m

21·3 The conservation of momentum. We now consider some further consequences of the equations of motion for a rigid body moving in a plane,

$$M\ddot{\xi} = X_0, \quad M\ddot{\eta} = Y_0, \quad M(\xi\ddot{\eta} - \eta\ddot{\xi}) + I_G\ddot{\theta} = N_0. \tag{1}$$

The notation is that used in § 21·2, and the equations just written are the equations (3), (4) and (5) of that paragraph; (ξ, η) are the coordinates of G, and the system of external forces has been reduced to a force (X_0, Y_0) at O and a couple N_0.

If we write

$$p = M\dot{\xi}, \quad q = M\dot{\eta}, \quad h = M(\xi\dot{\eta} - \eta\dot{\xi}) + I_G\dot{\theta}, \tag{2}$$

the equations (1) take the form

$$\dot{p} = X_0, \quad \dot{q} = Y_0, \quad \dot{h} = N_0. \tag{3}$$

The definitions (2) of $(p, q; h)$ suffice for our present purpose, but it is important to notice that if we take the momentum $(m\dot{x}, m\dot{y})$ of each particle, and reduce this system of localized vectors (as in § 19·1) to an $\mathscr{M}$-force at O and an $\mathscr{M}$-couple, the $\mathscr{M}$-force has components (p, q) and the $\mathscr{M}$-couple has moment h. To prove this we observe that the components of the $\mathscr{M}$-force are

$$(Sm\dot{x}, Sm\dot{y}) = (M\dot{\xi}, M\dot{\eta}).$$

The $\mathscr{M}$-couple has moment

$$Sm(x\dot{y}-y\dot{x}) = Sm\{(\xi+\alpha)(\dot{\eta}+\dot{\beta})-(\eta+\beta)(\dot{\xi}+\dot{\alpha})\}$$
$$= M(\xi\dot{\eta}-\eta\dot{\xi})+Sm(\alpha\dot{\beta}-\beta\dot{\alpha}), \quad (4)$$

where (α, β) are coordinates relative to G (as in § 19·6), and we have used the equations

$$Sm\alpha = Sm\beta = Sm\dot{\alpha} = Sm\dot{\beta} = 0.$$

In establishing the formula (4) for the $\mathscr{M}$-couple we have not yet used the fact that the particles are the constituent particles of a rigid body. In our present problem, where the particles are the constituent particles of a rigid body, we have

$$\dot{\beta} = \alpha\dot{\theta}, \quad \dot{\alpha} = -\beta\dot{\theta},$$

and the formula for the $\mathscr{M}$-couple becomes

$$M(\xi\dot{\eta}-\eta\dot{\xi})+Sm(\alpha^2+\beta^2)\,\dot{\theta} = M(\xi\dot{\eta}-\eta\dot{\xi})+I_G\,\dot{\theta}.$$

This completes the proof that the reduced momentum system, with O as origin, is $(p, q; h)$. We shall return to this reduction later (§ 22·2) in connexion with the theory of impulsive motion. The components (p, q) of the $\mathscr{M}$-force are called components of linear momentum, and the $\mathscr{M}$-couple h is called the angular momentum about O. This is the same notation that we used for the angular momentum in the theory of orbits, where we were concerned with a single moving particle.

Consider now a problem in which $X_0 = 0$ for all time. Then, from the first of the equations (3), we see that p remains constant. Explicitly, if the sum of the components of the external forces in a certain direction is zero for all time, the component of linear momentum in that direction remains constant: an equivalent statement is that the component of the velocity of G in that direction remains constant. This is the principle of *the conservation of linear momentum* for a single rigid body. It is a special case of a general principle which we shall establish later (§ 24·1) for a general dynamical system.

In the same way from the third of equations (3) we see that if $N_0 = 0$ for all time h is constant: if the moment of the system of external forces about a certain fixed point O is zero for all time the angular momentum about O is constant. This is the principle of *the conservation of angular momentum*. Again, it is a special case of a general principle which holds for a general dynamical system (§ 24·1).

Example. *The end A of a rod AB, not necessarily uniform, is freely attached to a small light ring which slides on a smooth horizontal rail. Prove that when the rod moves in the vertical plane through the rail the horizontal component of the velocity of the centre of gravity of the rod is constant.*

Since the ring is light the action of the ring on the rod is equal, in magnitude and direction, to the action of the rail on the ring; and this is vertical. Thus all the external forces on the rod are vertical, and from the principle of the conservation of linear momentum the horizontal component of the velocity of G is constant.

21·4 Motion of a solid.

Hitherto we have tacitly assumed that in the plane motion of a rigid body the equations of motion can be found by replacing the body by a lamina. We now consider this assumption more closely.

We consider then a plane motion of a rigid body, i.e. a motion in which the various particles of the body move in planes parallel to a given plane, the *plane of motion*. In many cases the fact that the motion is of this type is evident from symmetry. The theory also applies to a rigid body constrained to move in this way if the forces of constraint are all perpendicular to the plane of motion.

The kineta of the particles are not now coplanar, but are all in parallel planes. In most cases the external forces are also parallel to these planes; when they are not (as, for example, in the case of a body constrained in the way just mentioned) we must concentrate our attention on the components of these external forces parallel to the plane of motion, and the forces of constraint, which are perpendicular to the plane of motion, do not appear in the equation.

The kineton-system is equivalent to the external-force system; *the same is true of the systems obtained by projection on the plane of motion*. That is to say, if we take the orthogonal projections of the particles on the plane of motion, and think of the kineta and of the components of the external forces as acting at these projections, the two coplanar systems thus obtained are still equivalent. This is the principle we use when we replace the rigid body by a lamina in the plane of motion. As we have seen, the conditions expressing the equivalence of the two coplanar systems are sufficient to establish the equations of motion.

A complete proof of the statement that the projected systems are equivalent requires a little familiarity with the theory of a system of forces, or other localized vectors, in three dimensions. In the three-dimensional case the necessary and sufficient conditions for the equivalence of the external force-system and the kineton-system are

$$Sm\ddot{x} = SX, \tag{1}$$

$$Sm\ddot{y} = SY, \tag{2}$$

$$Sm\ddot{z} = SZ, \tag{3}$$

$$Sm(y\ddot{z} - z\ddot{y}) = S(yZ - zY), \tag{4}$$

$$Sm(z\ddot{x} - x\ddot{z}) = S(zX - xZ), \tag{5}$$

$$Sm(x\ddot{y} - y\ddot{x}) = S(xY - yX), \tag{6}$$

where (x, y, z) are the coordinates of a typical particle of mass m, and (X, Y, Z) are the components of the external force on it. If the motion is parallel to the plane $z = 0$ the equations (1), (2) and (6) constitute the result stated.

In the problem under discussion $\ddot{z} = 0$ for each particle, but this does not mean that the external force-system has zero moment about Ox and about Oy. For example, suppose the body to rotate steadily with angular velocity ω about an axis fixed in the body and coincident with Oz. In this case

$$\ddot{x} = -\omega^2 x, \quad \ddot{y} = -\omega^2 y,$$

and the moments about Ox and about Oy of the external-force system required to maintain the motion are (from equations (4) and (5))

$$-Smz\ddot{y} = \omega^2 Smyz, \quad Smz\ddot{x} = -\omega^2 Smzx,$$

and these are not zero in general. But they are zero if the plane $z = 0$ is a plane of symmetry for a uniform body, and this is true in most of the problems we shall meet in this book.

21·5 Kinetic energy, König's formula. We will suppose now that the rigid body considered is a rigid lamina moving in a plane; though in fact it is not difficult to see that the result we shall obtain is also applicable in the more general case of any rigid body moving parallel to a plane. We use the same notation as in § 21·3, (x, y) being the coordinates of a typical particle, of mass m, referred to fixed axes in the plane, and (α, β) its coordinates relative to G. The kinetic energy of the lamina, defined as the sum of the kinetic energies of the individual particles, is

$$\begin{aligned} T &= \tfrac{1}{2}Sm(\dot{x}^2+\dot{y}^2) \\ &= \tfrac{1}{2}Sm\{(\dot{\xi}+\dot{\alpha})^2+(\dot{\eta}+\dot{\beta})^2\} \\ &= \tfrac{1}{2}M(\dot{\xi}^2+\dot{\eta}^2)+\tfrac{1}{2}Sm(\dot{\alpha}^2+\dot{\beta}^2), \end{aligned} \tag{1}$$

where we have used the equations

$$Sm\dot{\alpha} = Sm\dot{\beta} = 0.$$

In establishing the formula (1) we have not yet used the fact that the particles are the constituent particles of a rigid body. In our present problem, where the particles are the constituent particles of a rigid body,

$$\dot{\alpha}^2+\dot{\beta}^2 = r^2\omega^2,$$

where $r^2 = \alpha^2+\beta^2$, and $\omega(=\dot{\theta})$ is the angular velocity of the lamina. (Actually $\dot{\alpha} = -\beta\omega, \dot{\beta} = \alpha\omega$.) The formula (1) thus becomes

$$\begin{aligned} T &= \tfrac{1}{2}M(\dot{\xi}^2+\dot{\eta}^2)+\tfrac{1}{2}Smr^2\omega^2 \\ &= \tfrac{1}{2}MW^2+\tfrac{1}{2}I_G\omega^2, \end{aligned}$$

where W is the velocity of G. This is König's formula for the kinetic energy of the moving body.

The kinetic energy can thus be expressed as the sum of two terms, the first $\tfrac{1}{2}MW^2$ as though all the mass were concentrated at G, the second $\tfrac{1}{2}I_G\omega^2$ as for rotation about a fixed axis at G.

21·6 The equation of energy. For the purpose of this paragraph it will be convenient to take G as the origin for the reduction of the external-force system. If the external forces are equivalent to a force (X_0, Y_0) at G and a couple N_G we have

$$X_0 = M\ddot{\xi}, \quad Y_0 = M\ddot{\eta}, \quad N_G = I_G\ddot{\theta},$$

whence
$$\begin{aligned} X_0\dot{\xi}+Y_0\dot{\eta}+N_G\dot{\theta} &= M(\dot{\xi}\ddot{\xi}+\dot{\eta}\ddot{\eta})+I_G\dot{\theta}\ddot{\theta} \\ &= \frac{d}{dt}\{\tfrac{1}{2}M(\dot{\xi}^2+\dot{\eta}^2)+\tfrac{1}{2}I_G\dot{\theta}^2\} \\ &= dT/dt. \end{aligned} \tag{1}$$

Thus, the rate of increase of kinetic energy is equal to the rate of working of the external forces.

Of course, as we should expect, the result is evident in other ways; the rate of increase of kinetic energy must be equal to the rate of working of *all* the forces, and the rate of working of the internal forces is zero.

Now consider the particular case when the external forces are derived from a field of force. We form the function $V(=Sm\phi)$ and from it the function $\Omega(t)$, as in § 19·9, and the rate of working of the external forces is $-d\Omega/dt$. Equation (1) above therefore leads to the equation

$$\frac{d}{dt}(T+\Omega)=0,$$

whence
$$T+\Omega = C. \tag{2}$$

This is the classical form of the equation of energy. As we have remarked in the simpler cases of the same theorem it is permissible to write it in the form

$$T+V=C, \tag{3}$$

the symbol V now referring, not to a general configuration, but to the particular configuration occupied at time t. The simplest case is that of the uniform field of gravity, for which

$$V=Mg\eta,$$

where the axis Oy is taken in the direction opposite to that of the field.

The equations (2) and (3) are valid in a much wider range of problems than those of free motion in the field. There may be other external forces on the body which, by their very nature, do no work as the body moves. Such forces are called *forces of constraint*, or *smooth constraints*. The intervention of such forces, in addition to the forces of the field, will not invalidate the classical equation of energy.

The commonest types of forces of constraint which occur in problems on the motion of a rigid body are these:

(i) The action on the body at a fixed point about which it rotates, e.g. the action on a pendulum at the point of support (§ 20·1). The action is of course here assumed to be a force through the point of support.

(ii) The action on the body at a point of it which is constrained to move on a smooth guiding curve; the force is perpendicular to the direction of motion of the point of the body on which it acts.

(iii) The action on the body where it slides on a fixed smooth surface; here again the force is perpendicular to the direction of motion of the point of the body on which it acts.

(iv) The action on the body where it rolls without sliding on a fixed rough surface; the velocity of the point of the body on which the force acts is zero.

The equation of energy is a *first integral* of the equations of motion. It is of great importance and usefulness in concrete applications of the theory, for example, in the solution of problems of the type we have already considered in § 21·2. The first two of the examples which follow have already been discussed there.

Example 1. *Consider again Example* 4 *of* § 21·2, *a cylinder rolling inside a fixed cylinder*. If v is the velocity of the centre of the rolling cylinder when OG makes an angle θ with the downward vertical (Fig. 21·2g) we have

$$T = \tfrac{1}{2}Mv^2 + \tfrac{1}{2}(\tfrac{1}{2}Ma^2)\frac{v^2}{a^2} = \tfrac{3}{4}Mv^2 = \tfrac{3}{4}Mc^2\dot{\theta}^2,$$

and

$$V = -Mgc\cos\theta.$$

Thus from the equation of energy

$$\dot{\theta}^2 = \frac{4g}{3c}(\cos\theta - \cos\alpha),$$

and we derive at once, by differentiation with respect to θ,

$$\ddot{\theta} = -\frac{2g}{3c}\sin\theta.$$

These are the equations (10) and (9) (of § 21·2) obtained previously.

In this problem there is another force acting on the rolling cylinder in addition to its weight, namely, the reaction of the fixed cylinder. But this additional force belongs to the category of forces of constraint, and the classical equation of energy

$$T+V = C$$

is still valid.

Example 2. *Consider again the loaded cylinder, Example 5 of* § 21·2. With the same notation as before (Fig. 21·2*j*) the velocity of the particle is the vector sum of $a\dot{\theta}$ horizontally and $a\dot{\theta}$ perpendicular to CP. Thus

$$T = \tfrac{1}{2}Ma^2\dot{\theta}^2 + \tfrac{1}{2}(\tfrac{1}{2}Ma^2)\,\dot{\theta}^2 + \tfrac{1}{2}mv^2,$$

where

$$v^2 = 2a^2\dot{\theta}^2(1-\cos\theta).$$

So we have

$$T = \tfrac{3}{4}Ma^2\dot{\theta}^2 + ma^2\dot{\theta}^2(1-\cos\theta),$$

and

$$V = -mga\cos\theta.$$

Therefore the equation of energy gives us

$$\tfrac{3}{4}Ma^2\dot{\theta}^2 + ma^2\dot{\theta}^2(1-\cos\theta) - mga\cos\theta = C.$$

Differentiating with respect to θ we find

$$(3M/2m)\,\ddot{\theta} + 2(1-\cos\theta)\,\ddot{\theta} + \sin\theta.\dot{\theta}^2 + (g/a)\sin\theta = 0,$$

as before.

In the two examples just considered the body has just one degree of freedom, and we only need one equation, such as the equation of energy, to determine the motion. The example following is one in which the body has two degrees of freedom.

Example 3. *The end A of a uniform rod AB, of mass M and length 2a, is freely attached to a small light ring, which is free to slide on a smooth horizontal rail. To study the motion when the rod moves in the vertical plane through the rail.*

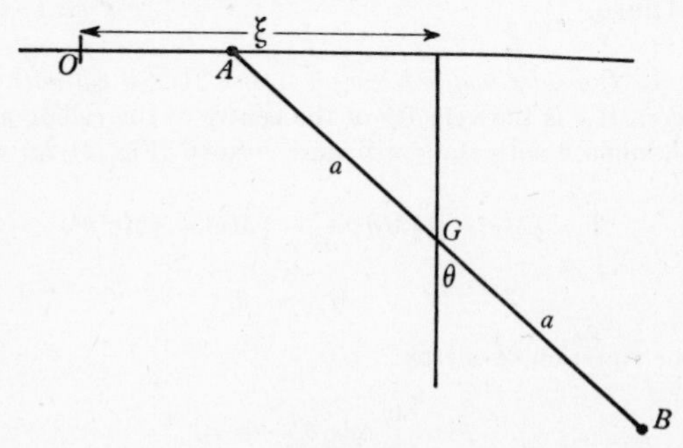

Fig. 21·6

The system has two degrees of freedom. We shall make use of the principles of the conservation of linear momentum (§ 21·3) and of energy (§ 21·6). Since all the external forces on the rod are vertical, the horizontal component of velocity of its centre of gravity G is constant (as we have noticed already in § 21·3, *Example*). Moreover, the force on the rod at A is a smooth constraint, i.e. it is a force of the kind that does not invalidate the equation of energy (the direction of motion of A is always at right angles to the force, and this force does no work) so the energy is conserved. We denote by ξ the horizontal distance of G from a fixed vertical line, and by θ the inclination of the rod to the vertical, as in Fig. 21·6.

From the conservation of momentum

$$\dot{\xi} = \text{constant},$$

and from the conservation of energy

$$\tfrac{1}{2}M(\dot{\xi}^2 + a^2 \sin^2 \theta . \dot{\theta}^2) + \tfrac{1}{2}M \frac{a^2}{3} \dot{\theta}^2 - Mga \cos\theta = C.$$

Hence $$\dot{\theta}^2(3\sin^2\theta + 1) - 6(g/a)\cos\theta = \text{constant},$$

and this is the equation controlling the angular motion; if $\dot{\theta} = \omega$ when $\theta = 0$ it becomes

$$\dot{\theta}^2(4 - 3\cos^2\theta) = \omega^2 - 6(g/a)(1 - \cos\theta). \tag{4}$$

The equation is of the familiar form $\tfrac{1}{2}\dot{\theta}^2 = f(\theta)$ which we have already encountered so often.

We notice that equation (4) does not involve the value of $\dot{\xi}$, and we could have foreseen a result of this form. The equation is the same as if $\dot{\xi}$ were zero. This must be so, because a base moving uniformly with velocity $\dot{\xi}$ is still a Newtonian base, and relative to this base the horizontal velocity of G is zero.

We will content ourselves with two simple applications of equation (4):

(i) To find the period of a small oscillation about $\theta = 0$. If we retain only terms of the second order in θ and $\dot{\theta}$ we obtain the approximate equation

$$\dot{\theta}^2 + 3(g/a)\theta^2 = \omega^2,$$

as for a simple pendulum of length $\tfrac{1}{3}a$ (§ 11·10).

(ii) To find the least value of ω in order that the rod may make complete revolutions. We need the condition that $\dot{\theta}^2 > 0$ for $0 \leqslant \theta \leqslant \pi$, i.e. $\omega^2 > 12g/a$.

21·7 Moments about the instantaneous centre.

We found, in § 20·1, that the equation of motion about a fixed axis is

$$N = I\dot{\omega}, \tag{1}$$

and the reader might be tempted to expect a similar equation to be valid for general motion in a plane if we take moments about the instantaneous centre. This expectation is not fulfilled! It is, however, easy to find the necessary correction.

In the figure (Fig. 21·7*a*) the path of G is shown, and the acceleration of G has components $\dot{W}$ and W^2/ρ along the tangent and normal to the path, ρ denoting the radius of curvature of the path of G. The instantaneous centre I lies on the normal to the path; we denote the distance IG by R, and we notice that $W = R\omega$. The kineton system is shown in the figure and this system is equivalent to the external-force system. If the moment of the external forces about the instantaneous centre I is N_1 we have

$$\begin{aligned} N_1 &= MR\dot{W} + Mk^2\dot{\omega} \\ &= MR(R\dot{\omega} + \dot{R}\omega) + Mk^2\dot{\omega} \\ &= M(R^2 + k^2)\dot{\omega} + MR\dot{R}\omega \\ &= I_1\dot{\omega} + MR\dot{R}\omega, \end{aligned} \tag{2}$$

where I_1 is the moment of inertia about the instantaneous centre. Equation (2) is the required equation of motion for moments about the instantaneous centre. We notice the additional term $MR\dot{R}\omega$ in the second member of equation (2) as compared with equation (1).

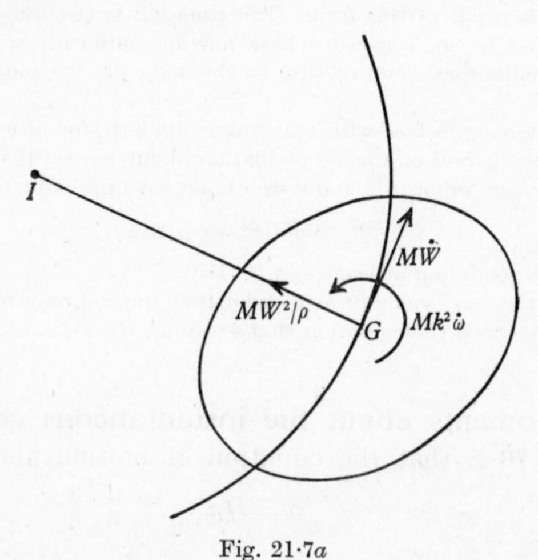

Fig. 21·7*a*

There is one conspicuous case where the additional term is absent, namely, when the body centrode is a circle with G as centre. In that case $\dot{R} = 0$, and I_1 is a constant. In particular, if a uniform circular disc rolls, without slipping, on a fixed curve, the equation of motion

$$I_1\dot{\omega} = N_1$$

is valid.

Another case in which the additional term is zero is that of motion from rest, where at the instant when the body is released, $\dot{R} = 0$ and $\dot{\omega} \neq 0$; but this of course only determines the initial value of $\dot{\omega}$.

A third case arises in a problem of small oscillations, when we need an equation correct only to the first order in θ (taking $\theta = 0$ in the position of equilibrium). The additional term is of the second order, and can be omitted; moreover, we can simplify the procedure still further, since (i), to the requisite order of approximation, I_1 may be given its equilibrium value, and (ii) we need only an approximation to N_1 correct to the order θ.

Example 1. *Consider still once again the problem of a cylinder rolling inside a fixed cylinder, which has already been dealt with in* §§ 21·2 *and* 21·6.

With the notation of Fig. 21·2*g*, $I_1 = \frac{3}{2}Ma^2$, $\omega = -\dot{\phi}$, and taking moments about B we obtain

$$Mga \sin\theta = -\tfrac{3}{2}Ma^2\ddot{\phi},$$

and, since

$$a\phi = c\theta,$$

this becomes

$$\ddot{\theta} + \frac{2g}{3c}\sin\theta = 0,$$

as before.

Example 2. *A uniform solid circular cylinder, of radius a, is cut in halves by a plane through its axis, and one half is placed with its curved surface in contact with a rough horizontal plane. To find the length of the equivalent simple pendulum when the half cylinder makes small oscillations about the position of equilibrium.*

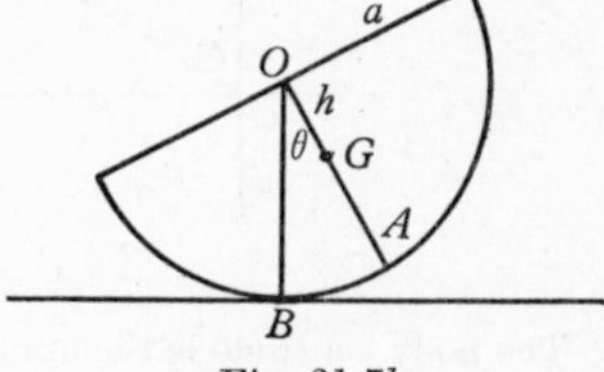

Fig. 21·7*b*

We use the notation of Fig. 21·7*b*, h denoting the distance of G from the axis. Taking moments about B, we have, correct to the first order in θ,

$$I_A\ddot{\theta} = -Mgh\theta,$$

where I_A is the moment of inertia of the half-cylinder about the generator through A.

To find h we use a uniform semi-circular disc, and we have, by Pappus' theorem,

$$\tfrac{1}{2}\pi a^2 . 2\pi h = \tfrac{4}{3}\pi a^3,$$

$$h = \frac{4}{3\pi}a.$$

To find I_A we notice that the moment of inertia about the axis (through O) is $\frac{1}{2}Ma^2$, whence

$$I_A = M\{\tfrac{1}{2}a^2 - h^2 + (a-h)^2\}$$
$$= M(\tfrac{3}{2}a^2 - 2ah).$$

The length of the equivalent simple pendulum is

$$\frac{I_A}{Mh} = \frac{3a^2}{2h} - 2a = \left(\frac{9\pi}{8} - 2\right)a,$$

which is approximately $1{\cdot}534a$.

Example 3. *To obtain the equation of energy (primitive form) from the equation of moments about the instantaneous centre.*

We have
$$N_1 = I_1\dot{\omega} + MR\dot{R}\omega. \tag{3}$$
Now
$$I_1 = M(k^2+R^2), \tag{4}$$
so that
$$\dot{I}_1 = 2MR\dot{R}.$$
Thus, from equations (3) and (4)
$$\begin{aligned} N_1\omega &= I_1\omega\dot{\omega} + \tfrac{1}{2}\dot{I}_1\omega^2 \\ &= \frac{d}{dt}(\tfrac{1}{2}I_1\omega^2) \\ &= dT/dt, \end{aligned}$$
and this proves that the rate of working of the external forces is equal to the rate of increase of the kinetic energy.

Example 4. *A rigid lamina moves in a vertical plane, one point A of the lamina being smoothly constrained to move on a horizontal line Ox and one point B of the lamina being smoothly constrained to move on a vertical line Oy; the centre of gravity of the lamina is the mid-point of AB. To discuss the motion.*

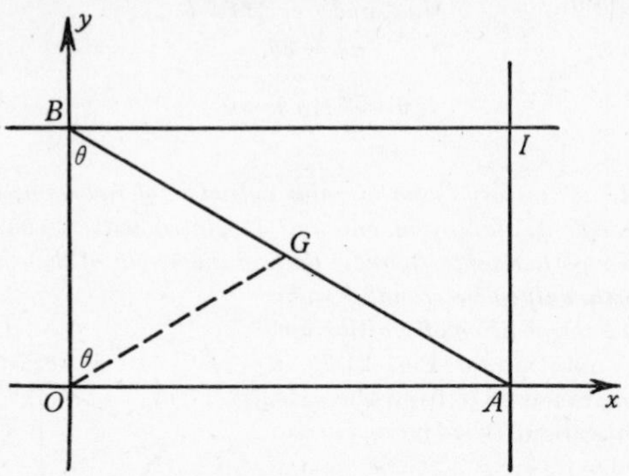

Fig. 21·7c

The body centrode is the circle on AB as diameter, with G as centre, so we can take moments about I without including the additional term. Thus, with the notation of the figure (Fig. 21.7c), taking Oy vertically downwards, and remembering that the reactions at A and B pass through I, we have, if $AB = 2a$,
$$M(a^2+k^2)\ddot{\theta} = -Mga\sin\theta.$$
Integrating, we have
$$\tfrac{1}{2}M(a^2+k^2)\dot{\theta}^2 - Mga\cos\theta = C,$$
which we could have written down from the conservation of energy. The θ-motion is similar to that of a simple pendulum; cf. equation (1) of § 11·9.

21·8 Reaction of pivot on pendulum.

In the problem of motion about a fixed axis (§ 20·2) we observed that among the external forces is the action **S** of the pivot on the lamina, but we did not then show how this reaction can be determined.

This is easily achieved from the general equations of motion. We will content ourselves with finding this reaction in the case of a rigid pendulum, oscillating through an angle α on either side of the vertical.

We denote the components of **S** perpendicular and parallel to GO by P and Q; the external force and kineton systems are shown in the figures. The equivalence of these systems enables us to find P and Q in terms of θ.

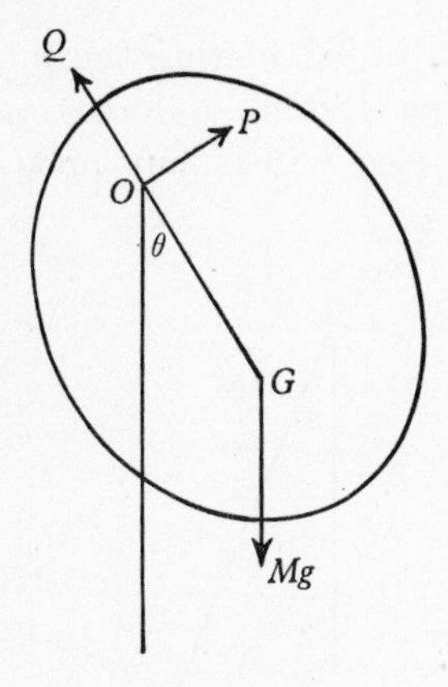

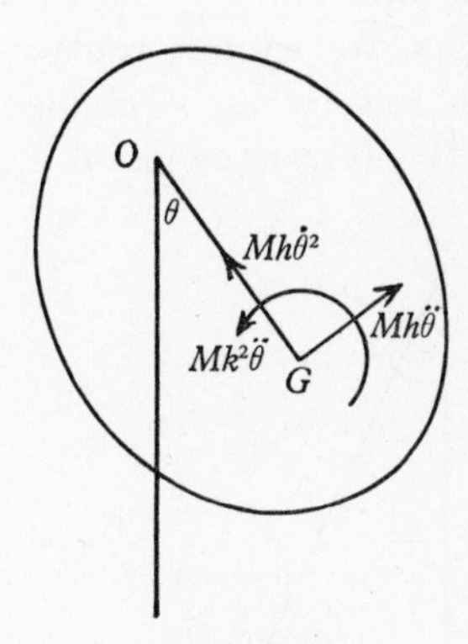

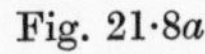
Fig. 21·8a

Fig. 21·8b

Taking moments about O we have the familiar equation

$$M(h^2+k^2)\,\ddot{\theta}+Mgh\sin\theta = 0, \tag{1}$$

whence
$$M(h^2+k^2)\,\dot{\theta}^2 = 2Mgh(\cos\theta-\cos\alpha), \tag{2}$$

as in § 20·3. Resolving perpendicular and parallel to OG we have

$$P-Mg\sin\theta = Mh\ddot{\theta},$$

$$Q-Mg\cos\theta = Mh\dot{\theta}^2,$$

and substituting for $\ddot{\theta}$ and $\dot{\theta}^2$ in terms of θ from equations (1) and (2) we find

$$P = \frac{k^2}{h^2+k^2}Mg\sin\theta,$$

$$Q = Mg\left\{\cos\theta+\frac{2h^2}{h^2+k^2}(\cos\theta-\cos\alpha)\right\}.$$

21·9 Internal stresses. Finally, we consider a problem involving the determination of internal stresses in a moving body. We take the simple case of a uniform rod AB, of mass M and length l, swinging as a pendulum about the end A which is fixed. The amplitude of the swing is α ($<\frac{1}{2}\pi$). The internal stress at a point C, taken as the action on the part CB, can be represented as a force (X, Y) at C, and a couple N; X is the thrust, Y the shearing force, and N the bending moment.

The first thing to notice is that, in calculating the internal stresses, we cannot replace a given system of forces on the whole rod by an equivalent one (see § 19·1) but must deal with the system as given.

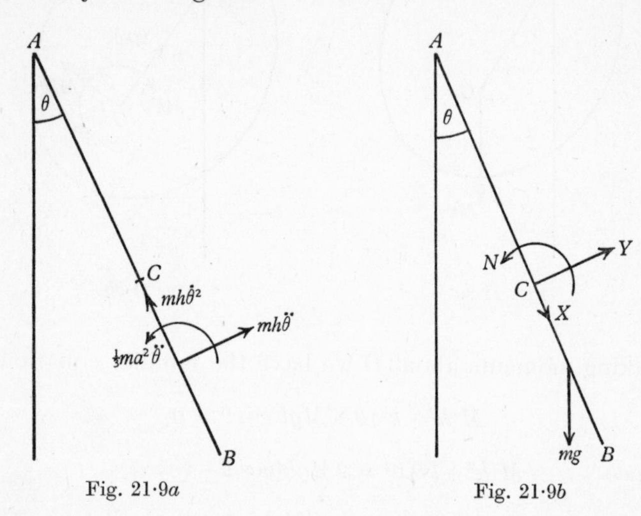

Fig. 21·9*a* Fig. 21·9*b*

Let us then show how X, Y, N may be determined. We denote the length CB by $2a$ ($<l$), and write h for $l-a$, so that h is the distance of the centre of CB from A.

Now consider the motion of the rigid body CB. Its kineton system is shown in Fig. 21·9*a* (m denotes the mass of CB, and $m = (2a/l)M$); and the external-force system in Fig. 21·9*b*. The equivalence of these two systems enables us to express X, Y, N in terms of $\ddot{\theta}$ and $\dot{\theta}^2$; and these in turn can be expressed as functions of θ from the equation of motion for the rod as a whole.

Thus, resolving along and perpendicular to CB, and taking moments about C, we find

$$X + mg\cos\theta \;= -mh\dot{\theta}^2,$$
$$Y - mg\sin\theta \;= mh\ddot{\theta},$$
$$N - mga\sin\theta = (\tfrac{1}{3}ma^2 + mah)\,\ddot{\theta}.$$

Now form the equation of motion for the whole rod AB, and we have

$$\tfrac{1}{3}Ml^2\ddot{\theta} = -\tfrac{1}{2}Mgl\sin\theta,$$
$$\ddot{\theta} = -\frac{3}{2}\frac{g}{l}\sin\theta,$$

whence
$$\dot{\theta}^2 = 3\frac{g}{l}(\cos\theta - \cos\alpha),$$

and we can now express X, Y, N in terms of θ and a. We find

$$X = -Mg\{2al\cos\theta + 6a(l-a)(\cos\theta - \cos\alpha)\}/l^2,$$

so that the rod is always in tension at all points, as we should expect; next
$$Y = Mg\sin\theta\{a(3a-l)/l^2\},$$

and the shear changes sign at a point K one-third of the way down the rod; finally,

$$N = -Mg\sin\theta\{a^2(l-2a)/l^2\},$$

so that the maximum value of $|N|$ which occurs at any point of the rod during the motion is $Mgl\sin\alpha/27$; this is obtained at K when $\theta = \alpha$.

EXAMPLES XXI

1. A uniform solid circular cylinder of mass m and radius a is rolled with its axis horizontal up a rough inclined plane by means of a constant couple L. Show that, for this to be possible, the coefficient of friction must be greater than

$$\frac{1}{3}\tan\theta + \frac{2}{3}\frac{L\sec\theta}{mag},$$

where θ is the inclination of the plane to the horizontal.

2. A uniform circular hoop of radius a is projected down a rough inclined plane in a vertical plane through the line of greatest slope, with forward velocity v_0 and backward spin with angular velocity $3v_0/a$. The inclination of the plane to the horizontal is $\tan^{-1}\frac{1}{10}$, and the coefficient of friction between the hoop and the plane is $\frac{1}{4}$. Show that after an interval of time t_1 the hoop ceases to skid, and find t_1.

3. A uniform right circular cylinder, of radius a, is projected up a rough inclined plane, of inclination α, so that its generators are perpendicular to the line of greatest slope. If the cylinder is projected with velocity u_0 and no spin, show that skidding ceases instantaneously after a time $u_0/g(3\mu\cos\alpha+\sin\alpha)$, where μ is the coefficient of friction between the plane and the cylinder. Show further that if $\mu>\frac{1}{3}\tan\alpha$ the cylinder subsequently rolls without skidding. What happens if $\mu<\frac{1}{3}\tan\alpha$?

4. A uniform solid sphere is projected up a line of greatest slope of a plane inclined at an angle α to the horizontal. Initially the sphere has no angular velocity and its centre is moving with velocity V up the plane. If the coefficient of friction between the sphere and the plane is μ, where $\mu>\frac{2}{7}\tan\alpha$, prove that the sphere will begin to move down the plane after a time $V/(g\sin\alpha)$.

5. A uniform solid circular cylinder of mass M and radius a rolls on the inside of a fixed circular cylindrical tube of radius $c+a$ whose axis is horizontal. The cylinder starts from the lowest position with a motion of pure rolling, the initial velocity of the centre being $\sqrt{(6gc)}$, and the coefficient of friction between the cylinder and the tube is greater than $\sqrt{3}/21$. Prove that in the subsequent motion the cylinder rolls right round the tube, and that the motion continues to be one of pure rolling throughout.

6. A uniform solid circular cylinder, of radius a and mass M, is placed with a generator in contact with the highest generator of a fixed horizontal circular cylinder of radius $c-a$, and is slightly displaced. If θ denotes the angle which the plane containing the axes of the cylinders makes with the vertical at time t, show that, so long as the motion is one of pure rolling,

$$\dot{\theta}^2 = \frac{4g}{3c}(1-\cos\theta).$$

If the coefficient of friction between the cylinders is $\frac{3}{8}$, show that slipping begins when $\theta=\alpha$, where $\cos\alpha=\frac{4}{5}$.

7. A uniform sphere of radius a rolls without slipping on the inside of a fixed rough circular cylinder of radius $b(=a+c)$ whose axis is horizontal. The sphere is started rolling at the lowest point so that the horizontal velocity u of its centre is perpendicular to the axis of the cylinder. Prove that the least value of u for which the sphere will roll right round without leaving the surface of the cylinder is given by $u^2 = 27gc/7$.

8. A uniform circular hoop, of radius b and mass M, hangs over a horizontal rod whose cross-section is a circle of radius a. The hoop swings, without slipping on the rod, in a vertical plane at right angles to the axis of the rod. If at time t the plane through the axis of the rod and the centre of the hoop makes an angle θ with the vertical, and if the maximum value of θ in the motion is $\alpha(<\frac{1}{2}\pi)$, prove that

$$c\dot{\theta}^2 = g(\cos\theta-\cos\alpha),$$

where $c=b-a$.

Find the least value of the coefficient of friction which will suffice to prevent slipping.

9. A rough symmetrical circular cylinder of radius a, and radius of gyration κ about its axis, is free to roll on a horizontal platform. Initially both the cylinder and the platform are at rest. The platform is then moved horizontally in a direction perpendicular to the axis of the cylinder. Show that, whatever the motion of the platform, the displacement of the axis of the cylinder at any instant is $\frac{\kappa^2}{\kappa^2+a^2}$ times the displacement of the platform, provided that the acceleration of the platform never exceeds $\mu g\left(1+\frac{a^2}{\kappa^2}\right)$ in absolute value, μ being the coefficient of friction between cylinder and platform.

If the total displacement of the platform is always small, show that the cylinder may be regarded as rotating about a fixed axis at height κ^2/a above its central axis.

10. A wheel, of radius a, with mass-centre at distance c from the geometric centre, rolls in a straight line with its plane vertical along a rough floor.

(i) If it be set rolling with angular velocity ω with the mass-centre in the lowest position, find the condition that the mass-centre should reach the highest position.

(ii) If the wheel performs small oscillations about the position of stable equilibrium, find the period.

11. A non-uniform circular cylinder of radius a has its centre of gravity displaced at a distance h from its central axis. The cylinder can roll with its axis horizontal on a rough plane inclined at an angle α to the horizontal, and is released from rest with its centre of gravity vertically below its axis. Show that the cylinder will perform complete revolutions if $(a/h)\sin\alpha$ exceeds the maximum value of the expression $\frac{1-\cos\theta}{\theta}$, and that if this condition is satisfied the angular velocity ω of the cylinder after one complete revolution is given by

$$\omega^2 = 4\pi ag\sin\alpha/(\kappa^2+a^2-2ha\cos\alpha),$$

where κ is the radius of gyration of the cylinder about its *central* axis.

12. A flywheel of radius a and mass M, uniformly distributed round the rim, is free to rotate about a horizontal axis through its centre. A mass $9M$ is eccentrically placed at a distance $\frac{1}{3}a$ from the centre. If ω is the angular velocity of the wheel when this mass is at its lowest point, show that the wheel will complete its revolutions if $\omega^2 > 6g/a$.

If, instead of the axis being fixed in position the wheel is free to roll along a horizontal plane, find the corresponding minimum value for ω which will allow the wheel to complete its revolutions.

*13. A uniform circular disc of mass M and radius a has a particle of mass m fixed in it at a point P at distance b from its centre C. An axis through C perpendicular to the disc can slide without friction horizontally, and the disc is free to rotate about the axis in a vertical plane. If the inclination of CP to the upward vertical at time t is denoted by θ, prove that

$$\tfrac{1}{2}\dot{\theta}^2 = \frac{\frac{1}{2}A\,\Omega^2+mgb(1-\cos\theta)}{A+B\sin^2\theta},$$

where Ω is the value of $\dot{\theta}$ when $\theta = 0$, and the constants A and B are defined by the equations

$$A = \tfrac{1}{2}Ma^2 + \frac{Mmb^2}{M+m}, \quad B = \frac{m^2b^2}{M+m}.$$

Show that the reaction on the axis at the same instant is

$$N = \frac{(M+m)g\{1+k(1-\cos\theta)^2\} - mb\Omega^2\cos\theta}{(1+k\sin^2\theta)^2},$$

where $k = B/A$.

Examine the special case in which $\Omega = 0$ and k is small. Find an approximation for $N/(M+m)g$ correct to order k, and prove that its greatest and least values are $1+4k$ and $1-\frac{4}{3}k$.

14. A uniform solid circular cylinder, of mass m and radius a, has a particle of mass m attached to a point of its surface. It is placed on a rough horizontal plane with the particle vertically over the axis and slightly displaced. Show that when the particle is at the level of the axis the velocity of the axis is $\sqrt{(4ga/7)}$.

Determine, for the same instant, the horizontal and vertical components of the reaction of the plane on the cylinder, and show that they are in the ratio 4 : 19.

(*Note.* To solve this problem it is not necessary to find the components of the reaction for the general position.)

15. A uniform rod is suspended in a horizontal position by two light vertical strings attached to its ends. Show that, when one string is suddenly cut, the tension in the other is instantaneously halved.

16. A uniform straight heavy rod AB of mass M is freely jointed about a smooth horizontal axis at A, and is supported at an inclination θ to the vertical by a light string which is perpendicular to the rod and attached to it at B. The string is suddenly cut. Find the pressures on the axis at A before and immediately afterwards.

17. A homogeneous sphere of mass M is placed on an imperfectly rough table, the coefficient of friction being μ. A particle of mass m is attached to the extremity of a horizontal diameter. Show that the sphere will begin to roll or slide according as μ is greater or less than

$$\frac{5(M+m)m}{7M^2+17Mm+5m^2}.$$

18. A particle of mass m is attached to one end of a uniform rod of mass M. The other end of the rod rests on a rough horizontal table. The system is released from rest with the rod inclined to the vertical at an angle $\frac{1}{3}\pi$. Prove that, if the rod does not slip initially, then the coefficient of friction is not less than

$$\frac{3\sqrt{3}(M+2m)^2}{7M^2+28Mm+12m^2}.$$

19. A uniform prism of mass M, whose cross-section is an equilateral triangle of side a, with plane ends at right angles to its length, is held with an edge on a horizontal plane and a rectangular face vertical and is let go.

Show that the initial angular acceleration is $2\sqrt{3g/5a}$ or $\sqrt{3g/a}$ according as the plane is rough or smooth.

20. A uniform rod, of mass M and length $2l$, is held vertically with its lower end in contact with a smooth plane which is inclined at an angle β to the horizontal. If the rod is released from rest, show that its initial angular acceleration is

$$\frac{3g\cos\beta\sin\beta}{l(1+3\sin^2\beta)}.$$

Find also the initial reaction on the plane.

21. A uniform rod AB of mass m and length $2a$ is turning in a vertical plane about the end A which has a constant horizontal acceleration f in this plane. Prove that, if the greatest value of the square of the angular velocity of the rod exceeds $3\sqrt{(f^2+g^2)}/a$, the rod is making complete revolutions; and prove that the difference between the greatest and least values of the pull along the rod of the attachment at A is $5m\sqrt{(f^2+g^2)}$.

22. A lamina moves in its own plane in such a way that the distance of the instantaneous centre of rotation from the centre of gravity of the lamina is constant. Prove that $Id\omega/dt = \Gamma$, where Γ is the moment of the external forces about the instantaneous centre, ω the angular velocity of the lamina, and I the (constant) moment of inertia of the lamina about the instantaneous centre.

Apply this result to the motion of a uniform circular disc of radius a which is free to move in a vertical plane and rolls under gravity on a curve in that plane. Show that the angular velocity of the disc is $\sqrt{(\frac{4}{3}gz)}/a$, where z is the depth of the centre of the disc below a certain level; and verify this result from the constancy of the energy.

23. A body of uniform density, having the form of a sector of a circle of radius $2a$, performs rolling oscillations of small amplitude with its circular edge in contact with a rough horizontal plane. Its centre of gravity is half-way between the edge and its centre. Prove that the length of the simple equivalent pendulum is $2a$.

24. A rigid body is moving in two dimensions with angular velocity ω at time t. Prove that

$$I\frac{d\omega}{dt}+\frac{1}{2}\frac{dI}{dt}\omega = L,$$

where I is the moment of inertia and L the moment of the external forces about the instantaneous centre of rotation.

The cross-section and end-sections of a uniform solid cylindrical body are semi-circles of radius 2 inches. The body is placed with its curved surface in contact with a horizontal plane rough enough to prevent slipping, and makes small oscillations about its position of equilibrium; find the length of the equivalent simple pendulum.

25. Three circular discs are fastened rigidly together as shown in the figure. The outer discs rest on two parallel horizontal rails, which are rough enough to prevent slipping. The discs are of the same thickness and density. The

radius of each outer disc is r, and that of the inner disc $2r$. Show that the periodic time of a small oscillation about the stable position of equilibrium is

$$2\pi\sqrt{\left(\frac{11}{4}\frac{r}{g}\right)}.$$

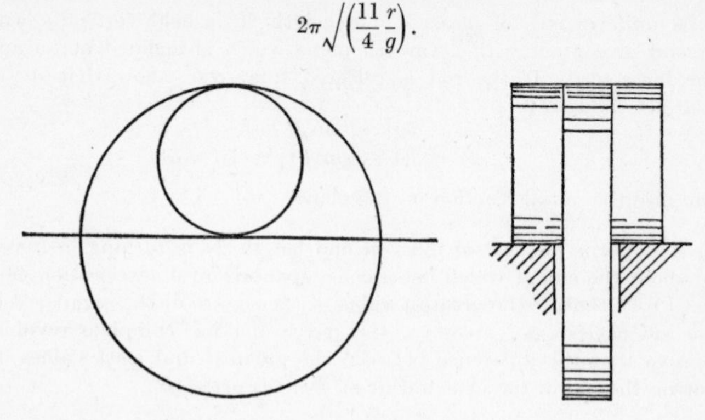

26. A uniform solid circular cylinder makes complete revolutions under gravity about a horizontal generator. Show that the supports must be able to bear at least 11/3 times the weight of the cylinder.

27. A compound pendulum of mass M oscillates with amplitude α about a fixed horizontal axis which is perpendicular to a plane of symmetry of the body; the axis meets this plane in A, the centre of gravity G is at a distance a from A, and the radius of gyration about a line through G parallel to the axis is k. Prove that when AG makes an angle θ with the downward vertical the components X and Y of the reaction of the axis on the body perpendicular to and along GA are given by

$$\frac{X}{Mg} = \frac{k^2}{a^2+k^2}\sin\theta, \quad \frac{Y}{Mg} = \cos\theta + \frac{2a^2}{a^2+k^2}(\cos\theta - \cos\alpha).$$

Consider the following applications of this result:

(i) The pendulum is drawn out of the vertical by means of a string in the plane of symmetry whose direction passes through the point on AG produced distant k^2/a from G. The system is at rest and the string is cut. Show that X remains unchanged immediately after the string is cut, but that Y changes unless the string is perpendicular to AG.

(ii) If the pendulum is a uniform rod suspended from one end, prove that $X = Y$ for just one positive value of θ if $\tan\alpha \geqslant 4$.

(iii) If the pendulum is a uniform rod with one end freely attached to a small light ring which is threaded on a rough horizontal wire, and if the rod is let go from rest when it is horizontal ($\alpha = \frac{1}{2}\pi$), prove that if the ring slips when the rod has turned through an angle β then the coefficient of friction is

$$\frac{9\tan\beta}{10\tan^2\beta+1}.$$

Prove that the ring will not slip if the coefficient of friction is greater than $(9\sqrt{10})/20$.

(iv) If the rigid body is a uniform solid sphere suspended from a point on its surface, and $\alpha = \frac{1}{2}\pi$, prove that the greatest value of the horizontal pull on the support during the motion is $15Mg/14$.

*28. A uniform rod is held nearly vertically with its lower end resting on an imperfectly rough table. It is released from rest and falls forwards. The inclination to the vertical at any subsequent time is θ. Prove (i) that if the coefficient of friction is less than a certain value (0·37 approximately) the lower end of the rod will slip backwards before $\cos\theta = \frac{2}{3}$, (ii) that however great the coefficient of friction may be the lower end will slip forward when $\cos\theta$ lies between $\frac{2}{3}$ and $\frac{1}{3}$.

29. A light straight rod AB of length $2l$ is smoothly hinged to a fixed point at A so that it can rotate in a vertical plane. Two particles, each of mass m, are rigidly attached to the rod, one at the end B and the other at the mid-point. The rod is released from rest at an angle α with the upward vertical. Show that, when it makes an angle θ with the upward vertical, its angular velocity $\dot\theta$ is given by

$$5l\dot\theta^2 = 6g(\cos\alpha - \cos\theta).$$

Calculate the shearing force and bending moment at a point of the rod at a distance h (less than l) from the upper end. For what range of values of $\cos\theta$ is the upper half of the rod in compression?

30. A rigid uniform ring, of mass $2\pi a\lambda$ and radius a, which is hanging over a smooth peg, is set spinning with angular velocity ω about its centre (which remains stationary) in its own vertical plane. It is completely fractured at one point A. Prove that the maximum bending moment experienced at the diametrically opposite point A' is

$$\lambda a^2\{2a\omega^2 + g\sqrt{(\pi^2+4)}\}.$$

What is the corresponding direction of AA'?

31. For a lamina moving freely under gravity in a vertical plane, prove that the space centrode is a parabola; and that the parametric equations, referred to axes moving with the lamina, of the body centrode may be written in the form

$$x = a\theta\cos\theta + b\sin\theta, \quad y = b\cos\theta - a\theta\sin\theta,$$

where a and b are constants.

*32. In Example 6, show that the cylinders separate when $\theta = \alpha + \beta$, where

$$\cos\beta = \tfrac{7}{9}e^{\frac{3}{4}\beta}.$$

CHAPTER XXII

IMPULSIVE MOTION OF A RIGID BODY

22·1 Impulsive motion. We saw that when an impulse acts on a particle there is a discontinuity in its velocity, and the study of this sudden change of velocity constitutes the impulse problem. In the same way when impulses act on a rigid body there is a discontinuity of velocity, and it is to the study of this sudden change of velocity that we now turn. In a problem of impulsive motion there is no change of position to worry about, and the coordinates (ξ, η, θ) defining the configuration are effectively constants. The variables in the problem are the velocity-components $(\dot{\xi}, \dot{\eta}, \dot{\theta})$. It is preferable in an impulse-problem to write say (U, V, ω) rather than $(\dot{\xi}, \dot{\eta}, \dot{\theta})$, since (ξ, η, θ) now have the role of mere constants.

Usually an impulse-problem is much simpler than a problem involving finite forces, since the equations to be solved are mere linear algebraic equations, not differential equations. We remind the reader that in a problem involving impulses the finite forces can be ignored. We begin with the case where the motion is set up from rest.

22·2 Motion set up from rest. There is a close analogy with the finite force problem already discussed. The momentum set up in a typical particle of the body is equal to the vector sum of the impulses, external and internal, acting on it. We assume that the body remains rigid when the impulses are applied, i.e. we assume that it can sustain not only internal forces but also internal impulses without distortion. Further, we make an assumption about the internal impulses analogous to that already made about internal forces, namely, that the internal impulses form a nul system. It follows, by an argument almost identical with that of § 19·10, that the external impulse system is equivalent to the momentum system set up. This is the analogue for impulses of d'Alembert's principle for forces, and it leads at once to the equations we require. As before, we shall deal in the first instance with a rigid lamina moving in a plane.

We reduce the momentum system, by the method of § 19·1, to an $\mathscr{M}$-force at G and an $\mathscr{M}$-couple. The momentum system set up is shown in Fig. 22·2a, where **W** is the velocity of G and ω the angular velocity of the body. The terms mW, all in the direction of **W**, are equivalent to an $\mathscr{M}$-force M**W** at G.

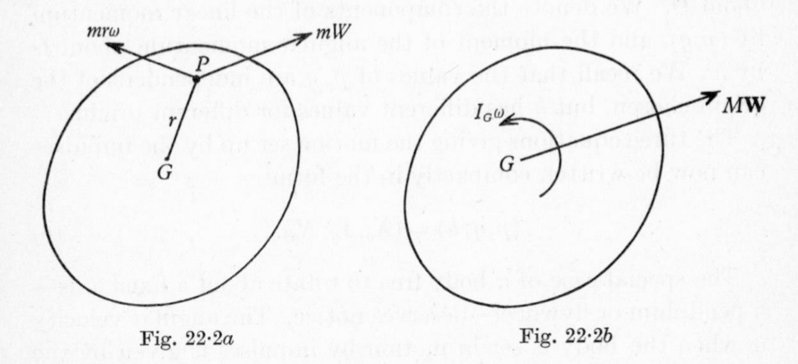

Fig. 22·2a Fig. 22·2b

The vector sum of the terms $mr\omega$ perpendicular to GP is zero, and they are equivalent to an $\mathscr{M}$-couple $Smr^2\omega = I_G\omega$. Thus the momentum set up when the body is jerked into motion by impulses equivalent to an $\mathscr{M}$-force M**W** at G and an $\mathscr{M}$-couple $I_G\omega$ (Fig. 22·2b). There is a striking similarity with the result we obtained for the reduced kineton system in the theory of motion under forces; cf. Fig. 21·1b with Fig. 22·2b.

If the external impulses are equivalent to an impulse (X_0, Y_0) at O and an impulse-couple N_0, the equations determining the motion immediately after the blows are

$$MU = X_0, \tag{1}$$

$$MV = Y_0, \tag{2}$$

$$M(\xi V - \eta U) + I_G\omega = N_0, \tag{3}$$

where (U, V) are the components of **W**. In practice, as in the force problem, we frequently obtain a third equation of simpler form by taking the moment N_G of the external impulses about G,

$$I_G\omega = N_G. \tag{4}$$

We have met the first members of the equations (1), (2), (3) already, in § 21·3, where we introduced the notation

$$p = MU, \quad q = MV, \quad h = M(\xi V - \eta U) + I_G \omega. \tag{5}$$

When we reduce the momentum system to an $\mathscr{M}$-force at O and an $\mathscr{M}$-couple, the $\mathscr{M}$-force is called the *linear momentum* of the body, and the $\mathscr{M}$-couple is called the *angular momentum* about O. We denote the components of the linear momentum by (p, q), and the moment of the angular momentum about O by h. We recall that the values of p, q are independent of the origin chosen, but h has different values for different origins.

The three equations giving the motion set up by the impulses can now be written compactly in the form

$$(p, q; h) = (X_0, Y_0; N_0). \tag{6}$$

The special case of a body free to rotate about a fixed axis—a pendulum or flywheel—deserves notice. The angular velocity ω when the body is set in motion by impulses is given by the equation

$$I\omega = N, \tag{7}$$

where I is the moment of inertia about the axis of rotation O, and N is the moment of the given impulses about the axis. This follows at once from the analogue of d'Alembert's principle, or as a special case of the third of the equations (5), which gives

$$h = Mb^2\omega + I_G\omega = I\omega,$$

where b denotes the distance of G from the axis of rotation O. The external impulses consist of the given impulses, and the impulsive action on the body at O, and the last has no moment about O.

Example 1. *A rigid lamina, free to rotate about a fixed point O of itself, is set in motion by an impulse acting at a point A in line with O and G; the magnitude of the impulse is P, its direction is perpendicular to OA. Find the angular velocity with which the lamina starts to move, and the impulsive reaction on the lamina at O.*

The angular velocity ω is given immediately by equation (7)

$$M(b^2 + k^2)\,\omega = aP,$$

where a denotes the length OA, and b the length OG. The reaction Q on the lamina at O is clearly at right angles to OA, and resolving perpendicular to

OA (equation (2)) we have

$$P+Q = Mb\omega,$$

giving $$Q = P\left(\frac{ab}{b^2+k^2}-1\right).$$

We see that Q vanishes if

$$a = b+(k^2/b),$$

and then A is called the *centre of percussion*; in this case the length OA is the length of the equivalent simple pendulum if the body is suspended from O.

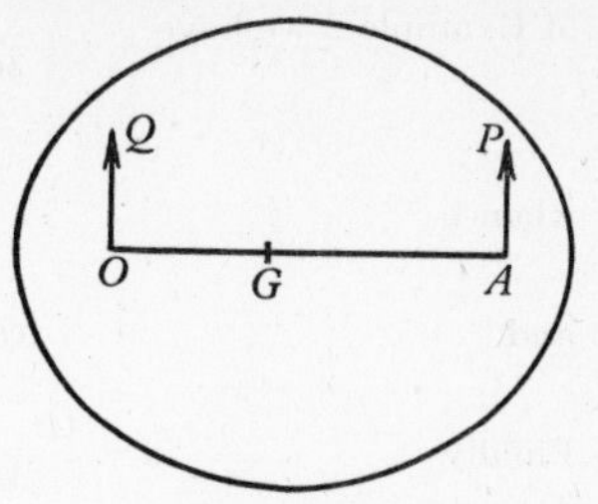

Fig. 22·2c

Example 2. *A uniform rod AB, of mass M and length 2a, is at rest, and is set in motion by parallel impulses P, Q applied at right angles to the rod at A and B. To find the velocities with which A and B start to move.*

The systems shown in Figs. 22·2d and 22·2e are equivalent, and we have

$$P+Q = MV,$$
$$(Q-P)a = (Ma^2/3)\,\omega,$$

Fig. 22·2d

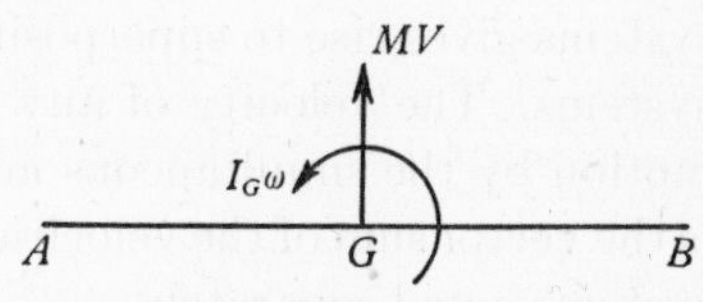

Fig. 22·2e

the first being found by resolving vertically and the second by taking moments about G. These equations determine V and ω, and the velocities of A and B are

$$V-a\omega = (4P-2Q)/M,$$
$$V+a\omega = (4Q-2P)/M.$$

Example 3. *Two uniform rods AB, BC, of masses M and m, freely jointed together at B, rest in line on a horizontal plane. The rods are set in motion by horizontal impulses P, R perpendicular to the rods at A and C. Find the initial velocities of A, B, C, and the internal impulse at B.*

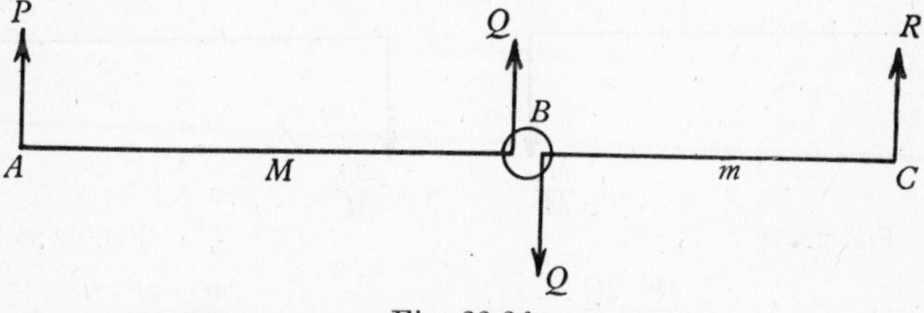

Fig. 22·2f

Taking the rods horizontal and P upwards, as in Fig. 22·2f we denote the internal impulse at B by Q (acting upwards on AB and downwards on BC), and the initial velocities upwards of A, B, C by u, v, w. From the result of Example 2 we have

$$v = \frac{4Q-2P}{M} = \frac{4(-Q)-2R}{m},$$

whence

$$v = \frac{-2(P+R)}{M+m},$$

and

$$2Q\left(\frac{1}{M}+\frac{1}{m}\right) = \frac{P}{M}-\frac{R}{m}.$$

Finally

$$u = \frac{4P-2Q}{M} = \frac{(4M+3m)P+MR}{M(M+m)},$$

$$w = \frac{4R+2Q}{m} = \frac{mP+(3M+4m)R}{M(M+m)}.$$

22·3 The principle of superposition.

The problem considered in Example 2 of § 22·2 can be solved even more simply. The equations determining the velocities in terms of the impulses when a rigid body is set in motion are homogeneous linear equations; it follows that superposition of two impulse systems gives rise to superposition of the effects of the separate systems. The velocity of any particle when the body is set in motion by the simultaneous action of the two impulse systems is the vector sum of the velocities it would acquire if the impulse systems acted separately.

Example. *As a trivial illustration of the principle of superposition, consider again the Example 2 of § 22·2.*

If only the impulse P at A acted on the rod the motion set up, determined as before, would be as in Fig. 22·3a. The motion set up by the impulse Q at B alone would be as shown in Fig. 22·3b. The effect of the two impulses applied simultaneously is found by superposition; it is shown in Fig. 22·3c, and this agrees of course with the result previously found.

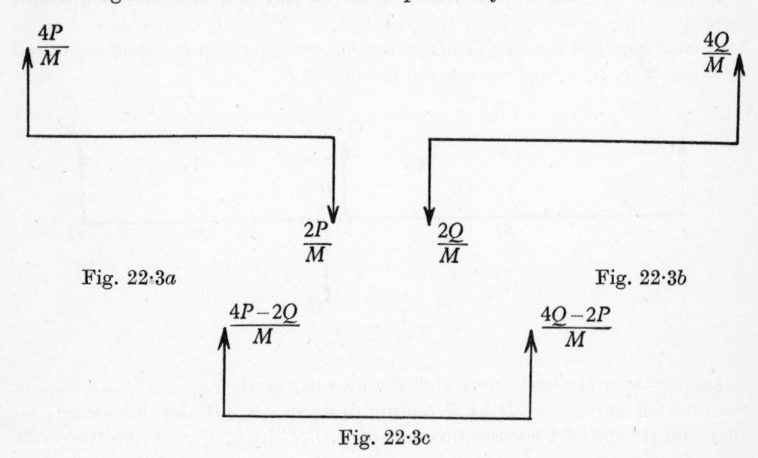

Fig. 22·3a

Fig. 22·3b

Fig. 22·3c

22·4 Energy communicated by impulses. When a body is set in motion by impulses the kinetic energy T it acquires is given by the equation

$$T = \tfrac{1}{2}\Sigma \mathbf{P}.\mathbf{v},$$

where $\mathbf{P}$ is one of the impulses, and $\mathbf{v}$ the velocity acquired by the point of the body at which the impulse is applied.

The proof is simple. If (U, V) is the velocity of G immediately after the impulses are applied, and ω the angular velocity, we have

$$MU = \Sigma X, \quad MV = \Sigma Y, \quad I_G\omega = \Sigma(xY - yX),$$

the origin being taken at G; X, Y are the components of the impulse $\mathbf{P}$ acting at (x, y). Now, by Konig's theorem (§ 21·5), the kinetic energy set up is

$$\tfrac{1}{2}M(U^2 + V^2) + \tfrac{1}{2}I_G\omega^2,$$

and this is equal to

$$\begin{aligned}\tfrac{1}{2}U(\Sigma X) + \tfrac{1}{2}V(\Sigma Y) + \tfrac{1}{2}\omega\Sigma(xY - yX)&\\ = \tfrac{1}{2}\{\Sigma X(U - y\omega) + \Sigma Y(V + x\omega)\}&\\ = \tfrac{1}{2}\Sigma \mathbf{P}.\mathbf{v}.&\end{aligned}$$

22·5 The general problem. We now consider the more general problem—the effect of impulses on a rigid body already in motion when the impulses are applied. The analogue of d'Alembert's principle now required is that the change-of-momentum system is equivalent to the impulse system; if (u_0, v_0) is the velocity of a typical particle immediately before the blow, and (u, v) immediately after, the system to be reduced consists of vector quantities $\{m(u - u_0), m(v - v_0)\}$ localized in the particles. The reduction is simple, and leads to an $\mathscr{M}$-force

$$M(U - U_0), M(V - V_0),$$

at G, with an $\mathscr{M}$-couple $I_G(\omega - \omega_0)$; the symbols $(U_0, V_0;\ \omega_0)$ describe the motion immediately before the blow, (U_0, V_0) being the velocity of G, and ω_0 the angular velocity; and similarly $(U, V;\ \omega)$ describe the motion immediately after the blow. The reduced change-of-momentum system is shown in Fig. 22·5*a*.

We see that the reduced change-of-momentum system is simply the difference between the reduced momentum systems after and before the blow. This we could have foreseen, since the relations involved in the reduction are homogeneous linear relations. Indeed, if we write the equations expressing the

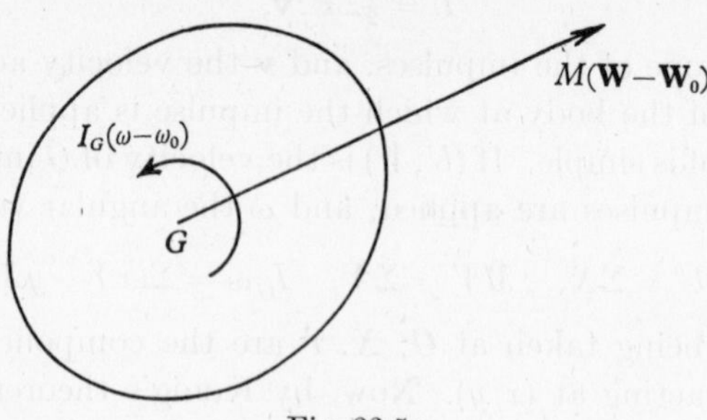

Fig. 22·5a

equivalence of the impulse system and the change-of-momentum system we have

$$X_0 = Sm(u-u_0) = p-p_0, \tag{1}$$

$$Y_0 = Sm(v-v_0) = q-q_0, \tag{2}$$

$$\begin{aligned} N_0 &= Sm\{x(v-v_0)-y(u-u_0)\} \\ &= Sm(xv-yu)-Sm(xv_0-yu_0) \\ &= h-h_0, \end{aligned} \tag{3}$$

where the external impulse system has been reduced to an impulse (X_0, Y_0) at O and an impulse-couple N_0. These equations give us as the equations determining (U, V, ω)

$$M(U-U_0) = X_0, \tag{4}$$

$$M(V-V_0) = Y_0, \tag{5}$$

$$M\{\xi(V-V_0)-\eta(U-U_0)\}+I_G(\omega-\omega_0) = N_0. \tag{6}$$

These are, of course, precisely the equations obtained from the reduced system shown in Fig. 22·5a.

As before, in practice we may replace the last equation (6) by the simpler one obtained by taking moments about G,

$$I_G(\omega-\omega_0) = N_G. \tag{7}$$

One or two special cases are worth notice. (i) If $U_0, V_0, \omega_0 = 0$ we recover the theory discussed in § 22·2. (ii) If $U, V, \omega = 0$ the body is brought to rest by the impulses. In this case the superposition of the impulse system and the initial momentum system gives a nul system. In this particular problem we may, if we like, show both systems in the same figure and treat it as an equilibrium problem! But the special technique is hardly worth while. (iii) If the body rotates about a fixed axis (a pendulum or flywheel) the equation determining the angular velocity after the blow is

$$I(\omega-\omega_0) = N,$$

where N is the moment of the external impulses about the axis, and I the moment of inertia of the lamina about the axis.

Example. *Door-stop. A uniform rod AB, of mass M and length $2a$, rotates with angular velocity ω_0 about the end A, which is fixed; it is brought to rest by colliding with an inelastic stop S at distance $a+x$ from A. To find the impulses on the rod at A and at S.*

We denote the impulse at A by P, and at S by Q, as shown in Fig. 22·5*b*; it is clear that both impulses are perpendicular to the rod. This is an example of (ii) above; we can safely mark the initial momentum system and the blows in the same figure and treat it as an equilibrium problem.

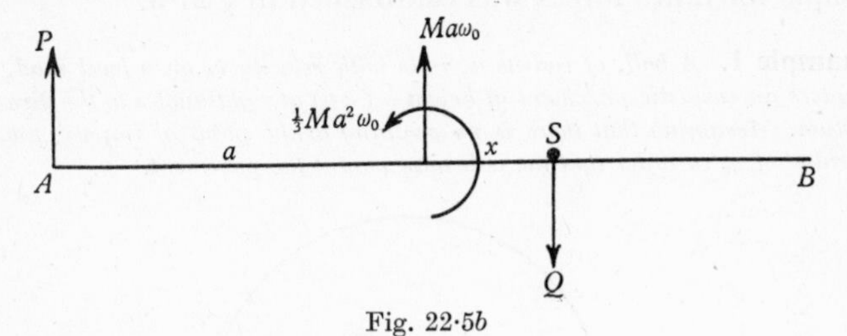

Fig. 22·5*b*

To find P: taking moments about S, we have

$$P(a+x) = M\frac{a^2}{3}\omega_0 - Ma\omega_0 x$$

$$= \tfrac{1}{3}Ma\omega_0(a-3x).$$

To find Q: taking moments about A, we have

$$Q(a+x) = \tfrac{4}{3}Ma^2\omega_0.$$

We notice that $P = 0$ if $x = \frac{1}{3}a$; there is no jar on the hinge if the stop is placed at the centre of percussion, i.e. at a distance from the hinge equal to two-thirds of the length of the rod.

22·6 The conservation of momentum.

Consider again the equation (1) of § 22·5,

$$X_0 = p - p_0.$$

If $X_0 = 0$ we have

$$p = p_0.$$

This means that if the sum of the components of the external impulses in a certain direction is zero the component of linear momentum in that direction is unchanged by the impulses. Another way of expressing the same law is that the component of the velocity of G in that direction remains unchanged. This is the principle of *the conservation of linear momentum* for impulsive motion of a rigid body. It is the analogue for impulses of the principle already established for finite forces in § 21·3.

Again, from the equation (3) of § 22·5,

$$N_0 = h - h_0,$$

we see that $N_0 = 0$ implies

$$h = h_0.$$

If the moment of the system of external impulses about a certain point O is zero, the angular momentum about O remains unchanged. This is the principle of *the conservation of angular momentum* for impulsive motion of a rigid body. The analogous principle for finite forces was established in § 21·3.

Example 1. *A ball, of radius a, rolls with velocity v_0 on a level road, and encounters an inelastic pavement of height b ($<a$) at right angles to the direction of motion. Assuming that there is no slipping at the point of impact, find the least value of v_0 in order that the ball may mount the pavement.*

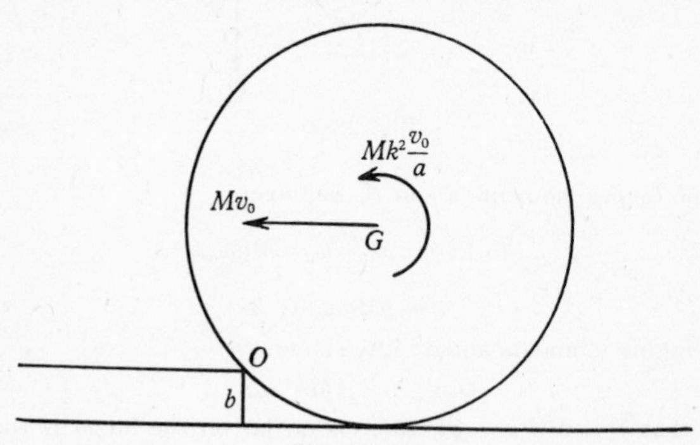

Fig. 22·6*a*

Our first task is to find the angular velocity ω immediately after impact. Now the momentum system before impact is as shown in Fig. 22·6*a*. There is a single impulse through O, and the conservation of the angular momentum about O,

$$h = h_0,$$

leads to

$$M(a^2+k^2)\,\omega = Mv_0(a-b)+Mk^2(v_0/a), \tag{1}$$

where Mk^2 is the moment of inertia about a diameter.

Next we ask what condition must be satisfied by ω in order that G may arrive with a finite velocity at the point vertically above O. Now in the motion subsequent to the impact there is no loss of energy, and the required condition is

$$\tfrac{1}{2}M(a^2+k^2)\,\omega^2 > Mgb.$$

Substituting for ω in terms of v_0 from equation (1) we obtain the required condition

$$(a^2-ab+k^2)^2 v_0^2 > 2ga^2 b(a^2+k^2).$$

Example 2. *A uniform rod AB, of mass M and length 2a, free to move in a plane, is turning about the end A with angular velocity ω_0 when it encounters an inelastic stop S at a distance $(a+x)$ from A. To find the fraction of the energy lost at the impact, and to examine how this varies with x.*

If immediately after the impact the rod is turning about S with angular velocity ω the momentum system before and after the impact is shown in Figs. 22·6*b* and 22·6*c*. Since the angular momentum about S is unchanged we have

$$M\left(\frac{a^2}{3}+x^2\right)\omega = M\left(\frac{a^2}{3}-ax\right)\omega_0,$$

so that

$$\frac{\omega}{\omega_0} = \frac{a(a-3x)}{a^2+3x^2}.$$

(We notice that $\omega = 0$ if $x = \frac{1}{3}a$, as we should expect from the example in § 22·5.)

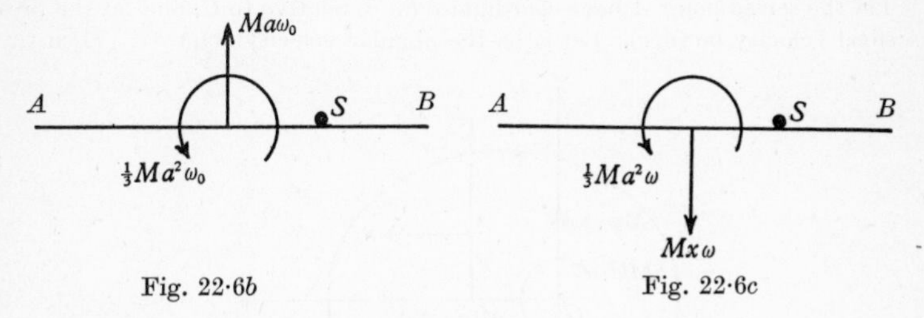

Fig. 22·6*b* Fig. 22·6*c*

The energies before and after are

$$T_0 = \tfrac{1}{2}(\tfrac{4}{3}Ma^2\omega_0^2),$$

$$T = \tfrac{1}{2}M\left(x^2+\frac{a^2}{3}\right)\omega^2,$$

and the fraction lost, say y, is

$$1-\frac{T}{T_0} = 1-\frac{a^2+3x^2}{4a^2}\frac{\omega^2}{\omega_0^2}.$$

Substituting for ω/ω_0 this is $\quad y = \dfrac{3(a+x)^2}{4(a^2+3x^2)},$

which we can also write in the form

$$y = \frac{3(a+x)^2}{3(a+x)^2+(a-3x)^2}.$$

The loss is zero when $x = -a$ (as we could anticipate) and has its maximum value 1 when $x = \frac{1}{3}a$. In plotting the graph of y, we notice that $y \to \frac{1}{4}$ as $x \to \pm\infty$, and that $y = \frac{3}{4}$ when $x = 0$ and when $x = a$. (Fig 22·6*d*).

For the practical problem, only the part of the graph from $x = -a$ to $x = a$ is relevant.

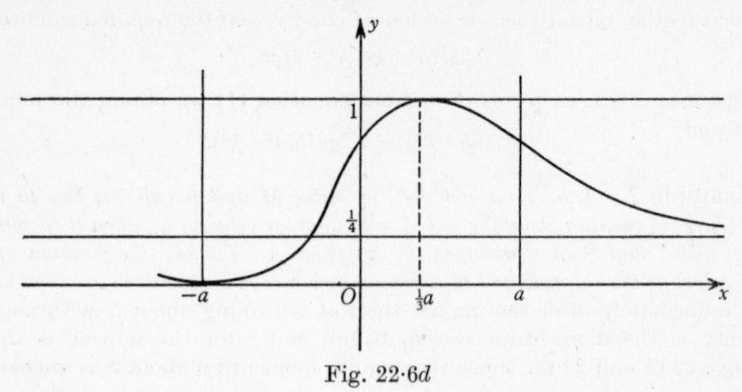

Fig. 22·6*d*

Example 3. *Kelvin's theorem. A lamina is lying at rest on a smooth table when one point A is seized and moved in a prescribed direction with prescribed speed. To show that the kinetic energy imparted to the lamina is as small as possible.*

We have only to show that any value of the angular velocity other than that acquired in the actual motion will give a greater value of T.

Let the seized point A have coordinates (α, β) relative to G, and let the prescribed velocity be (u, v). Let ω be the angular velocity acquired. Then the

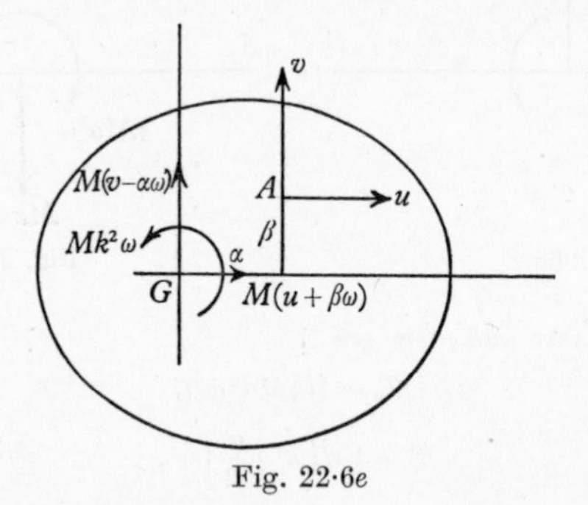

Fig. 22·6*e*

momentum system is as shown in Fig. 22·6*e*, and since the angular momentum about A is zero, we have

$$Mk^2\omega + M\beta(u+\beta\omega) - M\alpha(v-\alpha\omega) = 0,$$

whence
$$(k^2+\alpha^2+\beta^2)\,\omega = \alpha v - \beta u.$$

If the angular velocity were ω' the kinetic energy would be

$$T = \tfrac{1}{2}M(u+\beta\omega')^2 + \tfrac{1}{2}M(v-\alpha\omega')^2 + \tfrac{1}{2}Mk^2\omega'^2$$

$$= \frac{M}{2(k^2+\alpha^2+\beta^2)}\{[(k^2+\alpha^2+\beta^2)\,\omega' - (\alpha v - \beta u)]^2 + [k^2(u^2+v^2) + (\alpha u + \beta v)^2]\},$$

which is least when $\omega' = \omega$. Thus the kinetic energy acquired in the actual motion has the least possible value.

EXAMPLES XXII

1. A thin rod, whose mass is M and whose moment of inertia about its centre of gravity G is Mk^2, is free to turn about a fixed horizontal axis to which it is attached at a point P of the rod, the rod being perpendicular to the axis. Find the point Q of the rod such that, if the rod is set in motion from rest by an impulse at Q perpendicular to the rod, there is no impulsive pressure on the axis. Prove further that if the rod is rotating about P and is brought to rest by an inelastic stop at Q, there is again no impulsive pressure on the axis.

Prove that PQ is the length of the equivalent simple pendulum when the rod swings as a pendulum about P, and that the points P and Q are interchangeable.

2. A straight rod AB, whose mass-centre is at G, lies at rest on a smooth table. An impulse P perpendicular to the rod is applied horizontally at a point S. Determine the velocity with which any other point T begins to move. Show that if the impulse P is applied at T the point S begins to move with this same velocity.

3. A free lamina of any form is turning in its own plane about an instantaneous centre of rotation S. A point P in the line joining S to the centre of gravity G is brought to rest by an impulsive force passing through the point. Find the position of P in terms of SG and the radius of gyration about G (i) if the lamina is brought to rest, (ii) if the velocity of G has the same magnitude as before but is reversed in direction.

4. A uniform rectangular lamina of sides $2a, 2b$ and mass M swings freely through a small angle under gravity about a side $2a$. It is subject to a resistance, which for an element of area dA is $\mu v dA$, where v is the velocity of the element. Show that, if θ is the angle that the lamina makes with the vertical

$$\ddot{\theta}+\frac{4\mu ab\dot{\theta}}{M}+\frac{3g}{4b}\,\theta=0.$$

After every half-swing, when the lamina is vertical, an impulse is applied horizontally at the centre of percussion. If successive swings are to be through the same arc, find the magnitude of the impulse in terms of Ω, the angular velocity of the lamina at the bottom of its swing just after the impulse has been applied.

5. The balance-wheel of a watch consists of a symmetrical flywheel of moment of inertia I, pivoted with negligible friction about its axis. The hairspring provides a restoring couple of amount $-\lambda\theta$, where θ is the angular displacement of the balance-wheel from its position of rest. The balance-wheel is also subject to air resistance which exerts a couple of amount $-\sigma\dot{\theta}$, where $\dot{\theta}$ is its angular velocity. Obtain the equation of motion

$$\ddot{\theta}+2\kappa\dot{\theta}+n^2\theta=0$$

for the freely swinging balance-wheel, where $n^2=\lambda/I$, $\kappa=\sigma/2I$.

If ω_0 is its angular velocity at $\theta=0$, show that the angular displacement after time t is

$$\theta=\frac{\omega_0}{p}\,e^{-\kappa t}\sin pt,$$

where $p^2 = n^2 - \kappa^2$, and that its angular velocity first vanishes when

$$\theta = \frac{\omega_0}{n} e^{-\gamma \cot \gamma},$$

where $\gamma = \tan^{-1} p/\kappa$.

The balance-wheel receives an impulsive couple of magnitude C each time it passes through $\theta = 0$, in a direction always such as to assist the motion. Prove that the amplitudes θ_r of successive half-swings are connected by the relation

$$\theta_{r+1} = \theta_r e^{-\pi \cot \gamma} + \frac{C}{nI} e^{-\gamma \cot \gamma}.$$

Hence show that whatever its initial motion the balance-wheel will ultimately oscillate with a constant amplitude, which is approximately $\frac{2}{\pi}\frac{C}{\sigma}$ if $\cot \gamma$ is small, and that its period is independent of C.

6. A uniform rod of length $2a$ and mass m starts moving so that the ends have parallel velocities u and v normal to the rod and no velocity along the rod. Prove that the kinetic energy is

$$\tfrac{1}{6}m(u^2 + uv + v^2).$$

Find also by what impulses, acting at the ends, the rod can be so started.

7. A uniform rod AB, of mass M and length $2a$, is struck by an impulse ξ, acting at B in a direction BP which makes an angle θ with AB produced. Show that AB will become parallel to BP after a time t given by

$$t = \frac{Ma\theta}{3\xi \sin \theta}.$$

8. A uniform rod, of length $2a$ and mass M, moves with velocity U at right angles to its length on a smooth horizontal table. A point of the rod at distance c from the centre encounters a fixed inelastic obstacle. Determine the motion immediately after the impact, and show that a fraction $\frac{a^2}{a^2 + 3c^2}$ of the original energy is lost.

9. A uniform rectangular plate with sides $2a$, $2b$ is falling freely without rotation and with its plane vertical, with velocity V, when a string attached to the top corner becomes vertical and taut. Show that the velocity of the centre of mass immediately afterwards is

$$\frac{3Vp^2}{3p^2 + a^2 + b^2},$$

where p is the length of the perpendicular from the centre of mass to the string, and find the loss of kinetic energy.

10. A uniform rod AB is falling in a vertical plane and the end A is suddenly held fixed at an instant when the rod is horizontal and the vertical components of the velocities of A and B are v_1 downwards and v_2 upwards. Prove that the rod will begin to rise round the end A if $v_1 < 2v_2$.

11. A uniform circular disc of radius a is rolling without slipping along a smooth horizontal plane with velocity V when the highest point becomes

suddenly fixed. Prove that the disc will make a complete revolution round the point if $V^2 > 24ag$.

12. A uniform spherical ball of radius a is at rest on a rough horizontal table, and is set in motion by a horizontal blow in a vertical plane through the centre at a distance $\frac{1}{2}a$ above the table. Show that when the ball ceases to slip its linear velocity is 5/14 of its initial linear velocity.

13. A sphere, of radius a and of radius of gyration k about any axis through its centre, rolls with linear velocity v on a horizontal plane, the direction of motion being perpendicular to a vertical face of a fixed rectangular block of height h, where $h < a$. The sphere strikes the block; the sphere and the block are perfectly rough and perfectly inelastic. Show that the sphere will surmount the block if

$$(a^2 - ah + k^2)^2 v^2 > 2gha^2(a^2+k^2).$$

If $k^2 = \frac{2}{5}a^2$, $a = 25$ cm., $v = 500$ cm./sec., $g = 1000$ cm./sec./sec., show that the critical value of h is about 22·5 cm.

14. A cube of side $2a$ slides down a smooth plane inclined at an angle $2\tan^{-1}\frac{1}{5}$ to the horizontal, and meets a fixed horizontal bar placed perpendicular to the plane of motion and at a perpendicular distance $\frac{1}{4}a$ from the plane. Show that, if the cube is to have sufficient velocity to surmount the obstacle when it reaches it, it must be allowed first to slide down the plane through a distance $\frac{107}{60}a$. The obstacle may be taken to be inelastic and so rough that the cube does not slip on it.

15. A uniform straight rod AB, of length $2a$ and mass M, is lying at rest on a smooth table when the end B is seized and moved with velocity V horizontally at right angles to the rod. Find the initial velocity of the end A, the initial angular velocity of the rod, and the initial position of the instantaneous centre of rotation.

Prove that the energy acquired by the rod is the least possible consistent with the prescribed motion of B.

16. A uniform rectangular lamina $ABCD$ of mass M, and sides $2a, 2b$, initially at rest, is suddenly set in motion in its own plane by an impulse applied to the corner A. If the impulse is such that the velocity of A is of magnitude V in a direction making an angle θ with AB (of length $2a$), find the kinetic energy of the lamina, and show that it is less than the kinetic energy in any other possible motion in which A has the same velocity.

17. A square uniform lamina of side $2a$ has a concentric square hole of side $2b$ cut in it. Show that the square of the radius of gyration about a line through its centre perpendicular to its plane is $\frac{2}{3}(a^2+b^2)$.

The lamina can slide on a smooth horizontal table, and moves initially without rotation in a direction parallel to one of its edges with constant velocity V. It then strikes a small perfectly elastic fixed peg at a point close to one of its corners. Show that the lamina will never again strike the peg if $b > a/\sqrt{2}$.

If b is just less than $a/\sqrt{2}$, show that after the lamina strikes the peg a second time it moves without rotation with uniform velocity V at right angles to its direction of motion before it first struck the peg.

*18. A uniform circular disc of radius a and mass M is moved on a smooth horizontal table, with one plane face in contact with the table, by a rod which moves on the table and touches the circumference of the disc. At time $t = 0$ the rod is suddenly set rotating with constant angular velocity ω about one end O which is fixed, the system being previously at rest with the rod and disc in contact. The disc is inelastic and does not slip on the rod. If at time t the point of the rod in contact with the disc is at a distance x from O, prove that

$$3\ddot{x} - 2\omega^2 x = 0.$$

If $x = b$ when $t = 0$ show that at time t the component along the rod of the force exerted by the rod on the disc is

$$-\frac{M\omega^2}{3n}(a\omega \sinh nt + bn \cosh nt),$$

where $3n^2 = 2\omega^2$, and find the component perpendicular to the rod.

19. Two thin uniform rods AO, OB, of lengths $2a$ and $2b$ and of masses m and M, are smoothly jointed at O and lie in a straight line on a smooth table. An impulse P is applied at A perpendicular to AOB. Find the impulsive reaction at O and the initial velocity of O.

If equal and opposite impulses P are applied simultaneously at A and B, perpendicular to AOB, show that the impulse at O is numerically equal to $\frac{1}{2}P$ and that the initial velocity of O is zero.

20. Consider the extension of the theorem of § 22·4 to the more general case where the body is in motion when the impulses are applied; prove that the increase in kinetic energy is $\Sigma \mathbf{P} . \mathbf{v}$, where $\mathbf{v}$ is the mean of the velocities, immediately before and immediately after the blow, of the point at which the impulse $\mathbf{P}$ is applied.

CHAPTER XXIII

MOTION OF A SYSTEM

23·1 Dynamical systems. We now consider the motion of a system, built up of particles and rigid bodies connected in various ways. Since each rigid body can be thought of as a finite number of particles set rigidly in an imponderable framework, we can without loss of generality consider a system of particles. We shall confine our attention to problems in two dimensions, and for simplicity we shall suppose the system to lie in a plane. We denote the mass of a typical particle by m, its coordinates, relative to fixed rectangular axes Ox, Oy in the plane, by x, y.

Two simple examples of dynamical systems we have met already. The sun and planet, moving in a plane under their mutual gravitation, is a system containing two particles. The single rigid body is a system containing a large number of particles; in this particular case the geometrical configuration of the particles is maintained unchanged throughout the motion.

The enumeration of a few other concrete examples of dynamical systems may help to fix the ideas: (i) A particle sliding on the face of a wedge which itself slides on a horizontal table. (ii) Three particles moving in a plane under their mutual gravitation. (iii) A double pendulum, consisting of two rods, AB, BC, freely jointed together at B, the whole being freely suspended from A and swinging in a vertical plane through A. (iv) A framework consisting of four equal rods, AB, BC, CD, DA, freely jointed together at the ends to form a rhombus, the vertices A and C being connected by a light spring; the framework moves on a smooth horizontal table. (Notice that if we replace the spring by a string the equations of motion have different forms when the string is taut and when it is slack.) (v) A railway truck; this consists essentially of three rigid bodies, the box-like body of the truck, the front wheels-and-axle, constituting the second rigid body, and the back wheels-and-axle, constituting the third. (In examples (i) and (v) the particles do not actually lie in the same plane, and we must use the device, already mentioned in § 21·4, of projection on to one plane of motion.)

The attack on the problem of the motion of a dynamical system that immediately suggests itself is to start by forming the equations of motion for the separate particles and rigid bodies that constitute the system. These equations will involve the reactions between the component particles and rigid bodies, and the next step is to eliminate these unknown reactions and obtain equations connecting the configuration of the system with the time. For many simple systems this direct attack is quite practicable and expeditious. From the equations of motion from which we started we can also determine the unknown reactions; these will usually be determined in the first instance as functions of the configuration, and hence derivatively as functions of the time.

In other cases the calculations can be simplified by appealing to certain general principles which will be established in this chapter and the next. A really penetrating study of the motion of a dynamical system demands methods more powerful than those used in this book. Nevertheless, without going beyond those methods, we can make substantial progress, and establish theorems of great generality and importance. In the course of our work we shall meet the general form of many principles that we have met already in special cases. But we can hardly expect the theory of the motion of a system, as constructed with the tools available to us here, to possess the completeness and simplicity of the theories already constructed for a single particle and for a single rigid body.

23·2 Generalization of d'Alembert's principle; the three fundamental equations. The forces acting on the particles which constitute the system can be divided (just as in the special case when the system is a single rigid body) into two classes, internal and external; the internal forces are the interactions between different parts of the system, the external forces are forces imposed on the system from outside. For example, we class as internal forces the gravitational attraction between two particles of the system, the actions on the particles of a rigid body by the imponderable frame which maintains the rigidity, the pull on two particles of the system provided by the tension in a string joining them, the actions on the rods at the hinge where two rods are hinged together, the actions at the point of contact when one body rolls or slides on another.

We can find three equations *not involving the internal forces*.

Consider for a moment the analogous problem in statics, where we have a system of rigid bodies in equilibrium. The internal forces alone, *if they all acted in situ on a single rigid body*, would form a system in equilibrium. Therefore the external forces alone, *if they all acted in situ on a single rigid body*, would form a system in equilibrium, and this enables us to write (in the plane case) three equations, for example, by resolving in the directions Ox, Oy, and by taking moments about O. These equations do not involve the internal forces. If the system is just a single rigid body the equations express conditions which are sufficient, as well as necessary, for equilibrium. In the general case, the equations express necessary, but not sufficient, conditions for equilibrium. The finding of additional conditions sufficient to ensure equilibrium depends on the nature of the system, and the best attack may demand some ingenuity.

A similar situation appears in the dynamical problem. The kineton of a particle is equal to the vector sum of the forces, internal and external, acting on it. The internal forces, if they all acted *in situ* on a single rigid body, would form a system in equilibrium. Therefore the external force system and the kineton system are equivalent, if both are thought of as forces acting *in situ* on a single rigid body. This is the generalization for a dynamical system of d'Alembert's principle already encountered in the theory of the motion of a single rigid body. In that case the internal forces, conspicuous by their absence from the equations, consisted solely of the internal reactions in the body; here they consist partly of such reactions, but also partly of reactions between different bodies.

The exact significance of this principle must be carefully noticed. The equivalence of kineta and external forces is to be interpreted in terms of the theory of forces on a single rigid body. We have two sets of localized vectors, and they satisfy the conditions for equivalence familiar in the statics of a single rigid body. In particular, we can write three equations of motion, by resolving in the directions Ox, Oy, and by taking moments about O. The equations are

$$Sm\ddot{x} = SX, \tag{1}$$

$$Sm\ddot{y} = SY, \tag{2}$$

$$Sm(x\ddot{y} - y\ddot{x}) = S(xY - yX). \tag{3}$$

The symbol S denotes summation over the particles, and X, Y are the components of the external forces. If the system is a single rigid body these equations suffice to determine the motion, but this is not true for systems in general.

One important point should be noticed in connexion with the three equations. If the external forces acted on the particles of a single rigid body they would be equivalent to a force X_0, Y_0 at O (where $X_0 = SX, Y_0 = SY$) together with a couple N_0 (where $N_0 = S(xY - yX)$); and X_0, Y_0, N_0 are precisely the second members of the three equations. But when we are dealing with a dynamical system this reduction of the external forces has not the same physical validity that it had for a single rigid body, and should be regarded rather as a convenient artifice of thought.

The first members of the three equations are what we should obtain if we reduced the kineton system to a $\mathscr{K}$-force at O and a $\mathscr{K}$-couple in the same way as if the kineta were forces acting at points of a rigid body. In particular, the first member of the equation (3) is the moment of the kineton system about O. Now for one particular rigid body the kineton system is equivalent to $(M_r\ddot{\xi}_r, M_r\ddot{\eta}_r)$ at G_r, together with a couple $I_r\ddot{\theta}_r$, and the contribution to the first member of equation (3) arising from one rigid body can therefore be written down at once. (The notation is almost self-explanatory; (ξ_r, η_r) are the coordinates of G_r, the centre of gravity of the body, $\dot{\theta}_r$ is its angular velocity, M_r its mass, and I_r its moment of inertia about G_r.)

23·3 Motion of the centre of gravity.

Since

$$Smx = M\xi,$$

where M is the mass of the whole system, and (ξ, η) the centre of gravity G of the whole system, we have

$$Sm\ddot{x} = M\ddot{\xi},$$

and the first equation (of § 23·2) becomes

$$M\ddot{\xi} = X_0. \tag{1}$$

Similarly $$M\ddot{\eta} = Y_0. \tag{2}$$

The equations imply the classical theorem that the motion of G is the same as if all the external forces acted at G on all the mass concentrated in a single particle at G.

23·4 Transformation of the moment equation.

The original form of the equation obtained by taking moments about O, equation (3) of § 23·2, is not particularly useful in practice, but there are two other forms into which it can be changed which are important and useful.

(i) If we write
$$x = \xi + \alpha, \quad y = \eta + \beta,$$

so that (α, β) are coordinates of the particle relative to G, we can transform the first member of equation (3) of § 23·2 as follows:

$$Sm(x\ddot{y} - y\ddot{x}) = Sm\{(\xi + \alpha)(\ddot{\eta} + \ddot{\beta}) - (\eta + \beta)(\ddot{\xi} + \ddot{\alpha})\},$$

and the second member is equal to

$$M(\xi\ddot{\eta} - \eta\ddot{\xi}) + Sm(\alpha\ddot{\beta} - \beta\ddot{\alpha}),$$

since $Sm\alpha$, $Sm\beta$, $Sm\ddot{\alpha}$, $Sm\ddot{\beta}$ all vanish for all values of t. The moment equation becomes

$$M(\xi\ddot{\eta} - \eta\ddot{\xi}) + Sm(\alpha\ddot{\beta} - \beta\ddot{\alpha}) = N_0. \tag{1}$$

This is the first of the two forms mentioned. We notice in the first member the separation into two terms, the first involving only the motion of G, the second only the motion relative to G (i.e. relative to axes *in fixed directions* through G). The first term is the moment about O of a $\mathscr{K}$-force $(M\ddot{\xi}, M\ddot{\eta})$ at G. The second term is the moment about G of the *relative kineton system* $(m\ddot{\alpha}, m\ddot{\beta})$. In practice the contribution to this relative kineton system of one particular rigid body can be found as at the end of § 23·2, if we remember now that only the motion relative to G is involved. It must be emphasized that here the notion of relative motion does not involve any rotation of axes, and in particular that the angular velocity and angular acceleration of a body in the relative motion are the same as in the actual motion.

The special case when the system is a single rigid body deserves a moment's consideration. The term $Sm(\alpha\ddot{\beta}-\beta\ddot{\alpha})$ is the moment about G of the relative kineton system, and this reduces to the familiar $I_G\ddot{\theta}$. We thus recover an equation (equation (5) of § 21·2) already established in the theory of the motion of a single rigid body. A single rigid body is a particular example of a dynamical system, and we could have developed the theory in the reverse order, establishing first the three fundamental equations for a general system, and deducing the equations of motion for a single rigid body as a special case.

For a single rigid body the three fundamental equations (equations (1), (2), (3) of § 23·2) suffice to determine the motion. But, as we have noticed already, this is not true for a system; a similar situation occurs in statics. To determine the motion and the unknown reactions we usually require additional information, which we find, for example, by considering separately the motion of one of the bodies. Again, there is a close analogy with the corresponding situation in statics, and the finding of the neatest way to complete the solution may demand some ingenuity. The concrete illustration in the next paragraph (§ 23·5) illustrates this point.

Corollary. If G is at O we have $\xi=\eta=0$, and equation (1) becomes

$$Sm(\alpha\ddot{\beta}-\beta\ddot{\alpha}) = N_G. \tag{2}$$

The moment about G of the relative kineta is equal to the moment about G of the external forces.

(ii) The second of the two forms mentioned is obtained by reducing piecemeal the kineton system for each individual rigid body, as in the last sentences of § 23·2. We thus obtain the equation in the form

$$\Sigma M_r(\xi_r\ddot{\eta}_r-\eta_r\ddot{\xi}_r)+\Sigma I_r\ddot{\theta}_r = N_0, \tag{3}$$

where the symbol Σ denotes summation over the rigid bodies of the system. We can, of course, include the case where one of the rigid bodies is simply a single particle; for this term the rotation part $I_r \ddot{\theta}_r$ is missing.

Example. *A smooth wedge of mass M and angle α slides on a horizontal plane, and a particle of mass m slides on the face of the wedge; the motion is parallel to a vertical plane through a line of greatest slope. To find the motion and the reactions.*

The system contains one particle and one rigid body; the external forces are the weights and the reaction of the plane on the wedge. If at time t the acceleration of the wedge is f, the acceleration of the particle is the vector sum of f and the relative acceleration f' parallel to the face of the wedge. The kineton system is as shown in Fig. 23·4. From equation (1) of § 23·2 for the *system* we have

$$(M+m)f = mf' \cos\alpha,$$

and from the equation of motion of the *particle* in the direction of a line of greatest slope we have

$$f' - f\cos\alpha = g\sin\alpha.$$

Thus

$$\frac{f}{m\cos\alpha} = \frac{f'}{M+m} = \frac{g\sin\alpha}{M+m-m\cos^2\alpha} = \frac{g\sin\alpha}{M+m\sin^2\alpha}.$$

It turns out, as we could have foreseen, that f and f' are constants.

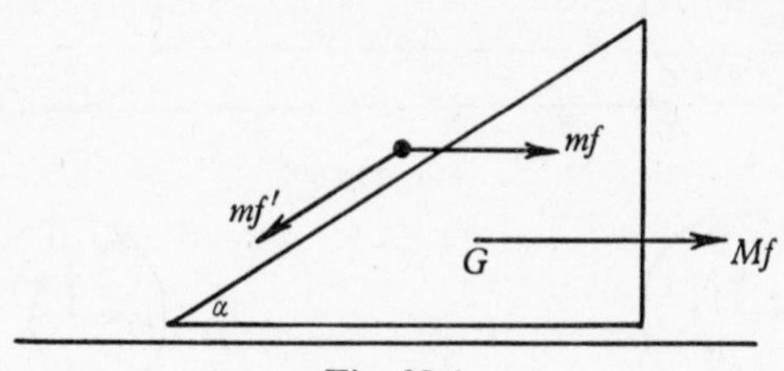

Fig. 23·4

It is easy to find the reactions. If R is the reaction between the wedge and the particle (normal to the face of the wedge) we have, from the horizontal motion of the wedge,

$$R\sin\alpha = Mf,$$

giving

$$R = \frac{m\cos\alpha}{M+m\sin^2\alpha} Mg;$$

and if N is the (vertical) reaction of the plane on the wedge we have

$$N = Mg + R\cos\alpha = \frac{M+m}{M+m\sin^2\alpha} Mg.$$

23·5 Motor car.

23·5 Motor car. The simplified car with which we deal consists of three rigid bodies, the box-like coachwork, the front wheels-and-axle, and the back wheels-and-axle. The motion of the parts of the engine is ignored, and the motion is maintained by a couple Γ on the back wheels; this can be thought of as provided by a spring of negligible mass. There is, of course, an equal and opposite couple on the coachwork. The frictional couple on the front wheels is denoted by N_1, that on the back wheels by N_2. If there were no friction at the bearings, N_1 and N_2 would vanish except when the brakes are applied. It is assumed that the external resistance to motion is represented to a sufficiently high degree of accuracy by a force R through G, the centre of gravity of the system. We shall calculate, on the assumption of no skidding, the acceleration of the car, and the reactions on the wheels by the road.

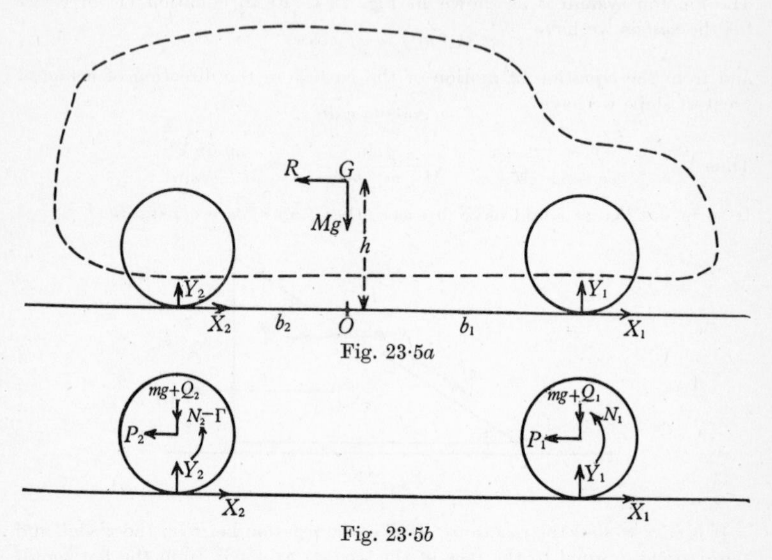

Fig. 23·5*a*

Fig. 23·5*b*

We denote the mass of the system by M, the mass of each pair of wheels and its axle by m, its moment of inertia about its axis by mk^2, and the radius of each wheel by a. The mass of the coachwork is thus $M-2m$. The height of G above the road is $h(>a)$, and the horizontal distances of G from the front and back axles are b_1, b_2. The reaction at the road on one of the front wheels has components $\frac{1}{2}X_1, \frac{1}{2}Y_1$; the corresponding components for one of the back wheels are $\frac{1}{2}X_2, \frac{1}{2}Y_2$. Our object is to determine the acceleration f and the components X_1, Y_1, X_2, Y_2.

The external forces on the system are shown in Fig. 23·5*a*. The external forces on the front wheels-and-axle, and those on the back wheels-and-axle, are shown in Fig. 23·5*b*; the direction of motion is to the right in the figures. The force (P_1, Q_1)

is the reaction of the front axle on the coachwork, (P_2, Q_2) the corresponding reaction of the back axle.

For the system we have, from equation (1) of § 23·3,

$$X_1 + X_2 - R = Mf. \tag{1}$$

For the front wheels, taking moments about the axis,

$$aX_1 + N_1 = -mk^2 f/a, \tag{2}$$

and for the back wheels

$$aX_2 + N_2 - \Gamma = -mk^2 f/a. \tag{3}$$

These equations determine f, X_1, and X_2:

$$(Ma^2 + 2mk^2) f/a = \Gamma - N_1 - N_2 - aR, \tag{4}$$

$$aX_1 = -N_1 - \theta(\Gamma - N_1 - N_2 - aR), \tag{5}$$

$$aX_2 = \Gamma - N_2 - \theta(\Gamma - N_1 - N_2 - aR), \tag{6}$$

where θ is a known constant of the system

$$\theta = \frac{mk^2}{Ma^2 + 2mk^2}.$$

Consider again the equation (4), which we can write in the form

$$M'f = X - F, \tag{7}$$

where

$$M' = M + 2m(k^2/a^2),$$

$$aX = \Gamma,$$

$$aF = aR + N_1 + N_2.$$

The equation (7) has already been anticipated (p. 125); it expresses the equation of motion of the car in the same form as the equation of motion for a single particle. The new mass-constant M' is larger than the mass M of the car; the propulsive force X is, as we should expect, such that aX is the driving couple; the resistance F comprises the air-resistance R, the friction at the axles and the effect of the brakes. When the brakes are not in action the resistance F is represented, to a high degree of accuracy, by a formula $A + Bv^2$ (see § 7·5), where v is the speed of the car.

We now return to the equations of motion. To determine the Y's we have, from equation (2) of § 23·3,

$$Y_1 + Y_2 = Mg, \tag{8}$$

and from equation (1) of § 23·4, taking moments about O,

$$Y_2 b_2 - Y_1 b_1 - Rh = Mfh + 2mk^2(f/a). \tag{9}$$

From (8) and (9), using the value of f in (4), we have

$$bY_1 = Mgb_2 - Rh - \phi(\Gamma - N_1 - N_2 - aR), \tag{10}$$

$$bY_2 = Mgb_1 + Rh + \phi(\Gamma - N_1 - N_2 - aR), \tag{11}$$

where $b = b_1 + b_2$, and represents the distance between the axles, and ϕ is another numerical constant of the system

$$\phi = \frac{Mah + 2mk^2}{Ma^2 + 2mk^2}.$$

This completes the solution. It can be recovered by writing the equations for the three separate rigid bodies, and eliminating the reactions P_1, Q_1, P_2, Q_2.

To establish equation (9) we used equation (1) of § 23·4, but we could equally well have used equation (2) or equation (3). The advantage of using equation (1) and taking moments about O is that X_1 and X_2 do not appear in the equation. If we use equation (2) of § 23·4, taking moments about G, we lose this advantage, but we gain in simplicity, and the equation only involves motion relative to G; if we take moments about G we find

$$Y_2 b_2 - Y_1 b_1 - (X_1 + X_2)h = 2mk^2(f/a), \tag{12}$$

and we recover equation (9) if we substitute for $X_1 + X_2$ from equation (1). If we use equation (3) of § 23·4, taking moments about O, we find

$$Y_2 b_2 - Y_1 b_1 - Rh = M_0 f h_0 + 2mfa + 2mk^2(f/a), \tag{13}$$

where $M_0 (= M - 2m)$ is the mass of the coachwork and h_0 the height of its centre of gravity. Since

$$M_0 h_0 + 2ma = Mh,$$

equation (13) is equivalent to equation (9).

Finally, a word about the numerical magnitudes of the constants θ and ϕ. We have $\theta < m/M$; more precisely, $\theta = O\left(\frac{mk^2}{Ma^2}\right)$; and $\frac{h}{a} > \phi > 1$. The two constants are connected by the relation

$$\phi - 1 = \left(\frac{h}{a} - 1\right)(1 - 2\theta).$$

23·6 Motor car, study of the motion. It may be of interest to examine the solution in greater detail for two special cases.

(i) If there is no braking, and all resistances to motion are neglected, we have $N_1 = N_2 = R = 0$, and the solution takes the simpler form

$$\begin{cases} M'f = \Gamma/a, \\ -aX_1 = \theta\Gamma, \\ aX_2 = (1-\theta)\Gamma, \\ bY_1 \ = Mgb_2 - \phi\Gamma, \\ bY_2 \ = Mgb_1 + \phi\Gamma. \end{cases}$$

We observe that f varies directly with Γ, as we should expect. We notice that, theoretically, Y_1 would vanish if Γ were large enough! But in practice for ordinary cars the maximum available value Γ_m of Γ is substantially less

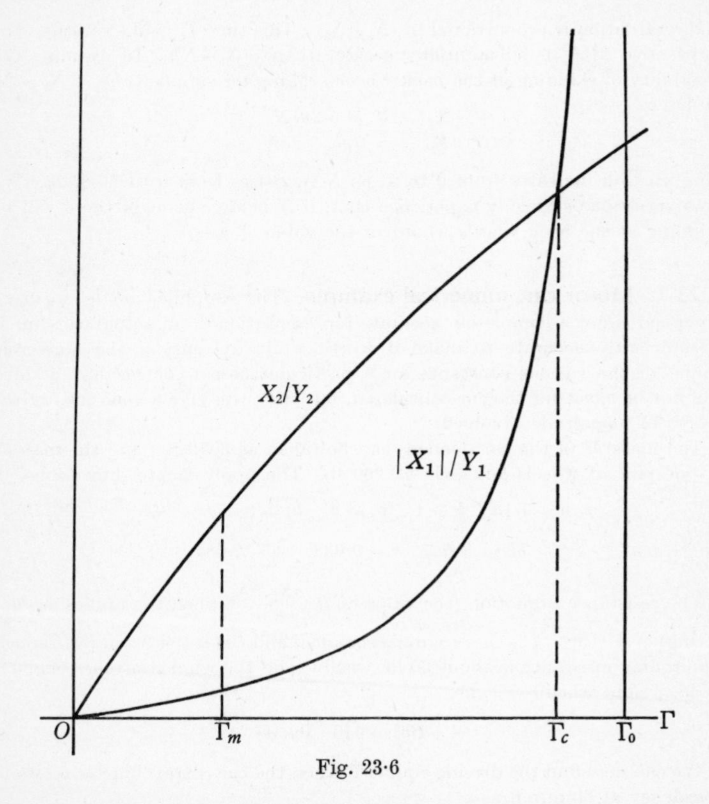

Fig. 23·6

than $\Gamma_0 = Mgb_2/\phi$. We suppose for simplicity that the coefficient of friction between the wheels and the road has the same value μ for front and back wheels. To avoid skidding the values of $|X_1|/Y_1$ and X_2/Y_2 must be less than μ. Now

$$\frac{|X_1|}{Y_1} = \frac{b\theta}{a\phi}\,\frac{\Gamma}{\Gamma_0 - \Gamma}, \quad \frac{X_2}{Y_2} = \frac{b(1-\theta)}{a\phi}\,\frac{\Gamma}{(b_1\Gamma_0/b_2)+\Gamma},$$

and the way these functions vary with Γ is shown in Fig. 23·6. So long as Γ is less than the critical value Γ_c (and this is true in practice, since usually Γ_c is greater than Γ_m) there is greater danger of skidding at the back than at the front wheels, as we should expect. (For the numerical case considered below the value of Γ_m/Γ_0 is about 0·273, and the value of Γ_c/Γ_0 is about 0·886.)

(ii) If $\Gamma = R = 0$, so that there is no driving couple and no resistance to motion other than the braking, the solution becomes

$$\begin{cases} M'f = -(N_1+N_2)/a, \\ -aX_1 = N_1 - \theta(N_1+N_2), \\ -aX_2 = N_2 - \theta(N_1+N_2), \\ bY_1 = Mgb_2 + \phi(N_1+N_2), \\ bY_2 = Mgb_1 - \phi(N_1+N_2). \end{cases}$$

The retardation is proportional to (N_1+N_2). This time Y_2 could vanish!—but in practice Mgb_1 is substantially greater than $\phi(N_1+N_2)$. To examine the possibility of skidding at the back wheels, taking for simplicity $N_1 = N_2 = N$, we have

$$\frac{|X_2|}{Y_2} = \frac{b}{a}\frac{(1-2\theta)N}{Mgb_1 - 2\phi N}.$$

This fraction increases from 0 to ∞ as N increases from 0 to $Mgb_1/2\phi$. We thus verify our everyday experience that, if N is large enough, there will be skidding at the back wheels whatever the value of μ.

23·7 Motor car, numerical example.

The simplified model we have discussed is not a completely accurate representation of an actual car; but it is sufficiently accurate to make it worth while to consider the numerical values of the various constants for a particular case. The results obtained will not be exact for the car considered, but they will give a good idea of the orders of magnitude involved.

The mass M of the car (loaded) can be taken as 3670 lb., and the mass m of one pair of wheels and axle as 200 lb. The approximate dimensions, in feet, are

$$a = 1{\cdot}15, \quad k = 1, \quad b_1 = 6, \quad b_2 = 3, \quad h = 2{\cdot}5.$$

These give

$$M' = 3972, \quad \theta = 0{\cdot}038, \quad \phi = 2{\cdot}084.$$

The resistance to motion (the value of $R + \dfrac{N_1+N_2}{a}$ when the brakes are not acting) is $A + Bv^2$; A is the 'rolling resistance' and Bv^2 is the 'wind resistance'. The rolling resistance is about 20 lb. weight, and the wind resistance is about $\frac{1}{50}$ lb. weight when $v = 1$, i.e.

$$A + Bv^2 = 640 + 0{\cdot}64v^2.$$

We can now find the driving couple Γ when the car is travelling at a steady speed, say at 60 m.p.h.:

$$\Gamma = 1{\cdot}15(640 + 0{\cdot}64 \times 88^2) = 6436,$$

or about 201 foot-pounds weight. (The maximum value Γ_m of the couple when the car is accelerating strongly in low gear is about 1440 foot-pounds weight.)

The power required at this speed is $\Gamma v/a$, or in horse-power

$$\frac{201\times 88}{1\cdot 15\times 550},$$

and this is about 28. The nominal horse-power of the car is 12, but the actual horse-power delivered can rise to three or four times this figure.

Next let us consider the distance in which the car can be brought to rest by the brakes, say from a speed of 30 m.p.h. The maximum braking couple obtainable is in the neighbourhood of 750 foot-pounds weight, so we will suppose each of the braking couples to have the value $N = 750g$. The equation of motion is

$$M'f = -R - \frac{N_1+N_2}{a},$$

$$= -A - Bv^2 - 2N/a,$$

giving
$$M'\frac{d}{dx}(\tfrac{1}{2}v^2) = -(Bv^2+C), \quad \text{say},$$

and the distance to rest from $v = v_0$ is

$$\frac{M'}{2B}\log\left(\frac{Bv_0^2+C}{C}\right).$$

With the given values this is about 90; the car can be brought to rest from 30 m.p.h. in a distance of about 30 yards. If we neglect all resistance to motion other than the braking, as in (ii) above, the distance is about 31 yards; in fact, as we should expect, the braking is of far more importance here than the other resistances, and the neglect of the terms $A+Bv^2$ in this context does not materially affect the result.

Finally, we ask what coefficient of friction is required to prevent slipping while the car is being brought to rest. To answer this precisely we should have to know how much of the resistance $A+Bv^2$ arises from air resistance and how much from axle friction. However, it will suffice for a first approximation to ignore again all resistance effects other than the braking couples. If we suppose as before that each of these has the value $750g$, we have, as in (ii) above,

$$\frac{|X_1|}{Y_1} = \frac{b}{a}\frac{N(1-2\theta)}{Mgb_2+2\phi N}, \quad \frac{|X_2|}{Y_2} = \frac{b}{a}\frac{N(1-2\theta)}{Mgb_1-2\phi N},$$

giving the approximate values 0·38 and 0·29. The danger of skidding is greater at the front wheels than at the back, and the coefficient of friction required is about 0·38. This is comfortably less than the actual value, which, for a rubber tyre on a concrete road, is in the neighbourhood of 0·6.

23·8 The equation of energy. We now return to the general theory of the motion of a dynamical system.

So far the theory has shown a close resemblance to the theory for the particular dynamical system we have studied in detail already, the single rigid body. We have seen that the principle of d'Alembert, which formed the basis of the theory for a single rigid body, can be extended in a certain sense to the general dynamical system. From this we were able to derive three equations of motion not containing the internal forces.

But when we come to considerations of work and energy the theory for a system diverges sharply from the theory for a single rigid body. The reason is as follows.

For a single rigid body the internal forces form a nul system, and the work done by them in any displacement of the body is zero. (The displacement contemplated here is of course a natural displacement in which the body remains rigid; we might wish, in some other connexion, to consider a displacement of a different kind, a non-natural displacement in which the body is distorted, but this non-natural displacement is not the kind contemplated here.)

For a system the internal forces form a nul system, in the sense that if they acted *in situ* on a single rigid body they would be in equilibrium. But it is no longer true that the work done by the internal forces in a displacement of the system is zero. (We have noticed this already in the case of a system of two particles, § 8·3.)

Thus, for a system, the work done in a displacement contains a contribution from the internal forces. When we consider the extension of the equation of energy, from the case of a single rigid body to that of a system, there is a new term in the equation arising from the work done by the internal forces. Explicitly the equations of motion for a typical particle are

$$m\ddot{x} = X, \quad m\ddot{y} = Y,$$

where (X, Y) is the resultant of all the forces acting on the particle. We multiply these equations by $\dot{x}, \dot{y}$ and add, and then sum over all the particles, and we obtain the equation

$$Sm(\dot{x}\ddot{x} + \dot{y}\ddot{y}) = S(X\dot{x} + Y\dot{y}),$$

i.e.

$$dT/dt = R, \tag{1}$$

where

$$T = \tfrac{1}{2}Sm(\dot{x}^2 + \dot{y}^2),$$

and R is the power, or rate of working, of all the forces. T, the sum of the kinetic energies of the particles, is the *kinetic energy of the system*. Equation (1) is the first, or primitive, form of the equation of energy: *the rate of increase of the kinetic energy of the system is equal to the power*.

The point we have to observe here is that *all* the forces, both internal and external, appear in the second member of equation (1). To this remark there is the obvious exception that we may omit the internal reactions in any one rigid body. These belong to the category of *forces of constraint* which we shall consider more fully in § 23·12. But the internal forces between different bodies of the system may not in general be omitted.

23·9 Conservative forces. A set of forces is said to be conservative if the work done by it when the system passes from one configuration to another is independent of the path. We have already met the idea of conservative forces in the theory of the motion of a single particle in a field of force.

Consider a set of forces of this type. If C_0 is some (arbitrary) standard configuration, and C any other configuration, we can define a function $V(C)$ by the equation

$$W_{C_0 C} = -V(C),$$

where $W_{C_1 C_2}$ is the work done by the given set of forces when the system moves from a configuration C_1 to a configuration C_2. If the forces of the given set are (X, Y) we have

$$W_{C_1 C_2} = \int_{C_1}^{C_2} S(Xdx + Ydy),$$

and we need specify no particular path from the configuration C_1 to the configuration C_2. The function $V(C)$—a function of the *configuration* C of the system—is called the *potential energy of the system*. The idea of the potential energy is a generalization of the idea of the potential function for a particle moving in a conservative field (§ 3·10). If we can find an explicit expression for V in terms of parameters (*generalized* or *Lagrangian* coordinates) defining the configuration, the work done by the given set of forces in a small displacement of the system is the perfect differential $-dV$.

We easily establish a generalization of the fundamental property of a conservative field of force (equation (4) of § 3·10). Since

$$W_{C_1C_2} + W_{C_2C_0} + W_{C_0C_1} = 0,$$

we have

$$W_{C_1C_2} = -W_{C_0C_1} + W_{C_0C_2}$$

$$= V(C_1) - V(C_2).$$

This is the fundamental property of the function $V(C)$.

Examples of conservative forces. If the system is a single particle the forces are conservative if the particle moves in a conservative field of force, and the potential V of the field is identical with the potential energy of the system. (But if ϕ denotes the potential per unit mass, a notation which we have used for example in the theory of orbits, the potential energy of the system is $m\phi$.)

For a more general type of system the property of being conservative is a property of the aggregate of all the forces of the set. The two cases that occur most frequently are these:

(i) If the particles of the system move in a conservative field whose potential function is ϕ per unit mass, then the set of forces arising from this field is conservative, and

$$V = Sm\phi.$$

The notation means that we form the sum of the products of the mass of each particle by the value of ϕ at the point occupied by it; this result has been noticed already (§ 19·8) for the special case of a single rigid body. In the particular case of the uniform field $(0, -g)$ we have

$$\phi = gy,$$

and

$$V = Smgy = Mg\eta,$$

where η denotes, as usual, the height of the centre of gravity of the system.

(ii) If there is a repulsion R between two particles, acting in the line joining them, this set of forces is conservative if R is a function $R(r)$ of r, the distance between the particles, and then (as in § 8·3)

$$V = -\int^r R(\xi)\, d\xi.$$

The result is almost evident, but a formal proof may be of interest. For this set of forces

$$W_{C_1C_2} = \int_{C_1}^{C_2} X_1 dx_1 + Y_1 dy_1 + X_2 dx_2 + Y_2 dy_2$$

$$= \int_{C_1}^{C_2} X_2 d(x_2 - x_1) + Y_2 d(y_2 - y_1)$$

$$= \int_{C_1}^{C_2} R\left\{\frac{x_2 - x_1}{r} d(x_2 - x_1) + \frac{y_2 - y_1}{r} d(y_2 - y_1)\right\},$$

and since $$r^2 = (x_2 - x_1)^2 + (y_2 - y_1)^2,$$

we have $$W_{C_1C_2} = \int_{C_1}^{C_2} R dr$$

$$= V(C_1) - V(C_2),$$

and the result follows. The extension to a number of particles is immediate. The internal actions between the pairs form a conservative set if the repulsion between every pair is a function only of their distance apart.

Notice that in (i) the forces of the set are external forces, and in (ii) they are internal.

23·10 The classical form of the equation of energy. Suppose that the aggregate of all the forces on the particles of the system form a conservative set. We form the function $\Omega(t)$ from the function $V(C)$ (as in §§ 5·3 and 11·4) by substituting for C the configuration occupied by the system at the instant t. Then

$$\frac{d\Omega}{dt} = -S(X\dot{x} + Y\dot{y})$$

$$= -R,$$

and the primitive form of the energy equation (equation (1) of § 23·8) becomes

$$\frac{dT}{dt} = -\frac{d\Omega}{dt},$$

whence $$T + \Omega = C.$$

This is the second, or classical, form of the equation of energy.

It is valid when the aggregate of all the forces on the system form a conservative set. It may be written, without fear of ambiguity, in the form

$$T + V = C,$$

provided that it is understood that the symbol V now refers not to a general configuration of the system, but to the particular configuration occupied by the system at time t.

23·11 The third form of the equation of energy.

There is still a third form of the equation of energy, and this third form embraces the other two as particular cases.

Suppose that some of the forces form a conservative set with the potential energy V, but that there are other forces as well; let us denote the forces of the conservative set by (X_1, Y_1), and the remaining forces by (X_2, Y_2). Then (equation (1) of § 23·8)

$$\frac{dT}{dt} = S(X_1\dot{x} + Y_1\dot{y}) + S(X_2\dot{x} + Y_2\dot{y})$$

$$= -\frac{d\Omega}{dt} + S(X_2\dot{x} + Y_2\dot{y}),$$

giving the third form, namely,

$$\frac{d}{dt}(T + \Omega) = S(X_2\dot{x} + Y_2\dot{y}). \qquad (1)$$

The second member is the power of the remaining forces, i.e. those forces not comprehended in the symbol V.

If $V = 0$ we recover the first form of the equation of energy, and if (X_2, Y_2) are absent we recover the second form.

We can also write the third form as

$$(T + V)\,| = W_2, \qquad (2)$$

where the first member is the increment in $T + V$, and the second is the work done by the remaining forces.

Example. *A flywheel has a horizontal shaft of radius a, the moment of inertia about the axis being Mk^2. A light string is wound round the shaft, and supports a mass M hanging vertically. The system is released from rest. The string does not slip on the axle, and the rotation is opposed by a constant frictional couple G ($< Mga$). After the flywheel has turned through an angle θ the mass M meets the inelastic floor. Find the angle ϕ through which the flywheel rotates further before coming to rest.*

If ω is the angular velocity when M meets the floor we have, using the third form of the equation of energy (equation (2) above),

$$\tfrac{1}{2}Mk^2\omega^2+\tfrac{1}{2}M(a\omega)^2-Mga\theta=-G\theta.$$

After this, applying the same principle to the motion of the flywheel alone, and, remembering that the angular velocity is continuous, we have

$$\tfrac{1}{2}Mk^2\omega^2=G\phi.$$

Eliminating ω we find
$$\phi=\frac{k^2(Mga-G)}{(a^2+k^2)G}\theta.$$

23·12 Forces of constraint. There may be forces, or sets of forces, which by their very nature do no work as the system moves. Such sets of forces are called *forces of constraint* or *smooth constraints*. (We have already met forces of this type in §§ 11·5 and 21·6.) If the rest of the forces belong to a conservative set with the potential energy V, *the classical form of the equation of energy is still valid*. This follows from equation (2) of § 23·11; the second member, representing the work done by forces of constraint only, is zero.

As we have noticed already the set of forces that are the internal reactions in a single rigid body belong to this category; the work done by these forces as a whole in any natural displacement of the body is zero.

Other common examples of forces of constraint are: (i) the reaction of a fixed smooth wire on a bead sliding on the wire (§ 11·5) or the reaction of a fixed smooth surface on a particle sliding on it; (ii) the internal reactions between two particles of the system *whose distance apart remains invariable*—in particular, the actions on two rods at a smooth pivot where the rods are hinged together; (iii) the reactions on two smooth surfaces sliding over one another; (iv) the reactions between two rough surfaces in contact when the relative motion is one of pure rolling.

It must be emphasized that when we say that the forces of constraint do no work we are contemplating a natural displacement compatible with the structure of the system. It may be useful in certain circumstances to contemplate a more general type of displacement—e.g. a displacement such as that mentioned in § 23·8 in which a rigid body is distorted like a jelly, or in which an inelastic string is stretched. For displacements of this type it is no longer true that the work done by the forces of constraint is zero. Thus the notion of forces of constraint is intimately connected with the notion of a natural displacement. In particular, the displacement that occurs in any interval of time in the actual motion of the system is a natural displacement, and in the actual motion the forces of constraint do no work.

The recognition of the existence of forces of this type is of great importance. Formerly, when we established the three fundamental equations (§ 23·2) the division of the forces that we needed was the division between *internal forces* and *external forces*. Here an entirely different division emerges, the division between *given forces* and *forces of constraint*. The given forces are prescribed in terms of position and velocity and time.

The new division, into given forces and forces of constraint, is of fundamental importance in the further developments, and in the subject known as Analytical Dynamics it is usual to assume at the outset that all the forces belong to one of these two classes or to the other. This assumption is valid for most of the systems we have studied in this book, *but not for all.* And the systems for which it is not valid are tacitly excluded from discussion in analytical dynamics. A simple example will make the point clear. If a bead slides on a fixed smooth wire in a vertical plane, all is well; the weight is a given force, the reaction of the wire on the bead is a force of constraint. But if the wire is rough the situation is quite different; the weight is a given force, but *the reaction is neither a given force nor a force of constraint.* The problem of the rough wire is of a type which is not studied in analytical dynamics. Of course this does not mean that the problem is intractable; indeed, we have dealt with it already in § 16·2.

Observe, finally, that the new division cuts right across the old; the external forces may be either given forces or forces of constraint, and the internal forces may be either given forces or forces of constraint. Thus, if the system is a bead sliding under gravity on a fixed smooth wire, the forces on it are all external forces, the weight being a given force, and the reaction of the wire a force of constraint. If the system consists of two particles the internal reaction between them is a given force if their distance apart varies under their mutual attraction, but it is a force of constraint if this distance is fixed, for example, if the particles are attached to the ends of a light rod.

23·13 Formulae for kinetic energy. We now establish the general forms of two results of which special cases have been discussed already (§§ 8·2, 21·5).

(i) *König's formula.* We have

$$\begin{aligned} T &= \tfrac{1}{2}Sm(\dot{x}^2+\dot{y}^2) \\ &= \tfrac{1}{2}Sm\{(\dot{\xi}+\dot{\alpha})^2+(\dot{\eta}+\dot{\beta})^2\} \\ &= \tfrac{1}{2}M(\dot{\xi}^2+\dot{\eta}^2)+\tfrac{1}{2}Sm(\dot{\alpha}^2+\dot{\beta}^2). \end{aligned}$$

This is the result we wished to establish; we have, in fact, met it already (equation (1) of § 21·5).

We notice the separation into two terms, the first depending on the motion of G, the other on the motion relative to G. The first of these terms is the same as the kinetic energy of a single particle which remains throughout coincident with G—the same as we should get if the whole mass of the system were concentrated at G. We have already used this theorem repeatedly in the special case of a single rigid body. And, indeed, when we are calculating T for a system it is often simpler to apply the theorem piecemeal to the separate rigid bodies of the system and take the sum, rather than to apply the theorem to the system as a whole.

(ii) *Lagrange's formula.* We start from the simple identity

$$(\Sigma m_r)(\Sigma m_r \dot{x}_r^2) = \Sigma m_r m_s(\dot{x}_s-\dot{x}_r)^2+(\Sigma m_r \dot{x}_r)^2,$$

where the summations run from 1 to ν except the first summation in the second member, which is over pairs of particles (containing therefore $\frac{1}{2}\nu(\nu-1)$ terms). Writing the corresponding equation for y, and adding, we obtain

$$M(2T) = \Sigma m_r m_s v_{rs}^2 + M^2(\dot{\xi}^2+\dot{\eta}^2),$$

where v_{rs} is the velocity of m_s relative to m_r. Thus

$$T = \tfrac{1}{2}M(\dot{\xi}^2+\dot{\eta}^2)+\tfrac{1}{2}\Sigma \frac{m_r m_s}{M} v_{rs}^2,$$

the summation being over pairs. The first term is the same as in König's formula. The second involves only the relative velocities of pairs of particles.

Of the two formulae for T that we have established, the first is the more generally useful.

23·14 Examples of the motion of a system. We now consider some concrete applications of the foregoing theory.

Example 1. *A car (§ 23·5) ascends a slope of inclination α; to find the acceleration.*

With the notation of § 23·5 we have, from the third form of the energy equation (§ 23·11),

$$\frac{d}{dt}(T+V) = (\Gamma - N_1 - N_2)\frac{v}{a} - Rv. \tag{1}$$

Now by König's formula (§ 23·13)

$$T = \tfrac{1}{2}Mv^2 + \tfrac{1}{2}\left(2mk^2\frac{v^2}{a^2}\right)$$

and

$$V = Mgx\sin\alpha,$$

where x is the distance up the slope, $\dot{x} = v$. The first member of equation (1) is therefore

$$\frac{d}{dt}\left(\tfrac{1}{2}Mv^2 + mk^2\frac{v^2}{a^2} + Mgx\sin\alpha\right) = \left(M + 2m\frac{k^2}{a^2}\right)vf + Mgv\sin\alpha.$$

Substituting in equation (1) we find

$$M'f = (\Gamma - N_1 - N_2)\frac{1}{a} - R - Mg\sin\alpha,$$

where

$$M' = M + 2m\frac{k^2}{a^2}$$

as before. For $\alpha = 0$ we recover the value of f already found by another method in § 23·5; and it is clear that to obtain the general case from the case $\alpha = 0$ we have only to replace g by $g\cos\alpha$ and to augment R by $Mg\sin\alpha$. If we require only the value of f the present method is the more expeditious.

The value of f is the same at whatever point of the car we suppose the resistance R to act; formerly we supposed R to act through G. But the values of Y_1 and Y_2 previously found depend essentially on the assumption that R acts through G.

Example 2. *Four uniform rods AB, BC, CD, DA, each of mass m and length 2a, are freely jointed together at their ends to form a rhombus, and the vertices A and C are joined by a light spring of natural length 2a and modulus λ. The framework moves on a smooth horizontal table, the vertices A and C being smoothly constrained to lie always on the straight line Ox. Discuss the motion; show that, if initially the length AC is* $2a(1+\alpha)$, $0<\alpha<1$, *and the angular velocity of the rods is zero, then the length AC varies between the limits* $2a(1\pm\alpha)$, *and find the period of the libration.*

The forces exerted by the external constraints at A and C, perpendicular to Ox, are in fact zero, and no external forces act on the framework. We denote the distance OG by ξ and the angle DAC by θ. For the kinetic energy we need, by König's theorem, the sum of $\tfrac{1}{2}(4m\dot{\xi}^2)$ and the kinetic energy of the motion relative to G. Now for one rod, say CD, the motion relative to G is such that C and D move on perpendicular straight lines through G, and the locus of the centre of the rod is a circle of radius a traversed with speed $a\dot{\theta}$. Thus the kinetic energy of CD in the relative motion is

$$\tfrac{1}{2}m(a^2\dot{\theta}^2 + \tfrac{1}{3}a^2\dot{\theta}^2),$$

and similarly for the other three rods. Thus for the whole system

$$T = 2m\dot{\xi}^2 + \tfrac{8}{3}ma^2\dot{\theta}^2.$$

The potential energy arises from the stretched or compressed spring, and is

$$V = \frac{\lambda}{4a}(4a\cos\theta - 2a)^2 = \lambda a(2\cos\theta - 1)^2.$$

Since there are no external forces $\dot{\xi}$ remains constant (§ 23·3), and from the conservation of energy

$$\tfrac{8}{3}ma^2\dot{\theta}^2 + \lambda a(2\cos\theta - 1)^2 = C. \tag{2}$$

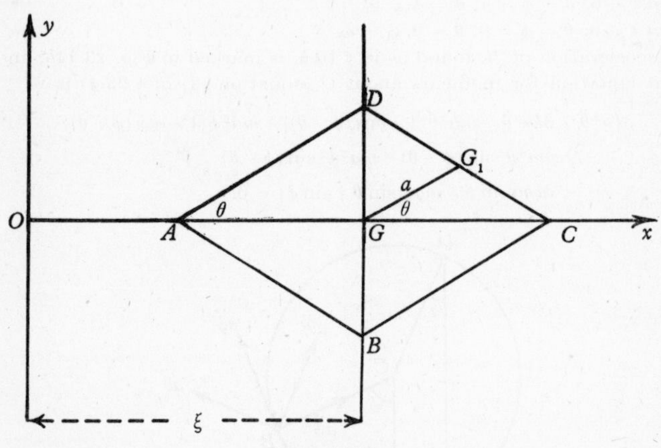

Fig. 23·14a

It is evident, as we could have foreseen, that the constant value of $\dot{\xi}$, prescribed once for all by the initial conditions, does not affect the way in which θ varies.

Since initially $\dot{\theta} = 0$ and $2\cos\theta - 1 = \alpha$ we have $C = \lambda a\alpha^2$, and the energy equation (2) leads to

$$\dot{\theta}^2 = \frac{3\lambda}{8ma}\{\alpha^2 - (2\cos\theta - 1)^2\}. \tag{3}$$

There is a libration in θ between the acute angles given by $\cos\theta = \frac{1}{2}(1 \pm \alpha)$, and this is equivalent to a libration in AC between the lengths $2a(1 \pm \alpha)$. This we can see from equation (3), or perhaps even more simply by rewriting the equation in terms of $x = 2\cos\theta - 1$. With this notation $AC = 2a(1+x)$, and equation (2) becomes

$$\dot{x}^2 = \frac{3\lambda}{8ma}(3+x)(1-x)(\alpha+x)(\alpha-x), \tag{4}$$

and the libration in x between $\pm\alpha$ is evident. We find a formula for the period σ from either of the equations (3) or (4); or a rather simpler formula by writing $x = \alpha\cos\phi$ (cf. § 6·5) giving

$$\sigma = 4\sqrt{\left(\frac{2ma}{3\lambda}\right)}\int_0^\pi \frac{d\phi}{\sqrt{\{(3+\alpha\cos\phi)(1-\alpha\cos\phi)\}}}.$$

Example 3. *A smooth uniform circular hoop, of mass M and radius a, swings in a vertical plane about a point O of itself which is freely hinged to a fixed support, and a bead B of mass m slides on the hoop. The inclination of OC (where C is the centre of the hoop) to the downward vertical is denoted by θ, and the inclination of CB to the downward vertical by ϕ. Establish equations of motion.*

Deduce the approximate equations for dealing with the small oscillations about the position of stable equilibrium, and show how to find the complete solution of these equations for given initial values, sufficiently small, of θ, ϕ, $\dot{\theta}$, $\dot{\phi}$. As concrete applications find the values of θ and ϕ at any subsequent time in the two following cases:

(i) *at* $t=0$, $\theta=\phi=\alpha$, $\dot{\theta}=\dot{\phi}=0$;

(ii) *at* $t=0$, $\theta=\phi=0$, $\dot{\theta}=0$, $\dot{\phi}=\omega$.

The acceleration of B, found as in § 10·5, is marked in Fig. 23·14*b*, and the moment equation for moments about O (equation (3) of § 23·4) is

$$\begin{aligned} Ma^2\ddot{\theta}+Ma^2\ddot{\theta}+ma^2\ddot{\theta}\{1+\cos(\phi-\theta)\}+ma^2\ddot{\phi}\{1+\cos(\phi-\theta)\} \\ +ma^2\dot{\theta}^2\sin(\phi-\theta)-ma^2\dot{\phi}^2\sin(\phi-\theta) \\ +Mga\sin\theta+mga(\sin\theta+\sin\phi)=0. \qquad (5) \end{aligned}$$

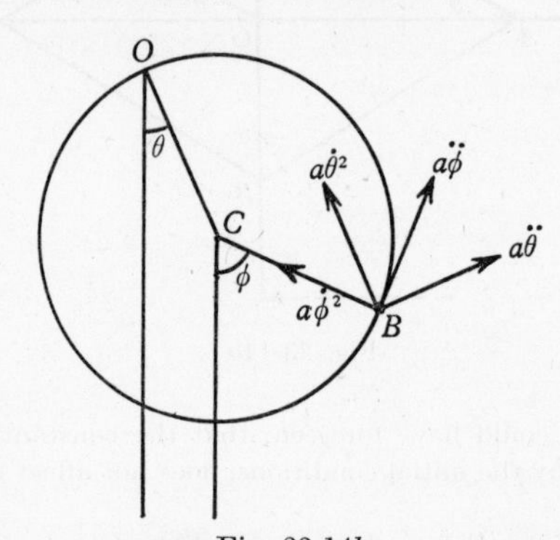

Fig. 23·14*b*

The equation of motion of the bead in the direction perpendicular to CB is

$$ma\ddot{\theta}\cos(\phi-\theta)+ma\ddot{\phi}+ma\dot{\theta}^2\sin(\phi-\theta)+mg\sin\phi=0. \qquad (6)$$

If we multiply equation (6) by a and subtract from (5) we get

$$(2M+m)\ddot{\theta}+m\ddot{\phi}\cos(\phi-\theta)-m\dot{\phi}^2\sin(\phi-\theta)+(M+m)n^2\sin\theta=0, \qquad (7)$$

where $n^2=g/a$. We rewrite equation (6) in the form

$$\ddot{\theta}\cos(\phi-\theta)+\ddot{\phi}+\dot{\theta}^2\sin(\phi-\theta)+n^2\sin\phi=0, \qquad (8)$$

and we take (7) and (8) as the equations of motion for the system.

The equation of energy

$$T+V=C$$

is a first integral of these equations, and we can take this in place of one of the others as the second equation for the study of the motion. But the forms (7) and (8) lead more readily to the equations we want for the problem of the small oscillations about the position $\theta=\phi=0$.

We now turn to this problem. If we take $V = 0$ in the position of equilibrium the motion takes place in the region $V \leqslant C$, since $T \geqslant 0$. Now

$$V = (M+m)ga(1-\cos\theta)+mga(1-\cos\phi),$$

and for small values of C the curve $V = C$ (in a subsidiary diagram with θ and ϕ as Cartesian coordinates) is a closed oval curve; if C is very small the curve is approximately an ellipse. The point (θ, ϕ) moves in the region $V \leqslant C$, which is the interior of the oval, and θ, ϕ remain small for all time; $\dot\theta$, $\dot\phi$, $\ddot\theta$, $\ddot\phi$ also remain small. We shall therefore obtain equations giving an approximation to the motion if, in the equations (7) and (8), we retain only terms of the first order in θ and ϕ and their derivatives. We thus obtain the approximate equations required for the study of the small oscillations, namely

$$(2M+m)\ddot\theta+m\ddot\phi+(M+m)n^2\theta = 0, \tag{9}$$

$$\ddot\theta+\ddot\phi+n^2\phi = 0. \tag{10}$$

We now introduce a technique for dealing with these equations. If we multiply the second of the equations by the constant number λ, and add to the first, we get

$$(2M+m+\lambda)\ddot\theta+(m+\lambda)\ddot\phi+n^2\{(M+m)\theta+\lambda\phi\} = 0. \tag{11}$$

We now choose λ so that we have an equation in only one variable. To achieve this end we need

$$\frac{M+m}{2M+m+\lambda} = \frac{\lambda}{m+\lambda},$$

and this is a quadratic equation for λ,

$$\lambda^2+M\lambda-m(M+m) = 0,$$

giving two values for λ, namely, $\lambda = m$ and $\lambda = -(M+m)$.

With the value $\lambda = m$ equation (11) becomes

$$2\{(M+m)\ddot\theta+m\ddot\phi\}+n^2\{(M+m)\theta+m\phi\} = 0, \tag{12}$$

and with the value $\lambda = -(M+m)$, equation (11) becomes

$$M(\ddot\theta-\ddot\phi)+n^2(M+m)(\theta-\phi) = 0. \tag{13}$$

We have thus replaced the equations (9) and (10) by the simpler equations (12) and (13), which we now write in the forms

$$\left.\begin{aligned}\ddot\xi+p^2\xi &= 0,\\ \ddot\eta+q^2\eta &= 0,\end{aligned}\right\} \tag{14}$$

where

$$\left.\begin{aligned}m\xi &= (M+m)\theta+m\phi,\\ \eta &= \theta-\phi,\end{aligned}\right\} \tag{15}$$

and $p^2 = \frac{1}{2}n^2$, $q^2 = (M+m)n^2/M$. We shall need also the expression of θ and ϕ in terms of ξ and η, namely,

$$\left.\begin{aligned}(M+2m)\theta &= m(\xi+\eta),\\ (M+2m)\phi &= m\xi-(M+m)\eta.\end{aligned}\right\} \tag{16}$$

The new coordinates ξ, η, which are the linear functions (15) of the original coordinates θ and ϕ, are called *normal coordinates* for the oscillating system.

They possess an interesting and important property; the ξ-motion and the η-motion are harmonic motions which are completely independent of one another. An oscillation in which only one normal coordinate varies, the other being zero, is called a *normal mode*. We have met the equations (14) already in § 10·12. The motion is periodic, whatever the initial conditions, if p/q is rational. If $p/q = r/s$, where r and s are positive integers with no common factor, the period is

$$\sigma = \frac{r}{p}2\pi = \frac{s}{q}2\pi.$$

In terms of the masses the condition for periodicity is that there are integers r and s such that

$$\frac{M}{m} = \frac{2r^2}{s^2-2r^2}.$$

This is true, for example, if $M = 8m$. If p/q is irrational the motion is never periodic, except when one normal mode is quiescent, i.e. unless either ξ or η vanishes for all values of t.

We can now find the complete solution of the equations (9) and (10) with any prescribed values for θ, ϕ, $\dot\theta$, $\dot\phi$ at $t = 0$. The equations (16) express θ and ϕ in terms of ξ and η. The equations (14) are of a familiar form from which the values of ξ and η at time t (>0) can be written down when the values of ξ, η, $\dot\xi$, $\dot\eta$ at $t = 0$ are known (§ 5·5). And these initial values are given in turn in terms of the initial values of θ, ϕ, $\dot\theta$, $\dot\phi$ by means of (15).

In the first of the two special cases mentioned the initial values are

$$\theta = \phi = \alpha, \quad \dot\theta = \dot\phi = 0.$$

$$\xi = \frac{M+2m}{m}\alpha, \quad \eta = 0, \quad \dot\xi = \dot\eta = 0.$$

Here $\eta = 0$ throughout; the oscillation is a normal oscillation in which only one normal coordinate is active. Since

$$\xi = \frac{M+2m}{m}\alpha\cos pt$$

the final solution is $\theta = \phi = \alpha\cos pt.$

In this normal mode the bead remains at rest relative to the hoop. (It is easy to verify that in the other normal mode the centre of gravity of the system remains vertically below O.)

In the second of the two special cases the initial values are

$$\theta = \phi = 0, \quad \dot\theta = 0, \quad \dot\phi = \omega,$$

$$\xi = \eta = 0, \quad \dot\xi = \omega, \quad \dot\eta = -\omega.$$

The solutions of (14) are

$$\xi = \frac{\omega}{p}\sin pt, \quad \eta = -\frac{\omega}{q}\sin qt,$$

whence

$$(M+2m)\,\theta = m\omega\left(\frac{1}{p}\sin pt - \frac{1}{q}\sin qt\right),$$

$$(M+2m)\,\phi = \omega\left(\frac{m}{p}\sin pt + \frac{(M+m)}{q}\sin qt\right).$$

23·15 Lagrangian coordinates. In Example 2 of § 23·14 the configuration of the system is defined by the values of ξ and θ. In Example 3 the configuration is defined by the values of θ and ϕ. Such parameters, already glanced at in § 23·9, are called *Lagrangian coordinates*.

The Lagrangian coordinates are generalized coordinates whose values define the configuration of the system. Their characteristic property is that the Cartesian coordinates of the ν particles that constitute the system can be expressed as functions of the Lagrangian coordinates and of t, i.e. we have equations of the type

$$x = \phi(q_1, q_2, \ldots, q_n; t),$$

where $q_1, q_2, \ldots, q_n$ are the Lagrangian coordinates, and x is one of the three Cartesian coordinates of one of the ν particles. In the simple cases the function ϕ does not involve t, but only $q_1, q_2, \ldots, q_n$.

For most systems we can arrange that the number n of Lagrangian coordinates is equal to the number of degrees of freedom of the system. For example, each of the systems mentioned at the beginning of this paragraph has two degrees of freedom. But there are also systems, called *non-holonomic systems*, for which it is necessary to use a number of Lagrangian coordinates in excess of the number of degrees of freedom; these systems, however, are somewhat outside the scope of this book.

EXAMPLES XXIII

1. A reel of mass M, consisting of a cylinder of radius a connecting two discs of radius b $(b > a)$, rolls without slipping on a horizontal plane. A light thread, wound on the cylinder, passes in a plane perpendicular to the axis of the reel, horizontally from its under side, over a smooth pulley, and thence vertically downwards; to its free end is attached a mass m. Show that the reel will move with acceleration

$$f = g \cdot \frac{mb(b-a)}{M(b^2+k^2)+m(b-a)^2},$$

where k is the radius of gyration of the reel about its axis.

Consider the corresponding problem when the reel rolls, with its axis horizontal, up a rough plane inclined at an angle α to the horizontal. The thread passes round and under the reel as before, then upwards parallel to a line of greatest slope of the plane, and then over a smooth pulley with a mass m hanging from the free end. Prove in particular that $f = 0$ in the critical case when $Mb \sin\alpha = m(b-a)$.

Consider also the corresponding problem when the inclined plane is smooth: it is assumed that the thread lies in the vertical plane of symmetry of the reel and that the motion is in two dimensions.

2. A body of mass M with two plane smooth faces inclined at an angle $\alpha (< 90°)$ is placed with one face in contact with a smooth horizontal table. A particle of mass $2M$ is placed on the other face and allowed to slide downwards under gravity. Show that the horizontal acceleration of the body and the pressure on the table are respectively

$$\frac{g \sin 2\alpha}{2-\cos 2\alpha} \quad \text{and} \quad \frac{3Mg}{2-\cos 2\alpha}.$$

3. The horizontal and inclined faces of a wedge of mass M meet at an angle α in a line AB. The horizontal face is on a smooth table, and a particle of mass m slides down the inclined face, which is rough, the coefficient of friction between particle and wedge being μ. The motion takes place in a plane perpendicular to AB. Show that if the system starts from rest the time taken by the particle to describe a given distance down the wedge bears to the time taken when the wedge is fixed the ratio

$$\left\{1-\frac{m \cos \alpha(\cos \alpha + \mu \sin \alpha)}{M+m}\right\}^{\frac{1}{2}} : 1.$$

4. A long rough plank of mass M rests on a horizontal table and a uniform sphere of mass m rests on the plank. The coefficient of friction between the sphere and the plank, and between the plank and the table, is μ. An impulse is applied to the plank which gives it an initial velocity V in the direction of its length. Explain why the sphere is at rest just after the impulse is applied to the plank.

Show that there will be no slipping between the sphere and the plank after a time

$$\frac{V}{\left(\frac{9}{2}+\frac{2m}{M}\right)\mu g}.$$

5. A uniform sphere of mass m and radius a, rotating about a horizontal diameter with angular velocity ω, is placed gently on the upper surface of a uniform rigid plane sheet of mass m which rests with its lower surface in contact with a rough horizontal table. The coefficient of friction between the sphere and the sheet is μ, and between the sheet and the table is $\frac{1}{4}\mu$. Prove that, if the sheet is held fixed, the sphere will travel a distance $\frac{2}{49}\cdot\frac{a^2\omega^2}{\mu g}$ before slipping ceases. If the sheet is free to slide, show that the distance relative to the sheet travelled by the sphere before it ceases to slip on the sheet is $\frac{3}{64}\cdot\frac{a^2\omega^2}{\mu g}$, and that in the same time the sheet has travelled a distance relative to the table one-third of this.

6. A uniform rectangular board of mass M is freely hinged along one edge which is horizontal; the breadth of the board at right angles to the hinge is $2a$. A particle of mass m is placed on the board at distance x from the hinge when the board is held at rest in a horizontal position, and the board is then released so that it can turn freely about the hinge under gravity.

There is a coefficient of friction μ between the board and the particle. Show that if $x > 4a/3$ the particle will leave the board immediately, and that if $x < 4a/3$ the particle will begin to slide on the board when it has fallen through an angle ϕ given by

$$\tan\phi = \frac{\mu Ma(4a-3x)}{2Ma(2a+3x)+9mx^2}.$$

7. A particle is attached to the end D of a light string CD of length b which is knotted at C to two equal light strings AC, CB fastened at points A, B at the same level; the distance of C from the line AB when the strings are taut is a. Find the equations of motion for small oscillations of the particle in the vertical plane through AB, and in the vertical plane through C perpendicular to AB, and integrate them.

If $a = 3b$, show that the particle may be made to describe an arc of a parabola. (Cf. § 10·12.)

8. Two particles, each of mass m, are joined by a light rod of length $2a$. The rod is placed in a vertical position on a smooth horizontal table and allowed to fall over. Assuming that the lower particle remains on the table prove that, when the inclination of the rod to the vertical is θ, (i) the speed of the lower particle is

$$\cos\theta\sqrt{\left(\frac{2ga(1-\cos\theta)}{2-\cos^2\theta}\right)},$$

and (ii) the acceleration of the mass-centre is vertically downwards and of magnitude

$$g(1-\cos\theta)(2+4\cos\theta-\cos^2\theta-\cos^3\theta)/(2-\cos^2\theta)^2.$$

Prove that the lower particle does remain on the table.

9. AB, BC are two uniform rods of masses m, m' and lengths $2a, 2b$ respectively. AB is pivoted freely at a fixed point A and is initially held in a horizontal position, while BC hangs freely from B. When AB is released and is free to rotate round A, show that $\ddot{\theta}_0$, its initial angular acceleration, is given by

$$\ddot{\theta}_0 = \frac{3g}{4a}\cdot\frac{m+2m'}{m+3m'}.$$

10. A uniform circular disc of mass $2m$ and radius a is free to rotate in a vertical plane about its centre, which is fixed. An insect of mass m stands on the rim of the disc at its lowest point, and the system is initially at rest. The insect then proceeds to walk round the rim, his motion relative to the rim having constant acceleration f. Prove that he will not reach the top if $f < g\sin 2\beta$, where β is the acute angle defined by the equation $\tan\beta = 2\beta$.

11. The speed of a railway truck, weighing 5 tons, is reduced uniformly from 25 to 20 m.p.h. on the level in a distance of $695\frac{3}{4}$ ft. by the brakes. Show that, if no slipping takes place between the wheels and the rails, the normal pressure between the rails and each of the front wheels is 50 lb. weight greater than the corresponding pressures on the back wheels, given that the distance between the axles is 12 ft. and that the centre of gravity of the truck is $4\frac{1}{2}$ ft. above the ground and equidistant from the axles, while the diameter of each wheel is 3 ft. and the moment of inertia of each pair of wheels and axle about its axis is 3600 lb.ft.^{2} units.

12. A bead P, of mass m, is free to slide on a smooth fixed circular hoop, of radius a, which is fixed in a horizontal plane. A light inextensible string attached to the bead passes through a small smooth ring fixed at a point A of the hoop and supports a particle of mass M hanging freely. Initially the system is at rest with the bead at B, the other end of the diameter through A. Prove that if the system is slightly disturbed, the velocity of the bead when the angle PAB has the value θ is v, where

$$v^2 = \frac{4Mga(1-\cos\theta)}{m+M\sin^2\theta}.$$

Prove also that the tension in the string at the same instant is

$$\frac{Mmg\{m+M(1-\cos\theta)^2\}}{(m+M\sin^2\theta)^2}.$$

13. A bead of mass m slides on a fixed smooth wire which has the form of a horizontal circle of radius a. The bead is connected to a hanging particle of the same mass by a light inextensible string, which passes through a small smooth ring which is fixed at a point in the plane of the circle at distance $\frac{1}{2}a$ from the centre. The system is released from rest, the horizontal part of the string being initially of length a. Show that when the bead reaches the point of the wire nearest the ring its velocity is $\sqrt{(ga)}$, and that the tension in the string at this instant is $2mg$. Find also the direction and magnitude of the reaction of the wire on the bead at this instant.

14. A heavy ring of mass m slides on a smooth vertical rod, and is attached to a light string which passes over a small light pulley at a distance a from the rod and is attached to a mass M ($>m$) which hangs freely. The system is released from rest when the string is taut, m level with the pulley, and M vertically beneath the pulley. Show that the ring drops a distance

$$\frac{2Mma}{M^2-m^2}$$

before coming to rest again. At this point the ring suddenly splits into two rings of equal mass, one of which falls freely while the string remains attached to the upper one. Show that the upper ring rises to a height

$$\frac{2M^2-m^2}{4M^2-m^2}\cdot\frac{2Mma}{M^2-m^2}$$

above the pulley before coming momentarily to rest.

15. A bead slides on a smooth parabolic wire in a horizontal plane and is connected to a hanging particle of the same mass as the bead by a light inelastic string passing through a small smooth hole at the focus. The bead is released from rest at a distance $5a/2$ from the focus, where $4a$ is the length of the latus rectum of the parabola. Show that the greatest velocity of the hanging particle in the subsequent motion is $\sqrt{(\frac{1}{2}ga)}$.

16. A bead of mass m slides on a smooth parabolic wire which is fixed with its axis vertical and vertex upwards, the latus rectum of the parabola being $4a$. It is attached to a particle of mass $2m$ by a light inelastic string passing through a smooth ring at the focus. The bead is held at the level

of the focus and released. Assuming that the string remains taut during the subsequent motion, show that, if z denotes the depth of the bead below the vertex at time t,

$$\dot{z}^2 = \frac{2gz(a-z)}{3z+a}.$$

Prove that the string does remain taut.

17. A bead of mass m slides on a smooth parabolic wire whose axis is vertical and vertex upwards. It is connected to a particle of mass $3m$ by a light inelastic string passing through a smooth ring at the focus. The system is let go from rest when the bead is at a depth $4a/9$ below the vertex, where $4a$ is the latus rectum of the parabola. Show that in the subsequent motion the greatest velocity of the particle is $\frac{1}{3}\sqrt{(ga)}$, and that the velocity of the bead when it passes through the vertex is $\frac{4}{3}\sqrt{(ga)}$.

18. A bead of mass m, free to slide on a fixed smooth vertical circular wire of radius a whose centre is O, is attached by a light elastic string of natural length $7a/12$ to a point at a distance $5a/12$ vertically below O. The string is such that a force mg would extend it to twice its unstretched length. When the bead is vertically below O, it is given a horizontal velocity $\sqrt{(17ga/7)}$ in the plane of the wire; show that it will just reach the level of O.

19. A bead of mass m slides on a smooth wire, which is bent into the form of a circle of radius a, and fixed in a vertical plane. The bead is joined to the highest point A of the circle by a light elastic string, whose natural length is a. The tension required to double the length of the string is $3mg$. Prove that, if the bead is let go from rest with the string just taut, it continually approaches the lowest point of the circle as t tends to infinity.

If the length of the string during the motion is denoted by $2as$, prove that

$$\dot{s}^2 = n^2(1+s)(1-s^2)(s-\tfrac{1}{2}),$$

where $n^2 = 2g/a$. Verify that the length of the string at time t after the start is

$$2a\frac{1+3\cosh nt}{5+3\cosh nt}.$$

20. Two equal particles are joined by a light inextensible string of length $2a\phi$ and rest in equilibrium on the surface of a smooth circular cylinder of radius a in a plane perpendicular to the axis of the cylinder, which is horizontal. Prove that, if the particles are slightly displaced in this plane and the lower particle leaves the surface when the other particle reaches the highest generator of the cylinder, then $\phi = \frac{1}{6}\pi$.

21. A tube ABC of mass m is bent at right-angles at B. The part AB is horizontal and slides freely through two fixed rings; the part BC is vertical. Particles P, Q, each of mass m, move without friction in AB, BC, and are connected by a string passing over a smooth pulley of negligible mass at B. The system is released from rest. Apply the principles of momentum and energy to show that, when Q has fallen a distance y from its initial position, its vertical velocity is $\sqrt{(6gy/5)}$.

Show that the vertical and horizontal components of the acceleration of Q are $3g/5$ and $g/5$.

22. A light string $ABCDE$, whose middle point is C, passes through smooth rings B, D, which are fixed in a horizontal plane at a distance $2a$ apart. To each of the points A, C, E is attached a mass m. Initially C is held at rest at O, the middle point of BD, and is then set free. Show that C will come instantaneously to rest when $OC = 4a/3$. (The total length of the string is greater than $10a/3$.)

Show that when C has fallen through $3a/4$ from O, its velocity is $\sqrt{(25ag/86)}$.

23. Two particles A, B, each of mass m, are constrained to move without friction on a circle of radius a, and they are joined by a light spring of natural length $2a \sin\alpha$ and modulus λ. The angle between the radii through A and B is denoted by 2ϕ. Show that during the motion

$$\dot{\phi}^2 + \frac{\lambda}{ma \sin\alpha}(\sin\phi - \sin\alpha)^2$$

remains constant; and that the difference between the normal reactions at A and B is proportional to $\dot{\phi}$.

If the maximum compression of the spring occurs when $\phi = \beta$, show that the maximum elongation occurs when

$$\sin\phi = 2\sin\alpha - \sin\beta,$$

provided that $2\sin\alpha - \sin\beta < 1$. What happens if $2\sin\alpha - \sin\beta > 1$?

24. A circular cylinder of mass m and radius r has mk^2 as moment of inertia about its axis. It rolls, without slipping, with its axis horizontal, on a plane making an angle α with the horizontal. Find the linear acceleration of the cylinder and the resultant reaction of the plane.

A bicycle of total mass 25 lb. has two equal wheels of 12 in. radius, each of mass 3 lb. When (i) the (riderless) bicycle and (ii) a wheel are allowed to run down a certain slope, the accelerations are 6·1 and 4·2 ft. sec.$^{-2}$ respectively. Show that the radius of gyration of each wheel is 10 in.

25. What purpose is served by a flywheel?

A flywheel of a pressing machine has 150,000 ft. lb. of kinetic energy stored in it when its speed is 250 r.p.m. What energy does it part with during a reduction in speed to 200 r.p.m.? If 82 per cent of this energy given out is imparted to the pressing rod during a stroke of 2 in., what is the average force exerted by the rod?

26. Two uniform rods AB and CD, each of mass M and length $4a$, are smoothly jointed together at a point O distant a from A and from C. The ends B and D are freely attached to small light rings which slide on a smooth horizontal wire. Initially the system is at rest, with O vertically below the wire at a depth a, and the ends B and D are then given velocities $2\sqrt{(\frac{1}{3}ag)}$ outwards along the wire. Show that O reaches the wire after a time

$$\frac{1}{2}\sqrt{\left(\frac{a}{g}\right)} \int_{\alpha}^{\frac{1}{2}\pi} (4 - 3\cos\theta)^{\frac{1}{2}}\, d\theta,$$

where α is the acute angle defined by $\cos\alpha = \frac{1}{3}$.

27. Three equal uniform rods AB, BC, CD each of length $2a$ and mass m are freely jointed at B and C, and at the ends A and D light rings are attached

which can slide freely on a fixed smooth horizontal rail. The system is released from rest when all the rods are horizontal. If in the subsequent motion θ is the angle of inclination of AB or CD to the horizontal, prove that

$$\dot{\theta}^2 = \frac{6g\sin\theta}{a(2+3\cos^2\theta)}.$$

Prove also that the total downward reaction on the rail is equal to $3mg$ when

$$\cos^2\theta = \sqrt{(\tfrac{7}{3})}-1.$$

28. A uniform circular disc, of mass M and radius r, rolls without slipping on a rough horizontal plane. A uniform rod, of mass M and length $2a$ ($>r$), has one end freely attached to the centre of the disc and the other in contact with a smooth vertical wall. The whole system moves in a vertical plane perpendicular to the wall. The inclination of the rod to the horizontal is denoted by θ, and the system starts from rest when $\theta=\alpha$. Show that

$$a\dot{\theta}^2(2+9\sin^2\theta) = 3g(\sin\alpha-\sin\theta).$$

Prove also that the ratio of the reaction of the wall to the frictional force on the disc is independent of the value of θ.

29. A uniform heavy chain of length l and mass $4m$ hangs symmetrically over a pulley of radius a and moment of inertia ma^2 which is free to turn in a vertical plane about a fixed horizontal axis. Bodies of mass m and $2m$ respectively are fixed to the ends of the chain and the system is released from rest. The chain does not slip on the pulley. Find from the equation of energy the velocity of the lighter body when it has risen a distance x, and deduce its acceleration. Verify the result for the acceleration by considering the moment of momentum about the axis of the pulley.

30. A mass M is suspended at the lower end of a vertical elastic wire of mass m and length L, suspended at its upper end. The system is caused to execute small vertical oscillations. Assuming that the wire can be treated as uniformly stretched throughout the motion, show that the kinetic energy of the system is $\frac{1}{2}\dot{x}^2(M+\frac{1}{3}m)$, where x is the displacement of the mass M from its equilibrium position. Hence show that the time of a complete oscillation is $2\pi\sqrt{\{kL(M+\frac{1}{3}m)\}}$, where k is such that unit force produces an extension of k units in unit length of the wire.

*31. Two uniform rods OA, AB, of the same length $2a$ and of the same mass m, are smoothly jointed at A; OA can turn freely about the fixed point O, while B is constrained by a small ring attached to it to remain on a smooth vertical wire passing through O. The system is released from rest, O and B being initially in contact. If OA is inclined to the vertical at an angle θ after a time t, prove that

$$\dot{\theta}^2 = \frac{g}{a}\frac{3\cos\theta}{4-3\cos^2\theta},$$

and find the initial action at the joint A. (We can express the components X, Y of the action at A on OA at time t as functions of θ. The difficulty is that $X\cos\theta$ tends to a finite limit as $\theta\to\frac{1}{2}\pi$. Therefore we cannot find the initial action at A from the position $\theta=\frac{1}{2}\pi$ alone, but must consider what happens as $\theta\to\frac{1}{2}\pi$.)

32. A uniform solid cylinder of mass M and radius a lies on a smooth horizontal plane, and a second uniform solid cylinder of mass m and radius b lies in contact with the first along a generator. The system moves from rest, and there is no slipping between the cylinders. Show that so long as the cylinders remain in contact

$$\frac{3M+2m\sin^2\theta}{2(M+m)}(a+b)\,\dot{\theta}^2 = 2g(\cos\alpha-\cos\theta),$$

where θ is the angle between the vertical and the plane through the axes of the cylinders, and α is its initial value.

*33. A uniform solid circular cylinder of radius a rests on a horizontal plane. A similar cylinder rests on the first, touching it along the highest generator, and the system is slightly disturbed. All the surfaces are perfectly rough. Show that, so long as the cylinders remain in contact, the angle θ which the plane containing the axes makes with the vertical satisfies the equation

$$\left(\frac{d\theta}{dt}\right)^2 = \frac{12g(1-\cos\theta)}{a(17+4\cos\theta-4\cos^2\theta)}.$$

Show also that, so long as the cylinders remain in contact, the path of a point on the axis of the upper cylinder, referred to the tangent and normal at the highest point as axes, may be expressed in the form

$$x = \tfrac{1}{3}a(\theta+4\sin\theta),$$

$$y = 2a(1-\cos\theta).$$

*34. A uniform circular cylinder of mass M is free to rotate about its axis which is smooth and horizontal and about which its radius of gyration is equal to its radius. A uniform solid sphere of mass m is placed with its lowest point in contact with the highest generator of the cylinder, the system being initially at rest. The sphere is then slightly disturbed and rolls down the cylinder, all points of the sphere moving in planes perpendicular to the axis of the cylinder. Show that slipping takes place before the sphere leaves contact with the cylinder, and begins when

$$2M\sin\theta = \mu[(17M+6m)\cos\theta-(10M+4m)],$$

μ being the coefficient of friction between sphere and cylinder, and θ the inclination to the vertical of the perpendicular from the centre of the sphere to the axis of the cylinder.

*35. One point O of a uniform circular hoop of radius a and mass m is fixed, and the hoop is free to rotate about O in a vertical plane. The centre of the hoop is A, and a bead B, of mass $\frac{1}{8}m$, slides on the hoop. Denoting the inclinations of OA and of AB to the downward vertical at time t by θ and ϕ, find the kinetic energy and the potential energy of the system.

If the system makes small oscillations about the position of stable equilibrium, find the periods of the normal modes, and show that the motion is periodic with period $2\pi/p$, where $p^2 = g/8a$.

If the motion is set up from rest in the equilibrium position by a small impulsive couple Γ applied to the hoop at the instant $t=0$, find the values of θ and of ϕ at any subsequent time.

CHAPTER XXIV

THE MOMENTUM OF A SYSTEM

24·1 The momentum. In the theory of impulsive motion we found it expedient to reduce the system of momenta of the particles of a rigid body to an $\mathscr{M}$-force at a given point and an $\mathscr{M}$-couple, the reduction being carried out in the same way as for a system of forces on a rigid body. We now consider a more general type of dynamical system. With each particle is associated, at a given instant, the momentum vector $m\mathbf{v}$, and we think of this vector as localized in the particle. We have thus a system of localized vectors, and we now consider this system and its significance in the theory of the motion.

If we reduce this momentum system by the same technique that we use for the reduction of a system of forces acting on a rigid body we obtain an $\mathscr{M}$-force (p, q) localized in the origin O, and an $\mathscr{M}$-couple h,

$$p = Sm\dot{x},$$
$$q = Sm\dot{y},$$
$$h = Sm(x\dot{y} - y\dot{x}).$$

The $\mathscr{M}$-force (p, q) is called the *linear momentum* of the system. The $\mathscr{M}$-couple h is called the *angular momentum about* O (cf. §§ 21·3, 22·2).

$$\text{Now} \quad \frac{dp}{dt} = Sm\ddot{x}, \quad \frac{dq}{dt} = Sm\ddot{y}, \quad \frac{dh}{dt} = Sm(x\ddot{y} - y\ddot{x}),$$

and the three fundamental equations for the system (§ 23·2) can be written compactly in the form

$$\frac{d}{dt}(p, q; h) = (X_0, Y_0; N_0), \tag{1}$$

where, as before,

$$X_0 = SX, \quad Y_0 = SY, \quad N_0 = S(xY - yX).$$

In particular, if $X_0 = 0$ for all values of t, p remains constant. This is the famous theorem of *the conservation of linear momentum*: if the sum of the components of the external forces in a given direction is zero, the component of linear momentum in that direction remains constant. If $N_0 = 0$ for all values of t, h remains constant. This is the theorem of *the conservation of angular momentum*: if the moment of the external forces about O is zero, the angular momentum about O remains constant.

24·2 Formulae involving the coordinates of *G*. With the usual notation (§ 23·4) we have

$$\left.\begin{aligned} p &= Sm\dot{x} = M\dot{\xi}, \quad q = Sm\dot{y} = M\dot{\eta}, \\ h &= Sm(x\dot{y} - y\dot{x}) \\ &= Sm\{(\xi+\alpha)(\dot{\eta}+\dot{\beta}) - (\eta+\beta)(\dot{\xi}+\dot{\alpha})\} \\ &= M(\xi\dot{\eta} - \eta\dot{\xi}) + Sm(\alpha\dot{\beta} - \beta\dot{\alpha}). \end{aligned}\right\} \tag{1}$$

If $X_0 = 0$ for all t, $\dot{\xi}$ remains constant; in other words, if the sum of the components of the external forces in a given direction is zero, the component of the velocity of G in that direction remains constant. This is a secondary form of the principle of the conservation of linear momentum. We have met already (§ 21·3) the special case of this principle when the system is a single rigid body.

If we substitute for p, q, h from the formulae (1) in the equations (1) of § 24·1 we recover, as we should expect, the equations (1), (2) of § 23·3 and the equation (1) of § 23·4.

For the special case of a single rigid body

$$h = M(\xi\dot{\eta} - \eta\dot{\xi}) + I_G\dot{\theta},$$

and substituting from the formulae (1), with this modified form for h, in the equations (1) of § 24·1, we find again the equations

$$M\ddot{\xi} = X_0, \quad M\ddot{\eta} = Y_0, \quad M(\xi\ddot{\eta} - \eta\ddot{\xi}) + I_G\ddot{\theta} = N_0,$$

already found in § 23·4, and originally in § 21·2. These are the fundamental equations for the motion of a single rigid body in a plane. The method originally used to establish them in § 21·2 is simpler, for two dimensions, than the present method; but the present method has the advantage of being readily extensible to three dimensions.

24·3 The case of no external forces. If there are no external forces G moves uniformly in a straight line, and the angular momentum about any point of the plane remains unchanged for all time. But we can go further than this; when there are no external forces the motion of the system relative to G is independent of the motion of G. This is evident because if we take a new base of reference with G as origin and axes in fixed directions, the new base is still Newtonian (cf. Example 3 of § 21·6). We have already met an example of this principle in Example 2 of § 23·14.

24·4 Angular momentum about G. Let us consider again the formula for h,

$$h = M(\xi\dot{\eta} - \eta\dot{\xi}) + Sm(\alpha\dot{\beta} - \beta\dot{\alpha}),$$

and the equation

$$N_0 = M(\xi\ddot{\eta} - \eta\ddot{\xi}) + Sm(\alpha\ddot{\beta} - \beta\ddot{\alpha}).$$

We see that the angular momentum about G at any instant is

$$h_G = Sm(\alpha\dot{\beta} - \beta\dot{\alpha}) \tag{1}$$

(since if G is at O, $\xi = \eta = 0$), so that the angular momentum about G can be calculated from the motion relative to G. Also if N_G is the moment about G of the external forces we have (as in equation (2) of § 23·4)

$$N_G = Sm(\alpha\ddot{\beta} - \beta\ddot{\alpha}), \tag{2}$$

so that

$$N_G = dh_G/dt. \tag{3}$$

This is a famous and remarkable result. It involves a new feature, for the point G about which the angular momentum is taken is a moving point, occupying different positions in space at different times. The equation (3) is a special case of a general theorem which we shall establish in § 24·6.

24·5 Piecemeal reduction. In finding p, q, h—for example, when applying the principles of the conservation of linear and of angular momentum—it is sometimes convenient to reduce the momentum system by effecting first the reduction for the individual rigid bodies. A similar procedure has already been used for the kineton system in § 23·2. With the notation used there the momentum system for one rigid body is equivalent to an $\mathscr{M}$-force $(M_r\dot{\xi}_r, M_r\dot{\eta}_r)$ at G_r, together with an $\mathscr{M}$-couple $I_r\dot{\theta}_r$. To find h we need the sum of the moments about O of the momenta of the separate rigid bodies,

$$h = \Sigma M_r(\xi_r\dot{\eta}_r - \eta_r\dot{\xi}_r) + \Sigma I_r\dot{\theta}_r, \tag{1}$$

where Σ indicates summation over the bodies. (If the sth body is a single particle the term $I_s\dot{\theta}_s$ is absent.) If we substitute the form (1) for h in the equation $\dot{h} = N_0$ we recover equation (3) of § 23·4.

Example 1. *A man of mass m stands at A on a horizontal lamina which can rotate freely about a fixed vertical axis which meets the lamina at O. Originally both man and lamina are at rest. The man proceeds to walk on the lamina, ultimately describes (relative to the lamina) a closed circle having OA, of length a, as diameter, and returns to the point of starting on the lamina. Show that the lamina has moved through an angle relative to the ground given by*

$$\pi\{1-\sqrt{[I/(I+ma^2)]}\},$$

where I is the moment of inertia of the lamina about the axis.

The angular momentum of the system about O remains zero throughout. (We may suppose that the man's velocity relative to the lamina varies continuously with t, so that there is no jerk at any time during the motion. But we shall see in the next chapter that the result is in fact unchanged even if the motion involves impulsive changes.) Therefore when the lamina has turned (clockwise) through an angle ϕ, and the man has reached the point P on the circle, where OP makes an angle θ with OM, the initial direction of OA, we have

$$mr^2\dot{\theta} = I\dot{\phi},$$

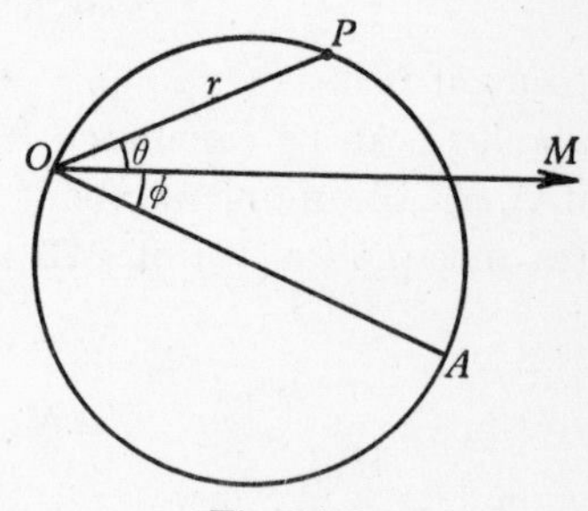

Fig. 24·5*a*

where the man has been treated as a particle! (Fig. 24·5*a*). If we write $\theta+\phi = \psi$ this gives

$$ma^2\cos^2\psi(\dot{\psi}-\dot{\phi}) = I\dot{\phi},$$

whence

$$\dot{\phi} = \frac{ma^2\cos^2\psi}{I+ma^2\cos^2\psi}\dot{\psi}.$$

The angle turned through by the lamina when the man reaches O is

$$\int_0^{\frac{1}{2}\pi}\frac{ma^2\cos^2\psi}{I+ma^2\cos^2\psi}\,d\psi$$

$$= \int_0^{\frac{1}{2}\pi}\left(1-\frac{I}{I+ma^2\cos^2\psi}\right)d\psi = \tfrac{1}{2}\pi\{1-\sqrt{[I/(I+ma^2)]}\}.$$

The final rotation, for the circuit of the complete circle, is twice as large. Notice that the angular velocity of the lamina is zero when the man is at O.

Example 2. *Two equal uniform rods AB, BC, each of mass m and length $2a$, are freely jointed at B. The end C moves on a smooth horizontal rail and the rod AB can turn freely about the end A which is fixed on the horizontal rail. The system is released from rest when the rods are in line, with B between A and C. Prove that when $B\widehat{A}C = \theta$,*

$$\dot{\theta}^2 = \frac{3g}{2a}\frac{\sin\theta}{(4-3\cos^2\theta)}.$$

Show that the angular momentum of the system about A is (when measured in the appropriate sense) $4ma^2\dot{\theta}$, and hence express as a function of θ the upward reaction of the rail on the rod BC at C. Prove that during the motion this reaction varies between $\frac{1}{4}mg$ and $\frac{35}{32}mg$.

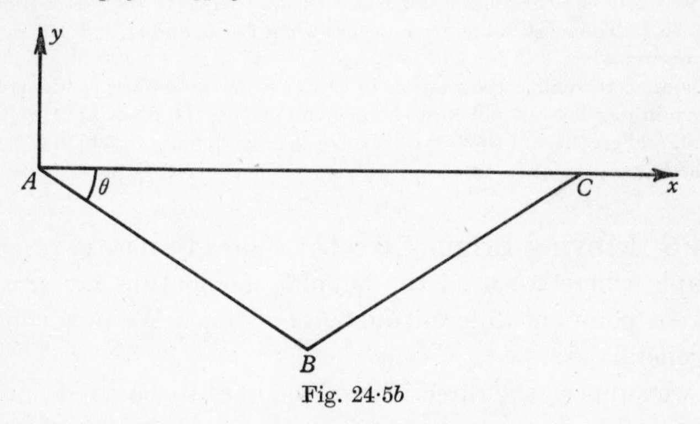

Fig. 24·5*b*

One way of finding T is to take one-half the value found in Example 2 of § 23·14, with $\xi = 2a\cos\theta$. Or directly, if (ξ_2, η_2) is the centre of BC with the axes indicated in Fig. 24·5*b*,

$$\xi_2 = 3a\cos\theta, \quad \eta_2 = -a\sin\theta$$

and
$$\begin{aligned} T &= \tfrac{1}{2}(\tfrac{4}{3}ma^2\dot{\theta}^2) + \tfrac{1}{2}m(\dot{\xi}_2^2 + \dot{\eta}_2^2) + \tfrac{1}{2}(\tfrac{1}{3}ma^2\dot{\theta}^2) \\ &= \tfrac{5}{6}ma^2\dot{\theta}^2 + \tfrac{1}{2}ma^2(9\sin^2\theta + \cos^2\theta)\dot{\theta}^2 \\ &= \tfrac{4}{3}ma^2(4-3\cos^2\theta)\dot{\theta}^2. \end{aligned}$$

The kinetic energy gained equals the potential energy lost, whence

$$\tfrac{4}{3}ma^2(4-3\cos^2\theta)\dot{\theta}^2 = 2mga\sin\theta,$$

giving
$$\dot{\theta}^2 = \frac{3g}{2a}\frac{\sin\theta}{(4-3\cos^2\theta)}.$$

Next, to find h, the angular momentum about A, we have (§ 24·5)

$$h = -\tfrac{4}{3}ma^2\dot{\theta} + m(\xi_2\dot{\eta}_2 - \eta_2\dot{\xi}_2) + \tfrac{1}{3}ma^2\dot{\theta} = -4ma^2\dot{\theta}.$$

If Y is the action (upwards) on BC at C we have

$$4a\cos\theta\, Y - 2mg\,.\,2a\cos\theta = \dot{h} = -4ma^2\ddot{\theta},$$

$$Y - mg = -\frac{ma}{\cos\theta}\ddot{\theta},$$

and
$$\ddot{\theta} = \frac{d}{d\theta}(\tfrac{1}{2}\dot{\theta}^2) = \frac{3g}{4a}\frac{d}{d\theta}\left(\frac{\sin\theta}{4-3\cos^2\theta}\right).$$

This gives us
$$\frac{Y}{mg} = \frac{70-105\cos^2\theta+36\cos^4\theta}{4(4-3\cos^2\theta)^2}.$$

The motion is a libration in θ between 0 and π, and to see how Y varies we must study the variation of the function

$$y = \frac{70-105x+36x^2}{4(4-3x)^2}$$

in the range from $x = 0$ to $x = 1$. A simple way of doing this is to express y in the form

$$y = \frac{35}{32} - \frac{27}{32}\left(\frac{x}{4-3x}\right)^2,$$

from which it is clear that y decreases steadily from $\frac{35}{32}$ to $\frac{1}{4}$ as x increases from 0 to 1. The least value for Y occurs when $\theta = 0$, and is $\frac{1}{4}mg$: the greatest value occurs when $\theta = \frac{1}{2}\pi$, and is $\frac{35}{32}mg$.

Of course, to realize the motion in practice the rods must be constrained to move in parallel but not quite coincident planes; the pivot at A is just off the rail, and on the rail there is a roller of negligible mass to which C is freely attached.

24·6 Moving origin.

We have already met (§ 24·4) one example where we used the angular momentum taken about different points at different instants of time. We now consider the general case.

If we suppose the directions of the axes to be fixed once for all, we see that the values of p and q do not depend on the choice of origin in the plane. This is evident in various ways; for example p is the sum of the momentum-components for the various particles in the direction Ox; or again, p is $M\dot{\xi}$, and $\dot{\xi}$ is the velocity of G in the given direction Ox, and is clearly independent of the position of O in the plane. The same is true of the resultant force in the reduction of any system of localized vectors.

But the same is not true of h; if h' is the angular momentum about a new origin O', whose coordinates with respect to the original axes are (x, y), we have (§ 19·3)

$$h = h' + xq - yp.$$

Thus, supposing O' to move in the plane, we have

$$N_0 = \dot{h} = \dot{h}' + \dot{x}q - \dot{y}p + x\dot{q} - y\dot{p},$$

N_0 denoting as usual the moment of the external forces about O.

Consider then an instant at which O' is passing through O, so that x and y vanish, but $\dot{x}$ and $\dot{y}$ do not. Then

$$\begin{aligned} N_0 &= \dot{h}' + \dot{x}q - \dot{y}p \\ &= \dot{h}' + M(\dot{x}\dot{\eta} - \dot{y}\dot{\xi}) \\ &= \dot{h}' + MV_0 V_G \sin\theta, \end{aligned}$$

where V_0 is the velocity of O' in the plane, V_G is the velocity of G in the plane, and θ is the angle between the directions of

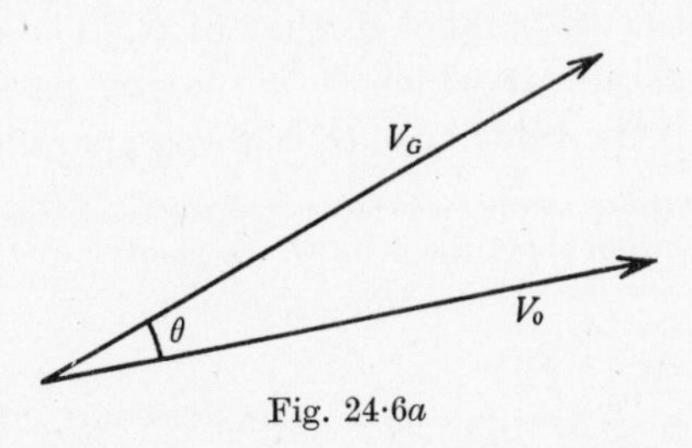

Fig. 24·6*a*

these two velocities; V_0 must be turned through an angle θ in the positive sense to bring it into the direction of V_G (Fig. 24·6*a*). If we use the notation of the vector product (§ 13·5) we have

$$\mathbf{N}_0 = \dot{\mathbf{h}}' + M(\mathbf{V}_0 \times \mathbf{V}_G),$$

where $\mathbf{N}_0$ and $\mathbf{h}'$ are now vectors in the direction Oz.

Example. *Suppose we have a single rigid body, and we take the moving origin O' to be, at each instant, the instantaneous centre I of rotation.* The velocity of G is $R\omega$ in the direction perpendicular to IG, and $\theta = \alpha + \frac{1}{2}\pi$ (see Fig. 24·6*b*). Thus, using the result just proved, we have

$$\begin{aligned} N_1 &= \dot{h}' + MV_0 V_G \sin\theta \\ &= \frac{d}{dt}(I_1\omega) + MV_0 V_G \cos\alpha \\ &= I_1\dot{\omega} + \dot{I}_1\omega + MR\omega(-\dot{R}), \end{aligned}$$

where I_1 is the moment of inertia about the instantaneous centre, and we have used the equation $V_0 \cos\alpha = -\dot{R}$. Thus, since $\dot{I}_1 = 2MR\dot{R}$, we find

$$N_1 = I_1\dot{\omega} + MR\dot{R}\omega,$$

and this is the equation already found in § 21·7.

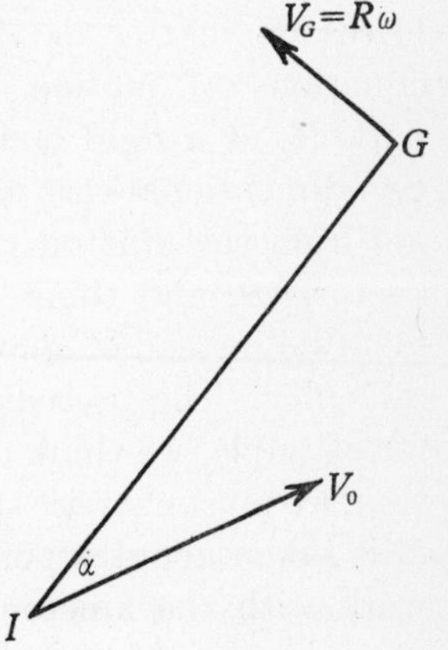

Fig. 24·6*b*

24·7 Special cases. The equation of motion found above in § 24·6 (involving h', the angular momentum about a moving point) takes the simpler form

$$N_0 = \dot{h}' \tag{1}$$

if $V_0 V_G \sin\theta = 0$. Two cases in which this condition is fulfilled are already familiar, the first when the origin O' is fixed (§ 24·1), the second when O' is identical with G (§ 24·4). There is still a third case, which is occasionally useful, namely, when the velocity of O' is parallel to that of G. The equation (1) is valid at any instant at which O' is moving parallel to G, and true generally if the velocity of O' is always parallel to that of G.

Example. *Consider again the motor-car problem of* § 23·5. If h' denotes the angular momentum about the point O, the point on the road immediately beneath G,

$$h' = -Mhv - 2mk^2(v/a),$$

and the equation $N_0 = \dot{h}'$ gives us

$$Y_1 b_1 - Y_2 b_2 + Rh = -(Mh + 2mk^2/a)f,$$

and this is equation (9) of § 23·5.

24·8 Moving axes. Theories of the type we have been discussing, with a moving origin, become vitally important later on in three-dimensional work; indeed, in three-dimensional problems it is frequently necessary to use axes not merely with a moving origin but also with varying orientation. But for two-dimensional systems, such as those we have dealt with in this book, it is hardly ever necessary to use even a moving origin. The fundamental vector, when we study the effects of forces on matter, is the kineton $m\mathbf{f}$. As we have seen, the equations of motion involving this vector—whether for a particle, or a rigid body, or a system—can be reduced in the two-dimensional case to very simple forms. If instead of starting from the kineton vector $m\mathbf{f}$ we start from the momentum vector $m\mathbf{v}$, and think primarily about rate of change of momentum instead of about kineton, we introduce an unnecessary difficulty. The preceding paragraphs illustrate this point. If, for example, we think of the rate of change of angular momentum, we must exercise special care when this angular momentum h' is taken about a point O' not fixed in space (§ 24·6). If we work with the kineton vector (§ 23·2) we can take moments about any point without hesitation.

EXAMPLES XXIV

1. A bead of mass m is free to slide on a smooth horizontal wire. A light rod of length a is freely attached to the bead, and carries a particle of mass m at the other end. The system rests in equilibrium, and the particle is struck a blow mu parallel to the wire. Show that in the subsequent motion the rod will just become horizontal if $u = 2\sqrt{(ga)}$. Show further that if u has this value the angular velocity of the rod when it makes an angle θ with the horizontal is

$$\sqrt{\left(\frac{4g\sin\theta}{a(2-\sin^2\theta)}\right)}.$$

2. A bead of mass M can slide on a smooth straight horizontal wire, and a particle of mass m is attached to the bead by a light string of length l. The system moves in the vertical plane through the wire, and the string remains taut. The horizontal displacement of the centre of gravity G at time t is denoted by x and the angle the string makes with the downward vertical by θ. Prove that during the motion $\dot{x}$ and

$$\tfrac{1}{2}(M+m\sin^2\theta)\,\dot{\theta}^2-(M+m)\,(g/l)\cos\theta$$

remain constant.

(i) If the system starts from rest with the string horizontal, find how far the bead has moved on the wire and the value of $\dot{\theta}$ when the inclination of the string to the vertical is θ.

(ii) If the system starts from rest with the string nearly vertical, find the period of the small oscillations.

3. Two particles A and B, each of mass m, are joined by a rigid straight uniform rod AB, of mass M and length $2a$. A is free to move without friction along a horizontal straight line l. The system is hanging at rest under gravity when an impulse I is applied at B parallel to l. Prove that the rod will make complete revolutions if

$$I^2 > \tfrac{4}{3}(M+2m)\,(M+6m)\,ga.$$

4. A uniform cubical block of mass M can slide on a smooth horizontal plane and a uniform rod, also of mass M, leans with one end on the plane and the other against a vertical cube face, which is smooth, the rod being in a vertical plane at right angles to the cube face and through its centre. The system is released from rest with the rod inclined at an angle α to the horizontal. Show that contact between the rod and the cube ceases when the inclination θ of the rod to the horizontal satisfies the equation

$$3\sin^3\theta-24\sin\theta+16\sin\alpha = 0.$$

5. Two equal uniform rods, AB, BC, each of length l, are smoothly hinged at B, and carry small rings, of negligible mass, at A and C so that the rods can slide with the ends A, C on a smooth horizontal rail. Initially the rods hang vertically with A and C side by side and B below the rail. The end C is now projected along the rod away from A with velocity U while A is initially at rest. Show by means of the equation of energy, or otherwise, that, if $U^2 < 12gl$, the angle ABC attains a maximum value 2θ given by $\sin^2\frac{1}{2}\theta = U^2/(24gl)$.

6. A wedge of mass M and angle α moves on a smooth horizontal plane, and a perfectly rough uniform sphere of mass m rolls on the inclined face of the wedge, the motion being in a vertical plane of symmetry containing a line of greatest slope of the wedge. If the system starts from rest, x' is the horizontal distance travelled by the wedge and x the distance on the wedge rolled over by the sphere in time t, prove that

$$(M+m)x' = mx\cos\alpha, \quad \tfrac{7}{5}x - x'\cos\alpha = \tfrac{1}{2}gt^2\sin\alpha.$$

7. The base of a solid hemisphere of mass M and radius a rests on a horizontal plane. A particle of mass m is placed on the highest point of the hemisphere and is slightly disturbed. Assuming the surfaces to be smooth, show that, so long as the particle remains in contact with the hemisphere, the path of the particle in space is an arc of an ellipse.

Show also that, if θ denotes the angle which the radius through the particle makes with the vertical at time t, then

$$\left(\frac{d\theta}{dt}\right)^2 = \frac{2g}{a}\frac{(1-\cos\theta)}{(1-k\cos^2\theta)},$$

where $k = m/(M+m)$.

8. A uniform solid sphere of mass m and radius b rolls down the rough curved surface of a uniform hemisphere of mass M and radius a, which is free to slide on its base in contact with a smooth horizontal plane, starting from rest when the common radius makes an angle α with the vertical. Show that, when this angle has increased to θ,

$$\{\tfrac{7}{5}(M+m) - m\cos^2\theta\}(a+b)\,\dot{\theta}^2 = 2(M+m)g(\cos\alpha - \cos\theta).$$

9. Two masses, $3m$ and m, are connected by a light inextensible string of length $2l$ which passes through a small hole in a smooth horizontal table on which the mass $3m$ can move while m hangs vertically. Initially m is released from rest at the hole and simultaneously $3m$ is projected with a horizontal velocity $\sqrt{(\frac{1}{3}gl)}$ at a distance $l/\sqrt{3}$ from the hole and at right angles to the line joining the mass to the hole. Show that just after the string becomes tight the mass m is instantaneously at rest, and show that subsequently it oscillates through a distance $\frac{1}{2}l$.

10. Two particles of masses m and M are connected by a light inextensible string of length $2l$ which passes through a smooth hole in a smooth horizontal table on which the mass M moves while m hangs vertically. Initially M is at rest, at a distance l from the hole, and it is then projected with velocity V at right angles to the string. Show that if $3MV^2 > 8mgl$, m will reach the hole with velocity

$$\sqrt{\left(\frac{3MV^2 - 8mgl}{4(M+m)}\right)}.$$

11. Two equal particles are joined by a smooth light string passing through a hole in a smooth horizontal table so that one particle rests on the table and the other hangs beneath. When held at rest on the table at a distance a from the hole the upper particle is given a horizontal velocity $\sqrt{(2nga)}$ at right angles to the string. Assuming that there is sufficient length of string to prevent the lower particle hitting the table, show that in the subsequent

motion the string remains taut and the lower particle moves between two levels at a vertical distance apart of

$$\tfrac{1}{2}a\,|\,(n^2+4n)^{\frac{1}{2}}+n-2\,|.$$

12. A particle of mass m is free to move in a thin smooth uniform straight tube of mass $3m$ and length a. The tube can turn freely in a horizontal plane about one end which is fixed. Initially the tube has angular velocity Ω and the particle is at rest relative to the tube at its mid-point. Find the velocity with which the particle leaves the tube.

13. A bead of mass m can slide freely on a rigid rod which is free to rotate in a horizontal plane about a vertical axis through a fixed point O of the rod. The moment of inertia of the rod about the axis of rotation is mc^2. The system is set in motion with the bead at rest relative to the rod at a distance a from O. Prove that the differential equation of the path of the bead is

$$(a^2+c^2)\left(\frac{dr}{d\theta}\right)^2 = (r^2-a^2)(r^2+c^2),$$

where r, θ are polar coordinates, with origin O, in the plane of motion.

14. Show that, for a system moving in one plane, the moment of momentum about a fixed axis perpendicular to the plane is constant, if the external forces acting on the system have no moment about that axis.

A horizontal wheel with buckets on its circumference revolves about a frictionless vertical axis. Water falls into the buckets at a uniform rate of mass m per unit of time. Treating the buckets as small compared with the wheel, find the angular velocity of the wheel after time t, if Ω be its initial value; and show that, if I be the moment of inertia of the wheel and buckets about the vertical axis and r the radius of the circumference on which the buckets are placed, the angle turned through by the wheel in time t is

$$\frac{I\Omega}{mr^2}\log_e\left(1+\frac{mr^2t}{I}\right).$$

15. A lamina is free to rotate in a horizontal plane, about a point O of the lamina which is fixed in space. An insect stands on the lamina at a point A of the lamina at a distance a from O. The mass of the insect is m, and the moment of inertia of the lamina about O is mk^2. The insect then walks on the lamina, his path relative to the lamina being the straight line through A at right angles to OA. Prove that the angle through which the lamina turns does not exceed $\pi/(2p)$, where

$$p^2 = (a^2+k^2)/a^2.$$

16. A bicycle is so geared that when the cranks turn through a radian the machine advances a distance k. The couple on the pedals being N and the power applied P, the rider works so that $P = AN(N_0-N)/N_0$, where A and N_0 are constants. Internal friction and road resistance together are equivalent to sliding friction with coefficient μ, and air resistance is neglected. Find what acceleration the rider can produce in climbing a hill of slope α with velocity v, and show that his maximum velocity is

$$Ak\left\{1-\frac{mkg}{N_0}(\sin\alpha+\mu\cos\alpha)\right\},$$

where m is the mass of the rider and machine together.

With k equal to 4·4 ft. and μ negligible, the rider can attain a speed of 22 ft./sec. on the flat, and 2·2 ft./sec. up a hill with $\sin\alpha = \frac{1}{20}$. Show that by adjusting his gear suitably he can climb a hill with $\sin\alpha = \frac{1}{10}$ with a velocity of 3·05 ft./sec.

17. A thin uniform rod OA, of length a and mass m, turns in a horizontal plane about a vertical axis through O; an equal rod AB is jointed at A to OA, and a smooth guide compels the end B to move along a horizontal straight line Ox. The angle $AOx = \theta$. No external forces act upon the rods except those due to the axis and the guide, and the motion takes place without friction. When $\theta = 0$, $d\theta/dt = \omega$. Show that

$$\left(\frac{d\theta}{dt}\right)^2 = \frac{\omega^2}{4-3\cos^2\theta}.$$

Prove that, if H be the angular momentum of the system about the vertical axis through O,

$$H = ma^2 d\theta/dt.$$

18. A running watch is placed on a smooth horizontal surface so that it may be regarded as free to rotate about its own centre of gravity. It is composed essentially of a balance-wheel of moment of inertia i, and a body of moment of inertia I, connected by a hair spring. The inertia of the hair spring and other parts of the mechanism may be ignored. Determine the angular motion, and show that in general in addition to its oscillations, the body of the watch rotates steadily with a small uniform spin.

If $I = 110$ g. (cm.)^2, $i = 0·05$ g. (cm.)^2, show that the watch gains nearly 20 sec. a day in this position, if its rate is correct when the body is rigidly held.

19. Two equal uniform rods AB, BC, each of mass M and length $2a$, are hinged freely at B and move freely on a smooth horizontal table with the centre of gravity of the system at rest. If 2θ is the angle between the rods and ϕ is the angle made by AC with a fixed direction in the plane, show that the kinetic energy T and the angular momentum h about the centre of gravity are given by

$$T = \tfrac{1}{3}Ma^2\{(1+3\cos^2\theta)\,\dot{\theta}^2+(1+3\sin^2\theta)\dot{\phi}^2\},$$

$$h = \tfrac{2}{3}Ma^2(1+3\sin^2\theta)\dot{\phi}.$$

The ends A and C are joined by a light inextensible string whose length is slightly less than $4a$. Initially the system rotates with angular velocity ω in the plane of the table. Show that if the string breaks the subsequent relative motion of the rods is approximately simply harmonic with period $2\pi/(\omega\sqrt{3})$.

20. A rigid body of mass M is in motion in two dimensions and at any given moment the perpendicular distance of the centre of mass from the instantaneous axis is denoted by r. If h denotes the angular momentum of the body and L the total moment of the external forces about the instantaneous axis, prove that

$$\frac{dh}{dt} = L + M\omega r\frac{dr}{dt},$$

where ω is the angular velocity.

If c is the distance of the centre of mass from the geometric centre and k the radius of gyration about a horizontal axis through the centre of mass of a circular wheel of radius a which rolls in a vertical plane on a horizontal table without slipping, calculate the period of the small oscillations about the position of stable equilibrium.

21. The mass of a railway truck, including the wheels, is 4 tons. The distance between the axles is 11 ft., the centre of gravity of the truck is 4 ft. above the rails and equidistant from the axles, the diameter of each wheel is 3 ft., and the moment of inertia of each pair of wheels and axle about its axis is 2688 lb. ft.^{2} Air resistance is negligible, and the wheels do not slip on the rails. By taking moments about the point on the ground vertically below the centre of gravity, or otherwise, show that, when the brakes produce a retardation of $\frac{1}{2}$ ft./sec./sec., the normal pressure between the rails and each of the front wheels is 56 lb. weight greater than the corresponding pressure for the back wheels.

CHAPTER XXV

IMPULSIVE MOTION OF A SYSTEM

25·1 The fundamental equations. We now consider the impulsive motion of a dynamical system. We have already discussed the theory of the impulsive motion of a single rigid body. The extension of this theory which we need in order to deal with simple problems for a system is very similar to that for finite forces. Indeed, the analogy is so close that the argument of § 23·2 can be applied forthwith *mutatis mutandis* to the corresponding impulse problem. But now it is impulses and momentum-changes with which we are concerned, instead of with forces and kineta. We recall that in these impulse problems there is no change of configuration to consider, only changes of velocity, and that finite forces can be neglected. In particular the effect of an elastic string joining two points of the system can be ignored, since such a string can only exert a finite tension, not an impulsive tension; only an inextensible string, which can sustain an impulsive tension, is relevant. In practice, of course, this is never actually attained; the inextensible string sustaining an impulsive tension is the theoretical limit of a nearly inextensible string sustaining a very large tension for a very short time.

One attack on the problem of the impulsive motion of a system is to form first the equations for the individual rigid bodies and then to eliminate the impulsive reactions between the bodies. As we have noticed already, a similar technique is applicable when the system moves under the action of finite forces. For impulses, as for finite forces, this elementary attack is quite effective in many simple cases.

But we can go further. Just as in the problem with finite forces we can establish certain important general principles which often greatly simplify the solution. The impulses acting on the particles of the system can be divided into the two classes, internal and external; such a division has already been made for finite forces. And, just as in the case of finite forces, the internal impulses, if they acted *in situ* on a single rigid body, would form a nul system. We have, therefore, the analogue for impulses of the generalization of d'Alembert's principle for finite forces (§ 23·2); the external impulse system and the change-of-momentum system are equivalent if both are thought of as acting *in situ* on a single rigid body.

We can now establish three fundamental equations analogous to those established in § 23·2 for finite forces. We consider the external impulses, and reduce them (by the same technique as for forces on a rigid body) to an impulse (X_0, Y_0) at O and an impulsive couple N_0. We notice again, as in § 23·2, that this reduction has not the same physical validity that it had when we were dealing with a single rigid body. (Two different impulse systems both having the same reduced form $(X_0, Y_0; N_0)$ will not in general generate the same change of motion; they would do so if the dynamical system were a single rigid body, and in that case the reduced impulse system would be equivalent to the original impulse system in the exact sense.) The equations expressing the equivalence of external impulses and changes-of-momentum are

$$X_0 = Sm(u-u_0) = p-p_0, \tag{1}$$

$$Y_0 = Sm(v-v_0) = q-q_0, \tag{2}$$

$$\begin{aligned} N_0 &= Sm\{x(v-v_0)-y(u-u_0)\} \\ &= Sm(xv-yu)-Sm(xv_0-yu_0) \\ &= h-h_0, \end{aligned} \tag{3}$$

where (u_0, v_0) is the velocity of the typical particle m immediately before the blow, and (u, v) is its velocity immediately afterwards. We see again, as in § 22·5, that the change-of-momentum system is the same thing as the difference of the momentum systems after and before the blows.

Another analogy with the theory of the motion under finite forces strikes us at this point. The three fundamental equations (1), (2) and (3) do not in general suffice to determine the motion after the blows. Usually we shall have to add to them some information obtained by dissecting the system, and considering for example the change in the motion of one rigid body (cf. § 23·4). This procedure will suffice for the simple cases. But, just as for the motion of a system under finite forces, more powerful methods than those used in this book are required for a really penetrating study.

The evaluation of $(p, q; h)$ has already been considered in the preceding chapter. We can use either the formulae of § 24·2, or the piecemeal reduction of § 24·5, whichever is more convenient for the particular problem.

If $X_0 = 0$, $p = p_0$. This means that if the sum of the components of the external impulses in a given direction is zero, the component of linear momentum in that direction remains unchanged. This is the theorem of the *conservation of linear momentum* for the impulsive motion of a system. In the same way if $N_0 = 0$, $h = h_0$. This means that if the moment about O of the system of external impulses is zero, the angular momentum about O remains unchanged. This is the theorem of the *conservation of angular momentum* for the impulsive motion of a system. The precisely analogous theorems for the motion of a system under finite forces have already been established in § 24·1.

Example 1. *Two light rigid rods AB, BC are freely jointed together at B, and particles, each of mass m, are attached at A, B and C. Initially the system is at rest on a smooth horizontal table, the angle ABC being 60°. An impulse P is applied to the particle B in the direction AB. Find the velocities with which the particles start to move.*

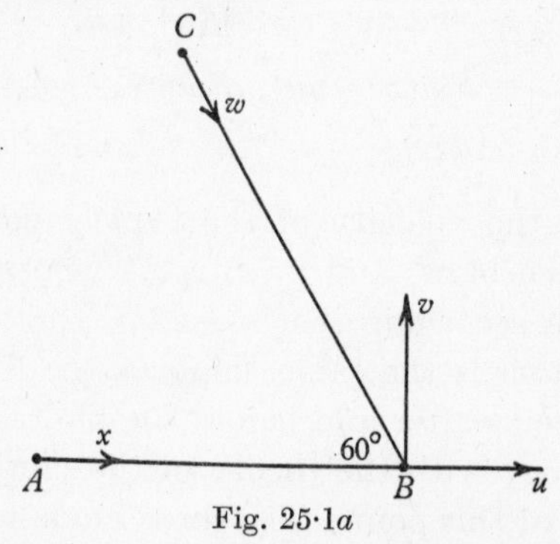

Fig. 25·1*a*

The initial velocity of A is in the direction AB, and that of C is in the direction CB, so we may use the notation indicated in Fig. 25·1*a*. Resolving in directions parallel and perpendicular to AB (equations (1) and (2) of the bookwork) we have

$$mx + mu + m(\tfrac{1}{2}w) = P,$$

$$v = w\sqrt{3}/2.$$

Now when a rod is in motion the components along the rod of the velocities of the two ends are equal, so

$$x = u,$$

$$w = \tfrac{1}{2}(u - \sqrt{3}v).$$

Thus finally
$$\frac{v}{\sqrt{3}} = \frac{w}{2} = \frac{2w + \sqrt{3}v}{7} = \frac{u}{7} = \frac{4u + w}{30} = \frac{P}{15m}.$$

Note. It is usual to assume (in a problem such as this in which the impulses in the rods are all tensions) that if the rods are replaced by inelastic strings, initially taut, the solution is unchanged. But this is not quite obvious, because the constraints are now one-sided, and the relation between the velocities of A and B for example should, strictly speaking, be $x \geqslant u$, to allow for the possibility that the string may slacken. But this would leave the solution indeterminate. The difficulty originates in the somewhat artificial concept of a perfectly inelastic string. Replacing the inequality $x \geqslant u$ by the equation $x = u$ amounts to introducing a new physical assumption—that the impulsive tension provided by the inelastic string is the least possible.

Example 2. *Two equal uniform rods AB and BC, each of mass M and length $2a$, are smoothly hinged together at B and hang freely from A. A horizontal impulse J is applied at C. Find the velocities with which B and C start to move.*

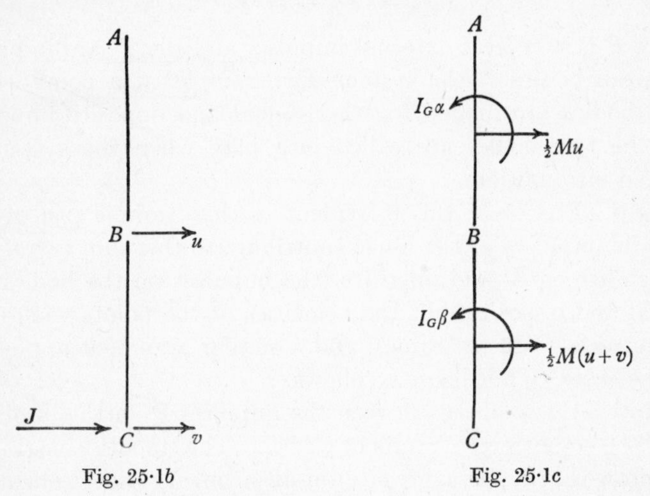

Fig. 25·1*b* Fig. 25·1*c*

If the required velocities of B and C are u and v, the momentum system, reducing for the separate rigid bodies, is as shown in Fig. 25·1*c*, where

$$\alpha = u/2a, \quad \beta = (v-u)/2a.$$

Taking moments about B for the rod BC alone we have

$$2aJ = M\frac{u+v}{2}a + M\frac{a^2}{3}\frac{v-u}{2a},$$

and taking moments about A for the system

$$4aJ = M\frac{u+v}{2}\,3a + M\frac{a^2}{3}\frac{v-u}{2a} + M\frac{u}{2}a + M\frac{a^2}{3}\frac{u}{2a}.$$

These equations can be written

$$\frac{2J}{M} = \frac{u+2v}{3}, \quad \frac{4J}{M} = \frac{6u+5v}{3},$$

whence
$$u = -\frac{6}{7}\frac{J}{M}, \quad v = \frac{24}{7}\frac{J}{M}.$$

25·2 Kinetic energy acquired by a system set in motion by impulses. The theorem proved for a single rigid body in § 22·4 is valid also for a system; if the system is set in motion by impulses the kinetic energy acquired is $\frac{1}{2}\Sigma\mathbf{P}\,.\,\mathbf{v}$, where $\mathbf{P}$ is a typical impulse, and $\mathbf{v}$ the velocity acquired by the point at which it is applied.

The proof is simple. By the theorem for a single rigid body the total energy acquired by the system is (by summing over the bodies)

$$\tfrac{1}{2}\Sigma\mathbf{P}\,.\,\mathbf{v} + \tfrac{1}{2}\Sigma\mathbf{P}'\,.\,\mathbf{v}'.$$

Here $\mathbf{P}$ is a typical external impulse, and the $\mathbf{P}'$ are impulses between bodies of the system; for example, at a point where two bodies are hinged together, equal and opposite impulses on the two bodies are called into play when the system is jerked into motion.

Now $\Sigma\mathbf{P}'\,.\,\mathbf{v}' = 0$; this is evident in the example just given, for the impulses at the hinge contribute to this sum two terms which are equal and opposite (the impulses on the bodies are equal and opposite, but the velocities of the points where the impulses act are the same); and a similar argument applies in other cases. The theorem follows.

Notice the analogy between the impulses $\mathbf{P}'$ in this problem and the smooth constraints of § 23·12. The smooth constraints do no work (in the aggregate) in an arbitrary displacement of the system, the impulses $\mathbf{P}'$ do no work (in the aggregate) when the system is given an arbitary motion. We therefore speak of the impulses $\mathbf{P}'$ as impulsive constraints.

Example. *A chain of equal rods, each of mass M, hinged end to end and infinite in one direction, is initially at rest with the rods in line. The system is set in motion by an impulse J applied in a direction perpendicular to the rods at the free end. Prove that the energy of the motion set up is $\sqrt{3}J^2/M$.*

If AB is the first rod, A being the free end, let the initial velocity of A (in the direction of J) be u, and of B (in the opposite direction) αu (Fig. 25·2a).

It is clear that u is proportional to J: so that, by considering the similar chain $BCD\ldots$, the downward impulse on the rod BC at B must be αJ. Thus the upward impulse on AB at B is αJ.

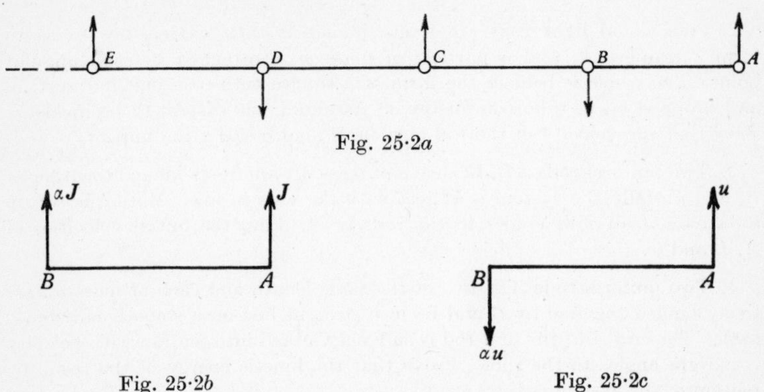

Fig. 25·2a

Fig. 25·2b

Fig. 25·2c

Thus impulses J and αJ on the rod AB at A and B give rise to a motion in which the velocities of A and B are u and $-\alpha u$ (Figs. 25·2b, c). Therefore (quoting the result of Example 2, § 22·2) we have

$$u = (4-2\alpha)J/M,$$

$$-\alpha u = (4\alpha-2)J/M,$$

whence $$\alpha^2-4\alpha+1 = 0,$$

and therefore (since $\alpha < 1$) $$\alpha = 2-\sqrt{3}.$$

Also $$u = 2\sqrt{3}J/M.$$

The energy communicated to the system is

$$\tfrac{1}{2}Ju = \sqrt{3}J^2/M.$$

Notice that here we have applied a theorem proved for a system with a finite number of bodies to a system with an infinite number; the extension is assumed to be valid on general physical grounds. Alternatively, we can find the energy of each rod, and find the total energy as the sum of an infinite series.

EXAMPLES XXV

1. A light inelastic string of length $2l$ has a particle of mass M attached to its mid-point and particles of mass m attached to each end. Initially the string is straight and at rest on a smooth horizontal table. The centre particle receives an impulse perpendicular to the string which gives it an initial velocity V in the plane of the table. If the angle between the two halves of the string is subsequently 2θ, show that the relative velocity of

either of the end particles with respect to the centre particle is of magnitude $V\sqrt{\left(\frac{M}{M+2m\cos^2\theta}\right)}$ until the end particles collide.

Assuming the end particles to be small spheres of coefficient of restitution e, show that when the string is once more straight they are moving with velocity eV relative to the centre particle.

2. Four equal light rods are freely joined at their extremities so as to form a framework, and a particle of mass m is attached to each angular point. The whole is held in the form of a square with one diagonal vertical and dropped on to a horizontal table. Assuming the system to be inelastic, show that the velocity of the top particle is unaltered by the impact.

3. Two uniform rods AB, BC, each of mass M, are freely hinged together at B, and initially the system is at rest with the rods in line. Motion is set up by a blow J at right angles to the rods at A. Find the initial velocities of A, B and C.

4. Two uniform rods AB, BC, of the same length and each of mass m, are freely hinged together at B and lie in a straight line on a smooth horizontal table. The end A of the first rod is suddenly jerked into motion with velocity v at right angles to the rods. Prove that the kinetic energy of the resulting motion of the system is $\frac{1}{7}mv^2$.

Show also that if the rods had been rigidly joined the kinetic energy produced would have been $\frac{1}{4}mv^2$.

5. Two equal uniform rods AB, BC are freely jointed at B and are initially at rest on a horizontal table, forming a right angle at B. An impulse is applied at A in the direction of the diagonal AC. Show that the initial angular velocities of the rods are in opposite senses and the ratio of their magnitudes is 3 : 1.

Show also that the initial motion of A makes an angle $\tan^{-1}\frac{3}{5}$ with the direction of the impulse.

6. The end B of a uniform rod AB of mass $2M$ and length $2a$ is smoothly hinged to a point on the rim of a uniform circular disc of mass M and radius r. The rod and disc are laid on a smooth horizontal table so that the direction of AB passes through the centre of the disc. An impulse P is applied at A at right angles to AB. Prove that the kinetic energy produced is $\frac{9}{10}\frac{P^2}{M}$ and that the impulsive reaction at the hinge is $\frac{1}{5}P$.

7. Three equal uniform rods, AB, BC, CD, are joined together and lie on a horizontal table in a straight line. Impulses J are applied perpendicular to the rods at A and D in the same sense. Show that the initial velocity of A, when the rods are freely jointed at B and C, is five times that which it would acquire if they were rigidly connected.

8. Three equal uniform rods AB, BC, CD, each of mass m, are freely jointed at B and C. The rods are at rest in a straight line on a horizontal table. If an impulsive couple is applied to the rod BC, show that its angular velocity is two-fifths of the value it would have had if both the rods AB, CD had been absent.

Show also that the initial angular velocity of each of the rods AB, CD is in magnitude three-quarters of that of the rod BC.

9. Three uniform rods AB, BC, CD, each of length $2a$ and mass M, are smoothly hinged at B and C, and are at rest on a smooth horizontal table with the free ends A and D in contact. An impulse is applied at the middle point of BC at right angles to its length and in the direction away from A. Find the magnitude of the impulse if BC starts to move with velocity V. Find also the initial angular velocities of AB and CD, and prove that when each of the angles ABC, BCD is θ,

$$a^2\dot{\theta}^2(2-\cos^2\theta) = \frac{63}{256}V^2.$$

10. A uniform rod AB of mass M and length $2a$ lies on a smooth table, and a particle of mass m is attached to B by a light inelastic string. The particle is projected from B along the table with velocity u at right angles to the rod. Find the velocity of the centre and the angular velocity of the rod immediately after the string tightens, and show that a fraction $\dfrac{M}{M+4m}$ of the original energy is lost. (See the Note on Example 1 of § 25·1.)

11. A uniform rod of length $2a$ is in motion in a plane with speed u at right angles to its length when it collides with a stationary small uniform elastic sphere of equal mass whose centre lies in the plane. If the sphere is free to move, prove that the angular velocity acquired by the rod cannot exceed

$$\frac{(1+e)u\sqrt{6}}{4a},$$

where e is the coefficient of restitution.

12. A symmetrical disc of mass M and radius of gyration κ about its centre is free to slide on a smooth horizontal table. Initially it spins in the plane of the table with angular velocity ω and with its centre at rest. A particle of mass m approaches it with velocity V in the plane of the table, along a line which is at distance h from the centre of the disc. The particle strikes the disc and embeds itself in it. By considering the angular momentum about the centre of gravity of the system, show that for the disc to move without rotation after the impact, V must be of magnitude $\dfrac{\kappa^2\omega}{h}\left(1+\dfrac{M}{m}\right)$.

Show also that the energy lost in the impact is then

$$\tfrac{1}{2}M\kappa^2\omega^2\left[1+\left(1+\frac{M}{m}\right)\frac{\kappa^2}{h^2}\right].$$

13. Two equal uniform rods AB, BC, each of mass M, are freely jointed at B and are on a smooth horizontal table, so that ABC is a straight line; the rods move horizontally with velocity U perpendicular to ABC. The rod AB impinges at its mid-point on an inelastic stop. Show that the loss of kinetic energy is $4MU^2/7$.

14. A flywheel whose moment of inertia is I is keyed to an axle of radius r and is rotating with angular velocity ω. As it rotates it winds up on the axle a light inextensible string which is attached to a mass M resting on the ground below. Show that at the instant the string becomes tight the angular velocity of the flywheel is reduced in the ratio of I to $I+Mr^2$, and that the kinetic energy of the system is reduced in the same ratio.

15. A uniform straight rod of length $2l$ and mass m falls vertically and strikes a rough circular cylinder of radius a, whose moment of inertia about its axis is I. The cylinder is free to turn about its axis, which is fixed and horizontal. The rod, when it strikes the cylinder, is at right angles to the axis, tangential to the cylinder, and inclined at an angle θ to the horizontal; it is moving vertically without rotation; and its centre of gravity is at a higher level than the point of impact and at a distance d from it. There is no rebound on impact. On the assumption that slipping does not occur on impact, write down equations to determine the components of the impulse along and at right angles to the rod, and the angular velocities of the rod and the cylinder immediately after the impact. Deduce that in order that no slipping should occur on impact the coefficient of friction between the rod and the cylinder must not be less than

$$\frac{1+3d^2/l^2}{1+ma^2/I}\tan\theta.$$

16. A flywheel A is rotating with an angular velocity ω about a horizontal axis. Attached to it is a drum of radius a, round which a light inextensible cord is wound. The cord passes round a pulley B of radius b, and is attached to a mass M which rests on the ground. Between A and B the cord is slack, and between B and M it is just tight and vertical. The moment of inertia of the flywheel and drum is I, and that of the pulley is I_1. The axles are free from friction and the cord does not slip. Find the impulsive tension in each part of the rope when it tightens, and show that the mass M rises through a distance

$$\frac{I\omega^2}{2gM}\frac{I}{I+I_1\dfrac{a^2}{b^2}+Ma^2}$$

before it comes to rest.

17. Two heavy circular discs, A and B, are mounted on parallel axles running in frictionless bearings in a fixed frame. Their radii and moments of inertia are r_1, r_2, I_1, I_2, respectively. Their rims touch at P, and the friction at P is such that if B is held fixed a couple L must be applied to A in order to cause slipping. The discs A and B are rotating without slip, A having an angular velocity ω_1. A third disc, whose moment of inertia is I, is mounted freely on the same axle as B, and is at rest. It is suddenly connected rigidly to B. Show that slipping occurs at P for a time

$$\frac{I_1\omega_1}{L}\frac{I/r_2^2}{I_1/r_1^2+(I_2+I)/r_2^2},$$

and find the final angular velocity of A.

18. Two cog-wheels A and B are mounted on parallel axes and are rotating with angular velocities of 2 revolutions per second and 1 revolution per second respectively in opposite directions. The distance between the axles is diminished so that the teeth of the wheels engage. Find the resulting angular velocity of A. The wheels may be treated as discs of the same material and thickness, with radius of B $1\frac{1}{2}$ times that of A.

Prove that a fraction 36/1885 of the energy is lost.

19. A uniform circular disc of mass m touches internally a uniform circular ring of mass M and they lie at rest on a smooth horizontal table. An impulse is applied to the ring, directed towards its centre, at a point whose angular distance from the point of contact of the two bodies is α ($<\frac{1}{2}\pi$). Show that, if the bodies are rough, the disc will at first roll or slide according as the coefficient of friction is greater or less than

$$\frac{(M+m)\tan\alpha}{3M+2m}.$$

20. Two inelastic balls lie in contact on a smooth horizontal plane. Each ball is a uniform solid sphere of mass M and radius a, the centres of the spheres being A and B. An impulse P is applied to the first ball in a horizontal line which passes through A and makes an acute angle α with the line AB. Prove that there is no slipping between the balls at the impact, if the coefficient of friction between them is greater than $\frac{2}{7}\tan\alpha$. Assuming this condition to be satisfied, find the initial direction of motion of the centre of the first ball.

*21. A circular hoop of radius a and mass $3m$ is free to move about its centre in a horizontal plane. An insect of mass m stands on the hoop, the system being initially at rest. The insect leaps with velocity w relative to the hoop, and alights again on the hoop. Show that when he alights the system comes to rest.

Show further that, provided $w^2 \geqslant \frac{5}{3}ga$, the greatest angle through which the hoop can be turned in this way is one-third of a radian. What is the greatest angle through which the hoop can be turned if $w^2 < \frac{5}{3}ga$?

CHAPTER XXVI

DIMENSIONS

26·1 Units. In this book we have used theoretical units; that is to say, the units have all been derived from the units of length, time, and mass. Thus, the unit of velocity is taken to be that velocity in which unit distance is traversed in unit time. Similarly, the unit of acceleration is taken to be that acceleration in which the velocity increases by one unit in unit time. Suppose we consider a particle moving on a straight line. Its distance from an origin O on the line at time t is x; i.e. x is the measure of the distance, and t is the measure of the time from a fixed origin of time. Then, with the conventions we have described, $\dot{x}$ is the measure of the velocity, and $\ddot{x}$ of the acceleration.

We choose the units of momentum and of kineton in a similar way. We get unit momentum when unit mass has unit velocity, and we get unit kineton when unit mass has unit acceleration. With these units the measure of the momentum is $m\dot{x}$, and the measure of the kineton is $m\ddot{x}$. (Actually, of course, velocity, acceleration, momentum, and kineton are vector quantities, but here we are only concerned with magnitudes or directed magnitudes.)

Next we choose as unit of impulse the impulse which sets up unit momentum, and as unit of force the force which sets up unit kineton. We then obtain the classical equations

$$I = mv, \quad P = mf.$$

The first of these determines the velocity v communicated to a particle of mass m, initially at rest, when it is acted on by an impulse I. The second is of course Newton's law of motion; it determines the acceleration of a particle of mass m acted on by a force P.

We derive in the same way the units of work and of power. The unit of work is the work done by unit force when the particle on which it acts moves through unit distance in the direction of the force. The unit of power is the power in which unit work is done in unit time. The unit of kinetic energy must be the same as the unit of work.

26·2 Change of units. We now consider how the measure of a quantity—such as a velocity (or more precisely a speed)—changes when we change to a new system of units.

It must be observed carefully that the change contemplated is the change to a new theoretical system based on new fundamental units of length, time, and mass. This must not be confused with the occasional use of units other than the theoretical ones as a practical convenience. Thus we have occasionally measured forces in pounds weight, because the theoretical unit, the poundal, is inconveniently small; and we have occasionally measured power as horse-power, instead of in foot-poundals per second, because again the theoretical unit is inconveniently small. This is merely a matter of comfort; it is not the kind of change of units now contemplated, which is the change from one theoretical system to another.

Let the new unit of length be L times the old, the new unit of time T times the old, and the new unit of mass M times the old. The numbers L, T, M are positive numbers characterizing the new system in relation to the old. A length whose measure is x in the old system has measure x' in the new system, where

$$x = Lx'.$$

And, with an analogous notation,

$$t = Tt', \quad m = Mm'.$$

Moreover, the new unit of velocity is L/T times the old. If v is the measure of a velocity in the old system, and v' its measure in the new, we have

$$v = \frac{L}{T}v'.$$

Similarly, using always the accented symbol to denote a measure in the new system of units, we find

$$f = \frac{L}{T^2}f', \quad I = \frac{ML}{T}I', \quad P = \frac{ML}{T^2}P'.$$

And generally, if q is the measure of a given physical quantity in the old system, and q' the measure of the same quantity in the new system, there is a relation

$$q = L^{\lambda} T^{\tau} M^{\mu} q',$$

where λ, τ, μ are constants, characteristic of the kind of quantity we are measuring. The multiplier $L^{\lambda} T^{\tau} M^{\mu}$ is called the *dimensional formula* of the quantity. Thus, for example, the dimensional formula of force is MLT^{-2}, of work or kinetic energy ML^2T^{-2}, of power ML^2T^{-3}.

The case in which $\lambda = \tau = \mu = 0$, and the dimensional formula is just 1, is of particular interest. Such a quantity is said to be dimensionless, and its measure is the same in all systems.

We have spoken above of *the* dimensional formula of a given physical quantity, thereby implying the assumption that this formula is necessarily unique. In fact this assumption is not universally valid; in some expositions of electrical theory it appears that the same quantity can have different dimensional formulae in different systems of measurement. Fortunately, however, this difficulty does not arise in the classical dynamics. It could arise if, for example, we introduced a new system in which the multiplier k in the equation $P = kmf$ was not dimensionless, but it does not arise so long as we stick to the system of measurement, with derived units, that we have used in this book. For our present purpose we may safely assume that each physical quantity with which we are concerned has a definite dimensional formula.

26·3 Change of measure. The first use we make of the dimensional formula is the practical one of finding the measure of a given quantity in the new system. Suppose, as a simple example, we change from feet-seconds-pounds to miles-hours-tons. Here

$$L = 5280, \quad T = 3600, \quad M = 2240.$$

For a velocity $$v = \frac{L}{T}v' = \frac{5280}{3600}v' = \frac{22}{15}v'.$$

The new unit of velocity, one mile per hour, is 88/60 ft./sec.; if $v' = 60$, $v = 88$. For a force

$$P = \frac{ML}{T^2}P' = 0{\cdot}9126P', \quad \text{approximately.}$$

The new unit is not very different from the old unit.

A more important application is the change from the c.g.s. to the m.k.s. system. In the m.k.s. system the unit of length is the metre and the unit of mass the kilogram, the unit of time, the second, being unchanged. Here

$$L = 100, \quad T = 1, \quad M = 1000.$$

For a force

$$P = 10^5 P',$$

and the new unit of force, called a newton, is 10^5 dynes. For work

$$W = 10^7 W',$$

and the new unit of work, called a joule, is 10^7 ergs.

26·4 Dimensional analysis.

We now come to a more far-reaching application. The use of dimensional formulae alone enables us to make some progress towards the solution of dynamical problems. Usually the solution so found is incomplete; sometimes unknown constants occur in the solution (as in the solution of an ordinary differential equation) and sometimes unknown functions (as in the solution of a partial differential equation). But in spite of this drawback the method is one of real value.

(i) Let us start with a very simple problem. Suppose a particle moves from rest with constant acceleration g—for example, a body falls freely under gravity *in vacuo*. What is the distance x through which it moves in time t?

We anticipate that x depends only on the acceleration g and the time t, and we assume tentatively

$$x = Ag^{\alpha} t^{\beta}, \tag{1}$$

where A, α, β are fixed numbers, independent of g and of t and of the system of units employed. Thus if we change to a new (theoretical) system we have

$$x' = Ag'^{\alpha} t'^{\beta}. \tag{2}$$

We now substitute in equation (1) the formulae

$$x = Lx', \quad g = (L/T^2)g', \quad t = Tt',$$

and we find

$$Lx' = A(L/T^2)^{\alpha} g'^{\alpha} T^{\beta} t'^{\beta}, \tag{3}$$

and comparing (2) and (3) we have

$$L = (L/T^2)^\alpha T^\beta. \tag{4}$$

Hence, since L and T are arbitrary independent numbers,

$$\alpha = 1, \quad \beta = 2.$$

The equation (1) thus becomes

$$x = Agt^2.$$

We know in fact that the relation *is* of this form, and that the value of the coefficient A is $\frac{1}{2}$; but this value is not determined by the present method.

But, having arrived at this point, it is clear that the matter can be put more concisely. The first and second members of equation (4) are simply the dimensional formulae of the first and second members of equation (1). The upshot of the argument is that the two members of equation (1) have the same dimensional formula. The argument is general. Whenever we have an equation connecting the measures of two physical quantities, the two quantities must have the same dimensional formula, or, briefly, the same dimensions.

The same is true of an equation containing three terms. The situation is more troublesome if the equation contains four terms; for example, in the problem just discussed, the equation

$$x + gx = \tfrac{1}{2}gt^2 + \tfrac{1}{2}v^2$$

is clearly true, and the four terms have not all the same dimensions. But the example is artificial, and in equations arising naturally we may expect all the terms to have the same dimensions, even when we are concerned with the terms of a finite or even of an infinite series.

(ii) We next consider a slightly more elaborate example. The particle moves in the uniform field as before, but this time it is projected, at the instant $t = 0$, with speed u in the direction of the field. What is the distance x covered in time t?

This time the matter is not quite so simple; x now depends on the *three* parameters g, u, t, and we are no longer justified in taking a single term $Ag^\alpha u^\beta t^\gamma$; instead we assume tentatively that x can be expressed as the sum of a series, finite or infinite, of the form

$$x = \Sigma A g^\alpha u^\beta t^\gamma. \tag{5}$$

If we assume that all the terms have the same dimensions we find

$$L = \left(\frac{L}{T^2}\right)^{\alpha}\left(\frac{L}{T}\right)^{\beta} T^{\gamma}, \tag{6}$$

giving $\qquad \alpha+\beta = 1, \quad 2\alpha+\beta = \gamma.$

The two equations do not suffice to determine the three numbers α, β, γ, and the best we can do is to express them all in terms of a single parameter, say in terms of α; thus

$$\beta = 1-\alpha, \quad \gamma = 1+\alpha,$$

and equation (5) becomes

$$x = \Sigma A g^{\alpha} u^{1-\alpha} t^{1+\alpha} = ut\Sigma A(gt/u)^{\alpha},$$

and we can write this more simply in the form

$$\frac{x}{ut} = f\left(\frac{gt}{u}\right).$$

The solution of this problem is therefore less complete than that of the previous one, since it contains the unknown function f. Actually, of course, we know on other grounds that the function $f(\theta)$ has the simple form

$$f(\theta) = 1+\tfrac{1}{2}\theta,$$

but this form is not determined by the present method.

26·5 Dimensionless quantities. If we examine a little more closely the solutions of the two problems just discussed we are struck by the importance of the dimensionless quantities which are products of powers of the quantities occurring in the problem. Thus in the first problem, involving x, g, and t, we find as the solution that x/gt^2 is constant, and x/gt^2 is dimensionless. Moreover, there is no dimensionless expression $x^{\alpha}g^{\beta}t^{\gamma}$ which is independent of x/gt^2; any such expression is just $(x/gt^2)^{\alpha}$.

We find a similar result in the second problem. Here there are four quantities, x, u, g, and t, and the solution turns out to be a functional relation connecting the dimensionless quantities x/ut and gt/u. These two dimensionless quantities are fundamental in the sense that there is no dimensionless quantity $x^\alpha u^\beta g^\gamma t^\delta$ which is independent of them; any such dimensionless quantity is a product of powers of x/ut and of gt/u. For example, gx/u^2 is a dimensionless quantity, but it is merely the product of the other two.

In these problems only L and T appear in the dimensional formulae, not M, and we can now foresee in a general way how the argument will run in similar problems. If, as in the first problem, only three quantities p, q, r are involved, there will in general be only one fundamental dimensionless combination $p^\alpha q^\beta r^\gamma$, because the fact that this is dimensionless leads to two equations for the ratios $\alpha : \beta : \gamma$. The solution of the problem will be that this quantity remains constant. In a problem with four quantities (all of which have dimensional formulae not involving M) there will be in general two independent dimensionless combinations, and the solution will be a functional relation between them. Similarly, in a problem with five quantities there will be in general three dimensionless combinations, and the solution will be a functional relation connecting them. For example, if in the second of the two problems just discussed (§ 26·4, (ii)) the motion takes place in a resisting medium producing a retardation k times the speed, we are concerned with the five quantities x, u, g, t, k. The dimensional formula of k is $1/T$, and we easily find three independent dimensionless quantities

$$kx/u, \quad g/ku, \quad kt.$$

We anticipate a solution of the form

$$\frac{kx}{u} = f\left(\frac{g}{ku}, kt\right),$$

and the reader will easily verify that the same solution is found by the method of § 26·4. Actually we know already that

$$f(\theta, \phi) = (1-\theta)(1-e^{-\phi}) + \theta\phi.$$

In a problem of a more elaborate type in which the dimensional formulae involve M as well as L and T, the appropriate modification is simple. Four quantities will in general provide one fundamental dimensionless combination, five will give two independent dimensionless combinations, and so on.

26·6 The simple pendulum. Suppose we have a simple pendulum oscillating with amplitude ψ. We denote by θ the inclination to the downward vertical at time t, and think of the motion as starting from rest with $\theta = \psi$; what can we find, by the method of dimensions, about (i) the time $\frac{1}{4}\sigma$ (quarter-period) required to reach the vertical, and (ii) the angular velocity ω when the vertical is reached? The two questions are very similar, and we can be content to discuss the first of them.

Now σ may depend on the mass m of the bob, the length a of the pendulum, the value of g, and on ψ. But here we meet a new point; we cannot hope to discover how σ depends on ψ by dimensional analysis, because ψ is dimensionless. We therefore make the tentative assumption

$$\sigma = Am^{\alpha}a^{\beta}g^{\gamma},$$

where A is not now a constant, but a function of ψ. Comparing the dimensional formulae of the two members of this equation, as in § 26·4, we find $\alpha = 0$ (as was almost evident at the start), and $\beta = -\gamma = \frac{1}{2}$. Thus

$$\sigma = f(\psi)\surd(a/g).$$

We could have arrived at this result equally simply by the method of § 26·5, since $g\sigma^2/a$ is dimensionless. We know in fact (§ 11·10) that, for small values of ψ, $f(\psi)$ differs only slightly from the constant value 2π. A similar argument leads to

$$\omega = F(\psi)\surd(g/a).$$

Now consider the relation between position and time in the motion. Let us now change our zero of time and measure from an instant when the pendulum is vertical, so that $\theta = 0, \dot{\theta} = \omega\ (>0)$ at $t = 0$. If θ is the inclination at time t we know that θ depends on the four arguments a, g, ω, t, and we assume

$$\theta = \Sigma A a^{\alpha} g^{\beta} \omega^{\gamma} t^{\delta},$$

where the A's are constants. The terms in the series in the second member are dimensionless,

$$1 = L^{\alpha}(L/T^2)^{\beta}(1/T)^{\gamma}T^{\delta},$$

so we have

$$\alpha + \beta = 0, \quad 2\beta + \gamma = \delta.$$

This time we have two equations with four unknowns, so two are arbitrary. If we leave in β and γ we have

$$\theta = \Sigma A(gt^2/a)^\beta(\omega t)^\gamma.$$

But
$$\omega t = F(\psi)\sqrt{(gt^2/a)},$$

so finally we find that θ is a function of the two arguments ψ and gt^2/a,

$$\theta = \chi(\psi, gt^2/a).$$

Of course ψ and gt^2/a are dimensionless, and we could have anticipated this result from the ideas of § 26·5.

Consider now two different pendulums, with lengths a_1 and a_2, swinging in fields g_1 and g_2, but with the same amplitude. We suppose both pendulums vertical at $t = 0$. If we consider times t_1 and t_2 such that

$$g_1 t_1^2/a_1 = g_2 t_2^2/a_2,$$

the inclination of the first pendulum at time t_1 is the same as the inclination of the second pendulum at time t_2. Two such systems, occupying similar configurations at corresponding times, are said to be *dynamically similar*.

26·7 Wind-tunnel experiments. We now consider the application of dimensional analysis to find the relation between the force-system, due to air pressure, on a model aeroplane at rest in a wind-tunnel and the force-system on the aeroplane in flight. We must think of the aeroplane itself as being at rest in a stream of air, a change which will not alter the force-system under discussion. The model is of course similar to the aeroplane, and we may expect that one particular component of force—say the lift—will depend on a, ρ, and v, where a is some linear measurement (say the span), ρ is the density of the air, and v the speed of the stream of air. We are supposing, as a first approximation, that the air is uniform and inviscid.

We assume then that the lift Y is given by a formula

$$Y = Aa^\alpha \rho^\beta v^\gamma.$$

Comparing the dimensional formulae, as in § 26·4, we find

$$\frac{ML}{T^2} = L^\alpha\left(\frac{M}{L^3}\right)^\beta\left(\frac{L}{T}\right)^\gamma,$$

giving
$$\alpha = 2, \quad \beta = 1, \quad \gamma = 2.$$

This is an example of the case mentioned at the end of § 26·5; we have four quantities Y, a, ρ, v, and one fundamental dimensionless combination.

Thus
$$Y = A\rho a^2 v^2,$$

where A is constant for models of the given shape, and we can infer from the lift in the wind-tunnel experiment, which can be measured directly, the lift on the aeroplane in flight. Explicitly

$$\frac{Y_1}{Y_2} = \frac{\rho_1 a_1^2 v_1^2}{\rho_2 a_2^2 v_2^2},$$

where the suffix 1 refers to the aeroplane in flight, the suffix 2 to the model in the wind-tunnel. The result is only approximately true, since we have neglected the effect of viscosity, and it fails still more seriously if one of the velocities is near that of sound, since in that case the assumption of uniform compression is no longer even approximately true.

In a typical experiment

$$\frac{\rho_1}{\rho_2} = \frac{1}{2}, \quad \frac{a_1}{a_2} = 10, \quad \frac{v_1}{v_2} = \frac{3}{2},$$

giving the lift on the aeroplane about 112 times that on the model.

26·8 Drops from a capillary tube. Finally, we consider a problem where dimensional analysis is combined with experimental evidence to establish an important physical law. The drops of liquid formed by slow dripping from a capillary tube are of nearly constant mass for a given liquid and a given tube. How does this mass vary with different liquids and different tubes?

If as a first approximation we suppose the liquids inviscid, we may assume that the mass m of the drop will depend on the radius a of the capillary tube, the density ρ of the liquid, the surface tension P, and on g. We have five quantities, and anticipate, as in § 26·5, a functional relation between two dimensionless combinations. The dimensional formula of P is M/T^2, and using the method of § 26·4 we write

$$m = \Sigma A a^{\alpha} \rho^{\beta} P^{\gamma} g^{\delta}.$$

Comparing the dimensional formulae we find

$$M = L^{\alpha}\left(\frac{M}{L^3}\right)^{\beta}\left(\frac{M}{T^2}\right)^{\gamma}\left(\frac{L}{T^2}\right)^{\delta},$$

whence $\alpha+\delta = 3\beta, \quad \gamma+\delta = 0, \quad \beta+\gamma = 1.$

Expressing the other numbers in terms of β we have

$$\alpha = 1+2\beta, \quad \gamma = 1-\beta, \quad \delta = -1+\beta,$$

so that
$$m = \frac{aP}{g}\Sigma A\left(\frac{g\rho a^2}{P}\right)^{\beta},$$

and we may express this in the form

$$\frac{mg}{aP} = f\left(\frac{g\rho a^2}{P}\right),$$

which is the expected functional relation between the dimensionless quantities mg/aP and $g\rho a^2/P$. The function f, which we can find experimentally by plotting values of the dependent variable mg/aP against those of the independent variable $g\rho a^2/P$, is absolute, independent of the tubes and liquids used in the experiments.

In fact, experiment shows that if the same liquid is used throughout, and only a is varied, then m is proportional to a; the function f is a constant! So finally

$$mg = kaP,$$

where k is constant. The weight of the drop is directly proportional to the radius a of the tube and to the surface tension P.

EXAMPLES XXVI

1. Give the dimensional formulae of: curvature, momentum, moment of inertia, angular velocity, impulsive couple, the gravitational constant γ.

2. (i) Express in pounds weight per square inch the pressure of 1020 millibars (1 millibar $= 10^3$ dynes/sq.cm.) (ii) Given that 1 lb. $= 453{\cdot}6$ g., $g = 32{\cdot}2$ ft./sec./sec. $= 981$ cm./sec./sec., prove that 1 h.p. $= 7{\cdot}46\times 10^7$ c.g.s. units.

3. A particle of unit mass moves under a central attractive force of magnitude μr^{-2}, where r is the distance from the centre of force. The angular momentum and total energy of the particle are denoted by h, C respectively.

Which (if any) of the following statements are plausible?

(i) The latus rectum of the orbit is $2h^2/\mu$.

(ii) The eccentricity of the orbit is $1+2Ch^2/\mu^2$.

(iii) If $C<O$, the orbit is closed and the period is $2\pi/n$, where $n^2=-8C/\mu^2$.

4. An examiner sets the following problem: 'A particle moves in a plane, subject to an attraction μr^{-5} per unit mass towards O. Its velocity at an apse A, distant a from O, is w. Find the differential equation of the orbit.' As a further exercise he sets a particular case. Show that this must be specified by an equation of the form $w^2 = k\mu a^{-4}$, where k is a number.

5. A ball of mass m is thrown vertically upwards with velocity u. The air resistance is kv^n, where v is the velocity of the ball. Show that the time taken to return to the point of projection is given by an equation of the form

$$t=\frac{u}{g}f\left(\frac{ku^n}{mg}\right).$$

6. A particle of mass m moves in a straight line, under the influence of a periodic force $P\sin nt$ and a damping force equal to k times the velocity. Show that the amplitude of the oscillation in the steady state is proportional to P.

7. In an experiment to find the surface tension P of a liquid of density ρ, ripples of frequency n are induced on the surface and their wave-length λ is measured. It is found that, if n is large, λ^3 is inversely proportional to n^2. Deduce that (for n large), $P \propto \lambda^3 n^2 \rho$.

8. A rotating mass of incompressible gravitating fluid becomes unstable when it attains a certain angular velocity. Show that this critical value is independent of the diameter, and is proportional to the square root of the density of the fluid.

9. The coefficient of viscosity η of a fluid has dimensions -1 in length, 1 in mass, and -1 in time. If for a certain fluid η has the value 0·23 in the centimetre-gram-second system, find its value in the metre-kilogram-minute system.

If a sphere of radius a and weight W falls through a fluid whose coefficient of viscosity is η, its speed approaches a terminal speed v. Assuming that no other variables are involved, find by a consideration of dimensions how v depends on a, W and η.

10. The mass of an electron is found to vary with the velocity according to the law

$$m=\frac{\lambda}{\sqrt{(1-v^2/\mu)}},$$

where m is the mass in grams, v is the velocity in centimetres per second, $\lambda = 9\times 10^{-28}$ and $\mu = 9\times 10^{20}$. Write down the law relating the mass M in kilograms with the velocity V in kilometres per minute.

Explain and justify the statement that the constants λ and μ have the dimensions of mass and (velocity)^2 respectively.

INDEX